Biodiesel Fuels

Biodiesel Fuels

Science, Technology, Health, and Environment

Edited by
Ozcan Konur

CRC Press is an imprint of the
Taylor & Francis Group, an **informa** business

First edition published 2021
by CRC Press
6000 Broken Sound Parkway NW, Suite 300, Boca Raton, FL 33487-2742

and by CRC Press
2 Park Square, Milton Park, Abingdon, Oxon, OX14 4RN

CRC Press is an imprint of Taylor & Francis Group, LLC

ISBN: 978-0-367-45614-6 (hbk)
ISBN: 978-0-367-70494-0 (pbk)
ISBN: 978-0-367-45623-8 (ebk)

Typeset in Times
by SPi Global, India

Contents

Preface

Crude oils have been primary sources of energy and fuels, such as petrodiesel. However, significant public concerns about the sustainability, price fluctuations, and adverse environmental impact of crude oils have emerged since the 1970s. Thus, biooils and biooil-based biodiesel fuels have emerged as an alternative to crude oils and crude-oil-based petrodiesel fuels, respectively, in recent decade. Nowadays, although petrodiesel fuels are still used extensively, biodiesel fuels are being used increasingly as petrodiesel–biodiesel blends in the transportation and power sectors. Therefore, there has been great public interest in the development of environment and human-friendly and sustainable petrodiesel and biodiesel fuels. However, it is necessary to reduce the total cost of biodiesel production by reducing the feedstock cost through the improvement of biomass and lipid productivity. It is also necessary to mitigate the adverse impact of petrodiesel fuels on the environment and human health.

Although there have been over 1,500 reviews and book chapters in this field, there has been no review of the research as a representative sample of all the population studies done in the field of both petrodiesel and biodiesel fuels. The first volume in this handbook provides such a sample for biodiesel fuels in general.

The major research fronts are determined from the sample and population paper-based scientometric studies. Table 1.1 presents information on the major and secondary research fronts emanating from these scientometric studies. There are four secondary research fronts for biodiesel fuels in general: biooils, biodiesel fuels in general, and glycerol; and, additionally, papers related to an overview of both biodiesel and petrodiesel fuels are included in this first volume.

As seen from Table 1.1 there is a substantial correlation between distribution of research fronts in this handbook and the 100 most-cited sample papers. Papers on biodiesel fuels in general are over-represented compared to the population papers: 34, 41 v. 27.7%.

Table 1.2 extends Table 1.1 and provides more information on the content of this handbook, including chapter numbers, paper references, primary and secondary research fronts, and finally the titles of the papers presented in the first volume.

The data presented in the tables and figures show that a small number of authors, institutions, funding bodies, journals, keywords, research fronts, subject categories, and countries have shaped the research in this field.

The findings show the importance of the development of efficient incentive structures for the development of research in this field, as in other fields. It further seems that although research funding is a significant element of these incentive structures, it might not be a sole solution for increasing the incentives for research. On the other hand, it seems there is more to do to reduce the significant gender deficit in this field as in other fields of science and technology.

The information provided on nanotechnology applications suggests that there is ample scope for the expansion of advanced applications in this research field.

The research on biooils is a fundamental part of the research on biodiesel fuels, which has intensified in recent years with the application of advanced catalytic technologies and nanotechnologies in both the production and upgrading of biooils.

The research on biodiesel fuels in general has progressed along the lines of production, properties, and emissions. As in the case of bio-oils, catalysts and additives play a crucial role for biodiesel fuels.

The research on glycerol has intensified in recent years with the increasing volume of biodiesel fuels creating eco-friendly solutions for the waste products of biodiesel fuels and for producing valuable biofuels and biochemicals from glycerol.

Thus, this first volume in this handbook is a valuable resource for stakeholders primarily in the research fields of Energy Fuels, Chemical Engineering, Environmental Sciences, Biotechnology and Applied Microbiology, Physical Chemistry, Petroleum Engineering, Environmental Engineering, Multidisciplinary Chemistry, Thermodynamics, Analytical Chemistry, Mechanical Engineering, Agricultural Engineering, Marine Freshwater Biology, Green Sustainable Science Technology, Applied Chemistry, Multidisciplinary Geosciences, Microbiology, Multidisciplinary Materials Science, Mechanics, Toxicology, Multidisciplinary Sciences, Biochemistry and Molecular Biology, Water Resources, Plant Sciences, Multidisciplinary Engineering, Transportation Science Technology, Geochemistry and Geophysics, Food Science Technology, Ecology, Public Environmental Occupational Health, Meteorology and Atmospheric Sciences, Electrochemistry, and Biochemical Research Methods.

This handbook is also particularly relevant in the context of biomedical sciences for Public and Environmental Occupational Health, Pharmacology, Immunology, Respiratory System, Allergy, Genetics Heredity, Oncology, Experimental Medical Research, Critical Care Medicine, General Internal Medicine, Cardiovascular Systems, Physiology, Medicinal Chemistry, and Endocrinology and Metabolism.

Ozcan Konur

Acknowledgements

This handbook was a multi-stakeholder project from its inception to its publication. CRC Press and Taylor & Francis Group were the major stakeholders in financing and executing it. Marc Gutierrez was the executive editor. Eighty-three authors have kindly contributed chapters despite the relatively low level of incentive, compared to journals. As stated in many chapters, a small number of highly cited scholars have shaped the research on biodiesel and petrodiesel fuels. The contribution of all these stakeholders is greatly acknowledged.

Editor

Ozcan Konur, as both a materials scientist and social scientist by training, has focused on the bibliometric evaluation of research in the innovative high-priority research areas of algal materials and nanomaterials for energy and fuel as well as for biomedicine at the level of researchers, journals, institutions, countries, and research areas, including the social implications of the research conducted in these areas.

He has also researched extensively in the development of social policies for disadvantaged people on the basis of disability, age, religious beliefs, race, gender, and sexuality at the interface of science and policy.

He has edited a book titled *Bioenergy and Biofuels* (CRC Press, January 2018) and a handbook titled *Handbook of Algal Science, Technology, and Medicine* (Elsevier, April 2020).

Contributors

Erika C. G. Aguieiras
Federal University of Rio de Janeiro, Brazil

May Ali Alsaffa
University of Technology, Iraq

Bamidele Victor Ayodele
Universiti Tenaga Nasional, Malaysia

Navneeta Bharadvaja
Delhi Technological University, India

Leichang Cao
Henan University; Fudan University, China

A. Chaube
Rajiv Gandhi Proudyogiki Vishwavidyalaya, India

Eliane P. Cipolatti
Federal University of Rio de Janeiro, Brazil

Madhu Sudan Reddy Dandu
Sree Vidyanikethan Engineering College, India

Hansham Dewal
Delhi Technological University, India

Jaqueline G. Duarte
Federal University of Rio de Janeiro, Brazil

Denise M. G. Freire
Federal University of Rio de Janeiro, Brazil

Celine Ming Hui Goh
Curtin University Malaysia, Malaysia

Anh Tuan Hoang
Ho Chi Minh City University of Transport; Ho Chi Minh City University of Technology, Vietnam

Mohd Lokman Ibrahim
Universiti Teknologi MARA, Malaysia

Shashi Kumar Jain
Technocrats Institute of Technology (Excellence), India

Md. Abul Kalam
University of Malaya, Malaysia

Jibrail Kansedo
Curtin University Malaysia, Malaysia

Ozcan Konur
(Former) Ankara Yildirim Beyazit University, Turkey

Lakhan Kumar
Delhi Technological University, India

N. Keerthi Kumar
BMS Institute of Technology and Management, India

Naveen Kumar
Delhi Technological University, India

Niraj Kumar
Amity University, India

Sunil Kumar
Rajiv Gandhi Proudyogiki Vishwavidyalaya, India

Evelin Andrade Manoel
Federal University of Rio de Janeiro, Brazil

N. M. Mubarak
Curtin University Malaysia, Malaysia

Siti Indati Mustapa
Universiti Tenaga Nasional, Malaysia

Xuan Phuong Nguyen
Ho Chi Minh City University of Transport, Vietnam

Van Viet Pham
Ho Chi Minh City University of Transport, Vietnam

Martina C. C. Pinto
Federal University of Rio de Janeiro, Brazil

Junming Sun
Washington State University, USA

Yie Hua Tan
Curtin University Malaysia, Malaysia

Mukul Tomar
Delhi Technological University, India

Daniel C. W. Tsang
Hong Kong Polytechnic University, China

Yong Wang
Pacific Northwest National Laboratory; Washington State University, USA

Jianghao Zhang
Washington State University, USA

Shicheng Zhang
Fudan University; Shanghai Technical Service Platform for Pollution Control and Resource Utilization of Organic Wastes; Shanghai Institute of Pollution Control and Ecological Security, China

Part I

Biodiesel and Petrodiesel Fuels

Science, Technology, Health, and the Environment

1 Biodiesel and Petrodiesel Fuels

Science, Technology, Health, and the Environment

Ozcan Konur

CONTENTS

1.1 INTRODUCTION

Crude oils have been primary sources of energy and fuels, such as petrodiesel. However, significant public concerns about their sustainability, price fluctuations, and adverse environmental impact have emerged since the 1970s (Ahmadun et al., 2009; Atlas, 1981; Babich and Moulijn, 2003; Haritash and Kaushik, 2009; Hazen et al., 2010; Kilian, 2009; Leahy and Colwell, 1990; Perron, 1989; Peters, 1986, Peterson et al., 2003). Thus, biooils (Bridgwater, 2012; Bridgwater et al., 1999; Bridgwater and Peacocke, 2000; Czernik and Bridgwater, 2004; di Blasi, 2008; Elliott, 2007; Mohan et al., 2006; Mortensen et al., 2011; Ragauskas et al., 2014; Tuck et al., 2012; Zakzeski et al., 2010; Zhang et al., 2007) and biooil-based biodiesel fuels (Agarwal, 2007; Brennan and Owende, 2010; Chen et al., 2011; Chisti, 2007, 2008; Gekko and Timasheff, 1981; Graboski and McCormick, 1998; Griffiths and Harrison, 2009; Grima et al., 2003; Hill et al., 2006; Hu et al., 2008; Knothe, 2005; Lapuerta et al., 2008; Lardon et al., 2009; Mata et al., 2010; Pagliaro et al., 2007; Rodolfi et al., 2009; Schenk et al., 2008; Srivastava and Prasad, 2000; Wijffels and Barbosa, 2010; Zhang et al., 2003a–2003b; Zhou et al., 2008) have emerged as an alternative to crude oils and crude oil-based petrodiesel fuels (Birch and Cary, 1996; Busca et al., 1998; McCreanor et al., 2007; Robinson et al., 2007; Rogge et al., 1993; Salvi et al., 1999; Schauer et al., 1999; Shelef, 1995; Song, 2003; Song and Ma, 2003), respectively, in recent decades. Nowadays, although petrodiesel fuels are still used extensively, biodiesel fuels are being used increasingly in the transportation and power sectors (Konur, 2021a–ag).

Therefore, there has been great public interest in the development of environment and human-friendly and sustainable petrodiesel and biodiesel fuels (Birch and Cary, 1996; Busca et al., 1998; Chisti, 2007; Hu et al., 2008; Lapuerta et al., 2008; Mata et al., 2010; McCreanor et al., 2007; Rodolfi et al., 2009; Schenk et al., 2008). However, it is necessary to reduce the total cost of the biodiesel production by reducing the feedstock cost through the improvement of biomass and lipid productivity (Brennan and Owende, 2010; Chen et al., 2011; Griffiths and Harrison, 2009; Grima et al., 2003; Hu et al., 2008; Lardon et al., 2009; Rodolfi et al., 2009; Wijffel and Barbosa, 2010) in a biorefinery context (John et al., 2011; Sialve et al., 2009). It is also necessary to mitigate the adverse impact of petrodiesel fuels on the environment and human health (Busca et al., 1998; Delfino et al., 2005; Diaz-Sanchez et al., 1997; Koebel et al., 2000; Lapuerta et al., 2008; McCreanor et al., 2007; Mills et al., 2007; Salvi et al., 1999; Schauer et al., 1999; Shelef, 1995; Song, 2003; Song and Ma, 2003; Stanislaus et al., 2010).

Furthermore, for the efficient development of research in this field, it is necessary to develop efficient incentive structures for the primary stakeholders and to inform these stakeholders about the research (Konur, 2000, 2002a–c, 2006a–b, 2007a–b); North, 1991a, 1991b)

Although there have been over 1,500 reviews and book chapters in this field (Brennan and Owende, 2010; Bridgwater and Peacocke, 2000; Busca et al., 1998; Chen et al., 2011; Chisti, 2007, 2008; di Blasi, 2008; Elliott, 2007; Grima et al., 2003; Hu et al., 2008; Lapuerta et al., 2008; Mata et al., 2010; Mohan et al., 2006; Mortensen et al., 2011; Ragauskas et al., 2014; Schenk et al., 2008; Tuck et al., 2012; Zakzeski et al., 2010), there has been no review of the research as a representative sample of whole-population studies done in the field of both petrodiesel and biodiesel fuels.

This handbook provides such a sample and this chapter introduces the key issues and findings.

1.2 MATERIALS AND METHODOLOGY

This book chapter provides a review of the representative sample of the whole population of research studies on both petrodiesel and biodiesel fuels, presented in this handbook in three volumes.

Table 1.1 provides information on the three key research fronts in this field and Table 1.2 provides information on the 58 chapters provided in this handbook. The related references are listed in the reference section to this chapter. The key issues and findings from these studies are provided and discussed briefly below in light of whole-population studies conducted in this field.

1.3 CONTENTS OF THE HANDBOOK

1.3.1 Volume 1: Biodiesel Fuels in General

There are four secondary research streams in the primary one of biodiesel fuels in general: an introduction to petrodiesel and biodiesel fuels (Part I), biooils (Part II), biodiesel fuels in general (Part III), and glycerol (Part IV) with four, six, seven, and three chapters, respectively, in the first volume.

1.3.1.1 Part I: Introduction to Petrodiesel and Biodiesel Fuels

There are four chapters introducing both petrodiesel and biodiesel fuels (Konur 2021e–h).

Chapter 1 (Konur, 2021e) provides an overview of the 58 chapters in this handbook under three primary research fronts: biodiesel fuels in general (volume 1), feedstock-based biodiesel fuels (volume 2), and petrodiesel fuels (volume 3). Table 1.1 provides brief information about the key research fronts covered. It provides data on the number of review papers and percentage of them presented, the percentage of the most-cited 100 sample papers in the field of biodiesel and petrodiesel fuels, and the percentage of over 121,000 population papers in the same field. Table 1.2 provides information on the content of the handbook with a focus on the individual chapters, paper references, research fronts, and chapter titles.

Chapter 2 (Konur, 2021f) provides a scientometric review of the research on both petrodiesel and biodiesel fuels using a population of over 121,000 papers. Konur uses two datasets for this study. The highly cited sample of 100 papers comprises the first dataset (n = 100 papers) whilst all the papers form the second dataset (n = over 121,000 papers). For this purpose Konur uses a carefully designed set of keywords for the research fields related to both biodiesel and petrodiesel fuels and the 'Web of Science' databases as of February 2020.

The data on the indices, document types, authors, institutions, funding bodies, source titles, 'Web of Science' subject categories, keywords, research fronts, and citation impact are presented and discussed.

TABLE 1.1
Research Fronts for Biodiesel and Petrodiesel Fuels: Science, Technology, Health, and Environment

	Research Fronts	No. of Reviews	No. of Reviews (%)	No. of Sample Papers (%)	Difference (%)	No. of Population Papers (%)
Vol. 1	Biodiesel fuels in general	20	34	41	–7	27.7
1.1	Introduction to petrodiesel and biodiesel fuels	4	7			
1.2	Biooils	6	10	14	–4	16.6
1.3	Biodiesel fuels in general	7	12	21	–9	5.4
1.4.	Glycerol	3	5	6	–1	5.7
Vol. 2	Feedstock-based biodiesel fuels	19	33	30	3	18.3
2.1	Edible oil-based biodiesel fuels	4	7	5	2	2.2
2.2	Nonedible oil-based biodiesel fuels	4	7	3	4	1.8
2.3	Waste oil-based biodiesel fuels	4	7	2	5	1.8
2.4	Algae-based biodiesel fuels	7	12	20	–8	12.5
Vol. 3	Petrodiesel fuels	19	33	34	–1	67.4
3.1	Crude oils	6	10	14	–4	43.8
3.2	Petrodiesel fuels in general	5	9	8	1	13.2
3.3	Emissions from petrodiesel fuels	4	7	11	–4	8.6
3.4	Health impact of the emissions from petrodiesel fuels	4	7	1	6	1.8

Notes: Sample size of review papers presented in this handbook: 58; of most-cited papers: 100; of population papers: over 121,000. Difference = no. of reviews – no. of sample papers.

TABLE 1.2
Biodiesel and Petrodiesel Fuels: Science, Technology, Health, and Environment

	Chapter No.	Research Front I	Research Front II	Chapter Title
Volume 1. Biodiesel Fuels: Science, Technology, Health, and Environment				
Part I Biodiesel and petrodiesel fuels: science, technology, health, and environment				
1	Konur (2021e)	Biodiesel and petrodiesel fuels	Overview	Biodiesel and petrodiesel fuels: Science, technology, health, and environment
2	Konur (2021f)	Biodiesel and petrodiesel fuels	Overview	Biodiesel and petrodiesel fuels: A scientometric review of the research
3	Konur (2021g)	Biodiesel and petrodiesel fuels	Overview	Biodiesel and petrodiesel fuels: A review of the research.
4	Konur (2021h)	Biodiesel and petrodiesel fuels	Nanotechnology applications	Nanotechnology applications in the diesel fuels and the related research fields: A review of the research
Part II Biooils				
5	Konur (2021i)	Biooils	Overview	Biooils: A scientometric review of the research
6	Konur (2021j)	Biooils	Characterization and properties	Characterization and properties of biooils: A review of the research
7	Konur (2021k)	Biooils	Pyrolysis	Biomass pyrolysis and pyrolysis oils: A review of the research
8	Zhang et al. (2021a)	Biooils	Biooil upgrading I	An overview of catalytic biooil upgrading. Part I: Processing aqueous-phase compounds
9	Zhang et al. (2021b)	Biooils	Biooil upgrading II	An overview of catalytic biooil upgrading. Part II: Processing oil-phase compounds and real biooil
10	Cao et al. (2021)	Biooils	Hydrothermal liquefaction (HTL)	Biooil production through hydrothermal liquefaction (HTL) of biomass: Recent development and future prospects
Part III Biodiesel fuels in general				
11	Konur (2021l)	Biodiesel fuels in general	Overview	Biodiesel fuels: A scientometric review of the research
12	Goh et al. (2021)	Biodiesel fuels in general	Biomass-based catalyst-assisted biodiesel production	Biomass-based catalyst-assisted biodiesel production
13	Aguieiras et al. (2021)	Biodiesel fuels in general	Enzymatic biodiesel production	Enzymatic biodiesel production: Challenges and future perspectives
14	Tomar et al. (2021)	Biodiesel fuels in general	Additives in biodiesel production	Biodiesel additives: Status and perspectives
15	Kumar and Kalam (2021)	Biodiesel fuels in general	Properties and characterization of biodiesel fuels	Qualitative characterization of biodiesel fuels: Basics and beyond

(Continued)

TABLE 1.2 (Continued)

	Chapter No.	Research Front I	Research Front II	Chapter Title
16	Hoang et al. (2021)	Biodiesel fuels in general	Performance of biodiesel fuels in diesel engines	Use of biodiesel fuels in diesel engines
17	Jain et al. (2021)	Biodiesel fuels in general	Biodiesel policies	Biodiesel promotion policies: A global perspective
Part IV Glycerol				
18	Konur (2021m)	Glycerol	Overview	Glycerol: A scientometric review of the research
19	Ayodele et al. (2021)	Glycerol	Biohydrogen production from glycerol: case study for biochemicals	Hydrogen-rich syngas production from biodiesel-derived glycerol: An overview of the modeling and optimization strategies
20	Konur (2021n)	Glycerol	Propanediol production from glycerol: case study for biochemicals	Propanediol production from glycerol: A review of the research
Volume 2. Biodiesel Fuels Based on Edible and Nonedible Feedstocks, Wastes, and Algae: Science, Technology, Health, and Environment				
Part V Edible oil-based biodiesel fuels				
21	Konur (2021o)	Edible oil-based biodiesel fuels	Overview	Edible oil-based biodiesel fuels: A scientometric review of the research
22	Westbrook (2021)	Edible oil-based biodiesel fuels	Soybean oil-based biodiesel fuels: case study I	Chemistry of biodiesel fuels based on soybean oil
23	Konur (2021p)	Edible oil-based biodiesel fuels	Palm oil-based biodiesel fuels: case study II	Palm oil-based biodiesel fuels: A review of the research
24	Konur (2021q)	Edible oil-based biodiesel fuels	Rapeseed oil-based biodiesel fuels: case study III	Rapeseed oil-based biodiesel fuels: A review of the research
Part VI Nonedible oil-based biodiesel fuels				
25	Konur (2021r)	Nonedible oil-based biodiesel fuels	Overview	Nonedible oil-based biodiesel fuels: A scientometric review of the research.
26	Banapurmath et al. (2021)	Nonedible oil-based biodiesel fuels	Jatropha oil-based biodiesel fuels: case study I	An exhaustive study on the use of Jatropha based biodiesel for modern diesel engine applications
27	Dandu et al. (2021)	Nonedible oil-based biodiesel fuels	Polanga oil-based biodiesel fuels: case study II	The effects of additives with *Calophyllum inophyllum* methyl ester (CIME) in CI engine applications
28	Niju and Janani (2021)	Nonedible oil-based biodiesel fuels	Moringa oil-based biodiesel fuels: case study III	*Moringa oleifera* oil as a potential feedstock for sustainable biodiesel production

Part VII Waste oil-based biodiesel fuels				
29	Konur (2021s)	Waste oil-based biodiesel fuels	Overview	Waste oil-based biodiesel fuels: A scientometric review of the research.
30	Siddique and Rohani (2021)	Waste oil-based biodiesel fuels	Wastewater sludge-based biodiesel: case study I	Biodiesel production from municipal wastewater sludge: Recent trends
31	Castanheiro (2021)	Waste oil-based biodiesel fuels	Waste cooking oil-based biodiesel fuels	Heterogeneous acid-catalysts for biodiesel production from waste cooking oil
32	Satyannarayana et al. (2021)	Waste oil-based biodiesel fuels	Microbial oil-based biodiesel fuels	Microbial biodiesel fuel production using microbes
Part VIII Algal biodiesel fuels				
33	Konur (2021t)	Algal oil-based biodiesel fuels	Overview	Algal biodiesel fuels: A scientometric review of the research
34	Konur (2021u)	Algal oil-based biodiesel fuels	Algal biomass production	Algal biomass production for biodiesel production: A review of the research
35	Konur (2021v)	Algal oil-based biodiesel fuels	Algal biomass production in wastewaters	Algal biomass production in wastewaters for biodiesel production: A review of the research
36	Konur (2021x)	Algal oil-based biodiesel fuels	Algal lipid production	Algal lipid production for biodiesel production: A review of the research
37	Koley et al. (2021)	Algal oil-based biodiesel fuels	Hydrothermal liquefaction of algal biomass	Biooil production from microalgae: Current status and future perspective
38	Krishnan et al. (2021)	Algal oil-based biodiesel fuels	Algal lipid extraction	Extraction of algal neutral lipids for biofuel production
39	Sivaramakrishnan and Incharoensakdi (2021)	Algal oil-based biodiesel fuels	Algal biodiesel production	Microalgal biodiesel production
Volume 3. Petrodiesel Fuels: Science, Technology, Health, and Environment				
Part IX Crude oils				
40	Konur (2021y)	Crude oils	Overview	Crude oils: A scientometric review of the research.
41	Fingas (2021a)	Crude oils	Crude oil spills	Introduction to oil spill behavior
42	Fingas (2021b)	Crude oils	Crude oil spills and their clean-up	Introduction to oil spills and their clean-up
43	Prince et al. (2021)	Crude oils	Crude oils properties and their removal	Crude oils and their fate in the environment
44	McFarlin and Prince (2021)	Crude oils	Biodegradation of crude oil-contaminated soils	Oil contaminated soil: Understanding biodegradation and remediation strategies
45	Jeong and Lee (2021)	Crude oils	Crude oil recovery	Biotechnology applications in oil recovery

(Continued)

TABLE 1.2 (Continued)

	Chapter No.	Research Front I	Research Front II	Chapter Title
Part X Petrodiesel fuels in general				
46	Konur (2021y)	Petrodiesel fuels in general	Overview	Petrodiesel fuels: A scientometric review of the research
47	Pandey and Sharma (2021)	Petrodiesel fuels in general	Combustion of biodiesel fuels in diesel engines	Combustion and formations of emission in compression ignition engines and emissions reduction techniques
48	Konur (2021aa)	Petrodiesel fuels in general	Bioremediation of biodiesel fuel-contaminated soils	Bioremediation of petroleum hydrocarbons in contaminated soils: A review of the research.
49	Issa and Ilinca (2021)	Petrodiesel fuels in general	Biodiesel power generation	Biodiesel and petrodiesel fuels for marine applications
50	Konur (2021ab)	Petrodiesel fuels in general	Desulfurization of diesel fuels	Desulfurization of diesel fuels: A review of the research
Part XI Emissions of petrodiesel fuels				
51	Konur (2021ac)	Petrodiesel fuel emissions	Overview	Diesel fuel exhaust emissions: A scientometric review of the research
52	Fattah et al. (2021)	Petrodiesel fuel emissions	Diesel emission mitigation	Diesel emissions and their mitigation approaches
53	Gao and Liu (2021)	Petrodiesel fuel emissions	Diesel particulate emissions	Particles from compression and spark ignition engines
54	Nuguid et al. (2021)	Petrodiesel fuel emissions	Diesel NO_x emissions	Selective catalytic reduction of NO_x emissions
Part XII Health impact of petrodiesel fuel emissions				
55	Konur (2021ad)	Health impact of petrodiesel fuel emissions	Overview	The adverse health and safety impact of diesel fuels: A scientometric review of the research
56	Konur (2021ae)	Health impact of petrodiesel fuel emissions	Respiratory illnesses	Respiratory illnesses caused by diesel fuel exhaust emissions: A review of the research
57	Konur (2021af)	Health impact of petrodiesel fuel emissions	Cancer	Cancer caused by diesel fuel exhaust emissions: A review of the research
58	Konur (2021ag)	Health impact of petrodiesel fuel emissions	Cardiovascular, brain, and reproductive system illnesses	Cardiovascular and other illnesses caused by diesel fuel exhaust emissions: A review of the research

The data presented in the tables and the figure show that a small number of authors, institutions, funding bodies, journals, keywords, research fronts, subject categories, and countries have shaped the research in this field.

The findings show the importance of the development of efficient incentive structures for the development of the research, as in other fields. It further seems that, although research funding is a significant element of these incentive structures, it might not be a sole solution. On the other hand, it seems there is more to do to reduce significant gender deficit in this field as in other fields of science and technology.

The data on the research fronts, keywords, source titles, and subject categories provide valuable evidence for the interdisciplinarity of the research. There is ample justification for the broad search strategy employed in this study due to the interdisciplinary nature of this research field as evidenced by the top subject categories.

There are three major topical research fronts with respect to these sample papers: 'biodiesel fuels in general', 'feedstock-based biodiesel fuels', and 'petrodiesel fuels'.

The detailed research fronts are 'biooils', 'biodiesel fuels in general', and 'glycerol' in the first group of sample papers. The detailed research fronts are 'edible oil-based biodiesel fuels', 'nonedible oil-based biodiesel fuels', 'waste oil-based biodiesel fuels', and 'algal oil-based biodiesel fuels', respectively, in the second group of sample papers. Finally, the detailed research fronts are 'crude oils', 'petrodiesel fuels in general', 'emissions from petrodiesel fuels', and 'health impact of the emissions from petrodiesel fuels', respectively, in the final group of sample papers.

Chapter 3 (Konur, 2021g) presents a review of the most-cited 50 article papers on both biodiesel and petrodiesel fuels, complementing Chapter 2 (Konur, 2021f). Table 3.1 provides information on the primary and secondary research fronts for both biodiesel fuels and petrodiesel fuels. It also provides information on the distribution of these research fronts in the article papers, 100 most-cited sample papers, and over 121,000 population papers.

As this table shows the research fronts of 'biodiesel fuels in general', 'feedstock-based biodiesel fuels', and 'petrodiesel fuels' comprise 18, 50, and 32% of these papers, respectively.

In the first group of article papers, those on 'biooils' constitute 14% of papers; in the second group, those on 'edible oil-based biodiesel fuels' and 'algae-based biodiesel fuels' constitute 18% each of these papers, and in the final group, those on 'crude oils' and 'emissions from petrodiesel fuels' constitute 16 and 10% of these papers, respectively – all of which emerge as the prolific research fronts.

Chapter 4 (Konur, 2021h) presents a review of the most-cited 25 article papers on nanotechnology applications in both biodiesel and petrodiesel fuels. Table 4.1 provides information on the primary and secondary research fronts for both biodiesel fuels and petrodiesel fuels. As this table shows the research fronts of 'biodiesel fuels in general', 'feedstock-based biodiesel fuels', and 'petrodiesel fuels' comprise 28, 24, and 48% of these papers, respectively, in this field.

In the first group of article papers, those on 'biooils', 'biodiesel fuels in general', and 'glycerol' comprise 4, 8, and 16% of these papers, respectively. In the second group of article papers, those on 'edible oil-based biodiesel fuels', 'nonedible oil-based biodiesel fuels', 'waste oil-based biodiesel fuels', and 'algal oil-based biodiesel fuels' comprise 16, 4, 4, and 0% of these papers, respectively. In the final group of article papers, those on 'crude oils', 'petrodiesel fuels in general', 'emissions from petrodiesel fuels', and 'health impact of the emissions from petrodiesel fuels' comprise 40, 8, 0, and 0% of these papers, respectively.

It is notable that there have been no prolific papers in the fields of 'biooil upgrading', 'biodiesel emissions', 'algal biodiesel fuels', 'crude oil properties', 'crude oil refining', 'petrodiesel desulfurization', 'power generation by petrodiesel fuels', 'petrodiesel fuel emissions', and 'health impact of petrodiesel emissions'.

These prolific studies on nanotechnology applications in the fields of biooils, biodiesel fuels, crude oils, and petrodiesel fuels, presented in this chapter, highlight its importance. These studies also show the importance of nanomaterials as nanocatalyts in these fields, ranging from 'crude oil recovery' and 'remediation of crude oils in the environment' to 'biooil production', 'biodiesel production', and 'production of biochemicals and bioenergy from glycerol', a by-product of biodiesel fuels.

1.3.1.2 Part II: Biooils

There are six chapters on the research front of biooils. Chapter 5 (Konur, 2021i) maps the research on biooils using a scientometric method comprising a sample of the 100 most-prolific sample papers and a population of over 20,000 papers.

Eight research fronts emerge from the examination of the sample papers: 'biomass pyrolysis', 'biofuels from biomass pyrolysis', 'pyrolysis oil property and characterization', 'biomass conversion for biofuel production', 'biomass torrefaction', 'biomass liquefaction', 'biooil upgrading', and 'lignin conversion' (Table 5.8).

Chapter 6 (Konur, 2021j) presents the key findings of the most-cited 25 article papers on the properties and characterization of biooils. Table 6.1 provides information on the research fronts in this field. As this table shows, the primary research fronts of 'characterization of biooils' and 'properties of biooils' comprise 48 and 52% of these papers, respectively. These prolific studies on two complementary research fronts provide valuable evidence of the characterization and properties of biooils obtained from a variety of feedstock such as algae and wood.

Chapter 7 (Konur, 2021k) presents the key findings of the most-cited 25 article papers on the pyrolysis of biooils. Table 7.1 provides information on the research fronts in this field. As this table shows, the primary research fronts of 'pyrolysis' and 'upgrading of pyrolysis oils' comprise 72 and 28% of these papers, respectively. 'Kinetic studies' and 'pyrolysis oils' comprise 52 and 20% of the papers in the first group, respectively. These prolific studies on two different research fronts provide valuable evidence of biomass pyrolysis and upgrading of pyrolysis oils.

Chapter 8 (Zhang et al., 2021a) provides an overview of catalytic biooil upgrading for processing aqueous-phase compounds. They first present the challenges and objectives in upgrading whole biooil. Then, they summarize the developed approaches for catalytic conversion of small oxygenates in the aqueous phase of biooil to produce valuable biofuels and biochemicals. They discuss the promising catalysts and proposed reaction mechanisms in an attempt to generate the relationships between catalyst properties (e.g. acid-base pairs) and catalytic performance which may guide the future rational design of catalysts in biooil upgrading.

Chapter 9 (Zhang et al., 2021b) provides an overview of catalytic biooil upgrading for processing oil-phase compounds and real biooils, complementing Chapter 8

(Zhang et al., 2021a). Specifically, they focus on hydrotreating and cracking the phenolics, furanics, or heavier oligomers, as well as the approaches discussed in Zhang et al. (2021a) (e.g. steam reforming), to produce valuable biofuels and biochemicals. In addition, they discuss the advances of catalyst development and reaction mechanisms to correlate catalyst structure with performance (e.g. oxophilicity vs hydrodeoxygenation), serving as a basis for the design of selective and durable catalysts for biooil upgrading.

Chapter 10 (Cao et al., 2021) reviews the research on the biooil production through catalytic 'hydrothermal liquefaction' (HTL) of biomass, focusing on recent developments and future prospects. They note that efficient hydrothermal conversion of biomass to produce biooil is in line with the current concepts of environmental conservation and sustainable development. However, they caution that the research in this field is still at the lab-scale or pilot stage, and they assert that future research should focus on the development of highly active, hydrothermally stable, green, and easily recoverable new catalysts to realize the production of: high-quality biooils through hydrothermal conversion; the cost-effective separation of the main biomass as the route to achieve value-added products with high yield and selectivity; green and environmentally friendly biphasic or multiphasic solvents that should gain more emphasis in terms of both efficient conversion and product purification.

1.3.1.3 Part III: Biodiesel Fuels in General

There are seven chapters in the research stream of biodiesel fuels in general. Chapter 11 (Konur, 2021l) maps the research on biodiesel fuels in general using a scientometric method employing a set of 100 most-cited sample papers and a population paper set of over 6,500. There are three major topical research fronts for these sample papers: 'production of biodiesel fuels', 'properties of biodiesel fuels', and 'emissions from biodiesel fuels'.

Chapter 12 (Goh et al., 2021) provides an overview of the research on biomass-based catalyst-assisted biodiesel production. Catalysts should be introduced in order to increase the overall rate of biodiesel production and reduce capital costs. They therefore illustrate the latest breakthroughs involved in the use of biomass-derived catalysts. In addition, they provide a better framework and methods to synthesize biomass-based catalysts for biodiesel production. This provides a solution to the catalyst separation problem in the current biodiesel field and enhances the economic viability of the industry, thus sustaining the environment while meeting energy demands.

Chapter 13 (Aguieiras et al., 2021) reviews the research on enzymatic biodiesel production focusing on the challenges and future prospects in this field. Many technologies have been developed in order to produce high performance biocatalysts and many of them have shown promising results towards the enzymatic synthesis of biodiesel. Alternative sources for enzyme and biodiesel production contribute to the advancement of this application. Among different raw materials, the use of low-cost agro-industrial waste as feedstock reinsert them in the production chain, reinforcing the circular economy bases, and approaching an economically feasible enzymatic biodiesel synthesis. Considering the economic aspects, Aguieiras et al. note that there are still some limitations in the industrial application of enzymes for biodiesel production and that this field still has numerous possibilities for development whilst enzymatic production becomes a more attractive and economically viable pathway.

Chapter 14 (Tomar et al., 2021) provides an overview of the research on biodiesel additives. In search of cleaner fuels and to make biodiesel usage economically viable, additives are becoming an indispensable tool in global trade. Additives covers a wide range of subjects and are categorized into various types according to their size, chemical compounds, state of matter and functionality. The amalgamation of additives in biodiesel and its blends has a significant effect on fuel properties such as viscosity, fire point, flash point, and calorific value, which furthermore influences the combustion, performance, and emission characteristics of biodiesel fuel. Various oxygenated fuel additives improve the combustion process and lower the in-cylinder pressure due to the higher latent heat of vaporization. Recent advancements in the field of biobased additives have also set a new benchmark in the era of fuel additives. Thus, as the energy sources are upgraded towards cleaner and renewable technology, the additives' share in the world market will likely increase over the next ten years.

Chapter 15 (Kumar and Kalam, 2021) reviews the research on the qualitative characterization of biodiesel fuels. The properties of biodiesel depends largely on the choice of feedstock which often is dependent on a domestic source. The fatty acid compositions of the parent oil or fat, the nature of the alcoholic head-group of biodiesel, and the contaminants present in the produced biodiesel determine the characteristics of the biodiesel produced. Intense research in biodiesel has outlined the significance of its characteristics as the expected performance and emission characteristics of engines are by and large governed by these properties. The performance of biodiesel fuel engines is comparable to that of petrodiesel. However, a significant improvement in emissions, except nitrogen oxides (NO_x), can be achieved by employing biodiesel as fuel. Hence, Kumar and Kalam propose that future research should focus on streamlining biodiesel properties and genetic modification of feedstock as a solution.

Chapter 16 (Hoang et al., 2021) reviews the research on the use of biodiesel fuels in diesel engines. The use of biodiesel or biodiesel–petrodiesel fuel blends for diesel engines may have significant effects on engine performance and emissions, deposit formation, problems relating to engine durability, and the corrosiveness of engine components. A small portion of biodiesel (<15% of volume) in blends with petrodiesel fuel has been found to have better engine performance and emissions, except for NO_x emissions, and it does not cause any noticeable issues for components, system, and lubricating oil in diesel engines. Higher NO_x emissions could be solved through the application of advanced technologies such as 'exhaust gas recirculation' (EGR) or exhaust gas-assisted fuel reforming. The use of additives and inhibitors could bring a high effectiveness in the reduction of deposit formation and lubricating oil degradation, as well as corrosion of engine components.

Chapter 17 (Jain et al., 2021) discusses biodiesel promotion policies from a global perspective. Most of the countries around the world are not able to follow the policy guidelines for production and use of biodiesel. This can be due to unrealistic estimates regarding cost, availability of cheap and abundant feedstock, tax benefits for parity with petroproducts, improper support from established petroproduct suppliers/producers and projections of high yields. Policy guidelines must be able to facilitate technology learning and the production scale-up necessary to reduce costs. Relevant policies should include advanced biofuel quotas and financial derisking measures, e.g. loan guarantees from development banks. These would be particularly effective in those countries which possess significant feedstock resources. Countries and regions should consider policies that specify reductions in the life-cycle carbon intensity of fuel and which are effective in boosting the demand for biodiesel from waste oil, fat and grease feedstocks.

1.3.1.4 Part IV: Glycerol

There are three book chapters on glycerol research. Chapter 18 (Konur, 2021m) maps the research on the production of biofuels and biochemicals from glycerol, a by-product of biodiesel fuels, using a scientometric method with 100 sample and over 6,900 population papers. There are eight major topical research fronts for these sample papers concerning glycerol, namely its: ‘catalytic conversion’, ‘microbial conversion’, ‘hydrogenolysis’, ‘oxidation’, ‘fermentation’, ‘dehydration’, ‘reforming’, and other methods of conversion such as ‘transesterification’.

There are eight primary research fronts regarding the products of glycerol conversion: ‘hydrogen’, ‘chemicals’ in general, ‘acrolein’, ‘fuels’ in general, ‘propanediol’, ‘carbonate’, ‘ethanol’, and ‘other products’ such as ‘hydroxybutyrate’ or ‘docosahexaenoic acids’. There are also eight papers related primarily to the characterization of glycerol.

Chapter 19 (Ayodele et al., 2021) reviews the research on hydrogen-rich syngas production from glycerol focusing on the overview of the modeling and optimization strategies as a case study for the production of biofuels from glycerol. The various modeling and strategies that have been investigated for hydrogen-rich syngas production from the thermo-catalysis and bioconversion of biodiesel-derived glycerol are discussed. These various strategies have been proven to be effective in modelling and optimizing the various glycerol conversion processes.

Chapter 20 (Konur, 2021n) presents the key findings of the most-cited 25 article papers in propanediol production from glycerol as a case study of biochemical production from glycerol. The primary research fronts of ‘catalytic propanediol production’ and ‘microbial propanediol production’ comprise 44 and 56% of these papers, respectively.

1.3.2 Volume 2: Feedstock-based Biodiesel Fuels

There are four primary research fronts in this section: ‘edible oil-based biodiesel fuels’, ‘nonedible oil-based biodiesel fuels’, ‘waste oil-based biodiesel fuels’, and ‘algal oil-based biodiesel fuels’ with five, four, four, and seven book chapters, respectively.

1.3.2.1 Part V: Edible Oil-based Biodiesel Fuels

Chapter 21 (Konur, 2021o) maps the research on ‘edible oil-based biodiesel fuels’ using a scientometric method using the 100 most-cited sample papers and 2,650 population papers. There are two major topical research fronts for these sample papers: ‘edible oil-based biodiesel fuel production’ and ‘properties and characterization of edible oil-based biodiesel fuels' (Table 21.8). There are nine research fronts for the edible oils used for biodiesel production. The most prolific research front is 'soybean oil-based biodiesel fuels' with 32 sample papers. This top research front is followed by 'palm oil-based biodiesel fuels', 'rapeseed oil-based biodiesel fuels', and 'sunflower oil-based biodiesel fuels' with 19, 16, and 16 sample papers, respectively. These top four research fronts form 83% of the sample papers in total.

Chapter 22 (Westbrook, 2021) provides an overview of the research on the chemistry of 'soybean oil-based biodiesel fuels'. He shows that the presence of only five unique methyl ester molecules in a biodiesel fuel is a major simplifying factor in understanding and using detailed chemical kinetic modeling to describe its combustion. He next shows that the number and location of C=C double bonds in these component molecules is a dominant factor in distinguishing soy biodiesel fuel from biodiesel fuels produced from other vegetable oils and for predicting the combustion properties of soybean oil diesel fuel. He finally shows that the monosaturated component of soybean oil, methyl oleate, has an optimal structure as a diesel fuel in terms of minimizing pollutant emissions, maximizing the diesel cetane number of the fuel, and limiting fuel thermal stability.

Chapter 23 (Konur, 2021p) presents the key findings of the most-cited 25 article papers on 'palm oil-based biodiesel fuels'. Table 23.1 provides information on the research fronts in this field. As this table shows, the primary research fronts of 'production of palm oil-based biodiesel fuels' and 'properties of palm oil-based biodiesel fuels' comprise 64 and 44% of these papers, respectively. In the first group of papers, 'production of crude palm oil-based biodiesel fuels', 'production of palm kernel oil-based biodiesel fuels', and 'production of waste palm oil-based biodiesel fuels' form 44, 12, and 8% of these papers, respectively.

Chapter 24 (Konur, 2021q) presents the key findings of the most-cited 25 article papers on 'rapeseed oil-based biodiesel fuels'. Table 24.1 provides information on the research fronts in this field. As this table shows the primary research fronts of 'production of rapeseed oil-based biodiesel fuels' and 'properties of rapeseed oil-based biodiesel fuels' comprise 72 and 28% of these papers, respectively.

1.3.2.2 Part VI: Nonedible Oil-based Biodiesel Fuels

There are four chapters in this section. Chapter 25 (Konur, 2021r) maps the research on 'nonedible oil-based biodiesel fuels' using a scientometric method. There are two major topical research fronts for these sample papers: 'nonedible oil-based biodiesel fuel production' and 'properties and characterization of nonedible oil-based biodiesel fuels' (Table 25.8). There are nine research fronts for the nonedible oils used for biodiesel production. The most-prolific research front is 'Jatropha oil-based biodiesel fuels' with 49 sample papers. This top research front is followed by 'karanja oil-based biodiesel fuels' and 'nonedible oil-based biodiesel fuels' in general, with 16 sample papers each. These top three research fronts form 91% of the sample papers in total.

The other prolific research fronts are 'castor oil-based biodiesel fuels', 'mahua oil-based biodiesel fuels', 'polanga oil-based biodiesel fuels', 'tobacco oil-based biodiesel fuels', 'rubber oil-based biodiesel fuels', and 'other nonedible oil-based biodiesel fuels' with five, five, four, four, three, and fourteen sample papers, respectively.

Chapter 26 (Banapurmath et al., 2021) presents and discusses the research on the use of 'Jatropha oil-based biodiesel fuels' for modern diesel engine applications. They focus on the approaches to nanofluid preparation using ultrasonic probes and bath sonication methods, the stability improvement of nanofluids by the addition of surfactants, numerous characterization techniques to estimate the nanoparticle shape and size, dispersion stability, and the chemical bonding of nanoadditives in liquid fuel. Jatropha oil was subsequently converted into its biodiesel called 'Jatropha oil methyl ester' (JOME) and blended with diesel to obtain its B20 blend. Further improvement in fuel properties of the B20 blend was realized with the addition of graphene as a fuel additive with the varying doses of 15, 20, and 30 mg respectively. Accordingly, JOMEB20Gr20 represents a B20 blend of JOME with a 20 mg dosage of graphene nanoparticles. The stability of the nanoadditive fuel blends resulted in their improved thermophysical properties. Further, their use in high-pressure-assisted 'common rail diesel injection' (CRDI) diesel engines resulted in a higher heat transfer rate and higher in-cylinder pressures, while 'brake thermal efficiency' (BTE) increased and 'brake specific fuel consumption' (BSFC) reduced, with the exhaust emissions of carbon monoxide, unburned hydrocarbons, and nitrogen oxide decreasing, depending on the dosage of graphene in the fuel.

Chapter 27 (Dandu et al., 2021) reviews the research on the effects of additives with '*Calophyllum inophyllum* methyl ester' (CIME) in CI engine applications. *Calophyllum inophyllum* have several advantages including abundant availability in nature and comprising superior properties with a modest extraction process. The application of CIME and their blends in diesel engines are reported to have better characteristics in all respects of the engine as compared to that of base diesel fuel. In order to enhance the properties of the CIME fuel, various additives are identified and added to achieve improved characteristics of the engine. The additives generally suitable for CIME blends are oxygenated additives like ethers, alcohols, and nanoparticles. All the experimental results projected slightly poor performance and better emission behaviors due to the reduction in net heat content. CIME fuel performance was highly improved on a par with the BTE of diesel fuel. On the other hand, the general drawback of CIME fuel, namely NO_x emission, is easily eradicated by the addition of more alcohols.

Chapter 28 (Niju and Janani, 2021) reviews the research on the *Moringa oleifera* oil as a potential feedstock for sustainable biodiesel production. *Moringa oleifera*, with a high cetane number, appreciable oxidation stability, and a low iodine value, along with wide availability, confers the potential to utilize it for biodiesel production. Though the pour point, cloud point, and 'cold filter plugging point' (CFPP) are higher, it is possible to reduce them appreciably with a ternary blend of diesel, bioethanol, and biodiesel; ultrasonication-mediated esterification has been experimentally proven to be a cutting edge technology that reduces the acid value from 81.5 mg KOH/g of oil to 2.78 mg KOH/g of oil with a volumetric ratio of 0.6:1 methanol to *Moringa oleifera* oil and a 1.5 vol% of concentrated H_2SO_4 for 40 min at 60°C.

1.3.2.3 Part VII: Waste Oil-based Biodiesel Fuels

There are four chapters in this section. Chapter 29 (Konur, 2021s) maps the research on 'waste oil-based biodiesel fuels' using a scientometric method through the use of the 100-most-cited sample papers and 2,150 population papers.

There are two major topical research fronts for these sample papers: 'waste oil-based biodiesel fuel production' and 'properties and characterization of waste oil-based biodiesel fuels' (Table 29.8). There are three primary research fronts for the waste oils used for biodiesel production. The most-prolific research front is 'waste oil-based biodiesel fuels'. This top research front is followed by 'animal fat-based biodiesel fuels' and 'other waste oil-based biodiesel fuels'.

Chapter 30 (Siddique and Rohani, 2021) reviews the research on biodiesel production from 'municipal wastewater sludge' (MWWS). Biodiesel production from MWWS is a promising alternative to reduce biodiesel production cost, to meet growing future energy demand, and to facilitate waste management for increasing sludge generation. Although MWWS is readily available free of cost, biodiesel production from it involves challenges in processing the sludge, extracting lipid, and producing biodiesel. However, there is significant progress being made in sludge processing, lipid extraction, and biodiesel production processes. The use of oleaginous substances to cultivate the secondary sludge along with the use of an ultrasonic technique would offer a promising improvement towards biodiesel production from MWWS. A mixture of polar and nonpolar solvents, especially hexane or chloroform and methanol, can greatly enhance lipid extraction efficiency and biodiesel yield. It is important to scale up biodiesel production and characterize the biodiesel produced following 'American Society for Testing and Materials' (ASTM) standards.

Chapter 31 (Castanheiro, 2021) reviews the research on heterogeneous acid catalysts for biodiesel production from 'waste cooking oils' (WCO). Biodiesel production is performed using homogeneous catalysts, which can cause some problems such as difficulty in removing the catalyst from the reaction mixture and possibility to soap production and corrosion. The use of heterogeneous catalysts can overcome these problems. Different heterogenous acid catalysts (zeolites, heteropolyacids, metal oxides, materials with sulfonic acid groups) have been employed in biodiesel production from waste cooking oil.

Chapter 32 (Satyannarayana et al., 2021) reviews the research on 'microbial biodiesel fuel production' using microbes such as algae, fungi, and yeasts. These feedstocks have their own advantages and disadvantages that are discussed in detail in this chapter. The selection of microbial species depends on the type of feedstock used and their culture conditions. These microbial sources promise effective biodiesel production in less time and the use of biodiesel which leads to a safer environment. The chapter also focuses on the methods to accumulate high lipids and lipid extraction techniques. Further, selection and analysis of various essential biodiesel parameters like pH, salinity, temperature, and feedstock are discussed. However, there is much scope for future research on the optimization of these parameters for obtaining high-quality biodiesel.

1.3.2.4 Part VIII: Algal Oil-based Biodiesel Fuels

There are seven chapters in this section. They include a scientometric review of the research on 'algal oil-based biodiesel fuels', 'algal biomass production for biodiesel production', 'algal biomass production in wastewater for biodiesel production', 'algal lipid production for biodiesel production', 'biooil production from microalgae', 'extraction of algal neutral lipids for biofuel production', and 'microalgal biodiesel production'.

Chapter 33 (Konur, 2021t) maps the research on 'algal biomass', 'algal biooils', and 'algal oil-based biodiesel fuels' using a scientometric method through the use of the 100-most-cited sample papers and over 15,000 population papers.

There are three major research fronts for these sample papers: 'algal biomass production', 'algal biooil production', and 'algal biodiesel production' (Table 33.8). There are three research subfronts for the top research front: 'algal biomass production in general', 'algal biomass production and nutrient removal in wastewaters', and 'CO_2 bioremediation by algal biomass', with 48, 14, and 7 sample papers, respectively.

There are six research subfronts for 'algal biooil production': 'algal lipid production in general', 'algal lipid production in wastewater', 'algal biomass pyrolysis', 'algal biomass liquefaction', 'algal lipid extraction', and 'algal lipid characterization'.

Chapter 34 (Konur, 2021u) reviews the research on 'algal biomass production for biodiesel production'. This chapter presents the key findings of the 25-most-cited article papers, in five pragmatically distinct research streams in this field.

These prolific studies provide valuable evidence on cultivation engineering in general, cultivation engineering in wastewater and nutrient removal, cultivation engineering and CO_2 bioremediation, life-cycle analysis of algal biomass production, and techno-economic studies of algal biomass production. There are also two more distinct research fields: photobioreactors and cell biology, and omics studies of algal biomass production, although there are no papers for these research streams in the table of the top 25 papers.

All these research streams contribute significantly towards improving algal biomass productivity, mostly in microalgae. It is apparent that it is highly desirable to enhance both biomass productivity and lipid productivity such as through the use of a two-phase cultivation strategy where, first, biomass growth is ensured in a nutrient (N, P) replete mode and then followed by a nutrient-deficient phase.

It is then apparent that the use of wastewater and flue gases is of critical importance to reduce the production costs as well as to reduce the adverse environmental impact of algal biomass production. It is also apparent that the studies on omics and cell biology studies complement the primary cultivation processes by providing valuable tools to understand and assess them.

The ecological, techno-economic, and environmental benefits from using wastewater for these purposes are significant. The use of wastewater would make biodiesel production more competitive in relation to petrodiesel fuels in reducing the feedstock costs and reducing environmental burdens significantly.

It would also help with the treatment of wastewater. Another beneficial effect of using wastewater for algal biomass accumulation and nutrient removal would be the lessening of the volume and harmful effects of algal blooms, which have created a significant ecological disaster in recent years.

The production of algal biodiesel fuels in competition with petrodiesel fuels would also help in shifting biodiesel production from edible-oil-based biodiesel fuels, which has been associated with public concern over 'food security' and the destruction of forests for the expansion of palm oil plantations, the destruction of ecological biodiversity, the significant increase in CO_2 emissions, and finally, the exploitation of local communities.

Chapter 35 (Konur, 2021v) presents the key findings of the 25-most-cited article papers in 'algal biomass production in wastewater for biodiesel production'. These prolific studies provide valuable evidence on algal biomass production and nutrient removal from wastewater by algae as well as lipid production and biodiesel production.

Chapter 36 (Konur, 2021x) reviews the research on algal lipid production for biodiesel production. This chapter presents the key findings of the 25-most-cited article papers, in five pragmatically distinct research streams in this field. These studies provide valuable evidence on cultivation engineering in general, cultivation engineering in wastewater and nutrient removal, cultivation engineering and CO_2 bioremediation, algal lipid analysis and imaging, and omics and cell biology studies in lipid production.

All these research streams contribute significantly towards improving lipid content and productivity in algae, mostly high-lipid microalgae. It is apparent that it is highly desirable to enhance both biomass productivity and lipid productivity, such as through the use of a two-phase cultivation strategy where, first, biomass growth is ensured in a nutrient (N, P) sufficient mode and then followed by a nutrient-deficient phase.

Chapter 37 (Koley et al., 2021) reviews the research on biooil production from microalgae through the hydrothermal liquefaction of algal biomass. The direct conversion of biomass to biocrude could be a viable solution. Alongside this, it enables the biooil produced to meet the characteristics of petrocrude, and to be refined in established petroleum refineries. To further enhance the economics of biofuel production, the by-products of direct conversion, i.e. the aqueous phase and biochar, are rich in nutrients and can be used as fertilizers. Biooil production can also be helpful if simultaneous production of biodiesel and biocrude is carried out from the same biomass. The defatted biomass following the extraction of lipid can be again subjected to hydrothermal liquefaction for the production of biocrude. Nevertheless, there is enough space to fill the gaps and improve the economics of biooil production from microalgal biomass for obtaining a sustainable substitute to fossil fuels.

Chapter 38 (Krishnan et al., 2021) provides an overview of the diverse conventional and advanced extraction methods used as efficient approaches for lipid extraction from algae for biofuel production. Lipids represent a major algal biochemical with an extensive range of applications, including biodiesel synthesis. Biodiesel constitutes 'fatty acid methyl esters' (FAMEs) from the transesterification of lipids. Among algal lipids, neutral or true lipids such as triacylglycerides are important for generating high quality and stable biodiesel fuel with enhanced calorific value. New and improved physicochemical lipid extraction methods are highly critical for enabling the selective generation of true lipids from algal biomass. The authors, finally, discuss the impact of nanotechnology on enhanced lipid extraction.

Chapter 39 (Sivaramakrishnan and Incharoensakdi, 2021) provides an overview of the research on algal biodiesel fuel production. Microalgae have a high growth rate and a high lipid content. Moreover, microalgal biodiesel production is less energy intensive and economically feasible. However, the commercialization of microalgal biodiesel production is still confronted with numerous challenges in practice. The selection of microalgae is very important for achieving high biomass and lipid yield. A microalgal cultivation system is one of the crucial factors to achieve sustainable microalgal production, which can undoubtedly affect the biomass and lipid yield. The authors focus on the lipid extraction and biodiesel production methods. The discussion and suggestions based on the findings could be well described as a roadmap for acquiring knowledge amendable for microalgal biodiesel production and future microalgal biodiesel development.

1.3.3 Volume 3: Petrodiesel Fuels

There are four primary research fronts in this section: 'crude oils', 'petrodiesel fuels in general', 'emissions from petrodiesel fuels', and the 'health impact of the emissions from petrodiesel fuels'; there are six, five, four, and four chapters, respectively.

1.3.3.1 Part IX: Crude Oils

There are six chapters in this field. Chapter 40 (Konur, 2021y) maps the research on crude oils using a scientometric method using a sample of the 100-most-cited papers and over 53,000 population papers.

Seven research fronts emerge from the examination of the sample papers: 'hydrocarbon contamination and bioremediation', 'crude oil spill bioremediation', 'crude oil properties and characterization', 'crude oil recovery', 'crude oil geology', 'crude oil refining', and 'crude oil pipelines' (Table 40.8).

Chapter 41 (Fingas, 2021a) provides an introduction to oil spill behavior. He provides summaries of the important processes such as evaporation, water-in-oil emulsification, natural dispersion, and dissolution and introduces other processes such as photooxidation, sedimentation, and oil–fines interaction. An important behavior process is that of evaporation. Evaporation of light oils can involve a loss of 30 to 60% of the oil's mass over a two-day period. He outlines the algorithms for calculating evaporation. Another important behavior is that of water uptake. Oil may take up water in one of five ways: as soluble water, as no significant uptake (unstable), as entrained water, as mesostable emulsions, or as stable emulsions. The latter three mechanisms are important as they have a large influence on the further behavior of the oil and, thus, a significant impact on countermeasures.

Chapter 42 (Fingas, 2021b) provides an introduction to oil spills and their clean-up as a follow-up to Fingas (2021a). The author notes that with the increasing use of these products in almost every facet of our lives, oil spills are also an accompanying reality. Despite a vast improvement in our technology to utilize oil, our ability to deal with oil spills has not kept pace. Experts still estimate that 30 to 50% of oil spills are either directly or indirectly caused by human error, with 20 to 40% of all spills caused by equipment failure or malfunction. Emerging spill risks include increased maritime activity in the Arctic, deep water exploration and development, and the rapid expansion of the rail transportation of crude oil. Oil spills have many adverse effects on the environment. However, efforts at spill containment and recovery are considered to be only moderately effective. Most often, spilled oil strands on the shoreline require clean-up efforts, though care is needed to minimize additional harm that can slow overall recovery. The understanding of the effects of oil spills is increasing, with the result that the consequences are much greater than once thought, as is their persistence.

Chapter 43 (Prince et al., 2021) provides an overview of the research on crude oils and their fate in the environment. They focus on the formation, production, and composition of crude oils, and the processes that change and eventually consume them if they are released into the environment. Most of these processes merely move components of the oil from one environmental compartment to another; only combustion and biodegradation remove it permanently. Fortunately microbes capable of consuming oil are ubiquitous, and under optimal conditions they can consume spilled hydrocarbons quite rapidly – with a half-life of one to three weeks when dispersed at sea, for example. The challenge is to get the oil into a form where such rapid biodegradation can occur.

Chapter 44 (McFarlin and Prince, 2021) reviews the research on crude-oil-contaminated soils, focusing on the biodegradation and remediation strategies. They highlight the environmental factors that affect the fate of petroleum hydrocarbons found in crude oil and refined products, and discuss the implementation and success of common bioremediation technologies that can significantly remove such hydrocarbons from the environment. Response options include '*ex situ* mechanical methods' involving haul and treat, and '*in situ* biological enhancement methods' that involve bioremediation. Chemical characterization of the spilled oil, combined with environmental characterization of the soil, helps to guide the most efficient remediation strategy. *Ex situ* bioremediation technologies, such as composting, landfarming, and biopiles, may offer advantages such as containment and process control. Other ameliorations, such as bulking agents and biochars, have been shown to enhance oil biodegradation in soil. Biostimulation and bioaugmentation techniques strive to optimize environmental conditions and microbial populations to enhance the bioavailability of hydrocarbons to oil-degrading microorganisms and increase biodegradation. A potential drawback to bioremediation is the amount of time required to achieve compliance compared to mechanical techniques that either combust the contaminant, or move it to a hazardous-waste landfill.

Chapter 45 (Jeong and Lee, 2021) discusses the biotechnology applications in oil recovery. This chapter deals with the various mechanisms used for oil production in the 'microbial enhanced oil recovery' (MEOR) process. The main strategies for the application of MEOR are determined by the main substances generated from microbial activity. In particular, selective plugging by biomass/biopolymers, and wettability alteration by biosurfactants, are described as main mechanisms for oil recovery. The effects of biosolvents, bioacids, and biogases are also engaged. Additionally, the degradation of hydrocarbons by microorganisms and the effect of improving oil recovery is described. This review aims to increase an understanding of how microbial technology can be applied to oil production engineering. Literature pertaining to laboratory experiments, numerical simulations, and field trials are included in the light of industrial applications.

1.3.3.2 Part X: Petrodiesel Fuels in General

There are five chapters in this section. Chapter 46 (Konur, 2021y) maps the research on petrodiesel fuels using a scientometric method. There are eight major topical research fronts for these sample papers. 'Diesel fuel exhaust emissions' is the most prolific research front. In this front, 'diesel fuel exhaust particle emissions' is the most-prolific research subfront. The other prolific subfronts are 'NO_x emissions', 'exhaust in general', and 'aerosol emissions'.

This top research front is followed by the 'health impact of diesel fuel exhaust emissions'. In this research front the most-prolific subfront is 'respiratory illnesses' caused by diesel fuel exhaust emissions. The other main subfronts are 'allergy', 'cancer', and 'vascular illnesses' caused by diesel fuel exhaust emissions, toxicity, and the general health impact of these emissions.

The other research fronts are 'desulfurization' of diesel fuels, 'combustion and properties of diesel fuels', 'power generation' by diesel fuels, 'diesel engines', 'environmental impact of diesel fuels', 'bioremediation of diesel-fuel-contaminated soils', and 'production of diesel fuels'.

Chapter 47 (Pandey and Sharma, 2021) reviews the research on the combustion and formation of emissions in 'compression ignition engines' (CIs) and emission reduction techniques. Thermal efficiency and fuel economy have improved in recent years, which has led to the development of 'common-rail fuel injection' (CRFI) systems. Recently, most of the research has been focused on emission control due to rigid new emission norms. Engine emissions can further be reduced by using after-treatment technology such as a 'diesel particulate filter' (DPF), 'selective catalytic reduction' (SCR), 'particulate matter' (PM), and NO_x reduction in exhausts. In diesel engines, vegetable oils and fatty acids can be used as alternative fuels. High viscosity and the tendency for NO_x emissions are the main barriers in the application of vegetable oils and fatty acids as alternative fuels. Alcohol has come out as an alternative to petrodiesel fuel, which could be facilitated by blending, emulsification, and fumigation. New combustion technology such as 'homogeneous charge compression ignition' (HCCI) can be used to improve diesel exhaust emissions. HCCI technology also improves performance while operating with alternate fuels or by using after-treatment technologies.

Chapter 48 (Konur, 2021aa) presents the key findings of the 20-most-cited article papers on the bioremediation of petroleum hydrocarbons in contaminated soil, the presence of which has been well established. This includes contamination by petrodiesel fuels and crude oils. The ecotoxicity of these soils contaminated with petroleum hydrocarbons to both terrestrial and aquatic organisms has also been well established.

Thus, the bioremediation of these toxic petroleum hydrocarbons in oils has been necessary to reduce the ecotoxicity of these compounds to terrestrial and aquatic organisms, meeting the large concerns of the public regarding this matter, such as the potential loss of biodiversity in soils.

The studies presented in this chapter provide the key findings on the bioremediation of the toxic petroleum compounds found in soils. These studies show that bioremediation of these toxic compounds has been efficiently achieved using a number of microbes and plants.

Chapter 49 (Issa and Ilinca, 2021) discusses petrodiesel and biodiesel fuels for marine applications. In the maritime market, fossil-derived fuels have been dominated by 'heavy fuel oil' (HFO), which is conventionally used in low-speed (main) engines, and more refined fuels such as 'marine diesel oil' (MDO), which is used in fast or medium-speed engines. Nonetheless, rising fuel costs and regulatory pressure, such as sulfur content restrictions, have increased interest in the use of alternate fuels. A variety of such fuels have been reported and which can be used in the maritime sector, including 'straight vegetable oil' (SVO) as an alternative to HFO in 'low-speed engines', biodiesel as a replacement for MDO/'marine gasoil' (MGO) in 'low-to medium-speed engines', and 'bio-liquid natural gas' (bio-LNG) in gas engines using LNG.

Chapter 50 (Konur, 2021ab) presents the key findings of the 25-most-cited article papers on the desulfurization of diesel fuels. Table 50.1 provides information on the research fronts in this field. As this table shows, the primary research fronts of 'hydrodesulfurization of sulfur compounds', 'adsorptive desulfurization of sulfur compounds', 'oxidative desulfurization of sulfur compounds', and 'extractive desulfurization of sulfur compounds', comprise 24, 32, 20, and 20% of these papers, respectively. Furthermore, 'biodesulfurization of sulfur compounds' comprises 4% of these papers.

These studies on five complementary research fronts provide valuable evidence on the desulfurization of sulfur compounds in petrodiesel fuels and other related fuels.

1.3.3.3 Part XI: Petrodiesel Fuel Exhaust Emissions

There are four chapters in this section. Chapter 51 (Konur, 2021ac) maps the research on 'diesel fuel exhaust emissions' using a scientometric method, complementing the chapters on the adverse health and safety impact of these emissions, through the use of the 100-most-cited sample papers and over 10,450 population papers.

Three research fronts emerge from the examination of the sample papers: 'diesel fuel exhaust emissions in general', 'diesel fuel exhaust particle emissions', and 'diesel fuel exhaust NO_x emissions'.

Chapter 52 (Fattah et al., 2021) provides an overview of the research on diesel emissions and their mitigation approaches. Diesel exhaust contains components of a complete air and carbon combustion (nitrogen, water vapor, and carbon dioxide), as well as the products of incomplete combustion (i.e. carbon monoxide, nitrogen oxides, hydrocarbons, and partially oxidized hydrocarbons such as particulates). This chapter begins with the introduction of the diesel combustion process and how those emissions are generated in diesel flames. After that, the criteria emissions are discussed in detail. Finally, some of the prominent exhaust emission reduction approaches and their efficiencies are discussed.

Chapter 53 (Gao and Lu, 2021) reviews the research on particles from compression (CI) and spark-ignition (SI) engines. In the first section, mass and number distributions, chemical compositions, and control methods of particles emitted by CI engines are introduced, whilst in the second section, physical properties of particles from 'port fuel injection' (PFI) and 'direct injection' (DI) engines are presented.

Chapter 54 (Nuguid et al., 2021) reviews the research on the SCR of NO_x emissions. SCR has emerged as the most efficient technology to curtail NO_x emissions from diesel-powered vehicles and power plants. SCR is an indispensable component of 'diesel exhaust after-treatment', where it is used to lower the 'NO_x emissions' to acceptable values. Although other reductants were employed in the past, only NH_3 achieved commercial success and widespread use due to its high efficiency and selectivity. Three families of materials have emerged as the leading SCR catalysts. Each offers a unique set of advantages and disadvantages that must be considered upon catalyst selection (Table 54.1).

1.3.3.4 Part XII: Health Impact of Petrodiesel Fuel Exhaust Emissions

There are four chapters in this part. Chapter 55 (Konur, 2021ad) maps the research on the 'adverse health and safety impact of diesel fuel exhausts' using a scientometric method through the use of the 100-most-cited papers and over 2,150 population papers.

Four research fronts emerge from the examination of the sample papers. The main research fronts are 'respiratory illnesses', 'cancer' and 'other illnesses' caused by diesel fuel exhausts, and the 'characterization of diesel exhaust particles' relevant to the determination of the health effect of these fuels.

Chapter 56 (Konur, 2021ae) presents a review of the 25-most-cited article papers on the respiratory illnesses caused by diesel fuel exhaust emissions. Table 56.1 provides information on the research fronts in this field. As this table shows, the research fronts of 'human studies', 'animal studies', and '*in vitro* studies' comprise 40, 20, and 44% of these papers, respectively'. These studies in three different research fronts provide valuable evidence on the causation of respiratory illnesses by 'diesel exhaust particles' (DEPs) with strong public policy implications.

Chapter 57 (Konur, 2021af) provides an overview of the research on cancer caused by diesel fuel exhaust emissions and the key findings of the 25-most-cited article papers in this field. Table 57.1 provides information on the research fronts in this field. As this table shows the research fronts of 'human studies', 'animal studies', and '*in vitro* studies' comprise 24, 32, and 44% of these papers, respectively. These studies on three different research fronts provide valuable evidence on the causation of cancer by the mutagenic fractions of DEPs with strong public policy implications.

Chapter 58 (Konur, 2021ag) presents the key findings of the 25-most-cited article papers on cardiovascular and other illnesses caused by diesel fuel exhaust emissions. Table 58.1 provides information on the research fronts in this field. As this table shows the research fronts of 'cardiovascular illnesses', 'brain illnesses', and 'reproductive system illnesses' comprise 64, 24, and 12% of these papers, respectively.

These studies on three different research fronts provide valuable evidence on the causation of 'cardiovascular illnesses', 'brain illnesses', and 'reproductive system illnesses' by DEPs with strong public policy implications.

1.4 DISCUSSION

This handbook aims to present a representative sample of the research on both biodiesel and petrodiesel fuels based on over 121,000 population papers as of January 2020. The major research fronts are determined from the sample and population paper-based scientometric studies. Table 1.1 presents information on the major and secondary research fronts emanating from these scientometric studies.

The chapters related to biodiesel fuels are divided into two separate major research fronts: biodiesel fuels in general, focusing on the processes, and feedstock-based biodiesel fuels, focusing on the feedstock-specific processes.

There are three secondary research fronts for the research front of biodiesel fuels in general: biooils, biodiesel fuels in general, and glycerol. Additionally, the papers related to an overview of both biodiesel and petrodiesel fuels are included in this first volume.

Similarly, there are four secondary research fronts for the second volume on feedstock-based biodiesel fuels: edible oil-based biodiesel fuels, nonedible oil-based biodiesel fuels, waste oil-based biodiesel fuels, and algal oil-based biodiesel fuels.

Finally, there are four secondary research fronts in the third volume on petrodiesel fuels: crude oils, petrodiesel fuels in general, petrodiesel fuel emissions, and the health impact of petrodiesel fuel emissions.

There are in total 58 chapters and 83 authors presented in this handbook in three volumes.

As seen from Table 1.1 there is a substantial correlation between the distribution of research fronts in this handbook and the 100-most-cited sample papers.

Papers on petrodiesel fuels (third volume) are under-represented significantly in both this handbook (33%) and the 100-most-cited sample papers (34%), compared to the population papers (67.4%). On the other hand, papers on feedstock-based biodiesel fuels (second volume) are over-represented compared to the population papers: 33, 30 v. 18.3%. Similarly, papers on the biodiesel fuels in general (first volume) are over-represented compared to the population papers: 34, 41 v. 27.7%.

Table 1.2 extends Table 1.1 and provides more information on the content of this handbook, including chapter numbers, paper references, primary and secondary research fronts, and finally the titles of the papers presented in this handbook.

The data presented in the tables and in the figures show that a small number of authors, institutions, funding bodies, journals, keywords, research fronts, subject categories, and countries have shaped the research in this field.

The findings show the importance of the development of efficient incentive structures for the progress of research in this field, as in other fields. It further seems that, although research funding is a significant element of these incentive structures, it might not be a sole solution for increasing incentives to perform research. On the other hand, it seems there is more to do to reduce significant the gender deficit in this field as in other fields of science and technology.

The information provided on the nanotechnology applications in both biodiesel and petrodiesel fuels suggests that there is ample scope for their expansion in this research field (Konur, 2021h).

The research on biooils is a fundamental part of the research on biodiesel fuels. Research in this field has intensified in recent years with the application of advanced catalytic technologies and nanotechnologies in both the production and upgrading of biooils.

The research on biodiesel fuels in general has progressed along the lines of their production, properties, and emissions. As in the case of biooils, catalysts and additives play a crucial role for biodiesel fuels.

The research on glycerol has intensified in recent years with the increasing volume of biodiesel fuels, creating eco-friendly solutions for biodiesel waste by producing valuable biofuels and biochemicals from glycerol.

The research on feedstocks for biodiesel fuels first focused on edible oils as first-generation biodiesel fuels. However, public concerns about the competition with foods based on these feedstocks and the adverse impact on ecological diversity and deforestation have resulted in the exploration of nonedible-oil-based biodiesel fuels as second-generation biodiesel fuels. Due to the ecological and cost benefits of treating waste, waste oil-based biodiesel fuels as third generation biodiesel fuels have emerged. Furthermore, following a seminal paper by Chisti (2007) and other influential review papers, the research has focused recently on algal oil-based biodiesel fuels.

Since the cost of feedstocks in general constitutes 85% of the total biodiesel production costs, the research has focused more on improving biomass and lipid productivity in this research field. Furthermore, since water, CO_2, and nutrients (primarily N and P) have been major ingredients in algal biomass and lipid production, research has also intensified into the use of wastewater and flue gasses for algal biomass production to reduce the ecological burdens and production costs.

Following substantial public concerns with the adverse impact of the emissions of petrodiesel fuels on the environment and human health, the research has intensified in the areas related to the reduction of these adverse effects. Thus, bioremediation of spills from crude oils and petrodiesel fuels at sea and soils as well as the desulfurization of petrodiesel fuels has emerged as publicly important research areas.

Similarly the emissions from diesel fuel exhausts due to their adverse effects on both human health and the environment have been researched more in recent years. These emissions cover particulate emissions, aerosol emissions, and NO_x emissions.

Research on the adverse health impact of petrodiesel fuel exhaust emissions on human health has progressed along the lines of respiratory illnesses, cancer, and other illnesses, such as cardiovascular, brain, and reproductive system illnesses, through human, animal, and *in vitro* studies.

It is clear that these illnesses caused by petrodiesel fuel exhaust emissions have been one of the most significant reasons for developing alternative biodiesel fuels.

1.5 CONCLUSIONS

This handbook presents a representative sample of the research on both biodiesel and petrodiesel fuels based on over 121,000 population papers (as of January 2020) in 3 volumes, 12 parts, and 58 chapters.

The major research fronts and issues have been determined through a number of scientometric studies and the handbook has been structured to cover these major research fronts and issues.

Thus, this handbook is a valuable resource for stakeholders primarily in the research fields of Energy Fuels, Chemical Engineering, Environmental Sciences, Biotechnology and Applied Microbiology, Physical Chemistry, Petroleum Engineering, Environmental Engineering, Multidisciplinary Chemistry, Thermodynamics, Analytical Chemistry, Mechanical Engineering, Agricultural Engineering, Marine Freshwater Biology, Green Sustainable Science Technology, Applied Chemistry, Multidisciplinary Geosciences, Microbiology, Multidisciplinary Materials Science, Mechanics, Toxicology, Multidisciplinary Sciences, Biochemistry and Molecular Biology, Water Resources, Plant Sciences, Multidisciplinary Engineering, Transportation Science Technology, Geochemistry and Geophysics, Food Science Technology, Ecology, Public Environmental Occupational Health, Meteorology and Atmospheric Sciences, Electrochemistry, and Biochemical Research Methods.

This handbook is also particularly relevant in the context of biomedical sciences for Public and Environmental Occupational Health, Pharmacology, Immunology, Respiratory System, Allergy, Genetics Heredity, Oncology, Experimental Medical Research, Critical Care Medicine, General Internal Medicine, Cardiovascular Systems, Physiology, Medicinal Chemistry, and Endocrinology and Metabolism.

It is recommended that similar handbooks are developed for other types of fuels such as hydrogen energy or ethanol fuels as they are the other most-studied fuels.

ACKNOWLEDGEMENTS

The contribution of the highly cited researchers in the fields of biodiesel and petrodiesel fuels ishas been greatly acknowledged.

REFERENCES

Agarwal, A. K. 2007. Biofuels (alcohols and biodiesel) applications as fuels for internal combustion engines. *Progress in Energy and Combustion Science* 33: 233–271.

Aguieiras, E. C. G., E. P. Cipolatti, M. C. C. Pinto, J. G. Duarte, E. A. Manoel, and D. M. G. G. Freire. 2021. Enzymatic biodiesel production: Challenges and future perspectives. In *Handbook of Biodiesel and Petrodiesel Fuels: Science, Technology, Health, and Environment. Volume 1. Biodiesel Fuels: Science, Technology, Health, and Environment*, ed. O. Konur. Boca Raton, FL: CRC Press.

Ahmadun, F. R., A. Pendashteh, and L. C. Abdullah, et al. 2009. Review of technologies for oil and gas produced water treatment. *Journal of Hazardous Materials* 170: 530–551.

Atlas, R. M. 1981. Microbial degradation of petroleum hydrocarbons: An environmental perspective. *Microbiological Reviews* 45: 180–209.

Ayodele, B. V., S. I. Mustapa, and M. A. Alsaffa. 2021. Hydrogen-rich syngas production from biodiesel-derived glycerol: An overview of modeling and optimization strategies. In *Handbook of Biodiesel and Petrodiesel Fuels: Science, Technology, Health, and Environment. Volume 1. Biodiesel Fuels: Science, Technology, Health, and Environment*, ed. O. Konur Boca Raton, FL: CRC Press.

Babich, I. V. and J. A. Moulijn. 2003. Science and technology of novel processes for deep desulfurization of oil refinery streams: A review. *Fuel* 82: 607–631.

Banapurmath, N. R., M. E. Soudagar, S. V. Ganachari, P. S. Kulkarni, N. K. Kumar, and V. S. Yaliwal. 2021. An exhaustive study on the use of Jatropha based biodiesel for modern diesel engine applications. In *Handbook of Biodiesel and Petrodiesel Fuels: Science, Technology, Health, and Environment. Volume 2. Biodiesel Fuels Based on the Edible and Nonedible Feedstocks, Wastes, and Algae: Science, Technology, Health, and Environment*, ed. O. Konur. Boca Raton, FL: CRC Press.

Birch, M. E. and R. A. Cary. 1996. Elemental carbon-based method for monitoring occupational exposures to particulate diesel exhaust. *Aerosol Science and Technology* 25: 221–241.

Brennan, L. and P. Owende. 2010. Biofuels from microalgae: A review of technologies for production, processing, and extractions of biofuels and co-products. *Renewable & Sustainable Energy Reviews* 14: 557–577.

Bridgwater, A. V. 2012. Review of fast pyrolysis of biomass and product upgrading. *Biomass & Bioenergy* 38: 68–94.

Bridgwater, A. V., D. Meier, and D. Radlein. 1999. An overview of fast pyrolysis of biomass. *Organic Geochemistry* 30: 1479–1493.

Bridgwater, A. V. and G. V. C. Peacocke. 2000. Fast pyrolysis processes for biomass. *Renewable & Sustainable Energy Reviews* 4: 1–73.

Busca, G., L. Lietti, G. Ramis, and F. Berti. 1998. Chemical and mechanistic aspects of the selective catalytic reduction of NO_x by ammonia over oxide catalysts: A review. *Applied Catalysis B-Environmental* 18: 1–36.

Cao, L. C., D. C. W. Tsang, and S. C. Zhang. 2021. Bio-oil production through hydrothermal liquefaction (HTL) of biomass: Recent development and future prospects. In *Handbook of Biodiesel and Petrodiesel Fuels: Science, Technology, Health, and Environment. Volume 1. Biodiesel Fuels: Science, Technology, Health, and Environment*, ed. O. Konur. Boca Raton, FL: CRC Press.

Castanheiro, J. E. 2021. Heterogeneous acid-catalysts for biodiesel production from waste cooking oil. In *Handbook of Biodiesel and Petrodiesel Fuels: Science, Technology, Health, and Environment. Volume 2. Biodiesel Fuels Based on the Edible and Nonedible Feedstocks, Wastes, and Algae: Science, Technology, Health, and Environment*, ed. O. Konur. Boca Raton, FL: CRC Press.

Chen, C. Y., K. L. Yeh, R. Aisyah, D. J. Lee, and J. S. Chang. 2011. Cultivation, photobioreactor design and harvesting of microalgae for biodiesel production: A critical review. *Bioresource Technology* 102: 71–81.

Chisti, Y. 2007. Biodiesel from microalgae. *Biotechnology Advances* 25: 294–306.

Chisti, Y. 2008. Biodiesel from microalgae beats bioethanol. *Trends in Biotechnology* 26: 126–131.

Czernik, S. and A. V. Bridgwater. 2004. Overview of applications of biomass fast pyrolysis oil. *Energy & Fuels* 18: 590–598.

Dandu, M. S. R., K. Nanthagopal, and B. Ashok. 2021. The effects of additives with *Calophyllum inophyllum* methyl ester (CIME) in CI engine applications. In *Handbook of Biodiesel and Petrodiesel Fuels: Science, Technology, Health, and Environment. Volume 2. Biodiesel Fuels based on the Edible and Nonedible Feedstocks, Wastes, and Algae: Science, Technology, Health, and Environment*, ed. O. Konur. Boca Raton, FL: CRC Press.

Delfino, R. J., C. Sioutas, and S. Malik. 2005. Potential role of ultrafine particles in associations between airborne particle mass and cardiovascular health. *Environmental Health Perspectives* 113: 934–946.

Di Blasi, C. 2008. Modeling chemical and physical processes of wood and biomass pyrolysis. *Progress in Energy and Combustion Science* 34: 47–90.

Diaz-Sanchez, D., A. Tsien, J. Fleming, and A. Saxon. 1997. Combined diesel exhaust particulate and ragweed allergen challenge markedly enhances human *in vivo* nasal ragweed-specific IgE and skews cytokine production to a T helper cell 2-type pattern. *Journal of Immunology* 158: 2406–2413.

Elliott, D. C. 2007. Historical developments in hydroprocessing bio-oils. *Energy & Fuels* 21: 1792–1815.

Fattah, I. M. R., H. C. Ong, T. M. I. Mahlia, and M. Mofijur. 2021. Diesel emissions and their mitigation approaches. In *Handbook of Biodiesel and Petrodiesel Fuels: Science, Technology, Health, and Environment. Volume 3. Petrodiesel Fuels: Science, Technology, Health, and Environment*, ed. O. Konur. Boca Raton, FL: CRC Press.

Fingas, M. 2021a. Introduction to oil spill behavior. In *Handbook of Biodiesel and Petrodiesel Fuels: Science, Technology, Health, and Environment. Volume 3. Petrodiesel Fuels: Science, Technology, Health, and Environment*, ed. O. Konur. Boca Raton, FL: CRC Press.

Fingas, M. 2021b. Introduction to oil spills and their cleanup. In *Handbook of Biodiesel and Petrodiesel Fuels: Science, Technology, Health, and Environment. Volume 3. Petrodiesel Fuels: Science, Technology, Health, and Environment*, ed. O. Konur. Boca Raton, FL: CRC Press.

Gao, J. and Y. Liu. 2021. Particles from compression and spark ignition engines. In *Handbook of Biodiesel and Petrodiesel Fuels: Science, Technology, Health, and Environment. Volume 3. Petrodiesel Fuels: Science, Technology, Health, and Environment*, ed. O. Konur. Boca Raton, FL: CRC Press.

Gekko, K. and S. N. Timasheff. 1981. Mechanism of protein stabilization by glycerol: Preferential hydration in glycerol-water mixtures. *Biochemistry* 20: 4667–4676.

Goh, C. M. H., Y. H. Tan, J. Kansedo, N. M. Mubarak, and M. Lokman. 2021. Biomass-based catalyst-assisted biodiesel production. In *Handbook of Biodiesel and Petrodiesel Fuels: Science, Technology, Health, and Environment. Volume 1. Biodiesel Fuels: Science, Technology, Health, and Environment*, ed. O. Konur. Boca Raton, FL: CRC Press.

Graboski, M. S. and R. L. McCormick. 1998. Combustion of fat and vegetable oil derived fuels in diesel engines. *Progress in Energy and Combustion Science* 24: 125–164.

Griffiths, M. J. and S. T. L. Harrison. 2009. Lipid productivity as a key characteristic for choosing algal species for biodiesel production. *Journal of Applied Phycology* 21: 493–507.

Grima, E. M., E. H. Belarbi, F. G. A. Fernandez, A. R. Medina, and Y. Chisti. 2003. Recovery of microalgal biomass and metabolites: Process options and economics. *Biotechnology Advances* 20: 491–515.

Haritash, A. K. and C. P. Kaushik. 2009. Biodegradation aspects of polycyclic aromatic hydrocarbons (PAHs): A review. *Journal of Hazardous Materials* 169: 1–15.

Hazen, T. C., E. A. Dubinsky, and T. Z. DeSantis, et al. 2010. Deep-sea oil plume enriches indigenous oil-degrading bacteria. *Science* 330: 204–208.

Hill, J., E. Nelson, D. Tilman, S. Polasky, and D. Tiffany. 2006. Environmental, economic, and energetic costs and benefits of biodiesel and ethanol biofuels. *Proceedings of the National Academy of Sciences of the United States of America* 103: 11206–11210.

Hoang, A. T., V. V. Pham, and X. P. Nguyen. 2021. Use of biodiesel fuels in diesel engines. In *Handbook of Biodiesel and Petrodiesel Fuels: Science, Technology, Health, and Environment. Volume 1. Biodiesel Fuels: Science, Technology, Health, and Environment*, ed. O. Konur. Boca Raton, FL: CRC Press.

Hu, Q., M. Sommerfeld, and E. Jarvis, et al. 2008. Microalgal triacylglycerols as feedstocks for biofuel production: Perspectives and advances. *Plant Journal* 54: 621–639.

Issa, M. and A. Ilinca. 2021. Petrodiesel and biodiesel fuels for marine applications. In *Handbook of Biodiesel and Petrodiesel Fuels: Science, Technology, Health, and Environment. Volume 3. Petrodiesel Fuels: Science, Technology, Health, and Environment*, ed. O. Konur. Boca Raton, FL: CRC Press.

Jain, S. K., S. Kumar, and A. Chaube. 2021. Biodiesel promotion policies: A global perspective. In *Handbook of Biodiesel and Petrodiesel Fuels: Science, Technology, Health, and Environment. Volume 1. Biodiesel Fuels: Science, Technology, Health, and Environment*, ed. O. Konur. Boca Raton, FL: CRC Press.

Jeong, M. S. and K. S. Lee. 2021. Biotechnology applications in oil recovery. In *Handbook of Biodiesel and Petrodiesel Fuels: Science, Technology, Health, and Environment. Volume 3. Petrodiesel Fuels: Science, Technology, Health, and Environment*, ed. O. Konur. Boca Raton, FL: CRC Press.

John, R. P., G. S. Anisha, K. M. Nampoothiri, and A. Pandey. 2011. Micro and macroalgal biomass: A renewable source for bioethanol. *Bioresource Technology* 102: 186–193.

Kilian, L. 2009. Not all oil price shocks are alike: Disentangling demand and supply shocks in the crude oil market. *American Economic Review* 99: 1053–1069.

Knothe, G. 2005. Dependence of biodiesel fuel properties on the structure of fatty acid alkyl esters. *Fuel Processing Technology* 86: 1059–1070.

Koebel, M., M. Elsener, M. Kleemann. 2000. Urea-SCR: A promising technique to reduce Nox emissions from automotive diesel engines. *Catalysis Today* 59: 335-345.

Koley, S., S. Sonkar, and N. Mallick. 2021. Bio-oil production from microalgae: Current status and future prospective. In *Handbook of Biodiesel and Petrodiesel Fuels: Science, Technology, Health, and Environment. Volume 2. Biodiesel Fuels based on the Edible and Nonedible Feedstocks, Wastes, and Algae*, ed. O. Konur. Boca Raton, FL: CRC Press.

Konur, O. 2000. Creating enforceable civil rights for disabled students in higher education: An institutional theory perspective. *Disability & Society* 15: 1041–1063.

Konur, O. 2002a. Access to Nursing Education by disabled students: Rights and duties of nursing programs. *Nurse Education Today* 22: 364–374.

Konur, O. 2002b. Assessment of disabled students in higher education: Current public policy issues. *Assessment and Evaluation in Higher Education* 27: 131–152.

Konur, O. 2002c. Access to employment by disabled people in the UK: Is the Disability Discrimination Act working? *International Journal of Discrimination and the Law* 5: 247–279.

Konur, O. 2006a. Participation of children with dyslexia in compulsory education: Current public policy issues. *Dyslexia* 12: 51–67.

Konur, O. 2006b. Teaching disabled students in Higher Education. *Teaching in Higher Education* 11: 351–363.

Konur, O. 2007a. A judicial outcome analysis of the Disability Discrimination Act: A windfall for the employers? *Disability & Society* 22: 187–204.

Konur, O. 2007b. Computer-assisted teaching and assessment of disabled students in higher education: The interface between academic standards and disability rights. *Journal of Computer Assisted Learning* 23: 207–219.

Konur, O., ed. 2021a. *Handbook of Biodiesel and Petrodiesel Fuels: Science, Technology, Health, and Environment*. Boca Raton, FL: CRC Press.

Konur, O., ed. 2021b. *Handbook of Biodiesel and Petrodiesel Fuels: Science, Technology, Health, and Environment. Volume 1. Biodiesel Fuels: Science, Technology, Health, and Environment.* Boca Raton, FL: CRC Press.

Konur, O., ed. 2021c. *Handbook of Biodiesel and Petrodiesel Fuels: Science, Technology, Health, and Environment. Volume 2. Biodiesel Fuels based on the Edible and Nonedible Feedstocks, Wastes, and Algae: Science, Technology, Health, and Environment.* Boca Raton, FL: CRC Press.

Konur, O., ed. 2021d. *Handbook of Biodiesel and Petrodiesel Fuels: Science, Technology, Health, and Environment. Volume 3. Petrodiesel Fuels: Science, Technology, Health, and Environment.* Boca Raton, FL: CRC Press.

Konur, O. 2021e. Biodiesel and petrodiesel fuels: Science, technology, health, and environment. In *Handbook of Biodiesel and Petrodiesel Fuels: Science, Technology, Health, and Environment. Volume 1. Biodiesel Fuels: Science, Technology, Health, and Environment*, ed. O. Konur. Boca Raton, FL: CRC Press.

Konur, O. 2021f. Biodiesel and petrodiesel fuels: A scientometric review of the research. In *Handbook of Biodiesel and Petrodiesel Fuels: Science, Technology, Health, and Environment. Volume 1. Biodiesel Fuels: Science, Technology, Health, and Environment*, ed. O. Konur. Boca Raton, FL: CRC Press.

Konur, O. 2021g. Biodiesel and petrodiesel fuels: A review of the research. In *Handbook of Biodiesel and Petrodiesel Fuels: Science, Technology, Health, and Environment. Volume 1. Biodiesel Fuels: Science, Technology, Health, and Environment*, ed. O. Konur. Boca Raton, FL: CRC Press.

Konur, O. 2021h Nanotechnology applications in the diesel fuels and the related research fields: A review of the research. In *Handbook of Biodiesel and Petrodiesel Fuels: Science, Technology, Health, and Environment. Volume 1. Biodiesel Fuels: Science, Technology, Health, and Environment*, ed. O. Konur. Boca Raton, FL: CRC Press.

Konur, O. 2021i. Biooils: A scientometric review of the research. In *Handbook of Biodiesel and Petrodiesel Fuels: Science, Technology, Health, and Environment. Volume 1. Biodiesel Fuels: Science, Technology, Health, and Environment*, ed. O. Konur. Boca Raton, FL: CRC Press.

Konur, O. 2021j. Characterization and properties of biooils: A review of the research. In *Handbook of Biodiesel and Petrodiesel Fuels: Science, Technology, Health, and Environment. Volume 1. Biodiesel Fuels: Science, Technology, Health, and Environment*, ed. O. Konur. Boca Raton, FL: CRC Press.

Konur, O. 2021k. Biomass pyrolysis and pyrolysis oils: A review of the research. In *Handbook of Biodiesel and Petrodiesel Fuels: Science, Technology, Health, and Environment. Volume 1. Biodiesel Fuels: Science, Technology, Health, and Environment*, ed. O. Konur. Boca Raton, FL: CRC Press.

Konur, O. 2021l. Biodiesel fuels: A scientometric review of the research. In *Handbook of Biodiesel and Petrodiesel Fuels: Science, Technology, Health, and Environment. Volume 1. Biodiesel Fuels: Science, Technology, Health, and Environment*, ed. O. Konur. Boca Raton, FL: CRC Press.

Konur, O. 2021m. Glycerol: A scientometric review of the research. In *Handbook of Biodiesel and Petrodiesel Fuels: Science, Technology, Health, and Environment. Volume 1. Biodiesel Fuels: Science, Technology, Health, and Environment*, ed. O. Konur. Boca Raton, FL: CRC Press.

Konur, O. 2021n. Propanediol production from glycerol: A review of the research. In *Handbook of Biodiesel and Petrodiesel Fuels: Science, Technology, Health, and Environment. Volume 1. Biodiesel Fuels: Science, Technology, Health, and Environment*, ed. O. Konur Boca Raton, FL: CRC Press,

Konur, O. 2021o. Edible oil-based biodiesel fuels: A scientometric review of the research. *In Handbook of Biodiesel and Petrodiesel Fuels: Science, Technology, Health, and Environment. Volume 2. Biodiesel Fuels based on the Edible and Nonedible Feedstocks, Wastes, and Algae: Science, Technology, Health, and Environment*, ed. O. Konur. Boca Raton, FL: CRC Press.

Konur, O. 2021p. Palm oil-based biodiesel fuels: A review of the research. In *Handbook of Biodiesel and Petrodiesel Fuels: Science, Technology, Health, and Environment. Volume 2. Biodiesel Fuels based on the Edible and Nonedible Feedstocks, Wastes, and Algae*, ed. O. Konur. Boca Raton, FL: CRC Press.

Konur, O. 2021q. Rapeseed oil-based biodiesel fuels: A review of the research. In *Handbook of Biodiesel and Petrodiesel Fuels: Science, Technology, Health, and Environment. Volume 2. Biodiesel Fuels based on the Edible and Nonedible Feedstocks, Wastes, and Algae*, ed. O. Konur. Boca Raton, FL: CRC Press.

Konur, O. 2021r. Nonedible oil-based biodiesel fuels: A scientometric review of the research. In *Handbook of Biodiesel and Petrodiesel Fuels: Science, Technology, Health, and Environment. Volume 2. Biodiesel Fuels based on the Edible and Nonedible Feedstocks, Wastes, and Algae: Science, Technology, Health, and Environment*, ed. O. Konur. Boca Raton, FL: CRC Press.

Konur, O. 2021s. Waste oil-based biodiesel fuels: A scientometric review of the research. In *Handbook of Biodiesel and Petrodiesel Fuels: Science, Technology, Health, and Environment. Volume 2. Biodiesel Fuels based on the Edible and Nonedible Feedstocks, Wastes, and Algae: Science, Technology, Health, and Environment*, ed. O. Konur. Boca Raton, FL: CRC Press.

Konur, O. 2021t. Algal biodiesel fuels: A scientometric review of the research. In *Handbook of Biodiesel and Petrodiesel Fuels: Science, Technology, Health, and Environment. Volume 2. Biodiesel Fuels based on the Edible and Nonedible Feedstocks, Wastes, and Algae: Science, Technology, Health, and Environment*, ed. O. Konur. Boca Raton, FL: CRC Press.

Konur, O. 2021u. Algal biomass production for biodiesel production: A review of the research. In *Handbook of Biodiesel and Petrodiesel Fuels: Science, Technology, Health, and Environment. Volume 2. Biodiesel Fuels based on the Edible and Nonedible Feedstocks, Wastes, and Algae*, Ed. O. Konur Boca Raton, FL: CRC Press.

Konur, O. 2021v. Algal biomass production in wastewaters for biodiesel production: A review of the research. In *Handbook of Biodiesel and Petrodiesel Fuels: Science, Technology, Health, and Environment. Volume 2. Biodiesel Fuels based on the Edible and Nonedible Feedstocks, Wastes, and Algae*, ed. O. Konur. Boca Raton, FL: CRC Press.

Konur, O. 2021x. Algal lipid production for biodiesel production: A review of the research. In *Handbook of Biodiesel and Petrodiesel Fuels: Science, Technology, Health, and Environment. Volume 2. Biodiesel Fuels based on the Edible and Nonedible Feedstocks, Wastes, and Algae*, Ed. O. Konur Boca Raton, FL: CRC Press.

Konur, O. 2021y. Crude oils: A scientometric review of the research. In *Handbook of Biodiesel and Petrodiesel Fuels: Science, Technology, Health, and Environment. Volume 3. Petrodiesel Fuels: Science, Technology, Health, and Environment*, ed. O. Konur. Boca Raton, FL: CRC Press.

Konur, O. 2021z. Petrodiesel fuels: A scientometric review of the research. In *Handbook of Biodiesel and Petrodiesel Fuels: Science, Technology, Health, and Environment. Volume 3. Petrodiesel Fuels: Science, Technology, Health, and Environment*, ed. O. Konur. Boca Raton, FL: CRC Press.

Konur, O. 2021aa. Bioremediation of petroleum hydrocarbons in the contaminated soils: A review of the research. In *Handbook of Biodiesel and Petrodiesel Fuels: Science, Technology, Health, and Environment. Volume 3. Petrodiesel Fuels: Science, Technology, Health, and Environment*, ed. O. Konur. Boca Raton, FL: CRC Press.

Konur, O. 2021ab. Desulfurization of diesel fuels: A review of the research. In *Handbook of Biodiesel and Petrodiesel Fuels: Science, Technology, Health, and Environment. Volume 3. Petrodiesel Fuels: Science, Technology, Health, and Environment*, ed. O. Konur. Boca Raton, FL: CRC Press.

Konur, O. 2021ac. Diesel fuel exhaust emissions: A scientometric review of the research. In *Handbook of Biodiesel and Petrodiesel Fuels: Science, Technology, Health, and Environment. Volume 3. Petrodiesel Fuels: Science, Technology, Health, and Environment*, ed. O. Konur. Boca Raton, FL: CRC Press.

Konur, O. 2021ad. The adverse health and safety impact of diesel fuels: A scientometric review of the research. In *Handbook of Biodiesel and Petrodiesel Fuels: Science, Technology, Health, and Environment. Volume 3. Petrodiesel Fuels: Science, Technology, Health, and Environment*, ed. O. Konur. Boca Raton, FL: CRC Press.

Konur, O. 2021ae. Respiratory illnesses caused by the diesel fuel exhaust emissions: A review of the research. In *Handbook of Biodiesel and Petrodiesel Fuels: Science, Technology, Health, and Environment. Volume 3. Petrodiesel Fuels: Science, Technology, Health, and Environment*, ed. O. Konur. Boca Raton, FL: CRC Press.

Konur, O. 2021af. Cancer caused by the diesel fuel exhaust emissions: A review of the research. In *Handbook of Biodiesel and Petrodiesel Fuels: Science, Technology, Health, and Environment. Volume 3. Petrodiesel Fuels: Science, Technology, Health, and Environment*, ed. O. Konur. Boca Raton, FL: CRC Press.

Konur, O. 2021ag. Cardiovascular and other illnesses caused by the diesel fuel exhaust emissions: A review of the research. In *Handbook of Biodiesel and Petrodiesel Fuels: Science, Technology, Health, and Environment. Volume 3. Petrodiesel Fuels: Science, Technology, Health, and Environment*, ed. O. Konur. Boca Raton, FL: CRC Press.

Krishnan, S., J. Jeevanandam, C. Acquah, and M. K. Danquah. 2021. Extraction of algal neutral lipids for biofuel production. In *Handbook of Biodiesel and Petrodiesel Fuels: Science, Technology, Health, and Environment. Volume 2. Biodiesel Fuels based on the Edible and Nonedible Feedstocks, Wastes, and Algae*, ed. O. Konur. Boca Raton, FL: CRC Press.

Kumar, N. and M. A. Kalam. 2021. Qualitative characterization of biodiesel fuels: Basics and beyond. In *Handbook of Biodiesel and Petrodiesel Fuels: Science, Technology, Health, and Environment. Volume 1. Biodiesel Fuels: Science, Technology, Health, and Environment*, ed. O. Konur. Boca Raton, FL: CRC Press.

Lapuerta, M., O. Armas, and J. Rodriguez-Fernandez. 2008. Effect of biodiesel fuels on diesel engine emissions. *Progress in Energy and Combustion Science* 34: 198–223.

Lardon, L., A. Helias, B. Sialve, J. P. Steyer, and O. Bernard. 2009. Life-cycle assessment of biodiesel production from microalgae. *Environmental Science & Technology* 43: 6475–6481.

Leahy, J. G. and R. R. Colwell. 1990. Microbial degradation of hydrocarbons in the environment. *Microbiological Reviews* 54: 305–315.

Mata, T. M., A. A. Martins, and N. S. Caetano. 2010. Microalgae for biodiesel production and other applications: A review. *Renewable & Sustainable Energy Reviews* 14: 217–232.

McCreanor, J., P. Cullinan, and M. J. Nieuwenhuijsen, et al. 2007. Respiratory effects of exposure to diesel traffic in persons with asthma. *New England Journal of Medicine* 357: 2348–2358.

McFarlin, K. M. and R. C. Prince. 2021. Oil contaminated soil: Understanding biodegradation and remediation strategies. In *Handbook of Biodiesel and Petrodiesel Fuels: Science, Technology, Health, and Environment. Volume 3. Petrodiesel Fuels: Science, Technology, Health, and Environment*, ed. O. Konur. Boca Raton, FL: CRC Press.

Mills, N. L., H. Tornqvist, and M. C. Gonzalez, et al. 2007. Ischemic and thrombotic effects of dilute diesel-exhaust inhalation in men with coronary heart disease. *New England Journal of Medicine* 357: 1075–1082.

Mohan, D., C. U. Pittman Jr, and P. H. Steele. 2006. Pyrolysis of wood/biomass for bio-oil: A critical review. *Energy & Fuels* 20: 848–889.

Mortensen, P. M., J. D. Grunwaldt, and P. A. Jensen, et al. 2011. A review of catalytic upgrading of bio-oil to engine fuels. *Applied Catalysis A-General* 407: 1–19.

Niju, S. and G. Janani. 2021. *Moringa oleifera* oil as a potential feedstock for sustainable biodiesel production. In *Handbook of Biodiesel and Petrodiesel Fuels: Science, Technology, Health, and Environment. Volume 2. Biodiesel Fuels based on the Edible and Nonedible Feedstocks, Wastes, and Algae,* ed. O. Konur. Boca Raton, FL: CRC Press.

North, D. C. 1991a. *Institutions, Institutional Change and Economic Performance*. Cambridge, MA: Cambridge University Press.

North, D.C. 1991b. Institutions. *Journal of Economic Perspectives* 5: 97–112.

Nuguid, R. J. G., F. Buttignol, A. Marberger, and O. Krocher. 2021. Selective catalytic reduction of NO_x emissions. In *Handbook of Biodiesel and Petrodiesel Fuels: Science, Technology, Health, and Environment. Volume 3. Petrodiesel Fuels: Science, Technology, Health, and Environment*, ed. O. Konur. Boca Raton, FL: CRC Press.

Pagliaro, M., R. Ciriminna, H. Kimura, M. Rossi, and C. della Pina. 2007. From glycerol to value-added products. *Angewandte Chemie-International Edition* 46: 4434–4440.

Pandey, S. and A. K. Sharma. 2021. Combustion and formations of emission in compression ignition engines and emissions reduction techniques. In *Handbook of Biodiesel and Petrodiesel Fuels: Science, Technology, Health, and Environment. Volume 3. Petrodiesel Fuels: Science, Technology, Health, and Environment*, ed. O. Konur. Boca Raton, FL: CRC Press.

Perron, P. 1989. The great crash, the oil price shock, and the unit root hypothesis. *Econometrica: Journal of the Econometric Society* 57: 1361–1401.

Peters, K. E. 1986. Guidelines for evaluating petroleum source rock using programmed pyrolysis. *AAPG Bulletin-American Association of Petroleum Geologists* 70: 318–329.

Peterson, C. H., S. D. Rice, and J. W. Short, et al. 2003. Long-term ecosystem response to the Exxon Valdez oil spill. *Science* 302: 2082–2086.

Prince, R. C., B. M. Hedgpeth, and K. M. McFarlin. 2021. Crude oils and their fate in the environment. In *Handbook of Biodiesel and Petrodiesel Fuels: Science, Technology, Health, and Environment. Volume 3. Petrodiesel Fuels: Science, Technology, Health, and Environment*, ed. O. Konur. Boca Raton, FL: CRC Press.

Ragauskas, A. J., G. T. Beckham, and M. J. Biddy, et al. 2014. Lignin valorization: Improving lignin processing in the biorefinery. *Science* 344: 1246843.

Robinson, A. L., N. M. Donahue, and M. K. Shrivastava, et al. 2007. Rethinking organic aerosols: Semivolatile emissions and photochemical aging. *Science* 315: 1259–1262.

Rodolfi, L., G. C. Zittelli, and N. Bassi, et al. 2009. Microalgae for oil: Strain selection, induction of lipid synthesis and outdoor mass cultivation in a low-cost photobioreactor. *Biotechnology and Bioengineering* 102: 100–112.

Rogge, W. F., L. M. Hildemann, M. A. Mazurek, G. R. Cass, and B. R. T. Simoneit. 1993. Sources of fine organic aerosol. 2. Noncatalyst and catalyst-equipped automobiles and heavy-duty diesel trucks. *Environmental Science & Technology* 27: 636–651.

Salvi, S., A. Blomberg, and B. Rudell, et al. 1999. Acute inflammatory responses in the airways and peripheral blood after short-term exposure to diesel exhaust in healthy human volunteers. *American Journal of Respiratory and Critical Care Medicine* 159: 702–709.

Satyannarayana, K. V. V., R. Singh, G. Moorthy, and R. V. Kumar. 2021. Microbial biodiesel production using microbes in general. In *Handbook of Biodiesel and Petrodiesel Fuels: Science, Technology, Health, and Environment. Volume 2. Biodiesel Fuels based on the Edible and Nonedible Feedstocks, Wastes, and Algae: Science, Technology, Health, and Environment*, ed. O. Konur. Boca Raton, FL: CRC Press.

Schauer, J. J., M. J. Kleeman, G. R. Cass, and B. R. T. Simoneit. 1999. Measurement of emissions from air pollution sources. 2. C_1 through C_{30} organic compounds from medium duty diesel trucks. *Environmental Science & Technology* 33: 1578–1587.

Schenk, P. M., S. R. Thomas-Hall, and E. Stephens, et al. 2008. Second generation biofuels: High-efficiency microalgae for biodiesel production. *Bioenergy Research* 1: 20–43.

Shelef, M. 1995. Selective catalytic reduction of NO_x with n-free reductants. *Chemical Reviews* 95: 209–225.

Sialve, B., N. Bernet, and O. Bernard. 2009. Anaerobic digestion of microalgae as a necessary step to make microalgal biodiesel sustainable. *Biotechnology Advances* 27: 409–416.

Siddique, M. N. and S. Rohani. 2021. Biodiesel production from municipal wastewater sludge: Recent trends. *Handbook of Biodiesel and Petrodiesel Fuels: Science, Technology, Health, and Environment. Volume 2. Biodiesel Fuels based on the Edible and Nonedible*

Feedstocks, Wastes, and Algae: Science, Technology, Health, and Environment, ed. O. Konur. Boca Raton, FL: CRC Press.

Sivaramakrishnan, R. and A. Incharoensakdi. 2021. Microalgal biodiesel production. In *Handbook of Biodiesel and Petrodiesel Fuels: Science, Technology, Health, and Environment. Volume 2. Biodiesel Fuels based on the Edible and Nonedible Feedstocks, Wastes, and Algae*, ed. O. Konur. Boca Raton, FL: CRC Press.

Song, C. S. 2003. An overview of new approaches to deep desulfurization for ultra-clean gasoline, diesel fuel and jet fuel. *Catalysis Today* 86: 211–263.

Song, C. and X. L. Ma. 2003. New design approaches to ultra-clean diesel fuels by deep desulfurization and deep dearomatization. *Applied Catalysis B-Environmental* 41: 207–238.

Srivastava, A. and R. Prasad. 2000. Triglycerides-based diesel fuels. *Renewable & Sustainable Energy Reviews* 4: 111–133.

Stanislaus, A., A. Marafi, and M. S. Rana. 2010. Recent advances in the science and technology of ultra low sulfur diesel (ULSD) production. *Catalysis Today* 153: 1–68.

Tomar, M., H. Dewal, L. Kumar, N. Kumar, and N. Bharadvaja. 2021. Biodiesel additives: Status and perspectives. In *Handbook of Biodiesel and Petrodiesel Fuels: Science, Technology, Health, and Environment. Volume 1. Biodiesel Fuels: Science, Technology, Health, and Environment*, ed. O. Konur. Boca Raton, FL: CRC Press.

Tuck, C. O., E. Perez, I. T. Horvath, R. A. Sheldon, and M. Poliakoff. 2012. Valorization of biomass: Deriving more value from waste. *Science* 337: 695–699.

Westbrook, C. K. 2021. Chemistry of biodiesel fuels based on soybean oil. In *Handbook of Biodiesel and Petrodiesel Fuels: Science, Technology, Health, and Environment. Volume 2. Biodiesel Fuels based on the Edible and Nonedible Feedstocks, Wastes, and Algae*, ed. O. Konur. Boca Raton, FL: CRC Press.

Wijffels, R. H. and M. J. Barbosa. 2010. An outlook on microalgal biofuels. *Science* 329: 796–799.

Zakzeski, J., P. C. A. Bruijnincx, A. L. Jongerius, and B. M. Weckhuysen. 2010. The catalytic valorization of lignin for the production of renewable chemicals. *Chemical Reviews* 110: 3552–3599.

Zhang, Q., J. Chang, T. J. Wang, and Y. Xu. 2007. Review of biomass pyrolysis oil properties and upgrading research. *Energy Conversion and Management* 48: 87–92.

Zhang, J., J. Sun, and Y. Wang. 2021a. An overview of catalytic bio-oil upgrading. Part I: Processing aqueous-phase compounds. In *Handbook of Biodiesel and Petrodiesel Fuels: Science, Technology, Health, and Environment. Volume 1. Biodiesel Fuels: Science, Technology, Health, and Environment*, ed. O. Konur. Boca Raton, FL: CRC Press.

Zhang, J., J. Sun, and Y. Wang. 2021b. An overview of catalytic bio-oil upgrading. Part II: Processing oil-phase compounds and real bio-oil. In *Handbook of Biodiesel and Petrodiesel Fuels: Science, Technology, Health, and Environment. Volume 1. Biodiesel Fuels: Science, Technology, Health, and Environment*, ed. O. Konur. Boca Raton, FL: CRC Press.

Zhang, Y., M. A. Dube, D. D. McLean, and M. Kates. 2003a. Biodiesel production from waste cooking oil: 1. Process design and technological assessment. *Bioresource Technology* 89: 1–16.

Zhang, Y., M. A. Dube, D. D. McLean, and M. Kates. 2003b. Biodiesel production from waste cooking oil. 2. Economic assessment and sensitivity analysis. *Bioresource Technology* 90: 229–240.

Zhou, C. H. C., J. N. Beltramini, Y. X. Fan, and G. Q. M. Lu. 2008. Chemoselective catalytic conversion of glycerol as a biorenewable source to valuable commodity chemicals. *Chemical Society Reviews* 37: 527–549.

2 A Scientometric Review of the Research
Biodiesel and Petrodiesel Fuels

Ozcan Konur

CONTENTS

2.1 INTRODUCTION

Crude oils have been primary sources of energy and fuels, such as petrodiesel. However, significant public concerns about their sustainability, price fluctuations, and adverse environmental impact of crude oils have emerged since the 1970s (Ahmadun et al., 2009; Atlas, 1981; Babich and Moulijn, 2003; Haritash and Kaushik, 2009; Kilian, 2009; Leahy and Colwell, 1990, Olah, 2005; Perron, 1989). Thus, biooils (Bridgwater and Peacocke, 2000; Czernik and Bridgwater, 2004; di Blasi, 2008; Gallezot, 2012; Mohan et al., 2006; Mortensen et al., 2011) and biooil-based biodiesel fuels (Agarwal, 2007; Chisti, 2007; Hill et al., 2006; Lapuerta et al., 2008; Ma and Hanna, 1999; Mata et al., 2010; Meher et al., 2006; Zhang et al., 2003a–b) have emerged as an alternative to crude oils and crude oil-based petrodiesel fuels in recent decades. Nowadays, although petrodiesel fuels are used extensively, biodiesel fuels are being used increasingly in the transportation and power sectors (Konur, 2021a–ag).

However, for the efficient progression of the research in this field, it is necessary to develop efficient incentive structures for the primary stakeholders and to inform these stakeholders about the research (Konur, 2000, 2002a–c, 2006a–b, 2007a–b; North, 1991a–b).

Scientometric analysis offers ways to evaluate the research in a certain field (Garfield, 1955, 1972). This method has been used to evaluate research in a number of fields (Konur, 2011, 2012a–n, 2015, 2016a–f, 2017a–f, 2018a–b, 2019a–b). However, there has been no current scientometric study of the research on both biodiesel and petrodiesel fuels in general.

This chapter presents a study of the scientometric evaluation of the research on both biodiesel and petrodiesel fuels using two datasets. The first dataset includes the 100-most-cited papers ($n = 100$ sample papers) whilst the second includes population papers ($n =$ over 121,000 population papers) published between 1980 and 2019.

The data on the indices, document types, authors, institutions, funding bodies, source titles, 'Web of Science' subject categories, keywords, research fronts, and citation impacts are presented and discussed.

2.2 MATERIALS AND METHODOLOGY

The search for the literature was carried out in the 'Web of Science' (WOS) database in February 2020. It contains the 'Science Citation Index Expanded' (SCI-E), the 'Social Sciences Citation Index' (SSCI), the 'Book Citation Index-Science' (BCI-S), the 'Conference Proceedings Citation Index-Science' (CPCI-S), the 'Emerging Sources Citation Index' (ESCI), the 'Book Citation Index-Social Sciences and Humanities' (BCI-SSH), the 'Conference Proceedings Citation Index-Social Sciences and Humanities' (CPCI-SSH), and the 'Arts and Humanities Citation Index' (A&HCI).

The keywords for the search of the literature were collated from the screening of the abstract pages for the first 1,000 highly cited papers in the related 11 research fields. These keyword sets are provided in the appendices of the related chapters (Konur, 2021e–ag).

Two datasets were used for this study. The 100-most-cited papers comprise the first dataset ($n = 100$ sample papers) whilst all the papers form the second dataset ($n =$ over 121,000 papers).

The data on the indices, document types, publication years, institutions, funding bodies, source titles, countries, 'Web of Science' subject categories, citation impact, keywords, and research fronts are collated from these datasets. The key findings are provided in the relevant tables and one figure, supplemented with explanatory notes in the text. The findings are discussed and a number of conclusions are drawn, along with recommendations for further study.

2.3 RESULTS

2.3.1 Indices and Documents

There are over 155,000 papers in this field in the 'Web of Science' as of February 2020. This original population dataset is refined by document type (article, review, book chapter, book, editorial material, note, and letter) and language (English), resulting in over 121,000 papers comprising over 77.9% of the original population dataset.

The primary index is the SCI-E for both the sample and population papers; 92.7% of the latter are indexed by this database. Additionally 3.7, 3.8, and 3.1% of these papers are indexed by the CPCI-S, ESCI, and BCI-S databases, respectively. The papers on the social and humanitarian aspects of this field are relatively negligible with only 1.9 and 0.1% of the population papers indexed by the SSCI and A&HCI, respectively.

Brief information on the document types for both datasets is provided in Table 2.1. The key finding is that articles are the primary documents for both sample and population papers, whilst reviews form 68% of the sample papers.

2.3.2 Authors

Brief information about the 13-most-prolific authors with at least two sample papers each is provided in Table 2.2. Around 330 authors contributed sample papers.

The most-prolific author is 'Yusuf Chisti' with three sample papers working primarily on 'algal biodiesel production'. The other authors have two papers each.

On the other hand, a number of authors have a significant presence in the population papers: 'Hassan H. Masjuki', 'Paul T. Williams', 'Jo-Shu Chang', 'Roger R. Ruan', 'Johan Sjoblom', 'Javier Bilbao', 'Paul Chen', 'M. Abul Kalam',

TABLE 2.1
Document Types

	Document Type	Sample Dataset (%)	Population Dataset (%)	Difference (%)
1	Article	32*	93.7	–61.7
2	Review	68*	3.1	64.9
3	Book chapter	0	1.6	–1.6
4	Proc. paper	4	5.3	–1.3
5	Editorial mat.	0	1.9	–1.9
6	Letter	0	0.5	–0.5
7	Book	0	0.1	–0.1
8	Note	0	0.7	–0.7

*Originally there were 53 articles and 47 reviews as classified by the database.

TABLE 2.2
Authors

	Authors	Sample Papers (%)	Population Papers (%)	Surplus (%)	Institution	Country	Research Front I	Research Front II
1	Chisti, Yusuf	3	0.1	2.9	Massey Un iv.	New Zealand	Biodiesel fuels	Algal biodiesel production, algal biomass production
2	Bernard, Olivier	2	0.1	1.9	INRIA	France	Biodiesel fuels	Algal biodiesel production
3	Bridgwater, Antony V.	2	0.1	1.9	Aston Univ.	UK	Biodiesel fuels	Biooil production
4	Cass, Glen R.	2	0.1	1.9	CALTECH	USA	Petrodiesel fuels	Emissions
5	Demirbas, Ayhan	2	0.1	1.9	Sila Sci.	Turkey	Biodiesel fuels	Biodiesel production
6	Dube, Marc A.	2	0.1	1.9	Univ. Ottawa	Canada	Biodiesel fuels	Biodiesel production-waste oils
7	Kates, Morris	2	0.1	1.9	Univ. Ottawa	Canada	Biodiesel fuels	Biodiesel production-waste oils
8	Knothe, Gerhard	2	0.1	1.9	Dept. Agric.	USA	Biodiesel fuels	Biodiesel properties
9	Mclean, David D.	2	0.1	1.9	Univ. Ottawa	Canada	Biodiesel fuels	Biodiesel production-waste oils
10	Sialve, Bruno	2	0.1	1.9	INRIA	France	Biodiesel fuels	Algal biodiesel production
11	Simoneit, Bernd R.T.	2	0.1	1.9	Oregon State Univ.	USA	Petrodiesel fuels	Emissions
12	Van Gerpen, Jon	2	0.1	1.9	Iowa State Univ.	USA	Biodiesel fuels	Biodiesel production
13	Zhang, Yongkui	2	0.1	1.9	Univ. Ottawa	Canada	Biodiesel fuels	Biodiesel production-waste oils

Source: 'Highly Cited Researchers' in 2019 (Clarivate Analytics, 2019).

'Young-Kwon Park', 'Chang Sik Lee', 'Jorge Ancheyta', 'Martin Olazar', 'Rene H. Wijffels', 'Riayz Kharrat', 'Tayfun Babadagli', 'Constantine Rakopoulos', 'Shahab Ayotollahi', 'Mohamed G. El-Din', 'Yun Hin Taufiq-Yap', 'Robert C. Brown', 'Avinash K. Agarwal', 'John V. Headley', 'Kefa Cen', 'G. Nagarajan', 'Rolf D. Reitz', 'Merv Fingas', 'Umer Rashid', 'Akwasi A. Boateng', 'Ajay K. Dalai', 'A. Mandal', 'Suzana Yusup', 'Manuel Garcia-Perez', 'Eilhann E. Kwon', 'Magin Lapuerta', 'Kerry M. Peru', 'Chun Shun Cheung', 'Christopher M. Reddy', 'Mustafa V. Kok', 'Ryan P. Rodgers', 'Gartzen Lopez', 'Antono Marcilla', 'Wei Du', 'Rafael Font', 'Mohammad H. Gazanfari', 'Hajime Takano', 'Barat Ghobadian', 'Hwai C. Ong', 'Oliver C. Mullins', 'Raul Payri', 'Bassim H. Hameed', 'Amir H. Mohammadi', 'Phillip M. Fedorak', 'Stephen R. Larter', and 'Masaru Sagai' with at least 0.55% of the population papers each.

The most-prolific institutions for these top authors are the 'University of Ottawa' of Canada and 'INRIA' of France with four and two sample authors, respectively. Thus, in total, nine institutions house these top authors.

It is notable that none of these top researchers is listed in the 'Highly Cited Researchers' (HCR) in 2019 (Clarivate Analytics, 2019; Docampo and Cram, 2019).

The most-prolific countries for these top authors are the USA and Canada with four authors each. These top countries are followed by France with three authors. In total, 6 countries contribute to these top papers.

There are two key topical research fronts for these top researchers: 'biodiesel fuels' and 'petrodiesel fuels', with 11 and 2 authors, respectively. At the secondary level, there are four and three authors with papers on 'biodiesel production from waste oils' and 'biodiesel production from algae', respectively. There are also two authors each with papers on 'biodiesel production in general' and 'petrodiesel exhaust emissions'. The other authors have papers on 'biooil production', 'algal biomass production', and 'biodiesel properties'.

It is further notable that there is a significant gender deficit among these top authors as all of these researchers are male (Lariviere et al., 2013; Xie and Shauman, 1998).

The author with the most impact is 'Yusuf Chisti' with a 2.9% publication surplus. The other authors have the same publication surplus of 1.9%.

2.3.3 Publication Years

The information about publication years for both datasets is provided in Figure 2.1. This figure shows that 6, 14, 64, and 16% of the sample papers and 7.5, 10.5, 19.1, and 63.4% of the population papers were published in the 1980s, 1990s, 2000s, and 2010s, respectively.

Similarly, the most-prolific publication years for the sample dataset are 2008, 2009, 2003, 2005, and 2010 with 11, 11, 9, 7, and 7 papers, respectively. On the other hand, the most-prolific publication years for the population dataset are 2019, 2018, 2017, 2016, and 2015 with 9.3, 8.2, 8.0, 7.6, and 6.9% of the population papers, respectively. It is notable that there is a sharply rising trend for population papers in the 2000s and 2010s.

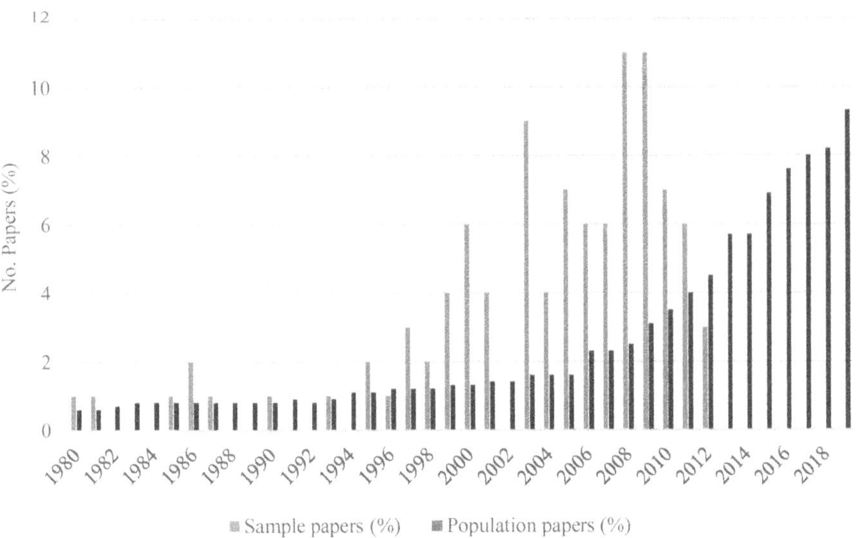

FIGURE 2.1 The research output between 1980 and 2019.

2.3.4 Institutions

Brief information on the top 24 institutions with at least 2% of the sample papers each is provided in Table 2.3. In total, around 160 and 27,200 institutions contribute to the sample and population papers, respectively. Additionally, 3.3% of the population papers have no institutional information on their abstract pages.

These top institutions publish 57 and 11.3% of the sample and population papers, respectively. The top institutions are the US 'Department of Energy' and 'Department of Agriculture' with five and four sample papers, respectively. These top institutions are followed by the 'Indian Institute of Technology' of India, 'Massey University' of New Zealand, the 'National Renewable Energy Laboratory' of the USA, and the 'University of Ottawa' of Canada with three sample papers each.

The most-prolific country for these top institutions is the USA with 11 institutions. The other prolific countries are France, Germany, and Italy with two institutions each. In total, 11 countries house these top institutions.

The institutions with the most impact are the US 'Department of Agriculture' and 'Department of Energy' with 3.5 and 3.3% publication surpluses, respectively. These top institutions are followed by the 'University of Ottawa', 'Massey University', and the 'National Renewable Energy Laboratory' with a 2.8% publication surplus each.

On the other hand, the institutions with the least impact are 'Rio de Janeiro Federal University' of Brazil, the 'Chinese Academy of Sciences', the 'Helmholtz Association' of Germany, the 'National Research Council' of Italy, and the 'Indian Institute of Technology' with –0.5, 1.2, 1.4, and 1.6% publication surpluses/deficits, respectively.

It is notable that some institutions have a heavy presence in the population papers: the 'National Scientific Research Center' of France, the 'China University of

TABLE 2.3
Institutions

	Institutions	Country	No. of Sample Papers (%)	No. of Population Papers (%)	Difference (%)
1	Dept. Energ.	USA	5	1.7	3.3
2	Dept. Agric.	USA	4	0.5	3.5
3	Indian Inst. Technol.	India	3	1.4	1.6
4	Massey Univ.	New Zealand	3	0.2	2.8
5	Natl. Renew. Ener. Lab.	USA	3	0.2	2.8
6	Univ. Ottawa	Canada	3	0.2	2.8
7	Chinese Acad. Sci.	China	2	2.5	–0.5
8	Helmholtz Assoc.	Germany	2	0.8	1.2
9	Natl. Res. Counc.	Italy	2	0.6	1.4
10	Penn. State Univ.	USA	2	0.3	1.7
11	Univ. Michigan	USA	2	0.3	1.7
12	Aston Univ.	UK	2	0.2	1.8
13	Calif. Inst. Technol.	USA	2	0.2	1.8
14	Chevron	USA	2	0.2	1.8
15	Colorado Sch. Mines	USA	2	0.2	1.8
16	INRAE	France	2	0.2	1.8
17	INRIA	France	2	0.2	1.8
18	Karlsruhe Inst. Technol.	Germany	2	0.2	1.8
19	Oregon State Univ.	USA	2	0.2	1.8
20	Dept. Interior	USA	2	0.2	1.8
21	Castilla La-Mancha Univ.	Spain	2	0.2	1.8
22	Univ. Genoa	Italy	2	0.2	1.8
23	Univ. Nebrasca Lincoln	USA	2	0.2	1.8
24	Univ. Queensland	Australia	2	0.2	1.8

Petroleum', the 'Russian Academy of Sciences', the 'University of Alberta' of Canada, the 'Council of Scientific Industrial Research' of India, the 'Superior Council of Scientific Investigations' of Spain, the 'University of Calgary' of Canada, 'Tsinghua University', the 'China National Petroleum Company', 'Tianjin University' of China, the 'University of Malaya' of Malaysia, 'Zhejiang University' of China, the 'Federal University of Rio de Janeiro' of Brazil, the 'İslamic Azad University' of Iran, and 'Shanghai Jia Tong University' of China with at least 0.5% of the population papers each.

2.3.5 Funding Bodies

Brief information about the top two funding bodies with at least 2% of the sample papers each is provided in Table 2.4. It is significant that only 20 and 46.9% of the sample and population papers declare any funding, respectively. Around 40 and 42,300 funding bodies fund the research for the sample and population papers, respectively.

The main funding bodies are the US 'Department of Energy' and the French 'National Research Agency' with two sample papers each.

TABLE 2.4
Funding Bodies

	Institutions	Country	No. of Sample Papers (%)	No. of Population Papers (%)	Difference (%)
1	French Natl. Res. Agcy.	France	2	0.2	1.8
2	Dept. Ener.	USA	2	1.2	0.8

It is notable that some top funding agencies have a heavy presence in the population studies. These include the 'National Natural Science Foundation' of China, the 'National Council for Scientific and Technological Development' of Brazil, the 'Natural Sciences and Engineering Research Council' of Canada, 'Fundamental Research Funds for the Central Universities' of China, the 'CAPES Foundation' of Brazil, the 'National Science Foundation' of the USA, the 'National Basic Research Program' of China, the 'European Union', the 'National High Technology Research and Development Program' of China, the US 'Department of Health Human Services', the 'China Postdoctoral Science Foundation', the 'National Institutes of Health' of the USA, the 'Ministry of Education Culture Sports Science and Technology' of Japan, the 'Engineering Physical Sciences Research Council' of the UK, the 'China Scholarship Council', and the 'Council of Scientific Industrial Research' of India with at least 0.5% of the population papers each. These funding bodies are from mainly Brazil, China, Canada, Europe, the USA, Japan, the UK, and India.

2.3.6 Source Titles

Brief information about the top 22 source titles with at least two sample papers each is provided in Table 2.5. In total, around 45 and 7,000 source titles publish the sample and population papers, respectively. On the other hand, these top 22 journals publish 75 and 18.6% of the sample and population papers, respectively.

The top journals are 'Bioresource Technology' and 'Renewable Sustainable Energy Reviews', publishing nine and eight sample papers, respectively. These top journals are followed by 'Energy Fuels' with five sample papers and 'Fuel', 'Energy Conversion and Management', 'Progress in Energy and Combustion Science', 'Biotechnology Advances', and 'Science' with four sample papers each.

Although these journals are indexed by 18 subject categories, the top categories are 'Energy Fuels', 'Engineering Chemicals', and 'Environmental Sciences' with eight, seven, and five journals, respectively. The other most-prolific subject categories are 'Biotechnology and Applied Microbiology', 'Engineering Environmental', 'Chemistry Physical', and 'Chemistry Applied' with three journals each. Furthermore, there are two journals for 'Agricultural Engineering' and 'Thermodynamics'.

TABLE 2.5
Source Titles

	Source Title	WOS Subject Category	No. of Sample Papers (%)	No. of Population Papers (%)	Difference (%)
1	Bioresource Technology	Agr. Eng., Biot. Appl. Microb., Ener. Fuels	9	2.8	6.2
2	Renewable Sustainable Energy Reviews	Green Sust. Sci. Technol., Ener. Fuels	8	0.5	7.5
3	Energy Fuels	Ener. Fuels, Eng. Chem.	5	3.2	1.8
4	Fuel	Ener. Fuels, Eng. Chem.	4	3.7	0.3
5	Energy Conversion and Management	Therm., Ener. Fuels, Mechs.	4	1.0	3.0
6	Progress in Energy and Combustion Science	Therm., Ener. Fuels, Eng. Chem., Eng. Mech.	4	0.1	3.9
7	Biotechnology Advances	Biot. Appl. Microb.	4	0.1	3.9
8	Science	Mult. Sci.	4	0.1	3.9
9	Environmental Science Technology	Eng. Env., Env. Sci.	3	1.0	2.0
10	Applied Catalysis B Environmental	Chem. Phys., Eng. Env., Eng. Chem.	3	0.7	2.3
11	Catalysis Today	Chem. Appl., Chem. Phys., Eng. Chem.	3	0.6	2.4
12	Applied Catalysis A General	Chem. Phys., Env. Sci.	3	0.5	2.5
13	AAPG Bulletin American Association of Petroleum Geologists	Geosci. Mult.	3	0.3	2.7
14	Industrial Engineering Chemistry Research	Eng. Chem.	2	1.1	0.9
15	Fuel Processing Technology	Chem. Appl., Ener. Fuels, Eng. Chem.	2	0.9	1.1
16	Biomass Bioenergy	Agr. Eng., Biot. Appl. Microb., Ener. Fuels	2	0.6	1.4
17	Journal of Hazardous Materials	Eng. Env., Env. Sci.	2	0.4	1.6
18	Environmental Pollution	Env. Sci.	2	0.3	1.7
19	Journal of the American Oil Chemists Society	Chem. Appl., Food Sci. Technol.	2	0.3	1.7
20	Atmospheric Environment	Env. Sci., Meteor. Atmosph. Sci.	2	0.2	1.8
21	Angewandte Chemie International Edition	Chem. Mult.	2	0.1	1.9
22	Microbiological Reviews	Microbiol.	2	0.1	1.9

The journals with the most impact are 'Renewable Sustainable Energy Reviews' and 'Bioresource Technology' with 7.5 and 6.2% publication surpluses, respectively. On the other hand, the journals with the least impact are 'Fuel', 'Industrial Engineering Chemistry Research', and 'Fuel Processing Technology' with 0.3, 0.9, and 1.1% publication surpluses, respectively.

It is notable that some journals have a heavy presence in the population papers. These include the 'Journal of Analytical and Applied Pyrolysis', 'Petroleum Science and Technology', the 'Oil Gas Journal', 'Energy', the 'Journal of Petroleum Science and Engineering', 'Energy Sources Part A Recovery Utilization and Environmental Effects', 'Marine Pollution Bulletin', 'Renewable Energy', the 'Journal of Applied Phycology', 'Applied Energy', 'Algal Research Biomass Biofuels and Bioproducts', 'Chemistry and Technology of Fuels and Oils', 'Chemosphere', the 'Chemical Engineering Journal', 'RSC Advances', and 'Science of the Total Environment' with at least 0.6% of the population papers.

2.3.7 Countries

Brief information about the top 17 countries with at least two sample papers each is provided in Table 2.6. In total, around 30 and 180 countries contribute to the sample and population papers, respectively. Additionally, 3.3% of the population papers have no country information.

TABLE 2.6
Countries

	Country	No. of Sample Papers (%)	No. of Population Papers (%)	Difference (%)
1	USA	38	18.9	19.1
2	India	8	6.3	1.7
3	China	7	16.7	–9.7
4	Canada	6	6.5	–0.5
5	Australia	6	2.5	3.5
6	Spain	5	3.8	1.2
7	Italy	5	2.6	2.4
8	UK	4	5.4	–1.4
9	Japan	4	4.0	0.0
10	Germany	4	3.6	0.4
11	France	3	3.5	–0.5
12	Turkey	3	1.7	1.3
13	Netherlands	3	1.6	1.4
14	Denmark	3	0.8	2.2
15	New Zealand	3	0.3	2.7
16	Malaysia	2	2.3	–0.3
17	Switzerland	2	0.7	1.3
	Europe-8	29	22.0	7.0
	Asia-5	24	29.6	–5.6

The top country is the USA, publishing 38 and 18.9% of the sample and population papers, respectively. India, China, Canada, Australia, Spain, and Italy follow the USA with at least five sample papers each.

On the other hand, the European and Asian countries represented in this table publish altogether 29 and 24% of the sample papers and 22.0 and 29.6% of the population papers, respectively.

It is notable that the publication surplus/deficit for the USA and these European and Asian countries is 19.1, 7.0, and –5.6%, respectively. On the other hand, the countries with the most impact are the USA, Australia, and New Zealand with 19.1, 3.5, and 2.7% publication surpluses, respectively. Furthermore, the countries with the least impact are the China, the UK, Canada, France, and Malaysia with –9.7, –1.4, –0.5, –0.5, and –0.3% publication deficits, respectively.

It is also notable that some countries have a heavy presence in the population papers. The major producers of population papers are Brazil, South Korea, Iran, Russia, Norway, Mexico, Taiwan, Poland, Sweden, Saudi Arabia, Egypt, Thailand, Nigeria, Greece, Portugal, Belgium, and Finland with at least 0.7% of the population papers each.

2.3.8 'Web of Science' Subject Categories

Brief information about the top 19 'Web of Science' subject categories with at least two sample papers each is provided in Table 2.7. The sample and population papers are indexed by around 25 and 230 subject categories, respectively.

For the sample papers, the top subject is 'Energy Fuels' with 43 and 30.1% of the sample and population papers, respectively. This top subject category is followed by 'Engineering Chemical', 'Biotechnology Applied Microbiology', 'Environmental Sciences', and 'Agricultural Engineering' with 26, 22, 16, and 12% of the sample papers, respectively. The other prolific subject categories are 'Chemistry Physical', 'Green Sustainable Science Technology', 'Engineering Environmental', 'Thermodynamics', and 'Chemistry Applied' with at least eight sample papers each.

It is notable that the publication surplus is most significant for 'Energy Fuels', 'Biotechnology Applied Microbiology', 'Agricultural Engineering', and 'Green Sustainable Science Technology' with 12.9, 11.6, 8.1, and 6.4% publication surpluses, respectively. On the other hand, the subjects with least impact are 'Engineering Chemical', 'Geosciences Multidisciplinary', 'Plant Sciences', and 'Chemistry Multidisciplinary' with –0.2, –0.2, 0.2, and 0.4% publication surpluses/deficits, respectively. This later group of subject categories are under-represented in the sample papers.

Additionally, some subject categories have also a heavy presence in the population papers: 'Engineering Petroleum', 'Chemistry Analytical', 'Marine Freshwater Biology', 'Materials Science Multidisciplinary', 'Toxicology', 'Biochemistry Molecular Biology', 'Water Resources', 'Engineering Multidisciplinary', 'Transportation Science Technology', 'Geochemistry Geophysics', 'Ecology', 'Public Environmental Occupational Health', 'Electrochemistry', 'Biochemical Research Methods', 'Engineering Electrical Electronic', and 'Engineering Civil' with at least 0.8% of the population papers each.

TABLE 2.7
'Web of Science' Subject Categories

	Subject	No. of Sample Papers (%)	No. of Population Papers (%)	Difference (%)
1	Energy Fuels	43	30.1	12.9
2	Engineering Chemical	26	26.2	–0.2
3	Biotechnology Applied Microbiology	22	10.4	11.6
4	Environmental Sciences	16	14.8	1.2
5	Agricultural Engineering	12	3.9	8.1
6	Chemistry Physical	10	9.1	0.9
7	Green Sustainable Science Technology	10	3.6	6.4
8	Engineering Environmental	9	6.6	2.4
9	Thermodynamics	9	5.2	3.8
10	Chemistry Applied	8	3.4	4.6
11	Chemistry Multidisciplinary	6	5.6	0.4
12	Engineering Mechanical	5	4.4	0.6
13	Multidisciplinary Sciences	5	1.9	3.1
14	Mechanics	4	2.1	1.9
15	Geosciences Multidisciplinary	3	3.2	–0.2
16	Microbiology	3	2.6	0.4
17	Food Science Technology	3	1.2	1.8
18	Meteorology Atmospheric Sciences	3	0.9	2.1
19	Plant Sciences	2	1.8	0.2

2.3.9 Citation Impact

These sample papers had received about 103,000 citations as of February 2020. Thus, the average number of citations per paper is about 1,030.

2.3.10 Keywords

Although a number of keywords are listed in the appendices of the related chapters for the datasets pertaining to this field, some of them are more significant for the sample papers.

The most-prolific keywords for the keyword set related to biodiesel and petrodiesel fuels are 'biodiesel' and 'diesel' with 37 and 11 occurrences, respectively. There is also one paper for 'bio-diesel'.

The prolific keywords for feedstocks are '*alga*' (20), 'oil*' (32), 'waste*' (5), 'biomass' (13), 'wood*' (3), 'bio-oil*' (3), 'triacylglycerol* (1), 'feedstock*' (4), 'lipid*' (4), 'microb*' (5), 'triglyceride*' (1), 'soybean*' (2), 'cooking oil*' (3), 'vegetable oil*' (3), 'rape*' (2), 'petroleum' (5), 'bacteria' (2), and '*seed*' (3). In these keyword, '*' allows for the different spellings. For example, '*alga*' covers 'microalgae' or 'microalgal'.

The prolific keywords related to the processes for biodiesel and petrodiesel production are 'transester*' (6), 'production' (27), 'engine*' (11), 'pyrolys*' (11), '*processing' (4), 'combustion' (3), '*fuel' (15), 'environment*' (5), 'cultivation' (2), 'photobioreactor*' (2), 'cataly*' (15), 'NOx' (5), 'desulfurization' (3), '*degradation' (7), 'hydrocarbon*' (8), 'pah*' (4), 'properties' (6), 'glycerol' (5), 'emission*' (5), 'upgrading' (3), 'pollut*' (4), 'composition' (4), 'soil' (2), and 'econom*' (5).

2.3.11 Research Fronts

Brief information about the key research fronts is provided in Table 2.8. There are three major topical research fronts for these sample papers: 'biodiesel fuels in general', 'feedstock-based biodiesel fuels', and 'petrodiesel fuels' with 41, 30, and 34 sample papers, respectively. The 'biodiesel fuels in general' and 'feedstock-based biodiesel fuels' have 13.8 and 11.7% publication surpluses, respectively, whilst 'petrodiesel fuels' have a 33.4% publication deficit.

There are 14, 21, and 6 papers for 'biooils' (Bridgwater and Peacocke, 2000; Czernik and Bridgwater, 2004; Di Blasi, 2008; Mohan et al., 2006; Mortensen et al., 2011), 'biodiesel fuels in general' (Agarwal, 2007; Hill et al., 2006; Ma and Hanna, 1999; Mata et al., 2010; Meher et al., 2006), and 'glycerol' (Behr et al., 2008; Dasari et al., 2005; Pagliaro et al., 2007; Zhou et al., 2008), respectively, in the first group of sample papers.

TABLE 2.8
Research Fronts

	Research Front	No. of Sample Papers (%)	No. of Population Papers (%)	Difference (%)
1	Biodiesel fuels in general	41	27.7	13.8
1.1	Biooils	14	16.6	–2.6
1.2	Biodiesel fuels in general	21	5.4	15.6
1.3	Glycerol	6	5.7	0.3
2	Feedstock-based biodiesel fuels	30	18.3	11.7
2.1	Edible oil-based biodiesel fuels	5	2.2	2.8
2.2	Nonedible oil-based biodiesel fuels	3	1.8	1.2
2.3	Waste oil-based biodiesel fuels	2	1.8	0.2
2.4	Algae-based biodiesel fuels	20	12.5	7.5
3.	Petrodiesel fuels	34	67.4	–33.4
3.1	Crude oils	14	43.8	–29.8
3.2	Petrodiesel fuels in general	8	13.2	–5.2
3.3	Emissions from petrodiesel fuels	11	8.6	2.4
3.4	Health impact of the emissions from petrodiesel fuels	1	1.8	–0.8

Similarly, there are 5, 3, 2, and 20 sample papers for 'edible oil-based biodiesel fuels' (Canakci and van Gerpen, 2001; Demirbas, 2003; Freedman et al., 1986; Graboski and McCormick, 1998; Saka and Kusdiana, 2001), 'nonedible oil-based biodiesel fuels' (Achten et al., 2008; Ramadhas et al., 2005), 'waste oil-based biodiesel fuels' (Kulkarni and Dalai, 2006; Zhang et al., 2003a–b), and 'algal oil-based biodiesel fuels' (Brennan and Owende, 2010; Chisti, 2007; Hu et al., 2008; Mata et al., 2010; Rodolfi et al., 2009, Schenk et al., 2008), respectively, in the second group of sample papers.

Finally, there are 14, 8, 11, and 1 papers for 'crude oils' (Ahmadun et al., 2009; Atlas, 1981; Haritash and Kaushik, 2009; Leahy and Colwell, 1990; Olah, 2005), 'petrodiesel fuels in general' (Babich and Moulijn, 2003; Budzinski et al., 1997; Peters, 1986; Song, 2003; Song and Ma, 2003), 'emissions from petrodiesel fuels' (Busca et al., 1998; Khalili et al., 1995; Rogge et al., 1993; Schauer et al., 1999; Shelef, 1995), and 'health impact of the emissions from petrodiesel fuels' (Birch and Cary, 1996), respectively, in the final group of sample papers.

'Biodiesel fuels in general', 'algal oil-based biodiesel fuels', and 'edible oil-based biodiesel fuels' have 15.7, 7.5, and 2.8% publication surpluses, respectively, whilst 'crude oils', 'petrodiesel fuels in general', and 'biooils' have –29.8, –5.2, and –2.6% publication deficits, respectively.

2.4 DISCUSSION

The size of the research in this field has increased to over 121,000 papers as of February 2020. It is expected that the number of population papers in this field will exceed 250,000 papers by the end of the 2020s.

The research has developed more in the technological aspects of the field, rather than the social and humanitarian pathways, as evidenced by the negligible number of population papers in the indices of the 'Web of Science', SSCI, and A&HCI.

The article types of documents are the primary documents: both datasets and reviews are over-represented by 64.9% in the sample papers whilst articles are under-represented by 61.7% (Table 2.1). Thus, the contribution of reviews by 68% of the sample papers in this field is highly exceptional (cf. Konur, 2011, 2012a–n, 2015, 2016a–f, 2017a–f, 2018a–b, 2019a–b).

Thirteen authors from nine institutions have at least two sample papers each (Table 2.2). Four, four, and three of these authors are from the USA, Canada, and France, respectively.

There are two key topical research fronts for these top researchers: 'biodiesel fuels' and 'petrodiesel fuels' with 11 and 2 authors, respectively. At the secondary level, there are four and three authors with papers on 'biodiesel production from waste oils' and 'biodiesel production from algae', respectively. There are also two authors each with papers on 'biodiesel production in general' and 'petrodiesel exhaust emissions'. The other authors have papers on 'biooil production', 'algal biomass production', and 'biodiesel properties'.

There is significant 'gender deficit' among these top authors as all of them are male (Lariviere et al., 2013; Xie and Shauman, 1998).

The population papers in the 2010s built on the sample papers primarily published in the 2000s (Figure 2.1). Following this rising trend, particularly in the 2000s and 2010s, it is expected that the number of papers will reach 250,000 by the end of the 2020s, more than doubling the current size.

The engagement of the institutions in this field at the global scale is significant as around 160 and 27,200 institutions contribute to the sample and population papers, respectively.

Twenty-four top institutions publish 57.0 and 11.3% of the sample and population papers, respectively (Table 2.3). The top institutions are the US 'Department of Energy' and 'Department of Agriculture'. These top institutions are followed by the 'Indian Institute of Technology', 'Massey University' of New Zealand, the 'National Renewable Energy Laboratory', and the 'University of Ottawa'.

The most-prolific country for these top institutions is the USA. The other prolific countries are France, Germany, and Italy. It is notable that some institutions with a heavy presence in the population papers are under-represented in the sample papers.

It is significant that only 20.0 and 46.9% of the sample and population papers declare any funding, respectively. These funding bodies are located in the USA and Brazil (Table 2.4). It is further notable that some top funding agencies for the population studies do not enter this top funding body list.

However, the substantial lack of Chinese funding bodies in Table 2.4 is notable. This finding is in contrast with studies showing heavy research funding in China, with the NSFC being the primary agency (Wang et al., 2012).

The sample and population papers are published by around 45 and 7,200 journals, respectively. It is significant that the top 22 journals publish 75.0 and 18.6% of the sample and population papers, respectively (Table 2.5).

The top journals are 'Renewable Sustainable Energy Reviews' and 'Bioresource Technology'. These are followed by 'Energy Fuels', 'Fuel', 'Energy Conversion and Management', 'Progress in Energy and Combustion Science', 'Biotechnology Advances', and 'Science'.

The top categories for these journals are 'Energy Fuels', 'Engineering Chemical', and 'Environmental Sciences'. The other most-prolific subject categories are 'Biotechnology and Applied Microbiology', 'Engineering Environmental', 'Chemistry Physical', and 'Chemistry Applied'. It is notable that some journals with a heavy presence in the population papers are relatively under-represented in the sample papers.

In total, around 30 and 180 countries contribute to the sample and population papers, respectively. The top country is the USA (Table 2.6). This finding is in line with the studies arguing that the USA is not losing ground in science and technology (Leydesdorff and Wagner, 2009).

The other prolific countries are India, China, Canada, Australia, Spain, and Italy. These findings are in line with the studies showing heavy research activity in these countries in recent decades (Bordons et al., 2015; Gorjiara and Baldock, 2014; Prathap, 2017; Tahmooresnejad et al., 2015; Zhou and Leydesdorff, 2006).

On the other hand, the European and Asian countries represented in this table publish altogether 29 and 24% of the sample papers and 22.0 and 29.6% of the population papers, respectively. These findings are in line with the studies showing that both European and Asian countries have superior publication performance in science and technology (Bordons et al., 2015; Glanzel and Schlemmer, 2007; Okubo et al., 1998; Youtie et al., 2008).

It is notable that the publication surplus/deficit for the US and these European and Asian countries is 19.1, 7.0, and –5.6%, respectively. On the other hand, the countries with the most impact are the US, Australia, and New Zealand. Furthermore, the countries with the least impact are the China, the UK, Canada, France, and Malaysia.

China's presence in this top table is notable. This finding is in line with China's efforts to be a leading nation in science and technology (Guan and Ma, 2007; Youtie et al., 2008; Zhou and Leydesdorff, 2006).

It is also notable that some countries have a heavy presence in the population papers. The major producers of these papers are Brazil, South Korea, Iran, Russia, Norway, Mexico, Taiwan, Poland, Sweden, Saudi Arabia, Egypt, Thailand, Nigeria, Greece, Portugal, Belgium, and Finland with at least 0.7% of the papers each (Glanzel et al., 2006; Hassan et al., 2012; Huang et al., 2006, Schmoch et al., 2016).

The sample and population papers are indexed by around 25 and 230 subject categories, respectively. For the sample papers, the top subject is 'Energy Fuels' with 43.0 and 30.1% of the sample and population papers, respectively (Table 2.7). This top subject category is followed by 'Engineering Chemical', 'Biotechnology Applied Microbiology', 'Environmental Sciences', and 'Agricultural Engineering'. The other main subject categories are 'Chemistry Physical', 'Green Sustainable Science Technology', 'Engineering Environmental', 'Thermodynamics', and 'Chemistry Applied'.

It is notable that the publication surplus is most significant for 'Energy Fuels', 'Biotechnology Applied Microbiology', 'Agricultural Engineering', and 'Green Sustainable Science Technology'. On the other hand, the subjects with least impact are 'Engineering Chemical', 'Geosciences Multidisciplinary', 'Plant Sciences', and 'Chemistry Multidisciplinary'. This latter group of subject categories are under-represented in the sample papers.

These sample papers receive about 103,000 citations, as of February 2020. Thus, the average number of citations per paper is about 1,030. Hence, the citation impact of these top 100 papers in this field is significant.

Although a number of keywords are listed in the appendix for the datasets related to this field, some of them are more significant for the sample papers.

The most-prolific keywords for the keyword set related to biodiesel and petrodiesel fuels are 'biodiesel*' and 'diesel'. There is also one paper for 'bio-diesel'.

The prolific keywords for feedstocks are '*alga*', 'oil*', 'waste*', 'biomass', 'wood*', 'bio-oil*', 'feedstock*', 'lipid*', 'microb*', 'soybean*', 'cooking oil*', 'vegetable oil*', 'rape*', 'petroleum', 'bacteria', and '*seed*'.

The prolific keywords related to the processes for biodiesel and petrodiesel production are 'transester*', 'production', 'engine*', 'pyrolys*', '*processing', 'combustion', '*fuel', 'environment*', 'cultivation', 'photobioreactor*', 'cataly*', 'NOx', 'desulfurization', '*degradation', 'hydrocarbon*', 'pah*', 'properties', 'glycerol', 'emission*', 'upgrading', 'pollut*', 'composition', 'soil', and 'econom*'.

As expected, these keywords provide valuable information about the pathways of the research in this field.

There are three major topical research fronts for these sample papers: 'biodiesel fuels in general', 'feedstock-based biodiesel fuels', and 'petrodiesel fuels'. The 'biodiesel fuels in general' and 'feedstock-based biodiesel fuels' have a publication surplus, respectively, whilst 'petrodiesel fuels' has a publication deficit.

There are 14, 21, and 6 papers for 'biooils', 'biodiesel fuels in general', and 'glycerol', respectively, in the first group of sample papers.

Similarly, there are 5, 3, 2, and 20 sample papers for 'edible oil-based biodiesel fuels', 'nonedible oil-based biodiesel fuels', 'waste oil-based biodiesel fuels', and 'algal oil-based biodiesel fuels', respectively, in the second group of sample papers.

Finally, there are 14, 8, 11, and 1 papers for 'crude oils', 'petrodiesel fuels in general', 'emissions from petrodiesel fuels', and 'health impact of the emissions from petrodiesel fuels', respectively, in the final group of sample papers.

'Biodiesel fuels in general', 'algae-based biodiesel fuels', and 'edible oil-based biodiesel fuels' have a publication surplus, respectively, whilst 'crude oils', 'petrodiesel fuels in general', and 'biooils' have publication deficits, respectively.

The key emphasis in these research fronts is the exploration of the structure–processing–property relationships of biodiesel and petrodiesel fuels (Cheng and Ma, 2011; Konur and Matthews, 1989; Rogers and Hopfinger, 1994; Scherf and List, 2002).

2.5 CONCLUSION

This chapter has mapped the research on biodiesel and petrodiesel fuels using a scientometric method.

The size of over 121,000 population papers shows the public importance of this interdisciplinary research field. However, it is significant that the research has developed more in its technological aspects than its social and humanitarian pathways.

Articles dominate both the sample and population papers. However, there are 68 review papers. The population papers, primarily published in the 2010s, build on these sample papers, primarily published in the 2000s.

The data presented in the tables and in the one figure show that a small number of authors, institutions, funding bodies, journals, keywords, research fronts, subject categories, and countries have shaped the research.

It is notable that the authors, institutions, and funding bodies from the USA, India, China, Canada, Australia, Spain, and Italy dominate the research. Furthermore, it is also notable that some countries have a heavy presence in the population papers. The major producers of the population papers are Brazil, South Korea, Iran, Russia, Norway, Mexico, Taiwan, Poland, Sweden, Saudi Arabia, Egypt, Thailand, Nigeria, Greece, Portugal, Belgium, and Finland. Additionally, China, the UK, and France are under-represented significantly in the sample papers.

These findings show the importance of the development of efficient incentive structures for the development of the research in this field, as in other fields. It seems that some countries (such as the USA, India, China, Canada, Australia, Spain, and

Italy) have efficient incentive structures for the development of the research in this field, contrary to Brazil, South Korea, Iran, Russia, Norway, Mexico, Taiwan, Poland, Sweden, Saudi Arabia, Egypt, Thailand, Nigeria, Greece, Portugal, Belgium, and Finland.

It further seems that, although research funding is a significant element of these incentive structures, it might not be a sole solution for increasing such incentives in this field, as is the case in Brazil, South Korea, Iran, Russia, Norway, Mexico, Taiwan, Poland, Sweden, Saudi Arabia, Egypt, Thailand, Nigeria, Greece, Portugal, Belgium, and Finland.

On the other hand, it seems there is more to do to reduce the significant gender deficit, as in other fields of science and technology (Lariviere et al., 2013; Xie and Shauman, 1998).

The data on the research fronts, keywords, source titles, and subject categories provide valuable evidence for the interdisciplinary nature of the research in this field (Lariviere and Gingras, 2010; Morillo et al., 2001).

There is ample justification for the broad search strategy employed in this study due to the interdisciplinary nature of the field, as evidenced by the top subject categories. The search strategy employed is in line with those employed in related and other research fields (Konur, 2011, 2012a–n, 2015, 2016a–f, 2017a–f, 2018a–b, 2019a–b; 2021a–ag). It is particularly noted that only 43.0 and 30.1% of the sample and population papers are indexed by the 'Energy Fuels' subject category, respectively.

There are three major topical research fronts for these sample papers. 'Biodiesel fuels in general', 'feedstock-based biodiesel fuels', and 'petrodiesel fuels'. The 'biodiesel fuels in general' and 'feedstock-based biodiesel fuels' have a publication surplus, respectively, whilst 'petrodiesel fuels' has a publication deficit.

The detailed research fronts are 'biooils', 'biodiesel fuels in general', and 'glycerol', respectively, in the first group of sample papers. Similarly, the detailed research fronts are 'edible oil-based biodiesel fuels', 'nonedible oil-based biodiesel fuels', 'waste oil-based biodiesel fuels', and 'algal oil-based biodiesel fuels', respectively, in the second group of sample papers. Finally, the detailed research fronts are 'crude oils', 'petrodiesel fuels in general', 'emissions from petrodiesel fuels', and 'health impact of the emissions from petrodiesel fuels', respectively, in the final group of sample papers.

'Biodiesel fuels in general', 'algal oil-based biodiesel fuels', and 'edible oil-based biodiesel fuels' have a publication surplus, respectively, whilst 'crude oils', 'petrodiesel fuels in general', and 'biooils' have publication deficits, respectively.

It is recommended that further scientometric studies are carried out for each of these research fronts, building on the pioneering studies in these fields ((Konur, 2021e–ag).

ACKNOWLEDGMENTS

The contribution of the highly cited researchers in the fields of biodiesel and petrodiesel fuels has been greatly acknowledged.

REFERENCES

Achten, W. M. J., L. Verchot, and Y. J. Franken, et al. 2008. Jatropha bio-diesel production and use. *Biomass & Bioenergy* 32: 1063–1084.

Agarwal, A. K. 2007. Biofuels (alcohols and biodiesel) applications as fuels for internal combustion engines. *Progress in Energy and Combustion Science* 33: 233–271.

Ahmadun, F. R., A. Pendashteh, and L. C. Abdullah, et al. 2009. Review of technologies for oil and gas produced water treatment. *Journal of Hazardous Materials* 170: 530–551.

Atlas, R. M. 1981. Microbial degradation of petroleum hydrocarbons: An environmental perspective. *Microbiological Reviews* 45: 180–209.

Babich, I. V. and J. A. Moulijn. 2003. Science and technology of novel processes for deep desulfurization of oil refinery streams: A review. *Fuel* 82: 607–631.

Behr, A., J. Eilting, K. Irawadi, J. Leschinski, and F. Lindner. 2008. Improved utilisation of renewable resources: New important derivatives of glycerol. *Green Chemistry* 10: 13–30.

Birch, M. E. and R. A. Cary. 1996. Elemental carbon-based method for monitoring occupational exposures to particulate diesel exhaust. *Aerosol Science and Technology* 25: 221–241.

Bordons, M., B. Gonzalez-Albo, J. Aparicio, and L. Moreno. 2015. The influence of R & D intensity of countries on the impact of international collaborative research: Evidence from Spain. *Scientometrics* 102: 1385–1400.

Brennan, L. and P. Owende. 2010. Biofuels from microalgae: A review of technologies for production, processing, and extractions of biofuels and co-products. *Renewable & Sustainable Energy Reviews* 14: 557–577.

Bridgwater, A. V. and G. V. C. Peacocke. 2000. Fast pyrolysis processes for biomass. *Renewable & Sustainable Energy Reviews* 4: 1–73.

Budzinski, H., I. Jones, J. Bellocq, C. Pierard, and P. Garrigues. 1997. Evaluation of sediment contamination by polycyclic aromatic hydrocarbons in the Gironde estuary. *Marine Chemistry* 58: 85–97.

Busca, G., L. Lietti, G. Ramis, and F. Berti. 1998. Chemical and mechanistic aspects of the selective catalytic reduction of NO_x by ammonia over oxide catalysts: A review. *Applied Catalysis B-Environmental* 18: 1–36.

Canakci, M. and J. Van Gerpen. 2001. Biodiesel production from oils and fats with high free fatty acids. *Transactions of the ASAE* 44: 1429–1436.

Cheng, Y. Q. and E. Ma. 2011. Atomic-level structure and structure–property relationship in metallic glasses. *Progress in Materials Science* 56: 379–473.

Chisti, Y. 2007. Biodiesel from microalgae. *Biotechnology Advances* 25: 294–306.

Clarivate Analytics. 2019. *Highly Cited Researchers: 2019 Recipients*. Philadelphia, PA: Clarivate Analytics. https://recognition. webofsciencegroup.com/awards/highly-cited/2019/ (accessed January, 3, 2020).

Czernik, S. and A. V. Bridgwater. 2004. Overview of applications of biomass fast pyrolysis oil. *Energy & Fuels* 18: 590–598.

Dasari, M. A., P. P. Kiatsimkul, W. R. Sutterlin, and G. J. Suppes. 2005. Low-pressure hydrogenolysis of glycerol to propylene glycol. *Applied Catalysis A-General* 281: 225–231.

Demirbas, A. 2003. Biodiesel fuels from vegetable oils via catalytic and non-catalytic supercritical alcohol transesterifications and other methods: A survey. *Energy Conversion and Management* 44: 2093–2109.

Di Blasi, C. 2008. Modeling chemical and physical processes of wood and biomass pyrolysis. *Progress in Energy and Combustion Science* 34: 47–90.

Docampo, D. and L. Cram. 2019. Highly cited researchers: A moving target. *Scientometrics* 118: 1011–1025.

Freedman, B., R. O. Butterfield, and E. H. Pryde. 1986. Transesterification kinetics of soybean oil. *Journal of the American Oil Chemists Society* 63: 1375–1380.

Gallezot, P. 2012. Conversion of biomass to selected chemical products. *Chemical Society Reviews* 41: 1538–1558.

Garfield, E. 1955. Citation indexes for science. *Science* 122: 108–111.

Garfield, E. 1972. Citation analysis as a tool in journal evaluation. *Science* 178: 471–479.

Glanzel, W. and B. Schlemmer. 2007. National research profiles in a changing Europe (1983–2003): An exploratory study of sectoral characteristics in the Triple Helix. *Scientometrics* 70: 267–275.

Glanzel, W., J. Leta, J., and B. Thijs. 2006. Science in Brazil. Part 1: A macro-level comparative study. *Scientometrics* 67: 67–86.

Gorjiara, T. and C. Baldock. 2014. Nanoscience and nanotechnology research publications: A comparison between Australia and the rest of the world. *Scientometrics* 100: 121–148.

Graboski, M. S. and R. L. McCormick. 1998. Combustion of fat and vegetable oil derived fuels in diesel engines. *Progress in Energy and Combustion Science* 24: 125–164.

Guan, J. C. and N. Ma. 2007. China's emerging presence in nanoscience and nanotechnology: A comparative bibliometric study of several nanoscience 'giants'. *Research Policy* 36: 880–886.

Haritash, A. K. and C. P. Kaushik. 2009. Biodegradation aspects of polycyclic aromatic hydrocarbons (PAHs): A review. *Journal of Hazardous Materials* 169: 1–15.

Hassan, S. U., P. Haddawy, P. Kuinkel, A. Degelsegger, and C. Blasy. 2012. A bibliometric study of research activity in ASEAN related to the EU in FP7 priority areas. *Scientometrics* 91: 1035–1051.

Hill, J., E. Nelson, D. Tilman, S. Polasky, and D. Tiffany. 2006. Environmental, economic, and energetic costs and benefits of biodiesel and ethanol biofuels. *Proceedings of the National Academy of Sciences of the United States of America* 103: 11206–11210.

Hu, Q., M. Sommerfeld, and E. Jarvis, et al. 2008. Microalgal triacylglycerols as feedstocks for biofuel production: Perspectives and advances. *Plant Journal* 54(4), 621–639.

Huang, M. H., H. W. Chang, and D. Z. Chen. 2006. Research evaluation of research-oriented universities in Taiwan from 1993 to 2003. *Scientometrics* 67: 419–435.

Khalili, N. R., P. A. Scheff, and T. M. Holsen. 1995. PAH source fingerprints for coke ovens, diesel and gasoline-engines, highway tunnels, and wood combustion emissions. *Atmospheric Environment* 29: 533–542.

Kilian, L. 2009. Not all oil price shocks are alike: Disentangling demand and supply shocks in the crude oil market. *American Economic Review* 99: 1053–1069.

Konur, O. 2000. Creating enforceable civil rights for disabled students in higher education: An institutional theory perspective. *Disability & Society* 15: 1041–1063.

Konur, O. 2002a. Access to Nursing Education by disabled students: Rights and duties of nursing programs. *Nurse Education Today* 22: 364–374.

Konur, O. 2002b. Assessment of disabled students in higher education: Current public policy issues. *Assessment and Evaluation in Higher Education* 27: 131–152.

Konur, O. 2002c. Access to employment by disabled people in the UK: Is the Disability Discrimination Act working? *International Journal of Discrimination and the Law* 5: 247–279.

Konur, O. 2006a. Participation of children with dyslexia in compulsory education: Current public policy issues. *Dyslexia* 12: 51–67.

Konur, O. 2006b. Teaching disabled students in Higher Education. *Teaching in Higher Education* 11: 351–363.

Konur, O. 2007a. A judicial outcome analysis of the Disability Discrimination Act: A windfall for the employers? *Disability & Society* 22: 187–204.

Konur, O. 2007b. Computer-assisted teaching and assessment of disabled students in higher education: The interface between academic standards and disability rights. *Journal of Computer Assisted Learning* 23: 207–219.

Konur, O. 2011. The scientometric evaluation of the research on the algae and bio-energy. *Applied Energy* 88: 3532–3540.

Konur, O. 2012a. Evaluation of the research on the social sciences in Turkey: A scientometric approach. *Energy Education Science and Technology Part B: Social and Educational Studies* 4: 1893–1908.

Konur, O. 2012b. Prof. Dr. Ayhan Demirbas' scientometric biography. *Energy Education Science and Technology Part A: Energy Science and Research* 28: 727–738.

Konur, O. 2012c. The evaluation of the biogas research: A scientometric approach. *Energy Education Science and Technology Part A: Energy Science and Research* 29: 1277–1292.

Konur, O. 2012d. The evaluation of the educational research: A scientometric approach. *Energy Education Science and Technology Part B: Social and Educational Studies* 4: 1935–1948.

Konur, O. 2012e. The evaluation of the global energy and fuels research: A scientometric approach. *Energy Education Science and Technology Part A: Energy Science and Research* 30: 613–628.

Konur, O. 2012f. The evaluation of the research on the Arts and Humanities in Turkey: A scientometric approach. *Energy Education Science and Technology Part B: Social and Educational Studies* 4: 1603–1618.

Konur, O. 2012g. The evaluation of the research on the biodiesel: A scientometric approach. *Energy Education Science and Technology Part A: Energy Science and Research* 28: 1003–1014.

Konur, O. 2012h. The evaluation of the research on the bioethanol: A scientometric approach. *Energy Education Science and Technology Part A: Energy Science and Research* 28: 1051–1064.

Konur, O. 2012i. The evaluation of the research on the biofuels: A scientometric approach. *Energy Education Science and Technology Part A: Energy Science and Research* 28: 903–916.

Konur, O. 2012j. The evaluation of the research on the biohydrogen: A scientometric approach. *Energy Education Science and Technology Part A: Energy Science and Research* 29: 323–338.

Konur, O. 2012k. The evaluation of the research on the microbial fuel cells: A scientometric approach. *Energy Education Science and Technology Part A: Energy Science and Research* 29: 309–322.

Konur, O. 2012l. The scientometric evaluation of the research on the production of bioenergy from biomass. *Biomass and Bioenergy* 47: 504–515.

Konur, O. 2012m. The scientometric evaluation of the research on the deaf students in higher education. *Energy Education Science and Technology Part B: Social and Educational Studies* 4: 1573–1588.

Konur, O. 2012n. The scientometric evaluation of the research on the students with ADHD in higher education. *Energy Education Science and Technology Part B: Social and Educational Studies* 4: 1547–1562.

Konur, O. 2015. Current state of research on algal biodiesel. In *Marine Bioenergy: Trends and Developments*, S. K. Kim, and C. G. Lee, ed., 487–512. Boca Raton, FL: CRC Press.

Konur, O. 2016a. Scientometric overview in nanobiodrugs. In *Nanoarchitectonics for Smart Delivery and Drug Targeting*, A. M. Holban and A.M. Grumezescu, ed., 405–428. Amsterdam: Elsevier.

Konur, O. 2016b. Scientometric overview regarding nanoemulsions used in the food industry. In *Emulsions: Nanotechnology in the Agri-Food Industry*, A. M. Grumezescu, ed., 689–711. Amsterdam: Elsevier.

Konur, O. 2016c. Scientometric overview regarding the nanobiomaterials in antimicrobial therapy. In *Nanobiomaterials in Antimicrobial Therapy,* A. M. Grumezescu, ed., 511–535. Amsterdam: Elsevier.

Konur, O. 2016d. Scientometric overview regarding the nanobiomaterials in dentistry. In *Nanobiomaterials in Dentistry,* A. M. Grumezescu, ed., 425–453. Amsterdam: Elsevier.

Konur, O. 2016e. Scientometric overview regarding the surface chemistry of nanobiomaterials. In *Surface Chemistry of Nanobiomaterials,* A. M. Grumezescu, ed., 463–486. Amsterdam: Elsevier.

Konur, O. 2016f. The scientometric overview in cancer targeting. In *Nanoarchitectonics for Smart Delivery and Drug Targeting,* A. M. Holban and A. Grumezescu, ed., 871–895. Amsterdam: Elsevier.

Konur, O. 2017a. Recent citation classics in antimicrobial nanobiomaterials. In *Nanostructures for Antimicrobial Therapy,* A. Ficai and A. M. Grumezescu, ed., 669–685. Amsterdam: Elsevier.

Konur, O. 2017b. Scientometric overview in nanopesticides. In *New Pesticides and Soil Sensors,* A. M. Grumezescu, ed. 719–744. Amsterdam: Elsevier.

Konur, O. 2017c. Scientometric overview regarding oral cancer nanomedicine. In *Nanostructures for Oral Medicine*, E. Andronescu, A. M. Grumezescu, ed., 939–962. Amsterdam: Elsevier.

Konur, O. 2017d. Scientometric overview regarding water nanopurification. In *Water Purification,* A. M. Grumezescu, ed., 693–716. Amsterdam: Elsevier.

Konur, O. 2017e. Scientometric overview in food nanopreservation. In *Food Preservation,* A. M. Grumezescu, ed., 703–729. Amsterdam: Elsevier.

Konur, O. 2017f. The top citation classics in alginates for biomedicine. In *Seaweed Polysaccharides: Isolation, Biological and Biomedical Applications*, J. Venkatesan, S. Anil, S. K. Kim, ed., 223–249. Amsterdam: Elsevier.

Konur, O. 2018a. Scientometric evaluation of the global research in spine: An update on the pioneering study by Wei et al. *European Spine Journal* 27: 525–529.

Konur, O. 2018b. Bioenergy and biofuels science and technology: Scientometric overview and citation classics. In *Bienergy and Biofuels*, O. Konur, ed., 3–63. Boca Raton: CRC Press.

Konur, O. 2019a. Cyanobacterial bioenergy and biofuels science and technology: A scientometric overview. In *Cyanobacteria: From Basic Science to Applications*, ed. A. K. Mishra, D. N. Tiwari and A. N. Rai, 419–442. Amsterdam: Elsevier.

Konur, O. 2019b. Nanotechnology applications in food: A scientometric overview. In *Nanoscience for Sustainable Agriculture*, R. N. Pudake, N. Chauhan, and C. Kole, ed., 683–711. Cham: Springer.

Konur, O., ed. 2021a. *Handbook of Biodiesel and Petrodiesel Fuels: Science, Technology, Health, and Environment.* Boca Raton, FL: CRC Press.

Konur, O., ed. 2021b. *Handbook of Biodiesel and Petrodiesel Fuels: Science, Technology, Health, and Environment. Volume 1. Biodiesel Fuels: Science, Technology, Health, and Environment.* Boca Raton, FL: CRC Press.

Konur, O., ed. 2021c. *Handbook of Biodiesel and Petrodiesel Fuels: Science, Technology, Health, and Environment. Volume 2. Biodiesel Fuels based on the Edible and Nonedible Feedstocks, Wastes, and Algae: Science, Technology, Health, and Environment.* Boca Raton, FL: CRC Press.

Konur, O., ed. 2021d. *Handbook of Biodiesel and Petrodiesel Fuels: Science, Technology, Health, and Environment. Volume 3. Petrodiesel Fuels: Science, Technology, Health, and Environment.* Boca Raton, FL: CRC Press.

Konur, O. 2021e. Biodiesel and petrodiesel fuels: Science, technology, health, and environment. In *Handbook of Biodiesel and Petrodiesel Fuels: Science, Technology, Health, and Environment. Volume 1. Biodiesel Fuels: Science, Technology, Health, and Environment*, ed. O. Konur. Boca Raton, FL: CRC Press.

Konur, O. 2021f. Biodiesel and petrodiesel fuels: A scientometric review of the research. In *Handbook of Biodiesel and Petrodiesel Fuels: Science, Technology, Health, and Environment. Volume 1. Biodiesel Fuels: Science, Technology, Health, and Environment*, ed. O. Konur. Boca Raton, FL: CRC Press.

Konur, O. 2021g. Biodiesel and petrodiesel fuels: A review of the research. In *Handbook of Biodiesel and Petrodiesel Fuels: Science, Technology, Health, and Environment. Volume 1. Biodiesel Fuels: Science, Technology, Health, and Environment*, ed. O. Konur. Boca Raton, FL: CRC Press.

Konur, O. 2021h Nanotechnology applications in the diesel fuels and the related research fields: A review of the research. In *Handbook of Biodiesel and Petrodiesel Fuels: Science, Technology, Health, and Environment. Volume 1. Biodiesel Fuels: Science, Technology, Health, and Environment*, ed. O. Konur. Boca Raton, FL: CRC Press.

Konur, O. 2021i. Biooils: A scientometric review of the research. In *Handbook of Biodiesel and Petrodiesel Fuels: Science, Technology, Health, and Environment. Volume 1. Biodiesel Fuels: Science, Technology, Health, and Environment*, ed. O. Konur. Boca Raton, FL: CRC Press.

Konur, O. 2021j. Characterization and properties of biooils: A review of the research. In *Handbook of Biodiesel and Petrodiesel Fuels: Science, Technology, Health, and Environment. Volume 1. Biodiesel Fuels: Science, Technology, Health, and Environment*, ed. O. Konur. Boca Raton, FL: CRC Press.

Konur, O. 2021k. Biomass pyrolysis and pyrolysis oils: A review of the research. In *Handbook of Biodiesel and Petrodiesel Fuels: Science, Technology, Health, and Environment. Volume 1. Biodiesel Fuels: Science, Technology, Health, and Environment*, ed. O. Konur. Boca Raton, FL: CRC Press.

Konur, O. 2021l. Biodiesel fuels: A scientometric review of the research. In *Handbook of Biodiesel and Petrodiesel Fuels: Science, Technology, Health, and Environment. Volume 1. Biodiesel Fuels: Science, Technology, Health, and Environment*, ed. O. Konur. Boca Raton, FL: CRC Press.

Konur, O. 2021m. Glycerol: A scientometric review of the research. In *Handbook of Biodiesel and Petrodiesel Fuels: Science, Technology, Health, and Environment. Volume 1. Biodiesel Fuels: Science, Technology, Health, and Environment*, ed. O. Konur. Boca Raton, FL: CRC Press.

Konur, O. 2021n. Propanediol production from glycerol: A review of the research. In *Handbook of Biodiesel and Petrodiesel Fuels: Science, Technology, Health, and Environment. Volume 1. Biodiesel Fuels: Science, Technology, Health, and Environment*, ed. O. Konur Boca Raton, FL: CRC Press.

Konur, O. 2021o. Edible oil-based biodiesel fuels: A scientometric review of the research. *In Handbook of Biodiesel and Petrodiesel Fuels: Science, Technology, Health, and Environment. Volume 2. Biodiesel Fuels based on the Edible and Nonedible Feedstocks, Wastes, and Algae: Science, Technology, Health, and Environment,* ed. O. Konur. Boca Raton, FL: CRC Press.

Konur, O. 2021p. Palm oil-based biodiesel fuels: A review of the research. In *Handbook of Biodiesel and Petrodiesel Fuels: Science, Technology, Health, and Environment. Volume 2. Biodiesel Fuels based on the Edible and Nonedible Feedstocks, Wastes, and Algae*, ed. O. Konur. Boca Raton, FL: CRC Press.

Konur, O. 2021q. Rapeseed oil-based biodiesel fuels: A review of the research. In *Handbook of Biodiesel and Petrodiesel Fuels: Science, Technology, Health, and Environment. Volume 2. Biodiesel Fuels based on the Edible and Nonedible Feedstocks, Wastes, and Algae,* ed. O. Konur. Boca Raton, FL: CRC Press.

Konur, O. 2021r. Nonedible oil-based biodiesel fuels: A scientometric review of the research. In *Handbook of Biodiesel and Petrodiesel Fuels: Science, Technology, Health, and Environment. Volume 2. Biodiesel Fuels based on the Edible and Nonedible Feedstocks, Wastes, and Algae: Science, Technology, Health, and Environment*, ed. O. Konur. Boca Raton, FL: CRC Press.

Konur, O. 2021s. Waste oil-based biodiesel fuels: A scientometric review of the research. In *Handbook of Biodiesel and Petrodiesel Fuels: Science, Technology, Health, and Environment. Volume 2. Biodiesel Fuels based on the Edible and Nonedible Feedstocks,*

Wastes, and Algae: Science, Technology, Health, and Environment, ed. O. Konur. Boca Raton, FL: CRC Press.

Konur, O. 2021t. Algal biodiesel fuels: A scientometric review of the research. In *Handbook of Biodiesel and Petrodiesel Fuels: Science, Technology, Health, and Environment. Volume 2. Biodiesel Fuels based on the Edible and Nonedible Feedstocks, Wastes, and Algae: Science, Technology, Health, and Environment*, ed. O. Konur. Boca Raton, FL: CRC Press.

Konur, O. 2021u. Algal biomass production for biodiesel production: A review of the research. In *Handbook of Biodiesel and Petrodiesel Fuels: Science, Technology, Health, and Environment. Volume 2. Biodiesel Fuels based on the Edible and Nonedible Feedstocks, Wastes, and Algae*, Ed. O. Konur Boca Raton, FL: CRC Press. February 23, 2020

Konur, O. 2021v. Algal biomass production in wastewaters for biodiesel production: A review of the research. In *Handbook of Biodiesel and Petrodiesel Fuels: Science, Technology, Health, and Environment. Volume 2. Biodiesel Fuels based on the Edible and Nonedible Feedstocks, Wastes, and Algae*, ed. O. Konur. Boca Raton, FL: CRC Press. February 23, 2020

Konur, O. 2021x. Algal lipid production for biodiesel production: A review of the research. In *Handbook of Biodiesel and Petrodiesel Fuels: Science, Technology, Health, and Environment. Volume 2. Biodiesel Fuels based on the Edible and Nonedible Feedstocks, Wastes, and Algae*, Ed. O. Konur Boca Raton, FL: CRC Press.

Konur, O. 2021y. Crude oils: A scientometric review of the research. In *Handbook of Biodiesel and Petrodiesel Fuels: Science, Technology, Health, and Environment. Volume 3. Petrodiesel Fuels: Science, Technology, Health, and Environment*, ed. O. Konur. Boca Raton, FL: CRC Press.

Konur, O. 2021z. Petrodiesel fuels: A scientometric review of the research. In *Handbook of Biodiesel and Petrodiesel Fuels: Science, Technology, Health, and Environment. Volume 3. Petrodiesel Fuels: Science, Technology, Health, and Environment*, ed. O. Konur. Boca Raton, FL: CRC Press.

Konur, O. 2021aa. Bioremediation of petroleum hydrocarbons in the contaminated soils: A review of the research. In *Handbook of Biodiesel and Petrodiesel Fuels: Science, Technology, Health, and Environment. Volume 3. Petrodiesel Fuels: Science, Technology, Health, and Environment*, ed. O. Konur. Boca Raton, FL: CRC Press.

Konur, O. 2021ab. Desulfurization of diesel fuels: A review of the research. In *Handbook of Biodiesel and Petrodiesel Fuels: Science, Technology, Health, and Environment. Volume 3. Petrodiesel Fuels: Science, Technology, Health, and Environment*, ed. O. Konur. Boca Raton, FL: CRC Press.

Konur, O. 2021ac. Diesel fuel exhaust emissions: A scientometric review of the research. In *Handbook of Biodiesel and Petrodiesel Fuels: Science, Technology, Health, and Environment. Volume 3. Petrodiesel Fuels: Science, Technology, Health, and Environment*, ed. O. Konur. Boca Raton, FL: CRC Press.

Konur, O. 2021ad. The adverse health and safety impact of diesel fuels: A scientometric review of the research. In *Handbook of Biodiesel and Petrodiesel Fuels: Science, Technology, Health, and Environment. Volume 3. Petrodiesel Fuels: Science, Technology, Health, and Environment*, ed. O. Konur. Boca Raton, FL: CRC Press.

Konur, O. 2021ae. Respiratory illnesses caused by the diesel fuel exhaust emissions: A review of the research. In *Handbook of Biodiesel and Petrodiesel Fuels: Science, Technology, Health, and Environment. Volume 3. Petrodiesel Fuels: Science, Technology, Health, and Environment*, ed. O. Konur. Boca Raton, FL: CRC Press.

Konur, O. 2021af. Cancer caused by the diesel fuel exhaust emissions: A review of the research. In *Handbook of Biodiesel and Petrodiesel Fuels: Science, Technology, Health, and Environment. Volume 3. Petrodiesel Fuels: Science, Technology, Health, and Environment*, ed. O. Konur. Boca Raton, FL: CRC Press.

Konur, O. 2021ag. Cardiovascular and other illnesses caused by the diesel fuel exhaust emissions: A review of the research. In *Handbook of Biodiesel and Petrodiesel Fuels: Science, Technology, Health, and Environment. Volume 3. Petrodiesel Fuels: Science, Technology, Health, and Environment*, ed. O. Konur. Boca Raton, FL: CRC Press.

Konur, O. and F. L. Matthews. 1989. Effect of the properties of the constituents on the fatigue performance of composites: A review. *Composites* 20: 317–328.

Kulkarni, M. G. and A. K. Dalai. 2006. Waste cooking oil-an economical source for biodiesel: A review. *Industrial & Engineering Chemistry Research* 45: 2901–2913.

Lapuerta, M., O. Armas, and J. Rodriguez-Fernandez. 2008. Effect of biodiesel fuels on diesel engine emissions. *Progress in Energy and Combustion Science* 34: 198–223.

Lariviere, V. and Y. Gingras. 2010. On the relationship between interdisciplinarity and scientific impact. *Journal of the American Society for Information Science and Technology* 61: 126–131.

Lariviere, V., C. Ni, Y. Gingras, B. Cronin, and C. R. Sugimoto. 2013. Bibliometrics: Global gender disparities in science. *Nature News* 504: 211–213.

Leahy, J. G. and R. R. Colwell. 1990. Microbial degradation of hydrocarbons in the environment. *Microbiological Reviews* 54: 305–315.

Leydesdorff, L. and C. Wagner. 2009. Is the United States losing ground in science? A global perspective on the world science system. *Scientometrics* 78: 23–36.

Ma, F. R. and M. A. Hanna. 1999. Biodiesel production: A review. *Bioresource Technology* 70: 1–15.

Mata, T. M., A. A. Martins, and N. S. Caetano. 2010. Microalgae for biodiesel production and other applications: A review. *Renewable & Sustainable Energy Reviews* 14: 217–232.

Meher, L. C., D. V. Sagar, and S. N. Naik. 2006. Technical aspects of biodiesel production by transesterification: A review. *Renewable & Sustainable Energy Reviews* 10: 248–268.

Mohan, D., C. U. Pittman, and P. H. Steele. 2006. Pyrolysis of wood/biomass for bio-oil: A critical review. *Energy & Fuels* 20: 848–889.

Morillo, F., M. Bordons, and I. Gomez. 2001. An approach to interdisciplinarity through bibliometric indicators. *Scientometrics* 51: 203–222.

Mortensen, P. M., J. D. Grunwaldt, P. A. Jensen, K. G. Knudsen, and A. D. Jensen. 2011. A review of catalytic upgrading of bio-oil to engine fuels. *Applied Catalysis A-General* 407: 1–19.

North, D. C. 1991a. *Institutions, Institutional Change and Economic Performance*. Cambridge, Mass: Cambridge University Press.

North, D.C. 1991b. Institutions. *Journal of Economic Perspectives* 5: 97–112.

Okubo, Y., J. C. Dore, T. Ojasoo, and J. F. Miquel. 1998. A multivariate analysis of publication trends in the 1980s with special reference to South-East Asia. *Scientometrics* 41: 273.

Olah, G. A. 2005. Beyond oil and gas: The methanol economy. *Angewandte Chemie-International Edition* 44: 2636–2639.

Pagliaro, M., R. Ciriminna, H. Kimura, H. M. Rossi, and C. Della Pina. 2007. From glycerol to value-added products. *Angewandte Chemie-International Edition* 46: 4434–4440.

Perron, P. 1989. The great crash, the oil price shock, and the unit root hypothesis. *Econometrica: Journal of the Econometric Society* 57: 1361–1401.

Peters, K. E. 1986. Guidelines for evaluating petroleum source rock using programmed pyrolysis. *AAPG Bulletin-American Association of Petroleum Geologists* 70: 318–329.

Prathap, G. 2017. A three-dimensional bibliometric evaluation of recent research in India. *Scientometrics* 110: 1085–1097.

Ramadhas, A. S., S. Jayaraj, and C. Muraleedharan. 2005. Biodiesel production from high FFA rubber seed oil. *Fuel* 84: 335–340.

Rodolfi, L., G. C. Zittelli, and N. Bassi, et al. 2009. Microalgae for oil: Strain selection, induction of lipid synthesis and outdoor mass cultivation in a low-cost photobioreactor. *Biotechnology and Bioengineering* 102: 100–112.

Rogers, D. and A. J. Hopfinger. 1994. Application of genetic function approximation to quantitative structure-activity relationships and quantitative structure-property relationships. *Journal of Chemical Information and Computer Sciences* 34: 854–866.

Rogge, W. F., L. M. Hildemann, M. A. Mazurek, G. R. Cass, and B. R. T. Simoneit. 1993. Sources of fine organic aerosol. 2. Noncatalyst and catalyst-equipped automobiles and heavy-duty diesel trucks. *Environmental Science & Technology* 27: 636–651.

Saka, S. and D. Kusdiana. 2001. Biodiesel fuel from rapeseed oil as prepared in supercritical methanol. *Fuel* 80: 225–231.

Schauer, J. J., M. J. Kleeman, G. R. Cass, and B. R. T. Simoneit. 1999. Measurement of emissions from air pollution sources. 2. C_1 through C_{30} organic compounds from medium duty diesel trucks. *Environmental Science & Technology* 33: 1578–1587.

Schenk, P. M., S. R. Thomas-Hall, and E. Stephens, et al. 2008. Second generation biofuels: High efficiency microalgae for biodiesel production. *Bioenergy Research* 1: 20–43.

Scherf, U. and E. J. List. 2002. Semiconducting polyfluorenes-towards reliable structure–property relationships. *Advanced Materials* 14: 477–487.

Schmoch, U., H. M. Fardoun, and A. S. Mashat. 2016. Establishing a world-class university in Saudi Arabia: intended and unintended effects. *Scientometrics* 109: 1191–1207.

Shelef, M. 1995. Selective catalytic reduction of NO_x with n-free reductants. *Chemical Reviews* 95: 209–225.

Song, C. and X. L. Ma. 2003. New design approaches to ultra-clean diesel fuels by deep desulfurization and deep dearomatization. *Applied Catalysis B-Environmental* 41: 207–238.

Song, C. S. 2003. An overview of new approaches to deep desulfurization for ultra-clean gasoline, diesel fuel and jet fuel. *Catalysis Today* 86: 211–263.

Tahmooresnejad, L., C. Beaudry, and A. Schiffauerova. 2015. The role of public funding in nanotechnology scientific production: Where Canada stands in comparison to the United States. *Scientometrics* 102: 753–787.

Wang, X., D. Liu, K. Ding, and X. Wang. 2012. Science funding and research output: A study on 10 countries. *Scientometrics* 91: 591–599.

Xie, Y. and K. A. Shauman. 1998. Sex differences in research productivity: New evidence about an old puzzle. *American Sociological Review* 63: 847–870.

Youtie, J, P. Shapira, and A. L. Porter. 2008. Nanotechnology publications and citations by leading countries and blocs. *Journal of Nanoparticle Research* 10: 981–986.

Zhang, Y., M. A. Dube, D. D. McLean, and M. Kates. 2003a. Biodiesel production from waste cooking oil: 1. Process design and technological assessment. *Bioresource Technology* 89: 1–16.

Zhang, Y., M. A. Dube, D. D. McLean, and M. Kates. 2003b. Biodiesel production from waste cooking oil: 2. Economic assessment and sensitivity analysis. *Bioresource Technology* 90: 229–240.

Zhou, C. H. C., J. N. Beltramini, Y. X. Fan, and G. Q. M. Lu. 2008. Chemoselective catalytic conversion of glycerol as a biorenewable source to valuable commodity chemicals. *Chemical Society Reviews* 37: 527–549.

Zhou, P. and L. Leydesdorff. 2006. The emergence of China as a leading nation in science. *Research Policy* 35: 83–104.

3 A Review of the Research

Biodiesel and Petrodiesel Fuels

Ozcan Konur

CONTENTS

3.1 INTRODUCTION

Crude oils have been primary sources of energy and fuels, such as petrodiesel. However, significant public concerns about their sustainability, price fluctuations, and adverse environmental impact have emerged since the 1970s (Ahmadun et al., 2009;

Atlas, 1981; Babich and Moulijn, 2003; Haritash and Kaushik, 2009; Kilian, 2009; Leahy and Colwell, 1990; Perron, 1989). Thus, biooils (Bridgwater and Peacocke, 2000; Czernik and Bridgwater, 2004; di Blasi, 2008; Gallezot, 2012; Mohan et al., 2006; Mortensen et al., 2011) and biooil-based biodiesel fuels (Agarwal, 2007; Chisti, 2007; Hill et al., 2006; Lapuerta et al., 2008; Ma and Hanna, 1999; Mata et al., 2010; Meher et al., 2006; Zhang et al., 2003a–b) have emerged as alternatives to crude oils and crude oil-based petrodiesel fuels in recent decades. Nowadays, although petrodiesel fuels are used extensively, biodiesel fuels are being used increasingly in the transportation and power sectors (Konur, 2021a–ag).

However, for the efficient progression of the research in this field, it is necessary to develop efficient incentive structures for the primary stakeholders and to inform these stakeholders about the research (Konur, 2000, 2002a–c, 2006a–b, 2007a–b; North, 1991a–b).

Although there have been over 5,500 reviews, book chapters, and books in this field, there has been no review of the 50-most-cited articles (cf. Konur, 2012, 2015). Thus, this chapter reviews these articles on both biodiesel and petrodiesel fuels. Then, it discusses the findings of the review.

3.2 MATERIALS AND METHODOLOGY

The search for the literature was carried out in the 'Web of Science' (WOS) database in February 2020. It contains the 'Science Citation Index-Expanded' (SCI-E), the Social Sciences Citation Index' (SSCI), the 'Book Citation Index-Science' (BCI-S), the 'Conference Proceedings Citation Index-Science' (CPCI-S), the 'Emerging Sources Citation Index' (ESCI), the 'Book Citation Index-Social Sciences and Humanities' (BCI-SSH), the 'Conference Proceedings Citation Index-Social Sciences and Humanities' (CPCI-SSH), and the 'Arts and Humanities Citation Index' (A&HCI).

The keywords for the search of the literature were collated from the screening of the abstract pages for the first 1,000 highly cited papers in the related 11 research fields. These keyword sets are provided in the appendices of the related chapters (Konur, 2021e–ag).

The 50-most-cited articles were selected for this review and the key findings are discussed briefly.

3.3 RESULTS

3.3.1 Biodiesel Fuels in General

3.3.1.1 Biooils

Van Zwieten et al. (2010) study the effects of biochar from the slow pyrolysis of paper-mill waste on agronomic performance and soil fertility in a paper with 722 citations. They modify two agricultural soils with two biochars and assess them in a glasshouse study. Both biochars had a high surface area and zones of a calcium mineral agglomeration, although they differed slightly in their liming values and carbon content. They find that both biochars significantly increased N uptake in wheat grown in a fertilizer-modified ferrosol. The concomitant increase in biomass production therefore suggested improved fertilizer-use efficiency. Likewise, biochar modification significantly increased biomass in soybean and radish in the ferrosol with fertilizer. There were no significant effects of biochar in the absence of fertilizer for wheat and soybean, while radish biomass increased significantly.

Vispute et al. (2010) study the production of chemicals from the integrated catalytic processing of pyrolysis oils in a paper with 672 citations. They combine hydroprocessing with zeolite catalysis. The hydroprocessing increased the intrinsic hydrogen content of the pyrolysis oil, producing polyols and alcohols. They find that the zeolite catalyst then converted these hydrogenated products into light olefins and aromatic hydrocarbons in a yield as much as three times higher than that produced with the pure pyrolysis oil. The yield of aromatic hydrocarbons and light olefins from the biomass conversion over zeolite was proportional to the intrinsic amount of hydrogen added to the biomass feedstock during hydroprocessing.

Evans and Milne (1987) carry out the molecular characterization of the pyrolysis of biomass in a paper with 633 citations. They apply the technique of 'molecular-beam, mass spectrometric' (MBMS) sampling to the elucidation of the molecular pathways in the fast pyrolysis of wood and its principal isolated constituents for the optimization of high-value fuel products by thermal and catalytic means. They find that the cellulose, lignin, and hemicellulose components of wood pyrolyze largely to monomer and monomer-related fragments and given characteristic mass spectral signatures. Whole wood behaves as the sum of its constituents, with few if any vapor species derived from interaction of the main polymer constituents. An important interaction, however, is the influence of mineral matter in the wood on the carbohydrate pyrolysis pathways.

Demirbas (2000) studies the mechanisms of liquefaction and the pyrolysis reactions of biomass in a paper with 617 citations. In the liquefaction process, the micellar-like broken-down fragments produced by hydrolysis are degraded into smaller compounds by dehydration, dehydrogenation, deoxygenation, and decarboxylation. These compounds once produced, rearrange through condensation, cyclization, and polymerization, leading to new compounds. He finds that thermal depolymerization and decomposition of biomass, cellulose, hemicelluloses, and products were formed, as well as a solid residue of charcoal. Cleavage of the aromatic C–O bond in lignin led to the formation of one oxygen atom product, and cleavage of the methyl C–O bond, to form two oxygen atom products, is the first reaction to occur in the thermolysis of 4-alkyl-guaiacol at 600–650 K.

Sharma et al. (2004) characterize biochars from the pyrolysis of lignin and its reactivity towards the formation of polycyclic aromatic hydrocarbons (PAHs) in a paper with 551 citations. They find that the biochar yield in pyrolysis decreased rapidly with an increase in temperature up to 400°C, after which there was a gradual decrease in the yield to ca. 40% at 750°C. In oxidative atmosphere, the char yield decreased to ca. 15% at 550°C. The pyrolysis led to the formation of melt, liquid phase, vesicles, precipitates of inorganic salts, and surface etching when these structures decomposed rapidly at high temperatures. There was a gradual decrease in the amounts of OH and CH_3 with increasing temperature. Both the H: C and O:C ratios of the biochar decreased with increase in temperature. The surface area, presence of inorganics, and aromaticity of char are important factors in PAH formation. These biochars have low reactivity, compared to chars from other biomass constituents probably due to the highly cross-linked and refractory nature of the lignin char.

Orfao et al. (1999) study the behavior of biomass components, cellulose, the xylan-representative of hemicelluloses, and lignin – thermogravimetrically with linear temperature programming, in N and air – in a paper with 546 citations. They find that the thermal decomposition of xylan and lignin could not be modeled with acceptable errors by means of simple reactions. They determine thermograms for pine and eucalyptus woods and pine bark, in an inert (N) or oxidizing (air) atmosphere. They model the pyrolysis of these lignocellulosic materials with good approximation by three first-order independent reactions. One of these reactions is associated with the primary pyrolysis of cellulose, its parameters being previously determined and fixed in the model. The model parameters are the activation energies and preexponential factors for the pyrolysis of the remaining two pseudo-components and two additional parameters related to the biomass composition.

Mohan et al. (2007) study biochar by-products from fast wood/bark pyrolysis as adsorbents for the removal of the toxic metals from water during biooil production in a paper with 500 citations. They obtain the biochars for oak bark, pine bark, oak wood, and pine wood at fast pyrolysis of 400 and 450°C in an auger-fed reactor. They find that maximum adsorption occurred over a pH range of 3–4 for arsenic and 4–5 for lead and cadmium. The optimum equilibrium time was 24 h with an adsorbent dose of 10 g/L and a concentration similar to 100 mg/L for lead and cadmium. Oak bark outperformed the other biochars and nearly mimicked Calgon F-400 adsorption for lead and cadmium. In an aqueous lead solution with initial concentration of 4.8×10^{-4} M, both oak bark and Calgon F-400 (10 g/L) removed nearly 100% of the heavy metal. Oak bark (10 g/L) also removed about 70% of arsenic and 50% of cadmium from aqueous solutions. Overall, the data are well fitted with both the models, with a slight advantage for the Langmuir model.

3.3.1.2 Biodiesel Fuels in General

Hill et al. (2006) compare environmental, economic, and energetic costs and benefits of biodiesel and bioethanol biofuels in a paper with 1,597 citations. They find that corn-based bioethanol yields 25% more energy than that invested in its production, whereas soybean-based biodiesel yields 93% more. Compared with bioethanol, biodiesel releases just 1.0, 8.3, and 13% of the agricultural N, P, and pesticide pollutants, respectively, per net energy gain. Relative to the fossil fuels they displace, greenhouse gas emissions are reduced by 12% by the production and combustion of bioethanol and by 41% by biodiesel. Biodiesel also releases less air pollutants per net energy gain than bioethanol. These advantages of biodiesel over bioethanol come from lower agricultural inputs and more efficient conversion of feedstocks to fuel. However, it is clear that neither biofuel can replace much petroleum without impacting food supplies. Even dedicating all US corn and soybean production to biofuels would meet only 12% of gasoline demand and 6% of diesel demand. Until recent increases in petroleum prices, high production costs made biofuels unprofitable without subsidies. Biodiesel provides sufficient environmental advantages to merit subsidy. Transportation biofuels such as synfuel hydrocarbons or cellulosic bioethanol, if produced from low-input biomass grown on agriculturally marginal land or from waste biomass, could provide much greater supplies and environmental benefits than food-based biofuels.

3.3.1.3 Glycerol

Dasari et al. (2005) study the low-pressure hydrogenolysis of glycerol to propylene glycol in a paper with 642 citations. They use nickel, palladium, platinum, copper, and copper-chromite catalysts. They find that at temperatures above 200°C and a hydrogen pressure of 200 psi, the selectivity to propylene glycol decreased due to excessive hydrogenolysis of the propylene glycol. At 200 psi and 200°C the pressures and temperatures were significantly lower than those reported in the literature while maintaining high selectivities and good conversions. The yield of propylene glycol increased with decreasing water content. They validate a new reaction pathway for converting glycerol to propylene glycol via an intermediate by isolating the acetol intermediate.

3.3.2 Feedstock-based Biodiesel Fuels

3.3.2.1 Edible Oil-based Biodiesel Fuels

Ramos et al. (2009) study the properties of vegetable-oil-based biodiesel fuels in a paper with 981 citations. They transesterify ten refined vegetable oils using potassium methoxide as a catalyst and standard reaction conditions (reaction time, 1 h; weight of catalyst, 1 wt.% of initial oil weight; molar ratio methanol:oil, 6:1; reaction temperature, 60°C). They find that some critical parameters, such as oxidation stability, cetane number, iodine value, and the cold filter plugging point, were correlated with the methyl ester composition of each biodiesel, according to two parameters: the degree of unsaturation and the long chain saturated factor.

Freedman et al. (1986) study the transesterification kinetics of soybean oils in an early paper with 813 citations. They find that transesterification of soybean oil and other triglycerides with alcohols, in the presence of a catalyst, yields fatty esters and glycerol, whereas di- and monoglycerides are intermediates and reactions are consecutive and reversible. They examine the effects of: the type of alcohol, 1-butanol or methanol (MeOH); the molar ratio of alcohol to soybean oil; the type and amount of catalyst; and the reaction temperature on rate constants and kinetic order. They find that forward reactions are pseudo-first order or second order, depending upon conditions used, whereas reverse reactions are second order. At a molar ratio of MeOH/soybean oil of 6:1, they observe a shunt reaction.

Vicente et al. (2004) study biodiesel production from sunflower oils in a paper with 789 citations. They compare different basic catalysts, sodium methoxide, potassium methoxide, sodium hydroxide, and potassium hydroxide for the methanolysis of sunflower oil. They carry out all the reactions under the same experimental conditions in a batch-stirred reactor and with the subsequent separation and purification stages in a decanter. They find that the biodiesel purity was near 100 wt% for all catalysts. However, near 100 wt% biodiesel yields were only obtained with the methoxide catalysts. Yield losses were due to triglyceride saponification and methyl ester dissolution in glycerol. The obtained biodiesel met the measured specifications, except for the iodine value, according to the German and EU draft standards. Although all the transesterification reactions were quite rapid and the biodiesel layers achieved nearly 100% methyl ester concentrations, the reactions using sodium hydroxide turned out to be the fastest.

Saka and Kusdiana (2001) study biodiesel production from rapeseed oil in a paper with 687 citations. They use supercritical methanol without using any catalyst. They carry out the experiment in a batch-type reaction vessel preheated to 350 and 400°C and at a pressure of 35–65 MPa, and with a molar ratio of 1:42 of the rapeseed oil to methanol. They find that, in a preheating temperature of 350°C, 240 s of supercritical treatment of methanol was sufficient to convert the rapeseed oil to methyl esters. Although the prepared methyl esters were basically the same as those of the common method with a basic catalyst, the yield of methyl esters by the former was higher than that by the latter. This new supercritical methanol process required the shorter reaction time and simpler purification procedure because of the unused catalyst.

Haas et al. (2006) estimate costs of biodiesel production from soybean oils in a paper with 604 citations. They develop a computer model to estimate the capital and operating costs of a moderately sized industrial biodiesel production facility. The major process operations in the plant were continuous-process vegetable oil transesterification, and ester and glycerol recovery. Crude, degummed soybean oil was specified as the feedstock. Annual production capacity of the plant was set around 37.8×10^6 1 (10×10^6 gal). Facility construction costs were US$11.3 million. The largest contributors to the equipment cost, accounting for nearly one-third of expenditures, were storage tanks to contain a 25-day capacity of feedstock and product. At a value of US$0.52/kg ($0.236/lb) for feedstock soybean oil, they predict a biodiesel production cost of US$0.53/l ($2.00/gal). The single greatest contributor to this value was the cost of the oil feedstock, which accounted for 88% of total estimated production costs. There was a direct linear relationship between the production costs and the cost of the feedstock, with a change of US$0.020/1 ($0.075/gal) in product cost per US$0.022/kg ($0.01/lb) change in biooil cost. Process economics included the recovery of glycerol, and its sale into the commercial glycerol market as an 80% w/w aqueous solution, which reduced production costs by approximately 6%. The production cost of biodiesel varied inversely and linearly with variations in the market value of glycerol, increasing by US$0.0022/1 ($0.0085/gal) for every US$0.022/kg ($0.01/lb) reduction in glycerol value.

Noureddini and Zhu (1997) study the kinetics of transesterification of soybean oil with methanol in a paper with 661 citations. They examine the effect of variations in mixing intensity (Reynolds number = 3,100–12,400) and temperature (30–70°C) on the rate of reaction, while the molar ratio of alcohol to triglycerol (6:1) and the concentration of catalyst (0.20 wt% based on soybean oil) were held constant. They find that the variations in mixing intensity impact the reaction, parallel to the variations in temperature. They propose a reaction mechanism consisting of an initial mass transfer-controlled region followed by a kinetically controlled region. The experimental data for the latter region are a good fit into a second-order kinetic mechanism. They finally determine the reaction rate constants and the activation energies for all the forward and reverse reactions.

Kusdiana and Saka (2001) study the kinetics of transesterification in rapeseed oil to biodiesel fuel as treated in supercritical methanol in a paper with 540 citations. They make runs in a bath-type reaction vessel ranging from 200°C at a subcritical temperature to 500°C at a supercritical state with different molar ratios of methanol to rapeseed oil. They find that the conversion rate of rapeseed oil to its methyl esters increased dramatically in the supercritical state, and that a reaction temperature of 350°C was the best condition, with the molar ratio of methanol in rapeseed oil being 42.

Granados et al. (2007) study biodiesel production from sunflower oil by using activated calcium oxide (CaO) in a paper with 507 citations. They use activated CaO as a catalyst in the production of biodiesel by transesterification of triglycerides with methanol. They find that CaO was rapidly hydrated and carbonated by contact with room air. A few minutes were enough to chemisorb a significant amount of H_2O and CO_2. The CO_2 was the main deactivating agent, whereas the negative effect of water was less important. The surface of the activated catalyst was an inner core of CaO particles covered by very few layers of $Ca(OH)_2$. The activation by outgassing at temperatures of at least 973 K were required to revert the CO_2 poisoning. The catalyst could be reused for several runs without significant deactivation. The catalytic reaction was the result of the heterogeneous and homogeneous contributions. Part of the reaction took place on basic sites at the surface of the catalyst, the rest was due to the dissolution of the activated CaO in methanol that created homogeneous leached active species.

Kim et al. (2004) study transesterification of vegetable oil to biodiesel using heterogeneous base catalysts in a paper with 507 citations. They develop an environmentally benign process for the production of biodiesel from vegetable oils using this catalyst. They adopt first an $Na/NaOH/\gamma{-}Al_2O_3$ heterogeneous base catalyst for the production of biodiesel. They optimize the reaction conditions, such as the reaction time, the stirring speed, the use of a co-solvent, the oil to methanol ratio, and the amount of catalyst. This heterogeneous base catalyst showed almost the same activity under the optimized reaction conditions compared to a conventional homogeneous NaOH catalyst. They estimate the basic strength of this catalyst and propose a correlation with the activity towards transesterification.

3.3.2.2 Nonedible Oil-based Biodiesel Fuels

Ramadhas et al. (2005) study biodiesel production from rubber seed oil with high 'free fatty acids' (FFAs) in a paper with 666 citations. They develop a two-step transesterification process to convert the high FFA oils to its monoesters. They find that the first step, acid catalyzed esterification, reduced the FFA content of the oil to less than 2%. The second step, an alkaline catalyzed transesterification process, converted the products of the first step to its monoesters and glycerol. The two-step esterification procedure converted rubber seed oil to its methyl esters. They find that the viscosity of biodiesel oil was nearer to that of diesel and that the calorific value was about 14% less than that of diesel.

Berchmans and Hirata (2008) develop a method to produce biodiesel from crude *Jatropha curcas* seed oil (JCO) having high FFAs (15%) in a paper with 564 citations.

They reduce the high FFA level of CJO to less than 1% by a two-step pretreatment process. The first step was carried out with a 0.60 w/w methanol-to-oil ratio in the presence of 1% w/w H_2SO_4 as an acid catalyst in a 1-h reaction at 50°C. After the reaction, the mixture was allowed to settle for 2 h and the methanol–water mixture separated at the top layer was removed. The second step was transesterified using 0.24 w/w methanol to oil and 1.4% w/w NaOH to oil as alkaline catalyst to produce biodiesel at 65°C. The final yield for methyl esters of fatty acids was ca. 90% in 2 h.

Azam et al. (2005) examine 'fatty acid methyl ester' (FAME) profiles of seed oils of 75 plant species having 30% or more fixed oil in their seed/kernel in a paper with 500 citations. They find that the 'saponification number' (SN), iodine value (IV), and 'cetane number' (CN) of FAMEs of oils varied from 169.2 to 312.5, 4.8 to 212, and 20.56 to 67.47, respectively. They use fatty acid compositions, the IV, and the CN to predict the quality of the FAMEs of oil for use as biodiesel. The FAME of oils of 26 species, including *Azadirachta indica*, *Calophyllum inophyllum*, *Jatropha curcas*, and *Pongamia pinnata*, were most suitable for use as biodiesel, and they met the major specification of biodiesel standards of the USA, Germany, and the European Standard Organization. The FAMEs of another 11 species met the specification of the biodiesel standard of the USA only.

3.3.2.3 Waste-Oil-based Biodiesel Fuels

Zhang et al. (2003a) study the process design and technological assessment of biodiesel production from waste cooking oil (WCO) and virgin vegetable oil in a paper with 997 citations. They consider four different continuous process flowsheets under alkaline or acidic conditions on a commercial scale. They find that the alkali-catalyzed process, using virgin vegetable oil as the raw material, required the fewest and smallest process equipment units but at a higher raw material cost than the other processes. The use of waste cooking oil to produce biodiesel reduced the raw material cost. The acid-catalyzed process using waste cooking oil was technically feasible with less complexity than the alkali-catalyzed process using waste cooking oil, thereby making it a competitive alternative to commercial biodiesel production by the alkali-catalyzed process.

Canakci and van Gerpen (2001) study biodiesel production from fats with high FFAs in a paper with 755 citations. They develop a technique to reduce the FFA content of these feedstocks using an acid-catalyzed pretreatment to esterify the FFAs before transesterifying the triglycerides with an alkaline catalyst to complete the reaction. They perform the initial process development with synthetic mixtures containing 20 and 40% FFAs, prepared using palmitic acid. They find that the acid level of the high FFA feedstocks could be reduced to less than 1% with a two-step pretreatment reaction. They allow the reaction mixture to settle between steps so that the water-containing alcohol phase can be removed. They show this two-step pretreatment reaction with actual feedstocks, including yellow grease with 12% FFAs and brown grease with 33% FFAs. After reducing the acid levels of these feedstocks to less than 1%, they complete the transesterification reaction with an alkaline catalyst to produce fuel-grade biodiesel.

Zhang et al. (2003b) carry out an economic assessment and sensitivity analysis of biodiesel production from waste cooking oil and virgin vegetable oils in a paper with

739 citations. They consider four continuous processes to produce biodiesel, including both alkali- and acid-catalyzed processes, using waste cooking oil and the 'standard' process using virgin vegetable oil as the raw material. They find that although the alkali-catalyzed process using virgin vegetable oil had the lowest fixed capital cost, the acid-catalyzed process using waste cooking oil was more economically feasible overall, providing a lower total manufacturing cost, a more attractive after-tax rate of return, and a lower biodiesel break-even price. They further find that plant capacity and prices of feedstock oils and biodiesel were the most significant factors affecting the economic viability of biodiesel production.

Leung and Guo (2006) study the transesterification of neat and used canola frying oil in a paper with 540 citations. They evaluate the characteristics and performance of three commonly used catalysts used for alkaline-catalyzed transesterification, i.e. sodium hydroxide, potassium hydroxide, and sodium methoxide. They find that with intermediate catalytic activity and a much lower cost, sodium hydroxide was more superior to the other two catalysts. They optimize the process variables that influenced the transesterification of triglycerides, such as catalyst concentration, the molar ratio of methanol to raw oil, the reaction time, the reaction temperature, and the FFA content of raw oil in the reaction system.

3.3.2.4 Algal-Oil-based Biodiesel Fuels

Rodolfi et al. (2009) study algal strain selection, the induction of algal lipid synthesis, and outdoor algal mass cultivation in a low-cost photobioreactor in a paper with 1,581 citations. They screen 30 microalgal strains for their biomass productivity and lipid content and select four strains (two marine and two freshwater). They find that only the two marine microalgae accumulated lipids under such conditions. One of them, *Nannochloropsis* sp. F&M-M24, which attained 60% lipid content after N starvation, was grown in a 20 L 'Flat Alveolar Panel photobioreactor'. They find that fatty acid content increased with high irradiances and following both N and P deprivation. This strain was grown outdoors in 110 L 'Green Wall Panel photobioreactors' under nutrient sufficient and deficient conditions. They then find that lipid productivity increased from nutrient sufficient media in N deprived media (117 v. 204 mg/L/day). In a two-phase cultivation process (a nutrient sufficient phase to produce the inoculum followed by an N deprived phase to boost lipid synthesis), they project the oil production potential as more than 90 kg per hectare per day. This *Nannochloropsis* sp. had the potential for an annual production of 20 tons of lipid per hectare in the Mediterranean climate and of more than 30 tons of lipid per hectare in sunny tropical areas.

Lardon et al. (2009) carry out a life-cycle assessment of biodiesel production from microalgae in a paper with 832 citations. They test two different culture conditions, nominal fertilizing or N starvation, as well as two different extraction options, dry or wet extraction. They then compare the best scenario to food-crop-based biodiesel and petrodiesel. They confirm the potential of microalgae as an energy source but highlight the imperative necessity of decreasing the energy and fertilizer consumption. Therefore, control of N stress during the culture and optimization of wet extraction were valuable options. They also emphasize the potential of the anaerobic digestion of oilcakes as a way to reduce external energy demand and to recycle a part of the mineral fertilizers.

Miao and Wu (2006) study biodiesel production from *Chlorella protothecoides* in a paper with 804 citations. They find that heterotrophic growth of *C. protothecoides* resulted in the accumulation of high lipid content in cells. They extract a large amount of microalgal oil efficiently from these heterotrophic cells by using n-hexane. They obtain biodiesel comparable to conventional diesel from heterotrophic microalgal oil by acidic transesterification. The best process combination was a 100% catalyst quantity (based on oil weight) with a 56:1 molar ratio of methanol to microalgal oil at a temperature of 30°C, which reduced the product's specific gravity in about 4 h of reaction time.

Li et al. (2008) study the effects of N sources on cell growth and lipid accumulation of *Neochloris oleoabundans* in a paper with 640 citations. They find that while the highest lipid cell content was obtained at the lowest sodium nitrate concentration, a remarkable lipid productivity was achieved at 5 mM with a lipid cell content of 0.34 g/g and a biomass productivity of 0.40 g/L/day. The highest biomass productivity was obtained at 10 mM of sodium nitrate, with a biomass concentration of 3.2 g/L and a biomass productivity of 0.63 g/L /day. They observe that cell growth continued after the exhaustion of the external N pool, hypothetically supported by the consumption of intracellular N pools such as chlorophyll molecules.

Clarens et al. (2010) compare the environmental life cycle of algae to other bioenergy feedstocks in a paper with 626 citations. They show that switchgrass, canola, and corn have lower environmental impacts than algae in energy use, greenhouse gas emissions, and water, regardless of cultivation location. Only in total land use and eutrophication potential, do algae perform favorably. The large environmental footprint of algae cultivation is driven predominantly by upstream impacts, such as the demand for CO_2 and fertilizer. To reduce these impacts, flue gas and, to a greater extent, wastewater could be used to offset most of the environmental burdens associated with algae. To demonstrate the benefits of algae production coupled with wastewater treatment, they expand the model to include three different municipal wastewater effluents as sources of nitrogen and phosphorus. Each provided a significant reduction in the burdens of algal cultivation, and the use of source-separated urine made algae more environmentally beneficial than the terrestrial crops.

Lee et al. (2010) compare several methods for effective lipid extraction from microalgae in a paper with 622 citations. They test various methods, including autoclaving, bead-beating, microwaves, sonication, and a 10% NaCl solution, to identify the most effective cell disruption method. They extract the total lipids from *Botryococcus* sp., *Chlorella vulgaris*, and *Scenedesmus* sp. using a mixture of chloroform and methanol (1:1). The lipid contents from the three species were 5.4–11.9, 7.9–8.1, 10.0–28.6, 6.1–8.8, and 6.8–10.9 g/L when using these methods, respectively. *Botryococcus* sp. showed the highest oleic acid productivity at 5.7 mg/L/d when the cells were disrupted using the microwave oven method. Thus, they conclude that among the tested methods, the microwave oven method was the most simple, easy, and effective for lipid extraction from microalgae.

Xu et al. (2006) study biodiesel production from *Chlorella protothecoides* in a paper with 610 citations. They apply the technique of metabolic controlling through the heterotrophic growth of *C. protothecoides* where the heterotrophic

C. protothecoides contained a crude lipid content of 55.2%. To increase the biomass and reduce the cost of alga, they use corn powder hydrolysate instead of glucose as the organic carbon source in a heterotrophic culture medium in fermenters. They find that cell density significantly increased under the heterotrophic condition, and the highest cell concentration reached 15.5 g/L. They extract a large amount of microalgal oil efficiently from the heterotrophic cells by using n-hexane, which is then transmuted into biodiesel by acidic transesterification. The biodiesel had a high heating value of 41 MJ/kg, a density of 0.864 kg/L, and a viscosity of 5.2×10^{-4} Pas (at 40°C).

Liang et al. (2009) study the biomass and lipid productivities of *Chlorella vulgaris* under autotrophic, heterotrophic, and mixotrophic growth conditions in a paper with 518 citations. They find that while autotrophic growth provided higher cellular lipid content (38%), lipid productivity was much lower compared with those from heterotrophic growth with acetate, glucose, or glycerol. They attain optimal cell growth (2 g/L) and lipid productivity (54 mg/L/day) using glucose at 1% (w/v), whereas higher concentrations were inhibitory. Growth of *C. vulgaris* on glycerol had similar dose effects as those from glucose. Overall, *C. vulgaris* was mixotrophic.

Biller and Ross (2011) study the hydrothermal liquefaction of microalgae with different biochemical contents in a paper with 451 citations. They liquefy them under hydrothermal conditions at 350°C, around 200 bar in water, 1 M Na_2CO_3, and 1 M formic acid. The model compounds included albumin and a soya protein, starch and glucose, the triglyceride from sunflower oil, and two amino acids. Microalgae included *Chlorella vulgaris*, *Nannochloropsis occulata*, *Porphyridium cruentum*, and *Spirulina*. They find that there was broad agreement between predictive yields and actual yields for the microalgae, based on their biochemical composition. The yields of biooil were 5–25 wt% higher than the lipid content of the algae, depending upon biochemical composition. The yields of biocrude followed the trend: lipids > proteins > carbohydrates.

3.3.3 Petrodiesel Fuels

3.3.3.1 Crude Oils

Budzinski et al. (1997) evaluate sediment contamination by 'polycyclic aromatic hydrocarbons' (PAHs) in the Gironde estuary in a paper with 785 citations. They analyze surface sediments for PAHs and find that total PAH concentrations ranged from 1000 to 2000 ng/g of dry sediment. They discuss the resulting distributions and molecular ratios of specific aromatic compounds in terms of sample location, origin of the organic matter, and seasonal variations. This estuary is moderately contaminated and the level of contamination and the relative distributions of PAHs are quite homogeneous throughout. They record a strong pyrolytic source fingerprint, but the detection of high amounts of perylene, which is always the most abundant PAH, indicates a dominant biogenic origin for the organic matter.

Moldowan et al. (1985) study the relationship between the petroleum composition and the depositional environment of petroleum source rocks in a paper with 698 citations. They distinguish crude oils of nonmarine source from those of marine shale source and from oils originating in marine carbonate sequences by using a battery of geochemical parameters, as demonstrated with a sample suite of nearly 40 oils. A novel parameter based on the presence of C_{30} steranes in the oil was a definitive indication of a contribution to the source from marine-derived organic matter. A second novel parameter based on monoaromatized steroid distributions was effective in helping to distinguish nonmarine from marine crudes and which can be used to gauge the relative amounts of a higher plant input to oils within a given basin. Sterane distributions were similarly useful for detecting a higher plant input but were less effective than monoaromatized steroid distributions for making marine versus nonmarine distinctions.

Scholle and Arthur (1980) study the carbon isotope fluctuations in Cretaceous pelagic limestones as a potential stratigraphic and petroleum-exploration tool in a paper with 662 citations. Significant short-term carbon isotope fluctuations are present in Cretaceous pelagic limestones from widely distributed onshore sections in the Circum-Atlantic-western Tethyan region. They analyze more than 1,000 closely spaced samples. At least seven major $\delta^{13}C$ excursions can be correlated from section to section. The most important heavy events occur near the Aptian–Albian and Cemonanian–Turonian boundaries, whereas light events are near the Jurassic–Cretaceous, Albian–Cemonanian, Turonian–Coniacian, and Cretaceous–Tertiary boundaries.

Hazen et al. (2010) study deep-sea oil plume enriching indigenous oil-degrading bacteria in a paper with 647 citations. They report that the dispersed hydrocarbon plume stimulated deep-sea indigenous γ-proteobacteria that are closely related to known petroleum degraders. Hydrocarbon-degrading genes coincided with the concentration of various oil contaminants. Changes in hydrocarbon composition with distance from the source and incubation experiments with environmental isolates demonstrated faster-than-expected hydrocarbon biodegradation rates at 5°C. Based on these results, the potential exists for intrinsic bioremediation of the oil plume in the deep-water column without substantial oxygen drawdown.

Zhang et al. (2013) study the superhydrophobic and superoleophilic 'poly(vinylidene fluoride)' (PVDF) membranes for the effective separation of water-in-oil emulsions in a paper with 588 citations. They fabricate this membrane via an inert solvent-induced phase inversion for the effective separation of both micrometer and nanometer-sized surfactant-free and surfactant-stabilized water-in-oil emulsions solely driven by gravity, with high separation efficiency and high flux, which is several times higher than those of commercial filtration membranes and reported materials with similar permeation properties.

Kubat et al. (1998) use machine learning for the detection of oil spills in satellite radar images in a paper with 577 citations. During a project examining the use of machine learning techniques for oil spill detection, they encounter several essential questions that they believe deserve the attention of the research community. They use their particular case study to illustrate such issues as problem formulation, selection of evaluation measures, and data preparation. They then relate these issues to properties of the oil spill application, such as its imbalanced class distribution, that are common to many applications. They note that their solutions to these issues are implemented in the Canadian Environmental Hazards Detection System, which underwent field testing.

Dojka et al. (1998) study the microbial diversity in a hydrocarbon- and chlorinated-solvent-contaminated aquifer undergoing intrinsic bioremediation in a paper with 555 citations. Of the 94 bacterial sequence types, they find that ten have no phylogenetic association with known taxonomic divisions and are phylogenetically grouped into six novel division levels; 21 belong to four recently described candidate divisions with no cultivated representatives; and 63 are phylogenetically associated with ten well-recognized divisions. One of these sequence types is associated with the genus *Syntrophus*; *Syntrophus* spp., produce energy from the anaerobic oxidation of organic acids, with the production of acetate and hydrogen. The organism represented by the other sequence type is closely related to *Methanosaeta* spp., which is capable of energy generation only through aceticlastic methanogenesis. They hypothesize, therefore, that the terminal step of hydrocarbon degradation in the methanogenic zone of the aquifer is aceticlastic methanogenesis and that the microorganisms represented by these two sequence types occur in syntrophic association.

He and Zhao (2005) study a new class of starch-stabilized bimetallic nanoparticles for the degradation of chlorinated hydrocarbons in water in a paper with 551 citations. The starched nanoparticles displayed much less agglomeration but greater dechlorination power than those prepared without a stabilizer. The starched nanoparticles were present as discrete particles, as opposed to dendritic flocs, for nonstarched particles. While starched nanoparticles remained suspended in water for days, nonstarched particles agglomerated and precipitated within minutes. The starched nanoparticles exhibited markedly greater reactivity when used for dechlorination of trichloroethylene (TCE) or polychlorinated biphenyls (PCBs) in water. The starched nanoparticles of around 1 g/L were able to transform over 80% of PCBs in less than 100 h, as compared to only 24% with nonstarched Fe-Pd nanoparticles.

3.3.3.2 Petrodiesel Fuels in General

There is no paper on petrodiesel fuels in general.

3.3.3.3 Petrodiesel Fuel Exhaust Emissions

Rogge et al. (1993) study particulate exhaust emissions in a paper with 1,125 citations. Internal combustion engines burning gasoline and diesel fuel contribute more than 21% of the primary fine particulate organic carbon emitted to the Los Angeles atmosphere. They examine particulate exhaust emissions from six noncatalyst automobiles, seven catalyst-equipped automobiles, and two heavy-duty diesel trucks. They quantify more than 100 organic compounds, including n-alkanes, n-alkanoic acids, benzoic acids, benzaldehydes, PAH, oxy-PAH, steranes, pentacyclic triterpanes, and azanaphthalenes.

Khalil et al. (1995) evaluate the chemical composition (source fingerprint) of the major sources of PAHs in the Chicago metropolitan area in a paper with 835 citations. The sources sampled were coke ovens, highway vehicles, heavy-duty diesel engines, gasoline engines, and wood combustion. They find that two and three ring PAHs were responsible for 98, 76, 92, 73, and 80% of the total concentration of measured 20 PAHs for coke ovens, diesel engines, highway tunnels, gasoline engines, and wood combustion samples, respectively. Six ring PAHs were mostly below the detection limit of this study and only found in the highway tunnel and diesel and gasoline engine samples. The source fingerprints were obtained by averaging the ratios of individual PAH concentrations to the total concentration of categorical pollutants.

Schauer et al. (1999) measure gaseous and particulate emissions from medium duty diesel trucks in a paper with 741 citations. They quantify emission rates of 52 gas-phase volatile hydrocarbons, 67 semivolatile and 28 particle-phase organic compounds, and 26 carbonyls, along with fine particle mass and chemical composition. When all carbonyls are combined, they accounted for 60% of the gas-phase organic compound mass emissions. They quantify fine particulate matter emission rates and chemical composition simultaneously, yielding an elemental carbon emission rate of 56 mg/km driven. They observe a significant enrichment in the ratio of unsubstituted PAHs to their methyl- and dimethyl-substituted homologues in the tailpipe emissions relative to the fuel. They finally quantify isoprenoids and tricyclic terpanes in the semivolatile organics emitted from diesel vehicles.

Qi et al. (2004) study the MnOx-CeO_2 mixed oxides prepared by co-precipitation for the 'selective catalytic reduction' (SCR) of nitrogen oxides (NO) with ammonia (NH_3) at low temperatures in a paper with 683 citations. They find that the best catalyst yielded 95% NO conversion at 150°C at a space velocity of 42,000/h. As the manganese content was increased from 0 to 40%, NO conversion increased significantly, but decreased at higher manganese contents. The most active catalyst was obtained with a molar Mn:(Mn + Ce) ratio of 0.4. Only N_2 rather than N_2O was present in the product when the temperature was below 150°C. The initial step was the adsorption of NH_3 on Lewis acid sites of the catalyst, followed by a reaction with nitrite species to produce N_2 and H_2O.

Zhu et al. (2002) study the ultrafine particles near a major highway with heavy-duty diesel traffic in a paper with 602 citations. They find that the range of average concentration of carbon monoxide (CO), black carbon (BC), and total particle number concentration at 17 m was 1.9–2.6 ppm, 20.3–24.8 μm/m^3, 1.8×10^5–3.5×10^5/cm^3, respectively. The relative concentration of CO, BC, and particle number decreased exponentially and tracked each other well as one moves away from the freeway. Both atmospheric dispersion and coagulation contribute to the rapid decrease in particle number concentration and change in particle size distribution with increasing distance from the freeway.

3.3.3.4 Health Impact of the Petrodiesel Exhaust Emissions

Birch and Cary (1996) present the results of an investigation into a thermal-optical technique for the analysis of the carbonaceous fraction of particulate diesel exhaust in a paper with 1,234 citations. Although they determine various carbon types, they find that elemental carbon is the superior marker of diesel particulate matter. This is because elemental carbon, which constitutes a large fraction of the particulate mass, can be quantified at low levels, and its only significant source in most workplaces is the diesel engine.

Salvi et al. (1999) study the acute inflammatory responses in the airways and peripheral blood after short-term exposure to diesel exhaust (DE) in healthy human volunteers in a paper with 608 citations. They expose 15 healthy human volunteers to air and diluted DE under controlled conditions for 1 h with intermittent exercise. They find that while standard lung function measures did not change following DE exposure, there was a significant increase in neutrophils and B lymphocytes in airway lavage, along with increases in histamine and fibronectin. There was a significant increase in neutrophils, mast cells, CD4+, and CD8+ T lymphocytes along with upregulation of the endothelial adhesion molecules ICAM−1 and VCAM−1, with increases in the numbers of LFA−1+ cells in the bronchial tissue. There were also significant increases in neutrophils and platelets in peripheral blood following DE exposure. They concluded that at high ambient concentrations, acute short-term DE exposure produced a well-defined and marked systemic and pulmonary inflammatory response in healthy human volunteers.

McCreanor et al. (2007) study respiratory effects of exposure to diesel traffic in persons with asthma in a paper with 531 citations. They find that participants had significantly higher exposures to fine particles, ultrafine particles, elemental carbon, and nitrogen dioxide (NO_2) on Oxford Street than in Hyde Park. Walking for two hours on Oxford Street induced asymptomatic but consistent reductions in the 'forced expiratory volume' in 1 second (FEV1) and the 'forced vital capacity' (FVC) that were significantly larger than the reductions in FEV1 and FVC after exposure in Hyde Park. The effects were greater in subjects with moderate asthma than in those with mild asthma. These changes were accompanied by increases in biomarkers of neutrophilic inflammation and airway acidification. The changes were associated most consistently with exposures to ultrafine particles and elemental carbon.

3.4 DISCUSSION

Table 3.1 provides information on the primary and secondary research fronts for both biodiesel fuels and petrodiesel fuels. It also provides information on the distribution of these research fronts in the article papers (sample size = 50), the 100-most-cited papers, and over 121,000 population papers.

As this table shows, the research fronts of 'biodiesel fuels in general', 'feedstock-based biodiesel fuels', and 'petrodiesel fuels' comprise 18, 50, and 32% of these papers, respectively.

In the first group of article papers, papers on 'biooils', 'biodiesel fuels in general', and 'glycerol' comprise 14, 2, and 2% of these papers, respectively. In the second group of article papers, 'edible oil-based biodiesel fuels', 'nonedible oil-based biodiesel fuels', 'waste oil-based biodiesel fuels', and 'algal oil-based biodiesel fuels' comprise 18, 6, 8, and 18% of these papers, respectively. In the final group of article papers, 'crude oils', 'petrodiesel fuels in general', 'emissions from petrodiesel fuels', and 'health impact of the emissions from petrodiesel fuels' comprise 16, 0, 10, and 6% of these papers, respectively.

All of the biooil papers are related to biomass pyrolysis (Demirbas, 2000; Evans and Milne, 1987, Mohan et al., 2007; Orfao et al., 1999; Sharma et al., 2004; Van Zwieten et al. (2010); Vispute et al., 2010) and one of the papers is related to biomass liquefaction (Demirbas, 2000). Evans and Milne (1987), Demirbas (2000), and Orfao et al. (1999) characterize the pyrolysis behavior of the biomass. Mohan et al. (2007)

TABLE 3.1
Research Fronts

	Research Front	No. Reviewed Papers (%)	No. of Sample Papers (%)	No. of Population Papers (%)
1	Biodiesel fuels in general	18	41	27.7
1.1	Biooils	14	14	16.6
1.2	Biodiesel fuels in general	2	21	5.4
1.3	Glycerol	2	6	5.7
2	Feedstock-based biodiesel fuels	50	30	18.3
2.1	Edible-oil-based biodiesel fuels	18	5	2.2
2.2	Nonedible-oil-based biodiesel fuels	6	3	1.8
2.3	Waste oil-based biodiesel fuels	8	2	1.8
2.4	Algal oil-based biodiesel fuels	18	20	12.5
3.	Petrodiesel fuels	32	34	67.4
3.1	Crude oils	16	14	43.8
3.2	Petrodiesel fuels in general	0	8	13.2
3.3	Emissions from petrodiesel fuels	10	11	8.6
3.4	Health impact of the emissions from petrodiesel fuels	6	1	1.8

Notes: No. of reviewed papers: sample size is 50 articles; no. of sample papers: sample size is 100 articles; no. of population papers: sample size is over 121,000 papers.

and Van Zwieten et al. (2010) study the applications of the biochars obtained through pyrolysis of the biomass in agronomics and heavy-metal bioremediation, respectively. Finally, Vispute et al. (2010) study biooil upgrading for the production of biochemicals, including biofuels.

The only paper in the group of the 'biodiesel fuels in general' compares environmental, economic, and energetic costs and benefits of soybean-based biodiesel and corn-based bioethanol biofuels (Hill et al., 2006). It is notable that there are no papers on the production, diesel engine performance, fuel properties and characterization, and emissions of biodiesel fuels.

The only paper in the group of 'glycerol', Dasari et al. (2005), studies the hydrogenolysis of glycerol to propylene glycol. It is notable that there are no papers on biohydrogen production or propanediol production from glycerol. As the glycerol is a by-product of biodiesel fuel production, it reduces the production cost of biodiesel fuels (Haas et al., 2006).

In the second group of papers, there are nine on 'edible oil-based biodiesel fuels'. Ramos et al. (2009) and Kim et al. (2004) study the production of biodiesel from vegetable oils in general. Freedman et al. (1986), Haas et al. (2006), and Noureddini and Zhu (1997) focus on the production and production costs of soybean-based biodiesel fuels. Vicente et al. (2004) and Granados et al. (2007) focus on biodiesel production from sunflower oils, whilst Saka and Kusdiana (2001) focus on the biodiesel production of biodiesel fuels from rapeseed.

It is significant that the single greatest contributor to biodiesel production cost is that of the oil feedstock, which accounts for 88% of total costs (Haas et al., 2006). It is also notable that there are no papers on palm oil-based biodiesel fuels. The adverse effect of edible feedstock production on biodiversity (Fitzherbert et al., 2008; Koh and Wilcove, 2007) and food security (Brown and Funk, 2008; Rosegrant and Cline, 2003) should be further noted.

In the second group of papers, there are three on 'nonedible oil-based biodiesel fuels'. Ramadhas et al. (2005), Berchmans et al. (2008), and Azam et al. (2005) focus on biodiesel production from rubber seed oil, jatropha, and nonedible seed oils in general, respectively. It is significant to note that a two-step production system is used for these nonedible oils with high FFA content: esterification and transesterification.

In the second group of papers, there are four on 'waste oil-based biodiesel fuels'. Zhang et al. (2003a–b) and Leung and Guo (2006) focus on the production of biodiesel from waste cooking oils in comparison with virgin cooking oils, whilst Canakci and van Gerpen (2001) study biodiesel production from waste animal fats. It is significant that Zhang et al. (2003b) find that biodiesel production from waste cooking oils is more economic in comparison with virgin cooking oils. This finding is not surprising as 88% of the total estimated production costs for biodiesel fuel production comes from the feedstock cost, in the light of the fact that waste cooking oil is chapter than virgin cooking oil (Haas et al., 2006; Hill et al., 2006).

In the second group of papers, there are nine on 'algal oil-based biodiesel fuels'. Rodolfi et al. (2009), Li et al. (2008), and Liang et al. (2009) study algal biomass and algal lipid production. Miao and Wu (2006) and Xu et al. (2006) focus on biodiesel production from microalgae. Biller and Ross (2011) study biooil production through the hydrothermal liquefaction of microalgal biomass, whilst Lee et al. (2010) focus on lipid extraction from microalgae. Finally, Clarens et al. (2010) and Lardon et al. (2009) carry out a life-cycle analysis of biodiesel fuel production from microalgae. Lardon et al. (2009) confirm the potential of microalgae as an energy source but highlight the imperative necessity of decreasing energy and fertilizer consumption. Similarly, Clarens et al. (2010) confirm the findings of Lardon et al. (2009) and assert that the large environmental footprint of algae cultivation is driven predominantly by upstream impacts, such as the demand for CO_2 and fertilizer. They note that the use of flue gas and wastewater would be an excellent choice to reduce this footprint.

In the final group of papers, there are eight on 'crude oils'. Budzinski et al. (1997) evaluate sediment contamination by polycyclic aromatic hydrocarbons (PAHs) whilst Hazen et al. (2010), Dojka et al. (1998), Kubat et al. (1998), and He and Zhao (2005) focus on the bioremediation of oil spills at sea. Moldowan et al. (1985) and Scholle and Arthur (1980) focus on crude oil exploration, whilst Zhang et al. (2013) study the separation of water-in-oil emulsions. However, there are no papers on the wastewater produced during crude oil production and oil refining, the bioremediation of crude oil-contaminated soils, the impact of crude oils on fish and other animals, and crude oil transportation through the pipelines.

In the final group of papers, there are none on 'petrodiesel fuels in general'. For example there are no papers on the production, desulfurization, characterization, and properties of petrodiesel fuels, power production by petrodiesel fuels, and the bioremediation of petrodiesel-contaminated soils.

In the final group of papers, there are five on 'petrodiesel fuel emissions in general'. Rogge et al. (1993), Khalil et al. (1995), Schauer et al. (1999), and Zhu et al.

(2002) focus on particulate diesel fuel exhaust emissions, whilst Qi et al. (2004) study nitrogen oxide (NOx) emissions.

In the final group of papers, there are three on the 'adverse health impact of petrodiesel fuel emissions'. Birch and Cary (1996) focus on the determination of elemental carbon in the workplaces whilst Salvi et al. (1999) and McCreanor et al. (2007) study the respiratory illnesses caused by diesel fuel exhaust emissions. It is notable that there are no papers on cancer and other illnesses such as cardiovascular illnesses caused by diesel fuel exhaust emissions.

3.5 CONCLUSION

This chapter has presented a review of the 50-most-cited article papers on both biodiesel and petrodiesel fuels.

Table 3.1 provides information on the primary and secondary research fronts for both biodiesel fuels and petrodiesel fuels. It also provides information on the distribution of these research fronts in the article papers (sample size = 50), the 100-most-cited papers, and over 121,000 population papers.

As this table shows the research fronts of 'biodiesel fuels in general', 'feedstock-based biodiesel fuels', and 'petrodiesel fuels' comprise 18, 50, and 32% of these papers, respectively.

In the first group of article papers, papers on 'biooils' comprise 14% of the papers, in the second group, 'edible oil-based biodiesel fuels' and 'algae-based biodiesel fuels' comprise 18 and 18% of these papers, and in the final group, 'crude oils' and 'emissions from petrodiesel fuels' comprise 16 and 10% of these papers that are emerging as the prolific research fronts.

It is recommended that similar studies are carried out for each research front as well.

ACKNOWLEDGMENTS

The contribution of the highly cited researchers in the fields of biodiesel and petrodiesel fuels has been greatly acknowledged.

REFERENCES

Agarwal, A. K. 2007. Biofuels (alcohols and biodiesel) applications as fuels for internal combustion engines. *Progress in Energy and Combustion Science* 33:233–271.

Ahmadun, F. R., A. Pendashteh, L. C. Abdullah, and et al. 2009. Review of technologies for oil and gas produced water treatment. *Journal of Hazardous Materials* 170: 530–551.

Atlas, R. M. 1981. Microbial degradation of petroleum hydrocarbons: An environmental perspective. *Microbiological Reviews* 45: 180–209.

Azam, M. M., A. Waris, and N. M. Nahar. 2005. Prospects and potential of fatty acid methyl esters of some non-traditional seed oils for use as biodiesel in India. *Biomass & Bioenergy* 29: 293–302.

Babich, I. V. and J. A. Moulijn. 2003. Science and technology of novel processes for deep desulfurization of oil refinery streams: A review. *Fuel* 82: 607–631.

Berchmans, H. J. and S. Hirata. 2008. Biodiesel production from crude *Jatropha curcas* L. seed oil with a high content of free fatty acids. *Bioresource Technology* 99: 1716–1721.

Biller, P. and A. B. Ross. 2011. Potential yields and properties of oil from the hydrothermal liquefaction of microalgae with different biochemical content. *Bioresource Technology* 102: 215–225.

Birch, M. E. and R. A. Cary. 1996. Elemental carbon-based method for monitoring occupational exposures to particulate diesel exhaust. *Aerosol Science and Technology* 25: 221–241.

Bridgwater, A. V. and G. V. C. Peacocke. 2000. Fast pyrolysis processes for biomass. *Renewable & Sustainable Energy Reviews* 4: 1–73.

Brown, M. E. and C. C. Funk. 2008. Food security under climate change. *Science* 319: 580–581.

Budzinski, H., I. Jones, J. Bellocq, C. Pierard, and P. Garrigues. 1997. Evaluation of sediment contamination by polycyclic aromatic hydrocarbons in the Gironde estuary. *Marine Chemistry* 58: 85–97.

Canakci, M. and J. van Gerpen. 2001. Biodiesel production from oils and fats with high free fatty acids. *Transactions of the ASAE* 44: 1429–1436.

Chisti, Y. 2007. Biodiesel from microalgae. *Biotechnology Advances* 25: 294–306.

Clarens, A. F., E. P. Resurreccion, M. A. White, and L. M. Colosi. 2010. Environmental life cycle comparison of algae to other bioenergy feedstocks. *Environmental Science & Technology* 44: 1813–1819.

Czernik, S. and A. V. Bridgwater. 2004. Overview of applications of biomass fast pyrolysis oil. *Energy & Fuels* 18: 590–598.

Dasari, M. A., P. P. Kiatsimkul, W. R. Sutterlin, and G. J. Suppes. 2005. Low-pressure hydrogenolysis of glycerol to propylene glycol. *Applied Catalysis A-General* 281: 225–231.

Demirbas, A. 2000. Mechanisms of liquefaction and pyrolysis reactions of biomass. *Energy Conversion and Management* 41: 633–646.

Di Blasi, C. 2008. Modeling chemical and physical processes of wood and biomass pyrolysis. *Progress in Energy and Combustion Science* 34: 47–90.

Dojka, M. A., P. Hugenholtz, S. K. Haack, and N. R. Pace. 1998. Microbial diversity in a hydrocarbon- and chlorinated-solvent-contaminated aquifer undergoing intrinsic bioremediation. *Applied and Environmental Microbiology* 64: 3869–3877.

Evans, R. J. and T. A. Milne. 1987. Molecular characterization of the pyrolysis of biomass. 1. Fundamentals. *Energy & Fuels* 1:123–137.

Fitzherbert, E. B., M. J. Struebig, A. Morel, and et al. 2008. How will oil palm expansion affect biodiversity? *Trends in Ecology & Evolution* 23: 538–545.

Freedman, B., R. O. Butterfield, and E. H. Pryde. 1986. Transesterification kinetics of soybean oil. *Journal of the American Oil Chemists Society* 63: 1375–1380.

Gallezot, P. 2012. Conversion of biomass to selected chemical products. *Chemical Society Reviews* 41:1538–1558.

Granados, M. L., M. D. Z. Poves, D. M. Alonso, and et al. 2007. Biodiesel from sunflower oil by using activated calcium oxide. *Applied Catalysis B-Environmental* 73: 317–326.

Haas, M. J., A. J. McAloon, W. C. Yee, and T. A. Foglia. 2006. A process model to estimate biodiesel production costs. *Bioresource Technology* 97: 671–678.

Haritash, A. K. and C. P. Kaushik. 2009. Biodegradation aspects of polycyclic aromatic hydrocarbons (PAHs): A review. *Journal of Hazardous Materials* 169: 1–15.

Hazen, T. C., E. A. Dubinsky, T. Z. DeSantis, and et al. 2010. Deep-sea oil plume enriches indigenous oil-degrading bacteria. *Science* 330: 204–208.

He, F. and D. Y. Zhao. 2005. Preparation and characterization of a new class of starch-stabilized bimetallic nanoparticles for degradation of chlorinated hydrocarbons in water. *Environmental Science & Technology* 39: 3314–3320.

Hill, J., E. Nelson, D. Tilman, S. Polasky, and D. Tiffany. 2006. Environmental, economic, and energetic costs and benefits of biodiesel and ethanol biofuels. *Proceedings of the National Academy of Sciences of the United States of America* 103: 11206–11210.

Khalili, N. R., P. A. Scheff, and T. M. Holsen. 1995. PAH source fingerprints for coke ovens, diesel and gasoline-engines, highway tunnels, and wood combustion emissions. *Atmospheric Environment* 29: 533–542.

Kilian, L. 2009. Not all oil price shocks are alike: Disentangling demand and supply shocks in the crude oil market. *American Economic Review* 99: 1053–1069.

Kim, H. J., B. S. Kang, M. J. Kim, and et al. 2004. Transesterification of vegetable oil to biodiesel using heterogeneous base catalyst. *Catalysis Today* 93–5: 315–320.

Koh, L. P. and D. S. Wilcove. 2007. Cashing in palm oil for conservation. *Nature* 448: 993–994.

Konur, O. 2000. Creating enforceable civil rights for disabled students in higher education: An institutional theory perspective. *Disability & Society* 15: 1041–1063.

Konur, O. 2002a. Access to Nursing Education by disabled students: Rights and duties of nursing programs. *Nurse Education Today* 22: 364–374.

Konur, O. 2002b. Assessment of disabled students in higher education: Current public policy issues. *Assessment and Evaluation in Higher Education* 27: 131–152.

Konur, O. 2002c. Access to employment by disabled people in the UK: Is the Disability Discrimination Act working? *International Journal of Discrimination and the Law* 5: 247–279.

Konur, O. 2006a. Participation of children with dyslexia in compulsory education: Current public policy issues. *Dyslexia* 12: 51–67.

Konur, O. 2006b. Teaching disabled students in Higher Education. *Teaching in Higher Education* 11: 351–363.

Konur, O. 2007a. A judicial outcome analysis of the Disability Discrimination Act: A windfall for the employers? *Disability & Society* 22: 187–204.

Konur, O. 2007b. Computer-assisted teaching and assessment of disabled students in higher education: The interface between academic standards and disability rights. *Journal of Computer Assisted Learning* 23:207–219.

Konur, O. 2012. The evaluation of the research on the biodiesel: A scientometric approach. *Energy Education Science and Technology Part A: Energy Science and Research* 28: 1003–1014.

Konur, O. 2015. Current state of research on algal biodiesel. In *Marine Bioenergy: Trends and Developments*, S. K. Kim, and C. G. Lee, ed., 487–512. Boca Raton, FL: CRC Press.

Konur, O., ed. 2021a. *Handbook of Biodiesel and Petrodiesel Fuels: Science, Technology, Health, and Environment*. Boca Raton, FL: CRC Press.

Konur, O., ed. 2021b. *Handbook of Biodiesel and Petrodiesel Fuels: Science, Technology, Health, and Environment. Volume 1. Biodiesel Fuels: Science, Technology, Health, and Environment.* Boca Raton, FL: CRC Press.

Konur, O., ed. 2021c. *Handbook of Biodiesel and Petrodiesel Fuels: Science, Technology, Health, and Environment. Volume 2. Biodiesel Fuels Based on the Edible and Nonedible Feedstocks, Wastes, and Algae: Science, Technology, Health, and Environment.* Boca Raton, FL: CRC Press.

Konur, O., ed. 2021d. *Handbook of Biodiesel and Petrodiesel Fuels: Science, Technology, Health, and Environment. Volume 3. Petrodiesel Fuels: Science, Technology, Health, and Environment.* Boca Raton, FL: CRC Press.

Konur, O. 2021e. Biodiesel and petrodiesel fuels: Science, technology, health, and environment. In *Handbook of Biodiesel and Petrodiesel Fuels: Science, Technology, Health, and Environment. Volume 1. Biodiesel Fuels: Science, Technology, Health, and Environment*, ed. O. Konur. Boca Raton, FL: CRC Press.

Konur, O. 2021f. Biodiesel and petrodiesel fuels: A scientometric review of the research. In *Handbook of Biodiesel and Petrodiesel Fuels: Science, Technology, Health, and*

Environment. Volume 1. Biodiesel Fuels: Science, Technology, Health, and Environment, ed. O. Konur. Boca Raton, FL: CRC Press.

Konur, O. 2021g. Biodiesel and petrodiesel fuels: A review of the research. In *Handbook of Biodiesel and Petrodiesel Fuels: Science, Technology, Health, and Environment. Volume 1. Biodiesel Fuels: Science, Technology, Health, and Environment*, ed. O. Konur. Boca Raton, FL: CRC Press.

Konur, O. 2021h Nanotechnology applications in the diesel fuels and the related research fields: A review of the research. In *Handbook of Biodiesel and Petrodiesel Fuels: Science, Technology, Health, and Environment. Volume 1. Biodiesel Fuels: Science, Technology, Health, and Environment*, ed. O. Konur. Boca Raton, FL: CRC Press.

Konur, O. 2021i. Biooils: A scientometric review of the research. In *Handbook of Biodiesel and Petrodiesel Fuels: Science, Technology, Health, and Environment. Volume 1. Biodiesel Fuels: Science, Technology, Health, and Environment*, ed. O. Konur. Boca Raton, FL: CRC Press.

Konur, O. 2021j. Characterization and properties of biooils: A review of the research. In *Handbook of Biodiesel and Petrodiesel Fuels: Science, Technology, Health, and Environment. Volume 1. Biodiesel Fuels: Science, Technology, Health, and Environment*, ed. O. Konur. Boca Raton, FL: CRC Press.

Konur, O. 2021k. Biomass pyrolysis and pyrolysis oils: A review of the research. In *Handbook of Biodiesel and Petrodiesel Fuels: Science, Technology, Health, and Environment. Volume 1. Biodiesel Fuels: Science, Technology, Health, and Environment*, ed. O. Konur. Boca Raton, FL: CRC Press.

Konur, O. 2021l. Biodiesel fuels: A scientometric review of the research. In *Handbook of Biodiesel and Petrodiesel Fuels: Science, Technology, Health, and Environment. Volume 1. Biodiesel Fuels: Science, Technology, Health, and Environment*, ed. O. Konur. Boca Raton, FL: CRC Press.

Konur, O. 2021m. Glycerol: A scientometric review of the research. In *Handbook of Biodiesel and Petrodiesel Fuels: Science, Technology, Health, and Environment. Volume 1. Biodiesel Fuels: Science, Technology, Health, and Environment*, ed. O. Konur. Boca Raton, FL: CRC Press.

Konur, O. 2021n. Propanediol production from glycerol: A review of the research. In *Handbook of Biodiesel and Petrodiesel Fuels: Science, Technology, Health, and Environment. Volume 1. Biodiesel Fuels: Science, Technology, Health, and Environment*, ed. O. Konur Boca Raton, FL: CRC Press.

Konur, O. 2021o. Edible oil-based biodiesel fuels: A scientometric review of the research. *In Handbook of Biodiesel and Petrodiesel Fuels: Science, Technology, Health, and Environment. Volume 2. Biodiesel Fuels based on the Edible and Nonedible Feedstocks, Wastes, and Algae: Science, Technology, Health, and Environment,* ed. O. Konur. Boca Raton, FL: CRC Press.

Konur, O. 2021p. Palm oil-based biodiesel fuels: A review of the research. In *Handbook of Biodiesel and Petrodiesel Fuels: Science, Technology, Health, and Environment. Volume 2. Biodiesel Fuels Based on the Edible and Nonedible Feedstocks, Wastes, and Algae*, ed. O. Konur. Boca Raton, FL: CRC Press.

Konur, O. 2021q. Rapeseed oil-based biodiesel fuels: A review of the research. In *Handbook of Biodiesel and Petrodiesel Fuels: Science, Technology, Health, and Environment. Volume 2. Biodiesel Fuels Based on the Edible and Nonedible Feedstocks, Wastes, and Algae,* ed. O. Konur. Boca Raton, FL: CRC Press.

Konur, O. 2021r. Nonedible oil-based biodiesel fuels: A scientometric review of the research. In *Handbook of Biodiesel and Petrodiesel Fuels: Science, Technology, Health, and Environment. Volume 2. Biodiesel Fuels Based on the Edible and Nonedible Feedstocks, Wastes, and Algae: Science, Technology, Health, and Environment*, ed. O. Konur. Boca Raton, FL: CRC Press.

Konur, O. 2021s. Waste oil-based biodiesel fuels: A scientometric review of the research. In *Handbook of Biodiesel and Petrodiesel Fuels: Science, Technology, Health, and Environment. Volume 2. Biodiesel Fuels Based on the Edible and Nonedible Feedstocks, Wastes, and Algae: Science, Technology, Health, and Environment*, ed. O. Konur. Boca Raton, FL: CRC Press.

Konur, O. 2021t. Algal biodiesel fuels: A scientometric review of the research. In *Handbook of Biodiesel and Petrodiesel Fuels: Science, Technology, Health, and Environment. Volume 2. Biodiesel Fuels Based on the Edible and Nonedible Feedstocks, Wastes, and Algae: Science, Technology, Health, and Environment*, ed. O. Konur. Boca Raton, FL: CRC Press.

Konur, O. 2021u. Algal biomass production for biodiesel production: A review of the research. In *Handbook of Biodiesel and Petrodiesel Fuels: Science, Technology, Health, and Environment. Volume 2. Biodiesel Fuels Based on the Edible and Nonedible Feedstocks, Wastes, and Algae*, ed. O. Konur Boca Raton, FL: CRC Press. February 23, 2020

Konur, O. 2021v. Algal biomass production in wastewaters for biodiesel production: A review of the research. In *Handbook of Biodiesel and Petrodiesel Fuels: Science, Technology, Health, and Environment. Volume 2. Biodiesel Fuels Based on the Edible and Nonedible Feedstocks, Wastes, and Algae*, ed. O. Konur. Boca Raton, FL: CRC Press. February 23, 2020

Konur, O. 2021x. Algal lipid production for biodiesel production: A review of the research. In *Handbook of Biodiesel and Petrodiesel Fuels: Science, Technology, Health, and Environment. Volume 2. Biodiesel Fuels Based on the Edible and Nonedible Feedstocks, Wastes, and Algae*, ed. O. Konur Boca Raton, FL: CRC Press. May 5, 2020

Konur, O. 2021y. Crude oils: A scientometric review of the research. In *Handbook of Biodiesel and Petrodiesel Fuels: Science, Technology, Health, and Environment. Volume 3. Petrodiesel Fuels: Science, Technology, Health, and Environment*, ed. O. Konur. Boca Raton, FL: CRC Press.

Konur, O. 2021z. Petrodiesel fuels: A scientometric review of the research. In *Handbook of Biodiesel and Petrodiesel Fuels: Science, Technology, Health, and Environment. Volume 3. Petrodiesel Fuels: Science, Technology, Health, and Environment*, ed. O. Konur. Boca Raton, FL: CRC Press.

Konur, O. 2021aa. Bioremediation of petroleum hydrocarbons in the contaminated soils: A review of the research. In *Handbook of Biodiesel and Petrodiesel Fuels: Science, Technology, Health, and Environment. Volume 3. Petrodiesel Fuels: Science, Technology, Health, and Environment*, ed. O. Konur. Boca Raton, FL: CRC Press.

Konur, O. 2021ab. Desulfurization of diesel fuels: A review of the research. In *Handbook of Biodiesel and Petrodiesel Fuels: Science, Technology, Health, and Environment. Volume 3. Petrodiesel Fuels: Science, Technology, Health, and Environment*, ed. O. Konur. Boca Raton, FL: CRC Press.

Konur, O. 2021ac. Diesel fuel exhaust emissions: A scientometric review of the research. In *Handbook of Biodiesel and Petrodiesel Fuels: Science, Technology, Health, and Environment. Volume 3. Petrodiesel Fuels: Science, Technology, Health, and Environment*, ed. O. Konur. Boca Raton, FL: CRC Press.

Konur, O. 2021ad. The adverse health and safety impact of diesel fuels: A scientometric review of the research. In *Handbook of Biodiesel and Petrodiesel Fuels: Science, Technology, Health, and Environment. Volume 3. Petrodiesel Fuels: Science, Technology, Health, and Environment*, ed. O. Konur. Boca Raton, FL: CRC Press.

Konur, O. 2021ae. Respiratory illnesses caused by the diesel fuel exhaust emissions: A review of the research. In *Handbook of Biodiesel and Petrodiesel Fuels: Science, Technology, Health, and Environment. Volume 3. Petrodiesel Fuels: Science, Technology, Health, and Environment*, ed. O. Konur. Boca Raton, FL: CRC Press.

Konur, O. 2021af. Cancer caused by the diesel fuel exhaust emissions: A review of the research. In *Handbook of Biodiesel and Petrodiesel Fuels: Science, Technology, Health, and Environment. Volume 3. Petrodiesel Fuels: Science, Technology, Health, and Environment*, ed. O. Konur. Boca Raton, FL: CRC Press.

Konur, O. 2021ag. Cardiovascular and other illnesses caused by the diesel fuel exhaust emissions: A review of the research. In *Handbook of Biodiesel and Petrodiesel Fuels: Science, Technology, Health, and Environment. Volume 3. Petrodiesel Fuels: Science, Technology, Health, and Environment*, ed. O. Konur. Boca Raton, FL: CRC Press.

Kubat, M., R. C. Holte, and S. Matwin. 1998. Machine learning for the detection of oil spills in satellite radar images. *Machine Learning* 30: 195–215.

Kusdiana, D. and S. Saka. 2001. Kinetics of transesterification in rapeseed oil to biodiesel fuel as treated in supercritical methanol. *Fuel* 80: 693–698.

Lapuerta, M., O. Armas, and J. Rodriguez-Fernandez. 2008. Effect of biodiesel fuels on diesel engine emissions. *Progress in Energy and Combustion Science* 34: 198–223.

Lardon, L., A. Helias, B. Sialve, J. P. Steyer, and O. Bernard. 2009. Life-cycle assessment of biodiesel production from microalgae. *Environmental Science & Technology* 43: 6475–6481.

Leahy, J. G. and R. R. Colwell. 1990. Microbial degradation of hydrocarbons in the environment. *Microbiological Reviews* 54: 305–315.

Lee, J. Y., Yoo, C., S. Y. Jun, C. Y. Ahn, and H. M. Oh. 2010. Comparison of several methods for effective lipid extraction from microalgae. *Bioresource Technology* 101: S75–S77.

Leung, D. Y. C. and Y. Guo. 2006. Transesterification of neat and used frying oil: Optimization for biodiesel production. *Fuel Processing Technology* 87: 883–890.

Li, Y. Q., M. Horsman, B. Wang, N. Wu, and C. Q. Lan. 2008. Effects of nitrogen sources on cell growth and lipid accumulation of green alga *Neochloris oleoabundans. Applied Microbiology And Biotechnology* 81: 629–636.

Liang, Y. N., N. Sarkany, and Y. Cui. 2009. Biomass and lipid productivities of *Chlorella vulgaris* under autotrophic, heterotrophic and mixotrophic growth conditions. *Biotechnology Letters* 31:1043–1049.

Ma, F. R. and M. A. Hanna. 1999. Biodiesel production: A review. *Bioresource Technology* 70: 1–15.

Mata, T. M., A. A. Martins, and N. S. Caetano. 2010. Microalgae for biodiesel production and other applications: A review. *Renewable & Sustainable Energy Reviews* 14: 217–232.

McCreanor, J., P. Cullinan, M. J. Nieuwenhuijsen, and et al. 2007. Respiratory effects of exposure to diesel traffic in persons with asthma. *New England Journal of Medicine* 357: 2348–2358.

Meher, L. C., D. V. Sagar, and S. N. Naik. 2006. Technical aspects of biodiesel production by transesterification: A review. *Renewable & Sustainable Energy Reviews* 10: 248–268.

Miao, X. L. and Q. Y. Wu. 2006. Biodiesel production from heterotrophic microalgal oil. *Bioresource Technology* 97: 841–846.

Mohan, D, C. U. Pittman, M. Bricka, and et al. 2007. Sorption of arsenic, cadmium, and lead by chars produced from fast pyrolysis of wood and bark during bio-oil production. *Journal of Colloid and Interface Science* 310: 57–73.

Mohan, D., C. U. Pittman, and P. H. Steele. 2006. Pyrolysis of wood/biomass for bio-oil: A critical review. *Energy & Fuels* 20: 848–889.

Moldowan, J. M., W. K. Seifert, E. J. Gallegos. 1985. Relationship between petroleum composition and depositional environment of petroleum source rocks. *AAPG Bulletin-American Association Of Petroleum Geologists* 69:1255–1268.

Mortensen, P. M., J. D. Grunwaldt, P. A. Jensen, K. G. Knudsen, and A. D. Jensen. 2011. A review of catalytic upgrading of bio-oil to engine fuels. *Applied Catalysis A-General* 407: 1–19.

North, D. C. 1991a. *Institutions, Institutional Change and Economic Performance*. Cambridge, Mass.: Cambridge University Press.

North, D.C. 1991b. Institutions. *Journal of Economic Perspectives* 5: 97–112.

Noureddini, H. and D. Zhu. 1997. Kinetics of transesterification of soybean oil. *Journal of the American Oil Chemists Society* 74: 1457–1463.

Orfao, J. J. M., F. J. A. Antunes, and J. L. Figueiredo. 1999. Pyrolysis kinetics of lignocellulosic materials: Three independent reactions model. *Fuel* 78: 349–358.

Perron, P. 1989. The great crash, the oil price shock, and the unit root hypothesis. *Econometrica: Journal of the Econometric Society* 57: 1361–1401.

Qi, G. S., R. T. Yang, and R. Chang. 2004. *MnOx-CeO2 mixed oxides prepared by co-precipitation for selective catalytic 51: 93–106.*

Ramadhas, A. S., S. Jayaraj, and C. Muraleedharan. 2005. Biodiesel production from high FFA rubber seed oil. *Fuel* 84: 335–340.

Ramos, M. J., C. M. Fernandez, A. Casas, L. Rodriguez, and A. Perez. 2009. Influence of fatty acid composition of raw materials on biodiesel properties. *Bioresource Technology* 100: 2618.

Rodolfi, L., G. C. Zittelli, N. Bassi, and et al. 2009. Microalgae for oil: Strain selection, induction of lipid synthesis and outdoor mass cultivation in a low-cost photobioreactor. *Biotechnology and Bioengineering* 102: 100–112.

Rogge, W. F., L. M. Hildemann, M. A. Mazurek, G. R. Cass, and B. R. T. Simoneit. 1993. Sources of fine organic aerosol. 2. Noncatalyst and catalyst-equipped automobiles and heavy-duty diesel trucks. *Environmental Science & Technology* 27: 636–651.

Rosegrant, M. W. and S. A. Cline. 2003. Global food security: Challenges and policies. *Science* 302: 1917–1919.

Saka, S. and D. Kusdiana. 2001. Biodiesel fuel from rapeseed oil as prepared in supercritical methanol. *Fuel* 80: 225–231.

Salvi, S., A. Blomberg, B. Rudell, F. Kelly, T. Sandstrom, S. T. Holgate, and A. Frew. 1999. Acute inflammatory responses in the airways and peripheral blood after short-term exposure to diesel exhaust in healthy human volunteers. *American Journal of Respiratory and Critical Care Medicine* 159: 702–709.

Schauer, J. J., M. J. Kleeman, G. R. Cass, and B. R. T. Simoneit. 1999. Measurement of emissions from air pollution sources. 2. C_1 through C_{30} organic compounds from medium duty diesel trucks. *Environmental Science & Technology* 33: 1578–1587.

Scholle, P. A. and M. A. Arthur. 1980. Carbon isotope fluctuations in cretaceous pelagic limestones: Potential stratigraphic and petroleum-exploration tool. *AAPG Bulletin-American Association Of Petroleum Geologists* 64: 67–87.

Sharma, R. K., J. B. Wooten, V. L. Baliga, and et al. 2004. Characterization of chars from pyrolysis of lignin. *Fuel* 83: 1469–1482.

Van Zwieten, L., S. Kimber, and S 2010. Effects of biochar from slow pyrolysis of papermill waste on agronomic performance and soil fertility. *Plant and Soil* 327: 235–246.

Vicente, G., M. Martinez, and J. Aracil. 2004. Integrated biodiesel production: A comparison of different homogeneous catalysts systems. *Bioresource Technology* 92: 297–305.

Vispute, T. P., H. Y. Zhang, A. Sanna, R. Xiao, and G. W. Huber. 2010. Renewable chemical commodity feedstocks from integrated catalytic processing of pyrolysis oils. *Science* 330: 1222–1227.

Xu, H., X. L. Miao, and Q. Y. Wu. 2006. High quality biodiesel production from a microalga *Chlorella protothecoides* by heterotrophic growth in fermenters. *Journal of Biotechnology* 126: 499–507.

Zhang, W. B., Z. Shi, F. Zhang, and et al. 2013. Superhydrophobic and superoleophilic PVDF membranes for effective separation of water-in-oil emulsions with high flux. *Advanced Materials* 25: 2071–2076.

Zhang, Y., M. A. Dube, D. D. McLean, and M. Kates. 2003a. Biodiesel production from waste cooking oil: 1. Process design and technological assessment. *Bioresource Technology* 89:1–16.

Zhang, Y., M. A. Dube, D. D. McLean, and M. Kates. 2003b. Biodiesel production from waste cooking oil: 2. Economic assessment and sensitivity analysis. *Bioresource Technology* 90: 229–240.

Zhu, Y. F., W. C. Hinds, S. Kim, S. Shen, and C. Sioutas. 2002. Study of ultrafine particles near a major highway with heavy-duty diesel traffic. *Atmospheric Environment* 36: 4323–4335.

4 Nanotechnology Applications in Diesel Fuels and Related Research Fields

A Review of the Research

Ozcan Konur

CONTENTS

4.1 INTRODUCTION

Crude oils have been primary sources of energy and fuels, such as petrodiesel. However, significant public concerns about their sustainability, price fluctuations, and adverse environmental impact have emerged since the 1970s (Ahmadun et al., 2009; Atlas, 1981; Babich and Moulijn, 2003; Perron, 1989). Thus, biooils (Bridgwater and Peacocke, 2000; Bridgwater et al. 1999; Czernik and Bridgwater, 2004) and biooil-based biodiesel fuels (Chisti, 2007; Hill et al., 2006; Ma and Hanna, 1999; Mata et al., 2010) have emerged as alternatives to crude oils and crude oil-based petrodiesel fuels in recent decades. Nowadays, although biodiesel fuels are being used increasingly, petrodiesel fuels (Birch and Cary, 1996; Khalili et al., 1995; Rogge et al., 1993; Song and Ma, 2003; Song, 2003) are still used extensively in the transportation and power sectors (Konur, 2021a–ag).

In the meantime, nanotechnology (Neto et al., 2009; Geim and Novoselov, 2007; Iijima, 1991; Novoselov et al., 2004, 2005) has emerged as an interdisciplinary research field in recent years with applications from bioenergy to biomedicine (Konur 2016a–f, 2017a–e, 2019).

However, for the efficient progression of research in this field, it is necessary to develop efficient incentive structures for the primary stakeholders and to inform these stakeholders about the research (Konur, 2000, 2002a–c, 2006a–b, 2007a–b; North, 1991a–b). Thus, it is crucial to provide an overview of the nanotechnology applications in the diesel fuels and related research fields.

Although there have been a number of reviews and book chapters in this field (Gao et al., 2013; Hashemi et al., 2014; Shamsi-Jazeyi et al., 2014; Verma et al., 2013), there has been no review of the 25-most-cited articles in this field. Thus, this chapter reviews these articles on nanotechnology applications in the diesel fuels and related research fields, highlighting the key findings from these prolific studies. Then, it discusses the findings of these studies.

4.2 MATERIALS AND METHODOLOGY

The search for the literature was carried out in the 'Web of Science' (WOS) database in February 2020. It contains the 'Science Citation Index Expanded' (SCI-E), the

Social Sciences Citation Index' (SSCI), the 'Book Citation Index-Science' (BCI-S), the 'Conference Proceedings Citation Index-Science' (CPCI-S), the 'Emerging Sources Citation Index' (ESCI), the 'Book Citation Index-Social Sciences and Humanities' (BCI-SSH), the 'Conference Proceedings Citation Index-Social Sciences and Humanities' (CPCI-SSH), and the 'Arts and Humanities Citation Index' (A&HCI).

The keywords for the diesel fuels and related research fields for the literature search were collated from the screening of abstract pages for the first 500 highly cited papers in the related 11 research fields (Konur, 2021e–ag). Similarly, the keywords for the nanotechnology literature search were collated from the screening of abstract pages for the first 500 highly cited papers in nanotechnology. These keyword sets are provided in the appendices of the related chapters (Konur, 2016a–f, 2017a–e, 2019, 2021e–ag).

The most-cited 25 articles are selected for this review and the key findings are highlighted and discussed briefly.

4.3 RESULTS

4.3.1 Biooils

Iliopoulou et al. (2007) test two mesoporous aluminosilicate Al-MCM-41 materials (Si/Al=30 or 50) as catalysts for the *in situ* upgrading of biomass pyrolysis vapors in comparison to a siliceous MCM-41 sample and to non-catalytic biomass pyrolysis in a paper with 167 citations. They observe that the product yields and the quality of the produced biooil were significantly affected by the use of all MCM-41 catalytic materials. They attribute this behavior mainly to the combination of the large surface area and tubular mesopores (pore diameter ~2–3 nm) of MCM-41 materials, with their mild acidity that led to the desired environment for controlled conversion of the high molecular weight lignocellulosic molecules. The major improvement in the quality of biooil with the use of Al-MCM-41 catalytic materials was the increase of phenol concentration and the reduction of corrosive acids. Higher Si/Al ratios of the Al-MCM-41 samples enhanced the production of the organic phase of the biooil, while lower Si/Al ratios favored the conversion of the hydrocarbons of the organic phase towards gases and coke. Moderate steaming of the Al-MCM-41 samples (at 550 and 750°C, 20% steam partial pressure) decreased their surface area and number of acid sites by 40–60%, depending on the Si/Al ratio of the samples and the steaming temperature. However, the steamed samples were still active in the *in situ* upgrading of biomass pyrolysis vapors, resulting in different product yields and biooil composition, compared to the parent calcined samples, mainly after higher-temperature steaming. They conclude that these mesoporous materials were instrumental in biooil production.

4.3.2 Biodiesel Fuels

4.3.2.1 Production

Liu et al. (2008) study the development of sulfonated 'ordered mesoporous carbon' (OMC) for the catalytic preparation of biodiesel in a paper with 149 citations. They synthesize OMC by covalent attachment of sulfonic-acid-containing aryl radicals on the surface of mesoporous carbons. Compared with other solid acid catalysts, the resulting materials showed stable and highly efficient catalytic performance in biodiesel production with the highest conversion reaching 73.59%. These catalysts could be reused several times without loss of activity. Furthermore, sulfonated OMC with a larger pore size made active sites easily accessible and exhibits higher catalytic ability. They conclude that these mesoporous catalysts were instrumental in biodiesel production.

4.3.2.2 Properties

Sajith et al. (2010) study the impact of the addition of cerium oxide in nanoparticle form on the major physicochemical properties and the performance of biodiesel in a paper with 194 citations. They measure the physicochemical properties of the base fuel and the modified fuel formed by dispersing the catalyst nanoparticles by ultrasonic agitation using ASTM standard test methods. They examine the effects of the additive nanoparticles on the individual fuel properties, the engine performance, and emissions, and optimize the dosing level of the additive. They compare the performance of the fuel with and without the additive. They observe that the flash point and the viscosity of biodiesel increased with the inclusion of the cerium oxide nanoparticles. The emission levels of hydrocarbon and NO_x were appreciably reduced with the addition of cerium oxide nanoparticles. They conclude that the addition of cerium oxide nanoparticles was instrumental in enhancing the major physicochemical properties and the performance of biodiesel.

4.3.2.3 Glycerol

Huang et al. (2008) prepare highly dispersed copper nanoparticles supported on silica by a simple and convenient precipitation-gel technique, and compare their physicochemical properties and activity to those of a catalyst prepared by the conventional impregnation method for glycerol hydrogenolysis in a paper with 254 citations. As a consequence of the preparation method, they observe that the texture (BET), dispersion (dissociative N_2O adsorption), morphology (TEM), reduction behavior (TPR, XRD), state of copper species (XPS), and catalytic performance (glycerol hydrogenolysis) differed between samples. Both samples showed high selectivity (>98%) toward '1,2-propanediol' in a glycerol reaction. Because of a much smaller particle size, a higher dispersion of copper species with a strong metal-support interaction, and more resistance to sintering, the CuO/SiO_2 catalyst prepared by the precipitation-gel method presented a much higher activity and remarkably better long-term stability in the glycerol reaction than did the catalyst prepared by the impregnation method. The catalytic behavior of calcined and reduced samples and the structure changes of these samples after reaction provided an understanding of the stability toward sintering as well as the possible mechanism of the reaction. They conclude that these nanoparticles were efficient and stable catalysts for glycerol hydrogenolysis.

Simoes et al. (2010) study the electrooxidation of glycerol with Pd-based nanocatalysts for an application in alkaline fuel cells for chemicals and energy cogeneration in a paper with 219 citations. They synthesize carbon-supported Pd, Pt, Au, and bimetallic PdAu and PdNi nanocatalysts with different compositions and evaluate their catalytic activity toward glycerol electrooxidation in an alkaline medium. They observe that the Pd_xAu_{1-x}/C catalysts were alloys, which presented an increase of crystallite (XRD) and particle (TEM) sizes with an increasing Au atomic fraction. Their surfaces were palladium rich whatever the Pd atomic ratio. The structure of the $Pd_{0.5}Ni_{0.5}/C$ catalyst was much more difficult to understand, but it was composed of a palladium phase in interaction with a $Ni(OH)_2$ phase. The onset potential of glycerol oxidation was ca. 0.15 V lower on Pt/C than on Pd/C and Au/C. All Pd_xMe_{1-x}/C catalysts presented a lower onset potential than monometallic Au/C and Pd/C ones, though higher than Pt/C. For bimetallic catalysts, the order of activity at low potentials was: $Pd_{0.3}Au_{0.7}/C > Pd_{0.5}Au_{0.5}/C > Pd_{0.5}Ni_{0.5}/C$. The glycerol electrooxidation mechanism was dependent on the catalyst, leading to different reaction products. They detect adsorbed CO species on monometallic Pt and on Pd-rich catalysts, but not on Au and $Pd_{0.3}Au_{0.7}$ catalysts, indicating that they were not able to break the C–C bond. They detect the formation of a hydroxypyruvate ion, which is a costly chemical product of a pure gold catalyst. They conclude that Pd-based nanocatalysts were instrumental in the electrooxidation of glycerol.

Hong et al. (2015) study the bimetallic PdPt 'nanowire networks' (NNWs) with enhanced electrocatalytic activity for ethylene glycol and glycerol oxidation in a paper with 161 citations. They synthesize NNWs with tunable compositions via a simple and efficient method. They observe that the catalytic activity of the as-prepared NNWs was related to their composition. They conclude that these synthesized NNWs displayed great potential as substitutes for commercial Pt/C catalysts for the effective catalysis of ethylene glycol and glycerol electrooxidation in an alkaline solution and among the prepared PdPt NNWs; Pd55Pt30 showed the best electrocatalytic activity. They conclude that these catalysts were instrumental in the efficient production of these biodiesel fuels.

Clacens et al. (2002) study the selective etherification of glycerol to polyglycerols over impregnated basic MCM-41 type mesoporous catalysts in a paper with 154 citations. They focus on the direct and selective synthesis, from glycerol and without a solvent, of polyglycerols having a low polymerization degree (di- and/or triglycerol), in the presence of solid mesoporous catalysts. This consists in the synthesis and the impregnation of mesoporous solids with different basic elements in order to make them active, selective, and stable for the target reaction. They observe that this impregnation method provided an important activity, which must be correlated to an important active species incorporation. Concerning the selectivity of the modified mesoporous catalysts, they obtain the best value for (di- and tri-) glycerol over solids prepared by caesium impregnation. The reuse of these caesium impregnated catalysts did not affect the selectivity of the (di- and tri-) glycerol fraction. In the presence of lanthanum or magnesium containing catalysts, the glycerol dehydration to acrolein was very significant, whereas this unwanted product was not formed when caesium was used as an impregnation promoter. They conclude that these mesoporous catalysts were instrumental in the selective etherification of glycerol to polyglycerols.

4.3.2.4 Edible Oil-based Biodiesel Fuels

Reddy et al. (2006) develop a route for the production of biodiesel via the transesterification of 'soybean oil' (SBO) and poultry fat with methanol in quantitative conversions at room temperature using nanocrystalline calcium oxides as catalysts in a paper with 230 citations. They observe that, under the same conditions, laboratory-grade CaO gave only a 2% conversion in the case of SBO, and there was no observable reaction with poultry fat. The soybean oil:methanol ratio was 1:27. With the most active catalyst, they observe deactivation after eight cycles with SBO and after three cycles with poultry fat. Deactivation might be associated with one or more of the following factors: the presence of organic impurities or adventitious moisture and enolate formation by the deprotonation of the carbon α to the carboxy group in triglyceride or biodiesel. These biodiesel fuels met the ASTM D-874 standard for sulfated ash for both substrates. They conclude that these nanocrystalline calcium oxides were instrumental in biodiesel production.

Verziu et al. (2008) study the catalytic activity for the production of sunflower and rapeseed oil-based biodiesel with three morphologically different nanocrystalline MgO materials prepared using simple, green, and reproducible methods in a paper with 161 citations. They consider MgO(111) nanosheets (MgOI), conventionally prepared MgO (MgO II), and aerogel prepared MgO (MgOIII). They use (a) a 4-methoxy-benzyl alcohol templated sol-gel process followed by supercritical drying and calcination in air at 773 K (MgOI), (b) from a commercial MgO that was boiled in water, followed by drying at 393 K, and dehydration under vacuum at 773 K (MgOII), and (c) *via* hydrolysis of $Mg(OCH_3)_2$ in a methanol–toluene mixture, followed by supercritical solvent removal with the formation of a $Mg(OH)_2$ aerogel that was dehydrated under vacuum at 773 K (MgOIII). They observe higher conversions and selectivities to methyl esters with these methods compared to autoclave or ultrasound conditions. Under ultrasound, they observe a leaching of the magnesium as a direct consequence of a saponification reaction. These systems also allow working with much lower ratios of methanol to vegetable oil than reported in the literature for other heterogeneous systems. The activation temperature providing the most active catalysts varied depending on the exposed facet: for MgOI this was 773 K, while for MgOII and MgOIII this was 583 K. They conclude that these catalysts were instrumental in the efficient production of these biodiesel fuels.

Wang and Yang (2007) study the transesterification reaction of soybean oil with supercritical/subcritical methanol by Nano-MgO catalysts in a paper with 157 citations. They examine the variables affecting the yield of methyl ester during the transesterification reaction, such as the catalyst content, reaction temperature, and the molar ratio of methanol to soybean oil, and compare them with those of a non-catalyst. When nano-MgO was added from 0.5 to 3 wt%, they observe that the transesterification rate clearly increased, while the catalyst content was further enhanced to 5 wt%, with little increase in yield. They then observe that increasing the reaction temperature had a favorable influence on the methyl ester yield. In addition, for molar ratios of methanol to soybean oil ranging from 6 to 36, the higher molar ratios of methanol to oil was charged, the faster transesterification rate was obtained. When the temperature was increased to 533 K, the transesterification reaction was essentially completed within 10 min with 3 wt% nano-MgO and a methanol/oil molar rate of 36:1. Such a high reaction rate with nano-MgO was mainly due to the lower activation energy (75.94 kJ/mol) and increased stirring. They conclude that these catalysts were instrumental in the efficient production of these biodiesel fuels.

Carmo et al. (2009) study the production of biodiesel by the esterification of palmitic acid over mesoporous aluminosilicate Al-MCM-41 in a paper with 145 citations. They obtain biodiesel by the esterification of palmitic acid with methanol, ethanol, and isopropanol in the presence of Al-MCM-41 mesoporous molecular sieves with Si/Al ratios of 8, 16, and 32. They synthesize the catalytic acids at room temperature and characterize them. They perform the reaction at 130°C whilst stirring at 500 rpm, with an alcohol:acid molar ratio of 60 and 0.6 wt% catalyst for 2 h. They observe that the alcohol reactivity followed the order: methanol > ethanol > isopropanol. The catalyst Al-MCM-41 with ratio Si/Al=8 produced the largest conversion values for the alcohols studied. The data followed a rather satisfactory approximation to first-order kinetics. They conclude that these nanocatalysts were instrumental in biodiesel production from palmitic acids.

4.3.2.5 Nonedible Oil-based Biodiesel Fuels

Deng et al. (2011) study the production of biodiesel from Jatropha oil catalyzed by a nanosized solid basic catalyst in a paper with 171 citations. They synthesize hydrotalcite-derived particles with an Mg:Al molar ratio of 3:1 by a co-precipitation method using urea as the precipitating agent, subsequently with a 'microwave-hydrothermal treatment', and followed by calcination at 773 K for 6 h. These particles were microsized mixed Mg/Al oxides, although actually they were nanosized. Because of their strong basicity, they use the nanoparticles further as a catalyst for biodiesel production from Jatropha oil after pretreatment. They perform the experiments with the solid basic catalyst in an ultrasonic reactor under different conditions. At the optimized condition, they obtain a biodiesel yield of 95.2%, and they observe that the biodiesel properties were close to those of the German standard. The catalyst could be reused eight times. They conclude that these nanocatalyts were instrumental for biodiesel production from Jatropha oils.

4.3.2.6 Waste Oil-based Biodiesel Fuels

Reddy et al. (2006) develop a route for the production of biodiesel via the transesterification of soybean oil (SBO) and poultry fat with methanol in quantitative conversions at room temperature using nanocrystalline calcium oxides as catalysts in a paper with 230 citations. They observe that under the same conditions, laboratory-grade CaO gave only a 2% conversion in the case of SBO, and there was no observable reaction with poultry fat. The soybean oil:methanol ratio was 1:27. With the most active catalyst, they observe deactivation after eight cycles with SBO and after three cycles with poultry fat. Deactivation might be associated with one or more of the following factors: the presence of organic impurities or adventitious moisture and enolate formation by the deprotonation of the carbon α to the carboxy group in triglyceride or biodiesel. These biodiesel fuels met the ASTM D-874 standard for sulfated ash for both substrates. They conclude that these nanocrystalline calcium oxides were instrumental in biodiesel production.

4.3.3 Crude Oils

4.3.3.1 Water-in Oil Emulsions

Si et al. (2015) report a strategy to create 'fibrous, isotropically bonded elastic reconstructed' (FIBER) aerogels with a hierarchical cellular structure and super-elasticity by combining electrospun nanofibers and the freeze-shaping technique for separating surfactant-stabilized water-in-oil emulsions in a paper with 333 citations. This approach allowed the intrinsically lamellar deposited electrospun nanofibers to assemble into elastic bulk aerogels with a tunable porous structure and wettability on a large scale. They observe that these FIBER aerogels exhibited the integrated properties of ultralow density (< 30 mg cm^{-3}), rapid recovery from 80% compression strain, superhydrophobic-superoleophilic wettability, and high pore tortuosity. More interestingly, the FIBER aerogels could effectively separate surfactant-stabilized water-in-oil emulsions, solely using gravity, with high flux (maximum of 8140 Lm-2h^{-1}) and high separation efficiency, which matched well with the requirements for treating the real emulsions. The synthesis of FIBER aerogels also provided a versatile platform for exploring the applications of nanofibers in a self-supporting, structurally adaptive, and 3D macroscopic form. They conclude that these aerogels were instrumental for the effective separation of oil/water emulsions.

Arriagada and Osseo-Asare (1999) study the effect of ammonia concentration on the region of existence of single-phase water-in-oil microemulsions for the system 'polyoxyethylene (5) nonylphenyl ether (NP-5)/cyclohexane/ammonium hydroxide' in a paper with 326 citations. They observe that the presence of ammonia decreased the size of the microemulsion region. A minimum concentration of surfactant (estimated at about 1.1 wt%) was required for solubilization of the aqueous phase, though this value was not significantly affected by ammonia concentration. The transition between bound and free water occurred when the water-to-surfactant molar ratio was about 1 and the presence of ammonium hydroxide did not have a significant effect on this. Ultrafine (30–70 nm diameter), monodisperse silica particles produced by hydrolysis of tetraethoxysilane (TEOS) in the microemulsion showed a complex dependence of the particle size on the 'water-to-surfactant molar ratio' (R) and on the concentration of ammonium hydroxide. At relatively low ammonia concentration in the aqueous pseudophase (1.6 wt% NH_3) the particle size decreased monotonically with an increase in R. However, for higher ammonia concentrations (6.3–29.6 wt% NH_3) a minimum in particle size occurred as R is increased. They rationalize these trends in terms of (a) the effects of the concentration, structure, and dynamics of the NP-5 reverse micelles on the hydrolysis and condensation reactions of TEOS, and (b) the effects of ammonia concentration on the stability of the microemulsion phase, the hydrolysis/condensation reactions of TEOS, and the depolymerization of siloxane bonds. They conclude that the effect of ammonia concentration on the region of existence of single-phase water-in-oil microemulsions for this system was significant.

4.3.3.2 Oil Recovery

Suleimanov et al. (2011) study nanofluids for 'enhanced oil recovery' (EOR) in a paper with 151 citations. They use an aqueous solution of anionic surface-active agents with the addition of light non-ferrous metal nanoparticles. They observe that the use of the nanofluid permitted a 70–90% reduction of surface tension on an oil boundary in comparison with a surface-active agent aqueous solution and was characterized by a shift in dilution. They conclude that the use of the developed nano-suspension results in considerably increased EOR.

Zhang et al. (2014) study EOR using nanoparticle dispersions in a paper with 149 citations. They focus on imbibition tests using a reservoir of crude oil and a reservoir of brine solution with a high salinity and a suitable nanofluid that displaces crude oil from Berea sandstone (water-wet) and single-glass capillaries. They formulate the Illinois Institute of Technology (IIT) nanofluid to survive in a high-salinity environment and which resulted in an efficiency of 50% for Berea sandstone, compared to 17% using the brine alone at a reservoir temperature of 55°C. They also present direct visual evidence of the underlying mechanism based on the structural disjoining pressure for the crude oil displacement using IIT nanofluid from the solid substrate in high-salinity brine. These findings aid the understanding of the role of the nanofluid in displacing crude oil from the rock, especially in a high-salinity environment containing Ca^{2+} and Mg^{2+} ions. They also use Berea sandstone and a nanofluid containing silica nanoparticles. They conclude that using nanoparticle dispersions was instrumental for EOR.

Hendraningrat et al. (2013) study nanofluid enhanced oil recovery in a paper with 143 citations focusing on the parameters involved in the structural disjoining pressure mechanism, such as lowering interfacial tensions (IFTs) and altering wettability. They perform laboratory coreflood experiments in water-wet Berea sandstone core plugs with permeability in the range 9–400 mD using different concentrations of nanofluids. They employ a crude oil from a field in the North Sea and synthesize three nanofluid concentrations at 0.01, 0.05, and 0.1 wt% with synthetic brine. They observe that IFT decreased when hydrophilic nanoparticles were introduced to brine. The IFT decreased as nanofluid concentration increased which indicates a potential for EOR. Increasing hydrophilic nanoparticles would also decrease the contact angle of the aqueous phase and increase water wetness. They also observe that the higher the concentrations of nanofluids, the more the impairment of porosity and permeability in Berea core plugs. Despite the fact that increasing nanofluid concentration shows decreasing IFT and alters wettability, additional recovery was not guaranteed. They conclude that application of nanofluid EOR was an alternative EOR method.

4.3.3.3 Remediation of Crude Oils

Calcagnile et al. (2012) present a novel composite material based on commercially available polyurethane foams functionalized with colloidal superparamagnetic iron oxide nanoparticles and submicrometer polytetrafluoroethylene particles, which can efficiently separate oil contaminants from water in a paper with 417 citations. Untreated foam surfaces are inherently hydrophobic and oleophobic, but they can be rendered water-repellent and oil-absorbing by a solvent-free, electrostatic polytetrafluoroethylene particle deposition technique. They observe that the combined functionalization of the polytetrafluoroethylene-treated foam surfaces with colloidal iron oxide nanoparticles significantly increased the speed of oil absorption. The combined effects of the surface morphology and of the chemistry of the functionalized foams greatly affected the oil-absorption dynamics. In particular, nanoparticle capping molecules played a major role in this mechanism. In addition to the water-repellent and oil-absorbing capabilities, the functionalized foams also exhibited magnetic responsivity. Finally, due to their light weight, they floated easily on water. Hence, they conclude that by simply moving them around oil-polluted waters using a magnet, they could absorb the floating oil from polluted regions, thereby purifying the water underneath. This low-cost process could easily be scaled up to clean large-area oil spills in water.

Toyoda and Inagaki (2000) study the sorption behaviors of four kinds of heavy oils into exfoliated graphites with different bulk densities in a paper with 270 citations. They observe that the maximum sorption capacity of exfoliated graphite with a bulk density of 6 kg/m^3 was surprisingly high: 86 g of A-grade heavy oil and 76 g of crude oil per 1 g of exfoliated graphite, respectively; its sorption occurred very rapidly, i.e. within 2 min. Sorption capacity depended strongly on the bulk density and pore volume of the exfoliated graphite and the time it took to reach maximum sorption, as well as on the sorption capacity, as exfoliated graphite depends strongly on the grade of heavy oil. Heavy oils sorbed into the exfoliated graphite could be recovered either by simple compression or suction filtration with a recovery ratio of 60–80%. Recovered oils showed no difference in molecular weight and hydrocarbon constituent from the original. They detected no increase of the water content in the recovered oils, suggesting preferential sorption of heavy oil into exfoliated graphite. The oils recovered from the exfoliated graphite could be recycled. They conclude that exfoliated graphite with low bulk density is a promising material for the sorption and recovery of spilled heavy oil.

Yang et al. (2011) show that 'fluorous metal-organic frameworks' (FMOFs) are highly hydrophobic porous materials with a high capacity and affinity to C_6-C_8 hydrocarbons of oil components in a paper with 240 citations. They observe that FMOF-1 exhibited reversible adsorption with a high capacity for n-hexane, cyclohexane, benzene, toluene, and p-xylene, with no detectable water adsorption even at near 100% relative humidity, drastically outperforming activated carbon and zeolite porous materials. FMOF-2, obtained from annealing FMOF-1, showed enlarged cages and channels with double toluene adsorption vs FMOF-1 based on crystal structures. They conclude that there is great promise for FMOFs in applications such as the removal of organic pollutants from oil spills or ambient humid air, hydrocarbon storage and transportation, and water purification, under practical working conditions.

Ge et al. (2017) report on a Joule-heated 'graphene-wrapped sponge' (GWS) to clean up viscous crude oil at a high sorption speed in a paper with 184 citations. They observe that the Joule heat of the GWS reduced *in situ* the viscosity of the crude oil, which prominently increased the oil-diffusion coefficient in the pores of the GWS and thus speeded up the oil-sorption rate. The oil-sorption time was reduced by 94.6% compared with that of non-heated GWS. In addition, the oil-recovery speed was increased because of the viscosity decrease of the crude oil. They conclude that this *in situ* Joule self-heated sorbent design would promote the practical application of hydrophobic and oleophilic oil sorbents in the clean-up of viscous crude-oil spills.

Lin et al. (2012) report on nanoporous polystyrene (PS) fibers, prepared via a one-step electrospinning process, used as oil sorbents for cleaning up oil spills in a paper with 155 citations. They observe that the oleophilic-hydrophobic PS oil sorbent with highly porous structures had a motor oil sorption capacity of 113.87 g/g, approximately three to four times that of natural sorbents and nonwoven polypropylene fibrous mats. Additionally, the sorbents also exhibited a relatively high sorption capacity for edible oils, such as bean oil (111.80 g/g) and sunflower seed oil (96.89 g/g). They examine the oil sorption mechanism of the PS sorbent and the sorption kinetics. They conclude that these nanoporous materials have great potential for use in wastewater treatment, oil accident remediation, and environmental protection.

4.3.4 Petrodiesel Fuels

4.3.4.1 Production

Kang et al. (2009) study the ruthenium nanoparticles supported on carbon nanotubes as efficient catalysts for the selective conversion of synthesis gas to diesel fuel in a paper with 190 citations. They observe that these catalysts were highly selective Fischer–Tropsch catalysts for the production of C_{10}-C_{20} hydrocarbons (diesel fuel). The C_{10}-C_{20} selectivity strongly depended on the mean size of the Ru nanoparticles. Nanoparticles with a mean size of around 7 nm exhibited the highest C_{10}-C_{20} selectivity (ca. 65%) and a relatively higher turnover frequency for CO conversion. They conclude that these catalysts were instrumental in biodiesel production.

4.3.4.2 Properties

Tyagi et al. (2008) improve the ignition properties of diesel fuels by investigating the influence of adding aluminum and aluminum oxide nanoparticles to diesel in a paper with 161 citations. They perform droplet ignition experiments on top of a heated plate. They use different types of fuel mixtures as they vary both particle size (15 and 50 nm) as well as the volume fraction (0, 0.1, and 0.5%) of nanoparticles added to diesel. For each type of fuel mixture, they put several droplets on the hot plate from a fixed height and under identical conditions, and they record the probability of ignition of that fuel based on the number of droplets that ignited. They repeat these experiments at several temperatures over the range of 688–768°C. They observe that the ignition probability for the fuel mixtures that contained nanoparticles was significantly higher than that of pure diesel. They conclude that the addition of these nanoparticles enhanced the ignition properties of petrodiesel fuels.

4.4 DISCUSSION

Table 4.1 provides information on the primary and secondary research fronts for both biodiesel fuels and petrodiesel fuels. It also provides information on the distribution of these research fronts in the article papers for nanotechnology application on diesel fuels, the 100-most-cited papers, and over 121,000 population papers in the broader fields of petrodiesel and biodiesel fuels.

As this table shows, the research fronts of 'biodiesel fuels in general', 'feedstock-based biodiesel fuels', and 'petrodiesel fuels' comprise 28, 24, and 48% of these papers, respectively, in this field.

TABLE 4.1
Research Fronts

	Research Front	No. of Papers Reviewed (%)*	No. of Sample Papers (%)**	No. of Population Papers (%)***
1	Biodiesel fuels in general	28	41	27.7
1.1	Biooils	4	14	16.6
1.2	Biodiesel fuels in general	8	21	5.4
1.3	Glycerol	16	6	5.7
2	Feedstock-based biodiesel fuels	24	30	18.3
2.1	Edible oil-based biodiesel fuels	16	5	2.2
2.2	Nonedible oil-based biodiesel fuels	4	3	1.8
2.3	Waste oil-based biodiesel fuels	4	2	1.8
2.4	Algal oil-based biodiesel fuels	0	20	12.5
3	Petrodiesel fuels	48	34	67.4
3.1	Crude oils	40	14	43.8
3.2	Petrodiesel fuels in general	8	8	13.2
3.3	Emissions from petrodiesel fuels	0	11	8.6
3.4	Health impact of the emissions from petrodiesel fuels	0	1	1.8

*Sample size: 25.
**Sample size: 100.
***Sample size: 121,000.

In the first group of article papers, those on 'biooils', 'biodiesel fuels in general', and 'glycerol' comprise 4, 8, and 16% of these papers, respectively. In the second group of article papers, 'edible oil-based biodiesel fuels', 'nonedible oil-based biodiesel fuels', 'waste oil-based biodiesel fuels', and 'algal oil-based biodiesel fuels' comprise 16, 4, 4, and 0% of these papers, respectively. In the final group of article papers, 'crude oils', 'petrodiesel fuels in general', 'emissions from petrodiesel fuels', and the 'health impact of the emissions from petrodiesel fuels' comprise 40, 8, 0, and 0% of these papers, respectively.

The data show that the most-prolific research front has concerned the applications of nanotechnology to 'crude oils', including 'crude oil recovery', 'remediation of crude oils in the environment', and 'water-in-oil emulsions'. The applications of nanotechnology in 'glycerol applications' and 'edible oil-based biodiesel fuels' have also been prolific.

It is notable that there have been no prolific papers in the fields of 'biooil upgrading', 'biodiesel emissions', 'algal biodiesel fuels', 'crude oil properties', 'crude oil refining', 'petrodiesel desulfurization', 'power generation by petrodiesel fuels', 'petrodiesel fuel emissions', and the 'health impact of petrodiesel emissions'.

These prolific studies on nanotechnology applications in the fields of biooils, biodiesel fuels, crude oils, and petrodiesel fuels, presented in this chapter, highlight the importance of nanotechnology in these fields. These studies also show the importance of nanomaterials as the nanocatalyts in these fields, ranging from 'crude oil recovery' and 'remediation of crude oils in the environment' to 'biooil production', 'biodiesel production', and the production of biochemical and bioenergy from glycerol, a by-product of biodiesel fuels.

4.4.1 Biooils

Iliopoulou et al. (2007) test two mesoporous aluminosilicate Al-MCM-41 materials (Si/Al=30 or 50) as catalysts for the *in situ* upgrading of biomass pyrolysis vapors in comparison to a siliceous MCM-41 sample and to a non-catalytic biomass pyrolysis in a paper with 167 citations. They conclude that these mesoporous materials were instrumental in biooil production.

4.4.2 Biodiesel Fuels

4.4.2.1 Production

Liu et al. (2008) study the development of sulfonated 'ordered mesoporous carbon' for the catalytic preparation of biodiesel in a paper with 149 citations. They conclude that these mesoporous catalysts were instrumental in biodiesel production.

4.4.2.2 Properties

Sajith et al. (2010) study the impact of the addition of cerium oxide in nanoparticle form on the major physicochemical properties and the performance of biodiesel in a paper with 194 citations. They conclude that such an addition of cerium oxide nanoparticles was instrumental in enhancing the major physicochemical properties and the performance of biodiesel.

4.4.2.3 Glycerol

Huang et al. (2008) prepare highly dispersed copper nanoparticles supported on silica by a simple and convenient precipitation-gel technique, and compare their physicochemical properties and activity to those of a catalyst prepared by the conventional impregnation method for glycerol hydrogenolysis in a paper with 254 citations. They conclude that these nanoparticles were efficient and stable catalysts for glycerol hydrogenolysis. Simoes et al. (2010) study the electrooxidation of glycerol on Pd-based nanocatalysts for application in alkaline fuel cells for chemicals and energy cogeneration in a paper with 219 citations. They conclude that Pd-based nanocatalysts were instrumental for the electrooxidation of glycerol.

Hong et al. (2015) study bimetallic PdPt nanowire networks with enhanced electrocatalytic activity for ethylene glycol and glycerol oxidation in a paper with 161 citations. They conclude that these catalysts were instrumental in the efficient production of these biodiesel fuels. Clacens et al. (2002) study the selective etherification of glycerol to polyglycerols over impregnated basic MCM-41 type mesoporous catalysts in a paper with 154 citations. They conclude that these mesoporous catalysts were instrumental in the selective etherification of glycerol to polyglycerols.

4.4.2.4 Edible Oil-based Biodiesel Fuels

Reddy et al. (2006) develop a route for the production of biodiesel via the transesterification of soybean oil (SBO) and poultry fat with methanol in quantitative conversions at room temperature using nanocrystalline calcium oxides as catalysts in a paper with 230 citations. They conclude that these nanocrystalline calcium oxides were instrumental in biodiesel production. Verziu et al. (2008) study the catalytic activity for the production of sunflower and rapeseed oil-based biodiesel with three morphologically different nanocrystalline MgO materials prepared using simple, green, and reproducible methods in a paper with 161 citations. They conclude that these catalysts were instrumental in the efficient production of these biodiesel fuels.

Wang and Yang (2007) study the transesterification reaction of soybean oil with supercritical/subcritical methanol by Nano-MgO catalysts in a paper with 157 citations. They conclude that these catalysts were instrumental in the efficient production of these biodiesel fuels. Carmo et al. (2009) study the production of biodiesel by esterification of palmitic acid over mesoporous aluminosilicate Al-MCM-41 in a paper with 145 citations. They conclude that these nanocatalysts were instrumental in biodiesel production from palmitic acids.

4.4.2.5 Nonedible Oil-based Biodiesel Fuels

Deng et al. (2011) study the production of biodiesel from Jatropha oil catalyzed by a nanosized solid basic catalyst in a paper with 171 citations. They conclude that these nanocatalyts were instrumental for biodiesel production from Jatropha oils.

4.4.2.6 Waste Oil-based Biodiesel Fuels

Reddy et al. (2006) develop a route for the production of biodiesel via transesterification of soybean oil and poultry fat with methanol in quantitative conversions at room temperature using nanocrystalline calcium oxides as catalysts in a paper with 230 citations. They conclude that these nanocrystalline calcium oxides were instrumental in biodiesel production.

4.4.3 Crude Oils

4.4.3.1 Water-in-Oil Emulsions

Si et al. (2015) report a strategy to create 'fibrous, isotropically bonded elastic reconstructed' aerogels with a hierarchical cellular structure and super-elasticity by combining electrospun nanofibers and the freeze-shaping technique for separating surfactant-stabilized water-in-oil emulsions in a paper with 333 citations. They conclude that these aerogels were instrumental in the effective separation of oil/water emulsions. Arriagada and Osseo-Asare (1999) study the effect of an ammonia concentration on the region of existence of single-phase water-in-oil microemulsions for the system 'polyoxyethylene (5) nonylphenyl ether (NP-5)/cyclohexane/ammonium hydroxide' in a paper with 326 citations. They conclude that the effect of an ammonia concentration on the region of existence of single-phase water-in-oil microemulsions for this system was significant.

4.4.3.2 Oil Recovery

Suleimanov et al. (2011) study the nanofluids for enhanced oil recovery (EOR) in a paper with 151 citations. They use an aqueous solution of anionic surface-active agents with the addition of light non-ferrous metal nanoparticles. They conclude that the use of a developed nano-suspension results in a considerably increased EOR. Zhang et al. (2014) study the EOR using nanoparticle dispersions in a paper with 149 citations. They conclude that using nanoparticle dispersions was instrumental for EOR. Hendraningrat et al. (2013) study nanofluid enhanced oil recovery in a paper with 143 citations focusing on the parameters involved in the structural disjoining pressure mechanism, such as lowering interfacial tensions (IFTs) and altering wettability. They conclude that application of nanofluid EOR was an alternative EOR method.

4.4.3.3 Remediation of Crude Oils

Calcagnile et al. (2012) present a novel composite material based on commercially available polyurethane foams functionalized with colloidal 'superparamagnetic iron oxide nanoparticles' and submicrometer polytetrafluoroethylene particles, which can efficiently separate oil contaminants from water in a paper with 417 citations. They conclude that by simply moving them around oil-polluted waters using a magnet, they could absorb the floating oil from polluted regions, thereby purifying the water underneath. This low-cost process could easily be scaled up to clean large-area oil spills in water. Toyoda and Inagaki (2000) study the sorption behaviors of four kinds of heavy oils into exfoliated graphites with different bulk densities in a paper with 270 citations. They conclude that exfoliated graphite with a low bulk density was a promising material for the sorption and recovery of spilled heavy oil.

Yang et al. (2011) show that fluorous 'metal-organic frameworks' are highly hydrophobic porous materials with a high capacity and affinity to C_6-C_8 hydrocarbons of oil components in a paper with 240 citations. They conclude that there is great promise for FMOFs in applications such as the removal of organic pollutants from oil spills or ambient humid air, hydrocarbon storage and transportation, and water purification, under practical working conditions. Ge et al. (2017) report on a Joule-heated 'graphene-wrapped sponge' to clean up viscous crude oil at a high sorption speed in a paper with 184 citations. They conclude that this *in situ* Joule self-heated sorbent design would promote the practical application of hydrophobic and oleophilic oil sorbents in the clean-up of viscous crude-oil spills. Lin et al. (2012) report on nanoporous polystyrene fibers prepared via a one-step electrospinning process used as oil sorbents for oil spill cleanup in a paper 155 citations. They conclude that these nanoporous materials have great potential for use in wastewater treatment, oil accident remediation, and environmental protection.

4.4.4 Petrodiesel Fuels

4.4.4.1 Production

Kang et al. (2009) study ruthenium nanoparticles supported on carbon nanotubes as efficient catalysts for the selective conversion of synthesis gas to diesel fuel in a paper with 190 citations. They conclude that these catalysts were instrumental in biodiesel production.

4.4.4.2 Properties

Tyagi et al. (2008) improve the ignition properties of diesel fuels by investigating the influence of adding aluminum and aluminum oxide nanoparticles to diesel in a paper with 161 citations. They conclude that the addition of these nanoparticles enhanced the ignition properties of petrodiesel fuels.

4.5 CONCLUSION

This chapter has presented a review of the 25-most-cited article papers on nanotechnology applications in both biodiesel and petrodiesel fuels.

Table 4.1 provides information on the primary and secondary research fronts for both biodiesel fuels and petrodiesel fuels. It also provides information on the distribution of these research fronts in the article papers for nanotechnology applications on diesel fuels, the 100-most-cited papers, and over 121,000 population papers in the broader fields of petrodiesel and biodiesel fuels.

As this table shows, the research fronts of 'biodiesel fuels in general', 'feedstock-based biodiesel fuels', and 'petrodiesel fuels' comprise 28, 24, and 48% of these papers, respectively in this field.

In the first group of article papers, those on 'biooils', 'biodiesel fuels in general', and 'glycerol' comprise 4, 8, and 16% of these papers. In the second group of article papers, 'edible oil-based biodiesel fuels', 'nonedible oil-based biodiesel fuels', 'waste oil-based biodiesel fuels', and 'algae-based biodiesel fuels' comprise 16, 4, 4, and 0% of these papers, respectively. In the final group of article papers, 'crude oils', 'petrodiesel fuels in general', 'emissions from petrodiesel fuels', and the 'health impact of the emissions from petrodiesel fuels' comprise 40, 8, 0, and 0% of these papers, respectively.

The data show that the most-prolific research front has concerned the applications of nanotechnology to crude oils, including 'crude oil recovery', 'remediation of crude oils in the environment', and 'water-in-oil emulsions'. The applications of nanotechnology in 'glycerol applications' and 'edible oil-based biodiesel fuels' have also been prolific.

It is notable that there have been no prolific papers in the fields of 'biooil upgrading', 'biodiesel emissions', 'algal biodiesel fuels', 'crude oil properties', 'crude oil refining', 'petrodiesel desulfurization', 'power generation by petrodiesel fuels', 'petrodiesel fuel emissions', and the 'health impact of petrodiesel emissions'.

These prolific studies on nanotechnology applications in the fields of biooils, biodiesel fuels, crude oils, and petrodiesel fuels, presented in this chapter highlight the importance of nanotechnology in these fields. These studies also show the importance of nanomaterials as the nanocatalyts in these fields, ranging from 'crude oil recovery' and 'remediation of crude oils in the environment' to 'biooil production', 'biodiesel production', and the production of biochemical and bioenergy from glycerol, a by-product of biodiesel fuels.

It is recommended that similar studies are carried out for each research front as well.

ACKNOWLEDGMENTS

The contribution of the highly cited researchers in this field is greatly acknowledged.

REFERENCES

Ahmadun, F. R., A. Pendashteh, L. C. Abdullah, and et al. 2009. Review of technologies for oil and gas produced water treatment. *Journal of Hazardous Materials* 170:530–551.

Arriagada, F. J. and K. Osseo-Asare. 1999. Synthesis of nanosize silica in a nonionic water-in-oil microemulsion: Effects of the water/surfactant molar ratio and ammonia concentration. *Journal of Colloid and Interface Science* 211:210–220.

Atlas, R. M. 1981. Microbial degradation of petroleum hydrocarbons: An environmental perspective. *Microbiological Reviews* 45: 180–209.

Babich, I. V. and J. A. Moulijn. 2003. Science and technology of novel processes for deep desulfurization of oil refinery streams: A review. *Fuel* 82: 607–631.

Birch, M. E. and R. A. Cary. 1996. Elemental carbon-based method for monitoring occupational exposures to particulate diesel exhaust. *Aerosol Science and Technology* 25: 221–241.

Bridgwater, A. V. and G. V. C. Peacocke. 2000. Fast pyrolysis processes for biomass. *Renewable & Sustainable Energy Reviews* 4: 1–73.

Bridgwater, A. V., D. Meier, and D. Radlein. 1999. An overview of fast pyrolysis of biomass. *Organic Geochemistry* 30: 1479–1493.

Calcagnile, P., D. Fragouli, I. S. Bayer, and et al. 2012. Magnetically driven floating foams for the removal of oil contaminants from water. *ACS Nano* 6: 5413–5419.

Carmo, A. C., L. K. C. de Souza, C. E. F. da Costa, and et al. 2009 Production of biodiesel by esterification of palmitic acid over mesoporous aluminosilicate Al-MCM-41. *Fuel* 88: 461–468.

Chisti, Y. 2007. Biodiesel from microalgae. *Biotechnology Advances* 25: 294–306.

Clacens, J. M., Y. Pouilloux, and J. Barrault. 2002. Selective etherification of glycerol to polyglycerols over impregnated basic MCM-41 type mesoporous catalysts. *Applied Catalysis A-General* 227: 181–190.

Czernik, S. and A. V. Bridgwater. 2004. Overview of applications of biomass fast pyrolysis oil. *Energy & Fuels* 18: 590–598.

Deng, X., Z. Fang, Y. H. Liu, and C. L. Yu. 2011. Production of biodiesel from Jatropha oil catalyzed by nanosized solid basic catalyst. *Energy* 36: 777–784.

Gao, F., J. H. Kwak, J. Szanyi, and C. H. F. Peden. 2013. Current understanding of Cu-exchanged chabazite molecular sieves for use as commercial diesel engine $DeNO_x$ catalysts. *Topics in Catalysis* 56:1441–1459.

Ge, J., L. A. Shi, Y. C. Wang, and et al. 2017. Joule-heated graphene-wrapped sponge enables fast clean-up of viscous crude-oil spill. *Nature Nanotechnology* 12: 434–440.

Geim, A. K. and K. S. Novoselov. 2007. The rise of graphene. *Nature Materials* 6:183–191.

Hashemi, R., N. N. Nassar, and P. P. Almao. 2014. Nanoparticle technology for heavy oil in-situ upgrading and recovery enhancement: Opportunities and challenges. *Applied Energy* 133: 374–387.

Hendraningrat, L., S. D. Li, and O. Torster. 2013. A coreflood investigation of nanofluid enhanced oil recovery. *Journal of Petroleum Science and Engineering* 111: 128–138.

Hill, J., E. Nelson, D. Tilman, S. Polasky, and D. Tiffany. 2006. Environmental, economic, and energetic costs and benefits of biodiesel and ethanol biofuels. *Proceedings of the National Academy of Sciences of the United States of America* 103:11206–11210.

Hong, W., C. S. Shang, J. Wang, and E. K. Wang. 2015. Bimetallic PdPt nanowire networks with enhanced electrocatalytic activity for ethylene glycol and glycerol oxidation. *Energy & Environmental Science* 8: 2910–2915.

Huang, Z. W., F. Cui, H. X. Kang, and et al. 2008. Highly dispersed silica-supported copper nanoparticles prepared by precipitation-gel method: A simple but efficient and stable catalyst for glycerol hydrogenolysis. *Chemistry of Materials* 20:5090–5099.

Iijima, S. 1991. Helical microtubules of graphitic carbon. *Nature* 354: 56–58.

Iliopoulou, E. F., E. V. Antonakou, S. A. Karakoulia, and et al. 2007. Catalytic conversion of biomass pyrolysis products by mesoporous materials: Effect of steam stability and acidity of Al-MCM-41 catalysts. *Chemical Engineering Journal* 134: 51–57.

Kang, J. C., S. L. Zhang, Q. H. Zhang, and Y. Wang. 2009. Ruthenium nanoparticles supported on carbon nanotubes as efficient catalysts for selective conversion of synthesis gas to diesel fuel. *Angewandte Chemie-International Edition* 48: 2565–2568.

Khalili, N. R., P. A. Scheff, and T. M. Holsen. 1995. PAH source fingerprints for coke ovens, diesel and gasoline-engines, highway tunnels, and wood combustion emissions. *Atmospheric Environment* 29: 533–542.

Konur, O. 2000. Creating enforceable civil rights for disabled students in higher education: An institutional theory perspective. *Disability & Society* 15: 1041–1063.

Konur, O. 2002a. Access to Nursing Education by disabled students: Rights and duties of nursing programs. *Nurse Education Today* 22: 364–374.

Konur, O. 2002b. Assessment of disabled students in higher education: Current public policy issues. *Assessment and Evaluation in Higher Education* 27: 131–152.

Konur, O. 2002c. Access to employment by disabled people in the UK: Is the Disability Discrimination Act working? *International Journal of Discrimination and the Law* 5: 247–279.

Konur, O. 2006a. Participation of children with dyslexia in compulsory education: Current public policy issues. *Dyslexia* 12: 51–67.

Konur, O. 2006b. Teaching disabled students in Higher Education. *Teaching in Higher Education* 11: 351–363.

Konur, O. 2007a. A judicial outcome analysis of the Disability Discrimination Act: A windfall for the employers? *Disability & Society* 22: 187–204.

Konur, O. 2007b. Computer-assisted teaching and assessment of disabled students in higher education: The interface between academic standards and disability rights. *Journal of Computer Assisted Learning* 23: 207–219.

Konur, O. 2016a. Scientometric overview in nanobiodrugs. In *Nanoarchitectonics for Smart Delivery and Drug Targeting*, A. M. Holban and A.M. Grumezescu, ed., 405–428. Amsterdam: Elsevier.

Konur, O. 2016b. Scientometric overview regarding nanoemulsions used in the food industry. In *Emulsions: Nanotechnology in the Agri-Food Industry*, A. M. Grumezescu, ed., 689–711. Amsterdam: Elsevier.

Konur, O. 2016c. Scientometric overview regarding the nanobiomaterials in antimicrobial therapy. In *Nanobiomaterials in Antimicrobial Therapy,* A. M. Grumezescu, ed., 511–535. Amsterdam: Elsevier.

Konur, O. 2016d. Scientometric overview regarding the nanobiomaterials in dentistry. In *Nanobiomaterials in Dentistry,* A. M. Grumezescu, ed., 425–453. Amsterdam: Elsevier.

Konur, O. 2016e. Scientometric overview regarding the surface chemistry of nanobiomaterials. In *Surface Chemistry of Nanobiomaterials,* A. M. Grumezescu, ed., 463–486. Amsterdam: Elsevier.

Konur, O. 2016f. The scientometric overview in cancer targeting. In *Nanoarchitectonics for Smart Delivery and Drug Targeting*, A. M. Holban and A. Grumezescu, ed., 871–895. Amsterdam: Elsevier.

Konur, O. 2017a. Recent citation classics in antimicrobial nanobiomaterials. In *Nanostructures for Antimicrobial Therapy,* A. Ficai and A. M. Grumezescu, ed., 669–685. Amsterdam: Elsevier.

Konur, O. 2017b. Scientometric overview in nanopesticides. In *New Pesticides and Soil Sensors,* A. M. Grumezescu, ed. 719–744. Amsterdam: Elsevier.

Konur, O. 2017c. Scientometric overview regarding oral cancer nanomedicine. In *Nanostructures for Oral Medicine*, E. Andronescu, A. M. Grumezescu, ed., 939–962. Amsterdam: Elsevier.

Konur, O. 2017d. Scientometric overview regarding water nanopurification. In *Water Purification,* A. M. Grumezescu, ed., 693–716. Amsterdam: Elsevier.

Konur, O. 2017e. Scientometric overview in food nanopreservation. In *Food Preservation,* A. M. Grumezescu, ed., 703–729. Amsterdam: Elsevier.

Konur, O. 2019. Nanotechnology applications in food: A scientometric overview. In *Nanoscience for Sustainable Agriculture*, R. N., Pudake, N. Chauhan, and C. Kole, ed., 683–711. Cham: Springer.

Konur, O., ed. 2021a. *Handbook of Biodiesel and Petrodiesel Fuels: Science, Technology, Health, and Environment.* Boca Raton, FL: CRC Press.

Konur, O., ed. 2021b. *Handbook of Biodiesel and Petrodiesel Fuels: Science, Technology, Health, and Environment. Volume 1. Biodiesel Fuels: Science, Technology, Health, and Environment.* Boca Raton, FL: CRC Press.

Konur, O., ed. 2021c. *Handbook of Biodiesel and Petrodiesel Fuels: Science, Technology, Health, and Environment. Volume 2. Biodiesel Fuels based on the Edible and Nonedible Feedstocks, Wastes, and Algae: Science, Technology, Health, and Environment.* Boca Raton, FL: CRC Press.

Konur, O., ed. 2021d. *Handbook of Biodiesel and Petrodiesel Fuels: Science, Technology, Health, and Environment. Volume 3. Petrodiesel Fuels: Science, Technology, Health, and Environment.* Boca Raton, FL: CRC Press.

Konur, O. 2021e. Biodiesel and petrodiesel fuels: Science, technology, health, and environment. In *Handbook of Biodiesel and Petrodiesel Fuels: Science, Technology, Health, and Environment. Volume 1. Biodiesel Fuels: Science, Technology, Health, and Environment*, ed. O. Konur. Boca Raton, FL: CRC Press.

Konur, O. 2021f. Biodiesel and petrodiesel fuels: A scientometric review of the research. In *Handbook of Biodiesel and Petrodiesel Fuels: Science, Technology, Health, and Environment. Volume 1. Biodiesel Fuels: Science, Technology, Health, and Environment*, ed. O. Konur. Boca Raton, FL: CRC Press.

Konur, O. 2021g. Biodiesel and petrodiesel fuels: A review of the research. In *Handbook of Biodiesel and Petrodiesel Fuels: Science, Technology, Health, and Environment. Volume 1. Biodiesel Fuels: Science, Technology, Health, and Environment*, ed. O. Konur. Boca Raton, FL: CRC Press.

Konur, O. 2021h Nanotechnology applications in the diesel fuels and the related research fields: A review of the research. In *Handbook of Biodiesel and Petrodiesel Fuels: Science, Technology, Health, and Environment. Volume 1. Biodiesel Fuels: Science, Technology, Health, and Environment*, ed. O. Konur. Boca Raton, FL: CRC Press.

Konur, O. 2021i. Biooils: A scientometric review of the research. In *Handbook of Biodiesel and Petrodiesel Fuels: Science, Technology, Health, and Environment. Volume 1. Biodiesel Fuels: Science, Technology, Health, and Environment*, ed. O. Konur. Boca Raton, FL: CRC Press.

Konur, O. 2021j. Characterization and properties of biooils: A review of the research. In *Handbook of Biodiesel and Petrodiesel Fuels: Science, Technology, Health, and Environment. Volume 1. Biodiesel Fuels: Science, Technology, Health, and Environment*, ed. O. Konur. Boca Raton, FL: CRC Press.

Konur, O. 2021k. Biomass pyrolysis and pyrolysis oils: A review of the research. In *Handbook of Biodiesel and Petrodiesel Fuels: Science, Technology, Health, and Environment. Volume 1. Biodiesel Fuels: Science, Technology, Health, and Environment*, ed. O. Konur. Boca Raton, FL: CRC Press.

Konur, O. 2021l. Biodiesel fuels: A scientometric review of the research. In *Handbook of Biodiesel and Petrodiesel Fuels: Science, Technology, Health, and Environment. Volume 1. Biodiesel Fuels: Science, Technology, Health, and Environment*, ed. O. Konur. Boca Raton, FL: CRC Press.

Konur, O. 2021m. Glycerol: A scientometric review of the research. In *Handbook of Biodiesel and Petrodiesel Fuels: Science, Technology, Health, and Environment. Volume 1. Biodiesel Fuels: Science, Technology, Health, and Environment*, ed. O. Konur. Boca Raton, FL: CRC Press.

Konur, O. 2021n. Propanediol production from glycerol: A review of the research. In *Handbook of Biodiesel and Petrodiesel Fuels: Science, Technology, Health, and Environment. Volume 1. Biodiesel Fuels: Science, Technology, Health, and Environment*, ed. O. Konur Boca Raton, FL: CRC Press.

Konur, O. 2021o. Edible oil-based biodiesel fuels: A scientometric review of the research. In *Handbook of Biodiesel and Petrodiesel Fuels: Science, Technology, Health, and Environment. Volume 2. Biodiesel Fuels Based on the Edible and Nonedible Feedstocks, Wastes, and Algae: Science, Technology, Health, and Environment,* ed. O. Konur. Boca Raton, FL: CRC Press.

Konur, O. 2021p. Palm oil-based biodiesel fuels: A review of the research. In *Handbook of Biodiesel and Petrodiesel Fuels: Science, Technology, Health, and Environment. Volume 2. Biodiesel Fuels based on the Edible and Nonedible Feedstocks, Wastes, and Algae*, ed. O. Konur. Boca Raton, FL: CRC Press.

Konur, O. 2021q. Rapeseed oil-based biodiesel fuels: A review of the research. In *Handbook of Biodiesel and Petrodiesel Fuels: Science, Technology, Health, and Environment. Volume 2. Biodiesel Fuels based on the Edible and Nonedible Feedstocks, Wastes, and Algae,* ed. O. Konur. Boca Raton, FL: CRC Press.

Konur, O. 2021r. Nonedible oil-based biodiesel fuels: A scientometric review of the research. In *Handbook of Biodiesel and Petrodiesel Fuels: Science, Technology, Health, and Environment. Volume 2. Biodiesel Fuels based on the Edible and Nonedible Feedstocks, Wastes, and Algae: Science, Technology, Health, and Environment*, ed. O. Konur. Boca Raton, FL: CRC Press.

Konur, O. 2021s. Waste oil-based biodiesel fuels: A scientometric review of the research. In *Handbook of Biodiesel and Petrodiesel Fuels: Science, Technology, Health, and Environment. Volume 2. Biodiesel Fuels Based on the Edible and Nonedible Feedstocks, Wastes, and Algae: Science, Technology, Health, and Environment*, ed. O. Konur. Boca Raton, FL: CRC Press.

Konur, O. 2021t. Algal biodiesel fuels: A scientometric review of the research. In *Handbook of Biodiesel and Petrodiesel Fuels: Science, Technology, Health, and Environment. Volume 2. Biodiesel Fuels Based on the Edible and Nonedible Feedstocks, Wastes, and Algae: Science, Technology, Health, and Environment*, ed. O. Konur. Boca Raton, FL: CRC Press.

Konur, O. 2021u. Algal biomass production for biodiesel production: A review of the research. In *Handbook of Biodiesel and Petrodiesel Fuels: Science, Technology, Health, and Environment. Volume 2. Biodiesel Fuels Based on the Edible and Nonedible Feedstocks, Wastes, and Algae*, ed. O. Konur Boca Raton, FL: CRC Press. February 23, 2020.

Konur, O. 2021v. Algal biomass production in wastewaters for biodiesel production: A review of the research. In *Handbook of Biodiesel and Petrodiesel Fuels: Science, Technology, Health, and Environment. Volume 2. Biodiesel Fuels based on the Edible and Nonedible Feedstocks, Wastes, and Algae*, ed. O. Konur. Boca Raton, FL: CRC Press. February 23, 2020.

Konur, O. 2021x. Algal lipid production for biodiesel production: A review of the research. In *Handbook of Biodiesel and Petrodiesel Fuels: Science, Technology, Health, and Environment. Volume 2. Biodiesel Fuels based on the Edible and Nonedible Feedstocks, Wastes, and Algae*, ed. O. Konur Boca Raton, FL: CRC Press.

Konur, O. 2021y. Crude oils: A scientometric review of the research. In *Handbook of Biodiesel and Petrodiesel Fuels: Science, Technology, Health, and Environment. Volume 3. Petrodiesel Fuels: Science, Technology, Health, and Environment*, ed. O. Konur. Boca Raton, FL: CRC Press.

Konur, O. 2021z. Petrodiesel fuels: A scientometric review of the research. In *Handbook of Biodiesel and Petrodiesel Fuels: Science, Technology, Health, and Environment. Volume 3. Petrodiesel Fuels: Science, Technology, Health, and Environment*, ed. O. Konur. Boca Raton, FL: CRC Press.

Konur, O. 2021aa. Bioremediation of petroleum hydrocarbons in the contaminated soils: A review of the research. In *Handbook of Biodiesel and Petrodiesel Fuels: Science, Technology, Health, and Environment. Volume 3. Petrodiesel Fuels: Science, Technology, Health, and Environment*, ed. O. Konur. Boca Raton, FL: CRC Press.

Konur, O. 2021ab. Desulfurization of diesel fuels: A review of the research. In *Handbook of Biodiesel and Petrodiesel Fuels: Science, Technology, Health, and Environment. Volume 3. Petrodiesel Fuels: Science, Technology, Health, and Environment*, ed. O. Konur. Boca Raton, FL: CRC Press.

Konur, O. 2021ac. Diesel fuel exhaust emissions: A scientometric review of the research. In *Handbook of Biodiesel and Petrodiesel Fuels: Science, Technology, Health, and Environment. Volume 3. Petrodiesel Fuels: Science, Technology, Health, and Environment*, ed. O. Konur. Boca Raton, FL: CRC Press.

Konur, O. 2021ad. The adverse health and safety impact of diesel fuels: A scientometric review of the research. In *Handbook of Biodiesel and Petrodiesel Fuels: Science, Technology, Health, and Environment. Volume 3. Petrodiesel Fuels: Science, Technology, Health, and Environment*, ed. O. Konur. Boca Raton, FL: CRC Press.

Konur, O. 2021ae. Respiratory illnesses caused by the diesel fuel exhaust emissions: A review of the research. In *Handbook of Biodiesel and Petrodiesel Fuels: Science, Technology, Health, and Environment. Volume 3. Petrodiesel Fuels: Science, Technology, Health, and Environment*, ed. O. Konur. Boca Raton, FL: CRC Press.

Konur, O. 2021af. Cancer caused by the diesel fuel exhaust emissions: A review of the research. In *Handbook of Biodiesel and Petrodiesel Fuels: Science, Technology, Health, and Environment. Volume 3. Petrodiesel Fuels: Science, Technology, Health, and Environment*, ed. O. Konur. Boca Raton, FL: CRC Press.

Konur, O. 2021ag. Cardiovascular and other illnesses caused by the diesel fuel exhaust emissions: A review of the research. In *Handbook of Biodiesel and Petrodiesel Fuels: Science, Technology, Health, and Environment. Volume 3. Petrodiesel Fuels: Science, Technology, Health, and Environment*, ed. O. Konur. Boca Raton, FL: CRC Press.

Lin, J. Y., Y. W. Shang, B. Ding, and et al. 2012. Nanoporous polystyrene fibers for oil spill cleanup. *Marine Pollution Bulletin* 64: 347–352.

Liu, R., X. Q. Wang, X. Zhao, and P. Y. Feng. 2008. Sulfonated ordered mesoporous carbon for catalytic preparation of biodiesel. *Carbon* 46:1664–1669.

Ma, F. R. and M. A. Hanna. 1999. Biodiesel production: A review. *Bioresource Technology* 70: 1–15.

Mata, T. M., A. A. Martins, and N. S. Caetano. 2010. Microalgae for biodiesel production and other applications: A review. *Renewable & Sustainable Energy Reviews* 14: 217–232.

Neto, A. H. C., F. Guinea, N. M. R. Peres, and et al. 2009. The electronic properties of graphene. *Reviews of Modern Physics* 81: 109–162.

North, D. C. 1991a. *Institutions, Institutional Change and Economic Performance*. Cambridge, Mass.: Cambridge University Press.

North, D. C. 1991b. Institutions. *Journal of Economic Perspectives* 5: 97–112.

Novoselov, K. S., A. K. Geim, S. V. Morozov, and et al. 2004. Electric field effect in atomically thin carbon films. *Science* 306: 666–669.

Novoselov, K. S., A. K. Geim, S. V. Morozov, and et al. 2005. Two-dimensional gas of massless Dirac fermions in graphene. *Nature* 438: 197–200.

Perron, P. 1989. The great crash, the oil price shock, and the unit root hypothesis. *Econometrica: Journal of the Econometric Society* 57: 1361–1401.

Reddy, C., V. Reddy, R. Oshel, J. G. Verkade. 2006. Room-temperature conversion of soybean oil and poultry fat to biodiesel catalyzed by nanocrystalline calcium oxides. *Energy & Fuels* 20: 1310–1314.

Rogge, W. F., L. M. Hildemann, M. A. Mazurek, G. R. Cass, B. R. T. Simoneit. 1993. Sources of fine organic aerosol. 2. Noncatalyst and catalyst-equipped automobiles and heavy-duty diesel trucks. *Environmental Science & Technology* 27: 636–651.

Sajith, V., C. B. Sobhan, G. P. Peterson. 2010. Experimental investigations on the effects of cerium oxide nanoparticle fuel additives on biodiesel. *Advances in Mechanical Engineering* 2: 581407.

Shamsi-Jazeyi, H., C. A. Miller, M. S. Wong, J. M. Tour, and R. Verduzco. 2014. Polymer-coated nanoparticles for enhanced oil recovery. *Journal of Applied Polymer Science* 131: 40576.

Si, Y., Q. X. Fu, X. Q. Wang, and et al. 2015. Superelastic and superhydrophobic nanofiber-assembled cellular aerogels for effective separation of oil/water emulsions. *ACS Nano* 9: 3791–3799.

Simoes, M., S. Baranton, and C. Coutanceau. 2010. Electro-oxidation of glycerol at Pd based nano-catalysts for an application in alkaline fuel cells for chemicals and energy cogeneration. *Applied Catalysis B-Environmental* 93: 354–362.

Song, C. and X. L. Ma. 2003. New design approaches to ultra-clean diesel fuels by deep desulfurization and deep dearomatization. *Applied Catalysis B-Environmental* 41:207–238.

Song, C. S. 2003. An overview of new approaches to deep desulfurization for ultra-clean gasoline, diesel fuel and jet fuel. *Catalysis Today* 86: 211–263.

Suleimanov, B. A., F. S. Ismailov, and E. F. Veliyev. 2011. Nanofluid for enhanced oil recovery. *Journal of Petroleum Science and Engineering* 78: 431–437.

Toyoda, M. and M. Inagaki. 2000. Heavy oil sorption using exfoliated graphite: New application of exfoliated graphite to protect heavy oil pollution. *Carbon* 38: 199–210.

Tyagi, H., P. E. Phelan, R. Prasher, and et al. 2008. Increased hot-plate ignition probability for nanoparticle-laden diesel fuel. *Nano Letters* 8: 1410–1416.

Verma, M. L., C. J. Barrow, and M. Puri. 2013. Nanobiotechnology as a novel paradigm for enzyme immobilisation and stabilisation with potential applications in biodiesel production. *Applied Microbiology and Biotechnology* 97: 23–39.

Verziu, M., B. Cojocaru, J. C. Hu, and et al. 2008. Sunflower and rapeseed oil transesterification to biodiesel over different nanocrystalline MgO catalysts. *Green Chemistry* 10: 373–381.

Wang, L. Y. and J. C. Yang. 2007. Transesterification of soybean oil with nano-MgO or not in supercritical and subcritical methanol. *Fuel* 86: 328–333.

Yang, C., U. Kaipa, Q. Z. Mather, and et al. 2011. Fluorous metal-organic frameworks with superior adsorption and hydrophobic properties toward oil spill cleanup and hydrocarbon storage. *Journal of the American Chemical Society* 133:18094-18097.

Zhang, H., A. Nikolov, and D. Wasan. 2014. Enhanced oil recovery (EOR) using nanoparticle dispersions: Underlying mechanism and imbibition experiments. *Energy & Fuels* 28: 3002–3009.

Part II

Biooils

5 Biooils
A Scientometric Review of the Research

Ozcan Konur

CONTENTS

5.1 INTRODUCTION

Crude oil-based fuels, such as diesel fuels, have been primary sources of energy and fuels (Chisti, 2007, 2008; Konur, 2012g, 2015; Lapuerta et al., 2008; van Gerpen, 2005). However, significant public concerns about the sustainability, price fluctuations, and adverse environmental impact of crude oils have emerged since the 1970s (Ahmadun et al., 2009; Atlas, 1981; Babich and Moulijn, 2003; Kilian, 2009; Moldowan et al., 1985; Perron, 1989). Thus, biooils have emerged as an alternative to crude oils in recent decades (Bridgwater and Peacocke, 2000; Czernik and Bridgwater, 2004; Evans and Milne, 1987; Gallezot, 2012; Mohan et al., 2006; Yaman, 2004; Zhang et al., 2007).

However, for the efficient progression of the research in this field, it is necessary to develop efficient incentive structures for the primary stakeholders and to inform these stakeholders about the research (Konur, 2000, 2002a–c, 2006a–b, 2007a–b; North, 1991a–b).

Scientometric analysis offers ways to evaluate the research in a respective field (Garfield, 1955, 1972). This method has been used to evaluate research in a number of fields (Konur, 2011, 2012a–n, 2015, 2016a–f, 2017a–f, 2018a–b, 2019a–b). However, there has been no scientometric study of this field.

This chapter presents a study on the scientometric evaluation of the research in this field using two datasets. The first data set includes the 100-most-cited papers (*n* = 100 sample papers) whilst the second set includes population papers (*n* = over 20,000 population papers) published between 1980 and 2019. It complements the chapter on crude oils as well as the other chapters in this handbook (Konur 2021a–ag).

The data on the indices, document types, authors, institutions, funding bodies, source titles, 'Web of Science' subject categories, keywords, research fronts, and citation impacts are presented and discussed.

5.2 MATERIALS AND METHODOLOGY

The search for the literature was carried out in the 'Web of Science' (WOS) database in January 2020. It contains the 'Science Citation Index-Expanded' (SCI-E), the Social Sciences Citation Index' (SSCI), the 'Book Citation Index-Science' (BCI-S), the 'Conference Proceedings Citation Index-Science' (CPCI-S), the 'Emerging Sources Citation Index' (ESCI), the 'Book Citation Index-Social Sciences and Humanities' (BCI-SSH), the 'Conference Proceedings Citation Index-Social Sciences and Humanities' (CPCI-SSH), and the 'Arts and Humanities Citation Index' (A&HCI).

The keywords for the search of the literature are collated from the screening of abstract pages for the first 1,000 highly cited papers. This keyword set is provided in the Appendix.

Two datasets are used for this study. The highly cited 100 papers comprise the first dataset (sample dataset, *n* = 100 papers) whilst all the papers form the second dataset (population dataset, *n* = over 20,000 papers).

The data on the indices, document types, publication years, institutions, funding bodies, source titles, countries, 'Web of Science' subject categories, citation impacts, keywords, and research fronts are collated from these datasets. The key findings are provided in the relevant tables and one figure, supplemented with explanatory notes in the text. The findings are discussed and a number of conclusions are drawn and recommendations for further study are made.

5.3 RESULTS

5.3.1 Indices and Documents

There are over 24,800 papers related to biooils in the 'Web of Science' as of January 2020. This original population dataset is refined by document type (article, review, book chapter, book, editorial material, note, and letter) and language (English), resulting in over 20,100 papers comprising over 81% of the original population dataset.

The primary index is SCI-E for both the sample and population papers. About 96% of the population papers are indexed by the SCI-E database. Additionally 6.3%, 2.1%, and 2.1% of these papers are indexed by CPCI-S, ESCI, and BCI-S databases, respectively. The papers on the social and humanitarian aspects of this field are relatively negligible with 0.5% and 0.1% of the population papers indexed by the SSCI and A&HCI, respectively.

Brief information on the document types for both datasets is provided in Table 5.1. The key finding is that article types of documents are the primary documents for the population papers whilst reviews form 42% of the sample papers. Articles are under-represented by –35.1% whilst reviews are over-represented by 37.4% in the sample papers.

5.3.2 Authors

Brief information about the most-prolific eight authors with at least three sample papers each is provided in Table 5.2. Around 310 and 37,000 authors contribute to sample and population papers, respectively.

The most-prolific author is 'George W. Huber' with seven sample papers, working primarily on 'biomass pyrolysis' and 'biooil upgrading'. The other prolific researchers are 'Stefan Czernik' and 'Anthony V. Bridgwater' with five and four sample papers, respectively. These top three authors have the most impact with a 15.4% publication surplus altogether.

The most-prolific institution for these top authors is the 'University of Massachusetts' of the USA with two authors. In total, 16 institutions house these top authors.

It is notable that four of these top researchers are listed in 'Highly Cited Researchers' (HCR) in 2019 (Clarivate Analytics, 2019; Docampo and Cram, 2019).

The most-prolific country for these top authors is the USA with five authors. The other countries are Spain, Turkey, and the UK. In total, four countries contribute to these top papers.

TABLE 5.1
Document Types

	Document Type	Sample Dataset (%)	Population Dataset (%)	Difference (%)
1	Article	58	93.1	–35.1
2	Review	42	4.6	37.4
3	Book chapter	0	1.9	–1.9
4	Proceeding paper	5	6.2	–1.2
5	Editorial material	0	0.8	–0.8
6	Letter	0	0.3	–0.3
7	Book	0	0.1	–0.1
8	Note	0	1.1	–1.1

TABLE 5.2
Authors

	Author	Sample Papers (%)	Population Papers (%)	Surplus (%)	Institution	Country	Research Front
1	Huber, George W.*	7	0.2	6.8	Univ. Massachusetts	USA	Pyrolysis and upgrading
2	Czernik, Stefan	5	0.1	4.9	Natl. Renew. Energ. Lab.	USA	Pyrolysis
3	Bridgwater, Anthony V.	4	0.3	3.7	Aston Univ.	UK	Pyrolysis
4	Corma, Avelino*	3	0.1	2.9	Univ. Polytech. Valencia	Spain	Upgrading
5	Demirbas, Ayhan	3	0.3	2.7	Selcuk Univ.	Turkey	Pyrolysis
6	Dumesic, James A.*	3	0.1	2.9	Univ. Wisconsin	USA	Biomass conversion, upgrading
7	Mohan, Dinesh*	3	0.1	2.9	Mississippi Univ.	USA	Pyrolysis
8	Tompsett, Geoffrey A.	3	0.1	2.9	Univ. Massachusetts	USA	Pyrolysis

*Highly cited researchers in 2019 (Clarivate Analytics, 2019).

There are three key research fronts for these top researchers: 'biomass conversion to biooil in general', 'biomass pyrolysis', and 'biooil upgrading to fuels'. The top research front is 'biomass pyrolysis' with six authors. The other prolific research front is 'biooil upgrading' with three authors.

It is further notable that there is a significant gender deficit among these top authors as they are all male (Lariviere et al., 2013; Xie and Shauman, 1998).

5.3.3 Publication Years

Information about publication years for both datasets is provided in Figure 5.1.

This figure shows that 14%, 33%, 40%, and 13% of the sample papers and 10.3%, 13.0%, 21.1%, and 55.1% of the population papers were published in the 1980s, 1990s, 2000s, and 2010s, respectively.

Similarly, the most-prolific publication years for the sample dataset are 1998, 2000, and 2003 with seven papers each. On the other hand, the most-prolific publication years for the population dataset are 2015, 2016, 2017, 2018, and 2019 with at least 6% of the population papers each. It is notable that there is a sharply rising trend for the population papers, particularly in the 2010s.

5.3.4 Institutions

Brief information on the top ten institutions with at least 3% of the sample papers each is provided in Table 5.3. In total, around 140 and 6,400 institutions contribute to the sample and population papers, respectively.

These top institutions publish 44% and 12.4% of the sample and population papers, respectively. The top institution is the 'National Renewable Energy

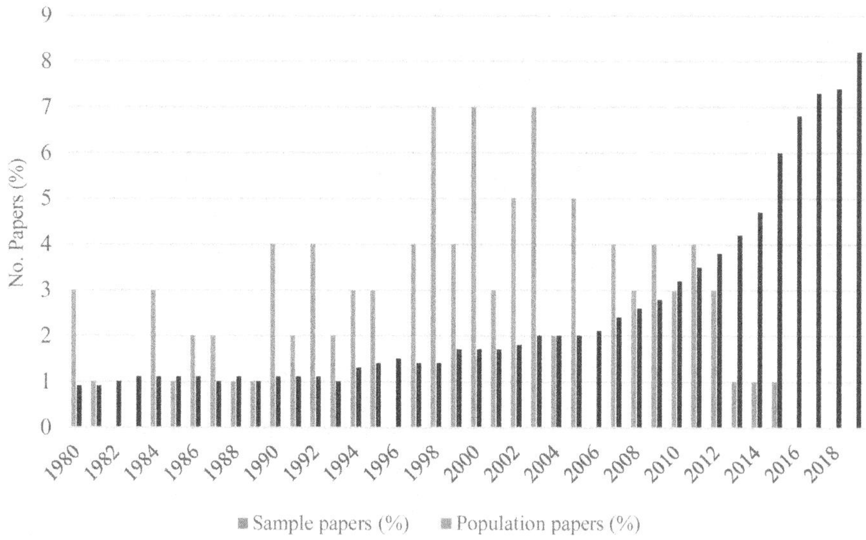

FIGURE 5.1 The research output between 1980 and 2019.

TABLE 5.3
Institutions

	Institution	Country	No. of Sample Papers (%)	No. of Population Papers (%)	Difference (%)
1	Natl. Renew. Energ. Lab.	USA	7	1.0	6.0
2	Univ. Massachusetts	USA	6	0.3	5.7
3	Super. Counc. Sci. Invest.	Spain	5	1.5	3.5
4	Univ. Wisconsin	USA	5	0.4	4.6
5	Aston Univ.	UK	4	0.7	3.3
6	Zhejiang Univ.	China	4	1.1	2.9
7	Sci. Res. Natl. Ctr.	France	4	2.0	2.0
8	Chinese Acad. Sci.	China	3	4.0	–1.0
9	Counc. Sci. Ind. Res.	India	3	1.1	1.9
10	Univ. Polytech. Valencia	Spain	3	0.3	2.7

Laboratory' of the USA with seven sample papers and a 6% publication surplus. The other top institutions are the 'University of Massachusetts' of the USA, the 'Superior Council of Scientific Investigations' of Spain, and the 'University of Wisconsin' of the USA with 6, 5, and 5 sample papers, respectively.

The most-prolific country for these top institutions is the USA with three. The other prolific countries are China and Spain with two institutions each. France, India, and the UK are the other contributing countries with one institution each.

The institutions with the most impact are the 'National Renewable Energy Laboratory', the 'University of Massachusetts', and the 'University of Wisconsin' with at least a 4.6% publication surplus each. On the other hand, the institutions with the least impact are the 'Chinese Academy of Sciences', the 'Council of Scientific Industrial Research', and the 'Scientific Research National Center' with at least a –1% publication surplus/deficit each.

It is notable that some institutions with a heavy presence in the population papers are under-represented in the sample papers: the 'Russian Academy of Sciences', the 'University of Science Technology of China', the 'Indian Institute of Technology', 'Huazhong University of Science Technology', the 'University of Basque Country', the US 'Department of Agriculture', and 'Southeast University China' with at least a 1% presence in the population papers each.

5.3.5 Funding Bodies

Brief information about the top eight funding bodies with at least 2% of the sample papers each is provided in Table 5.4. It is significant that only 38.0% and 54.7% of the sample and population papers declare any funding, respectively.

The top funding body is the US 'Department of Energy', funding 10.0% and 2.7% of the sample and population papers, respectively, with a 7.3% publication surplus. This top funding body is followed by the US 'Natural Science Foundation' with seven sample papers. The other prolific funding bodies are the US 'Defense Advanced

TABLE 5.4
Funding Bodies

	Institution	Country	No. of Sample Papers (%)	No. of Population Papers (%)	Difference (%)
1	Dept. Energy	USA	10	2.7	7.3
2	Natl. Science Found.	USA	7	1.7	5.3
3	Defense Adv. Res. Proj.	USA	4	0.1	3.9
4	Natl. Nat. Sci. Found. China	China	4	12.2	–8.2
5	Dept. Defense	USA	4	0.2	3.8
6	Natl. Basic Res. Prog. China	China	3	2.4	0.6
7	Eng. Phys. Sci. Res. Counc.	UK	2	1.0	1.0
8	Min. Ed. Cult. Sci.	Netherlands	2	0.7	1.3

Research Projects', the 'National Natural Science Foundation of China', and the US 'Department of Defense' with four sample papers each.

It is notable that some top funding agencies for the population studies do not enter this top funding body list. Some of them are the 'Fundamental Research Funds for the Central Universities' of China, the 'European Union', the 'National Council for Scientific and Technological Development' and 'CAPES' of Brazil, the 'China Postdoctoral Science Foundation', and the 'China Scholarship Council' with at least 1% of the population papers each.

It is notable that the most-prolific country for these top funding bodies is the USA with four bodies, whilst China has two.

The US 'Department of Energy' and the US 'National Science Foundation' are the funding bodies with the most impact, whilst the 'National Natural Science Foundation' and the 'National Basic Research Program of China' are those with the least impact.

5.3.6 Source Titles

Brief information about the top 13 source titles with at least three sample papers each is provided in Table 5.5. In total, 40 and over 1,380 source titles publish the sample and population papers, respectively. On the other hand, these top 13 journals publish 67% and 7.6% of the sample and population papers, respectively.

The top journal is 'Bioresource Technology', publishing 11 sample papers with a 5.2% publication surplus. The other top journals are 'Energy Conversion and Management', the 'Journal of Analytical and Applied Pyrolysis', and 'Energy Fuels' with seven, seven, and six sample papers, respectively.

Although these journals are indexed by 13 subject categories, the top categories are 'Energy Fuels', 'Chemical Engineering', and 'Thermodynamics' with nine, five, and three journals, respectively.

The journals with the most impact are 'Energy Conversion and Management', 'Bioresource Technology', and 'Chemical Society Reviews' with at least a 4.9%

TABLE 5.5
Source Titles

	Source Title	WOS Subject Category	No. of Sample Papers (%)	No. of Population Papers (%)	Difference (%)
1	Bioresource Technology	Agr. Eng., Biot. Appl. Microb., Ener. Fuels	11	5.8	5.2
2	Energy Conversion and Management	Therm., Ener. Fuels, Mechs.	7	1.5	5.5
3	Journal of Analytical and Applied Pyrolysis	Chem. Anal., Ener. Fuels, Eng. Chem.	7	10.5	–3.5
4	Energy Fuels	Ener. Fuels, Eng. Chem.	6	5.4	0.6
5	Applied Catalysis A General	Chem. Phys., Env. Sci.	5	0.8	4.2
6	Chemical Society Reviews	Chem. Mult.	5	0.1	4.9
7	Fuel	Ener. Fuels, Eng. Chem.	5	4.7	0.3
8	Renewable Sustainable Energy Reviews	Green Sust. Sci. Technol., Ener. Fuels	5	0.8	4.2
9	Science	Mult. Sci.	4	0.1	3.9
10	Angewandte Chemie International Edition	Chem. Mult.	3	0.2	2.8
11	Energy	Therm., Ener. Fuels	3	1.4	1.6
12	Energy Environmental Science	Chem. Mult., Ener. Fuels, Eng. Chem., Env. Sci.	3	0.2	2.8
13	Progress in Energy and Combustion Science	Therm., Ener. Fuels, Eng. Chem., Eng. Mech.	3	0.2	2.8
			67	7.6	59.4

publication surplus each. On the other hand, the journals with the least impact are 'Fuel', 'Energy Fuels', and 'Energy' with at least a 0.3 publication surplus each.

It is notable that some journals are relatively under-represented in the sample papers. Some of them are 'Industrial Engineering Chemistry Research', 'Fuel Processing Technology', 'Biomass Bioenergy', 'ACS Sustainable Chemistry Engineering', 'Waste Management', the 'Journal of Thermal Analysis and Calorimetry', 'Green Chemistry', 'Energy Sources Part A Recovery Utilization and Environmental Effects', the 'Journal of Cleaner Production', and 'Renewable Energy' with at least a 1% presence in the population papers each.

5.3.7 Countries

Brief information about the top 13 countries with at least three sample papers each is provided in Table 5.6. In total, 24 and over 135 countries contribute to the sample and population papers, respectively.

The top country is the USA, publishing 36% and 18.3% of the sample and population papers, respectively. China follows the USA with 14% and 21.4% of the sample and population papers, respectively. The other prolific countries are Canada, the UK, and Spain, publishing ten, ten, and nine sample papers, respectively.

On the other hand, the European and Asian countries represented in this table publish altogether 42% and 27% of the sample papers and 25% and 28.7% of the population papers, respectively.

It is notable that the publication surplus for the USA and these European and Asian countries is 17.7%, 17.0%, and –1.7%, respectively.

It is also notable that some countries do not have a presence in this top country table. Some of them are Japan, Italy, South Korea, Russia, Brazil, Malaysia, Sweden,

TABLE 5.6
Countries

	Country	No. of Sample Papers (%)	No. of Population Papers (%)	Difference (%)
1	USA	36	18.3	17.7
2	China	14	21.4	–7.4
3	Canada	10	4.5	5.5
4	UK	10	6.6	3.4
5	Spain	9	5.9	3.1
6	Germany	7	4.5	2.5
7	India	7	4.5	2.5
8	Netherlands	7	2.6	4.4
9	Australia	6	2.8	3.2
10	Turkey	4	2.6	1.4
11	Denmark	3	1.2	1.8
12	France	3	3.6	–0.6
13	Norway	3	0.6	2.4
	Europe-7	42	25	17
	Asia-3	27	28.7	–1.7

Taiwan, Poland, Finland, Belgium, Thailand, Greece, and Portugal with at least a 1% presence in the population papers each.

5.3.8 'Web of Science' Subject Categories

Brief information about the top 15 'Web of Science' subject categories with at least three sample papers each is provided in Table 5.7. The sample and population papers are indexed by 26 and 104 subject categories, respectively.

For the sample papers, the top subject is 'Energy Fuels' with 54% and 46% of the sample and population papers, respectively. This top subject category is followed by 'Engineering Chemical' with 32% and 41% of the sample and population papers, respectively. The other prolific subjects are 'Biotechnology Applied Microbiology', 'Chemistry Multidisciplinary', 'Agricultural Engineering', 'Environmental Sciences', and 'Thermodynamics' with at least 13 papers each.

It is notable that the publication surplus is most significant for 'Energy Fuels', 'Thermodynamics', 'Mechanics', and 'Biotechnology Applied Microbiology' with at least a 4.7% surplus each. On the other hand, the subjects with least impact are 'Chemistry Analytical', 'Engineering Chemical', 'Engineering Environmental', 'Chemistry Physical', and 'Chemistry Applied' with at least a –9.5% publication deficit each. These subject categories are under-represented in the sample papers.

Additionally, some subject categories do not have a place in this top subject table: 'Biochemical Research Methods', 'Agronomy', 'Engineering Multidisciplinary', 'Chemistry Inorganic Nuclear', and 'Electrochemistry', which have at least a 1.2% presence in the population papers each.

TABLE 5.7
Web of Science Subject Categories

	Subject	No. of Sample Papers (%)	No. of Population Papers (%)	Difference (%)
1	Energy Fuels	54	46.0	8.0
2	Engineering Chemical	32	41.0	–9.0
3	Biotechnology Applied Microbiology	16	11.3	4.7
4	Chemistry Multidisciplinary	16	12.9	3.1
5	Agricultural Engineering	13	9.0	4.0
6	Environmental Sciences	13	12.2	0.8
7	Thermodynamics	13	7.6	5.4
8	Chemistry Physical	10	11.6	–1.6
9	Green Sustainable Science Technology	9	8.0	1.0
10	Chemistry Analytical	7	16.5	–9.5
11	Mechanics	7	2.0	5.0
12	Multidisciplinary Sciences	5	1.1	3.9
13	Chemistry Applied	3	4.5	–1.5
14	Engineering Environmental	3	7.7	–4.7
15	Engineering Mechanical	3	1.9	1.1

5.3.9 Citation Impact

These sample papers received about 60,300 citations as of January 2010. Thus, the average number of citations per paper is 603.

5.3.10 Keywords

Although a number of keywords are listed in the Appendix for the datasets related to this field, some of them are more significant for the sample papers. The most-prolific keyword is 'pyroly*' with 59 occurrences. This top keyword is followed by 'biomass' with 52 occurrences. The other prolific keywords are '*fuel*' and '*conversion' with 36 and 31 citations, respectively. 'Lignin', 'biochar', '*cellulos*', '*wood*', 'upgrading', 'waste', and 'torrefaction' are the other prolific keywords with 16, 10, 12, 8, 5, 5, and 5 citations, respectively.

5.3.11 Research Fronts

Brief information about the key research fronts is provided in Table 5.8. There are eight research fronts for these sample papers: 'biomass pyrolysis' (Bridgwater and Peacocke, 2000; Mohan et al., 2006), 'biofuels from biomass pyrolysis' (Czernik and Bridgwater, 2004; Yaman, 2004), 'pyrolysis oil property and characterization' (Evans and Milne, 1987; Zhang et al., 2007), 'biomass conversion for biofuel production' (Alonso et al., 2010; Gallezot, 2012), 'biomass torrefaction' (Bridgeman et al., 2008; van der Stelt, 2011), 'biomass liquefaction' (Akhtar and Amin, 2011; Toor et al., 2011), 'biooil upgrading' (Saidi et al., 2014; Xiu and Shahbazi, 2012), and 'lignin conversion' (Ragauskas et al., 2014; Zakzeski et al., 2010).

The most-prolific research front is 'biomass conversion for biofuel production' with 46 sample papers. The other prolific research fronts are 'biomass pyrolysis', 'biofuels from biomass pyrolysis', 'lignin conversion, and 'pyrolysis oil property and characterization' with 19, 22, 16, and 12 sample papers, respectively.

TABLE 5.8
Research Fronts

	Research Front	No. of Sample Papers (%)
1	Biomass pyrolysis	19
2	Biofuels from biomass pyrolysis	22
3	Pyrolysis oil property and characterization	12
4	Biomass conversion for biofuel production	46
5	Biomass torrefaction	7
6	Biomass liquefaction	4
7	Biooil upgrading	9
8	Lignin conversion	16

5.4 DISCUSSION

The size of the research on biooils has increased to over 20,000 papers as of January 2020. It is expected that the number of population papers in this field will exceed 50,000 by the end of the 2020s.

The research has developed more in the technological aspects of this field, rather than the social and humanitarian pathways, as evidenced by the negligible number of population papers in the indices of the 'Web of Science', SSCI, and A&HCI.

The article types of documents are the primary documents for both datasets and reviews are over-represented by 37.4% in the sample papers (Table 5.1). Thus, the contribution of reviews by 42% of the sample papers in this field is highly exceptional (cf. Konur, 2011, 2012a–n, 2015, 2016a–f, 2017a–f, 2018a–b, 2019a–b).

Seven authors from 16 institutions have at least three sample papers each (Table 5.2). Five of these authors are from the USA and the remaining ones are from Spain, Turkey, and the UK.

These authors focus on 'biomass conversion to biooil in general', 'biomass pyrolysis', and 'biooil upgrading to biofuels'.

It is significant that there is ample 'gender deficit' among these top authors as all researchers are male (Lariviere et al., 2013; Xie and Shauman, 1998).

About 14%, 33%, 40%, and 13% of the sample papers and 10.3%, 13.0%, 21.1%, and 55.1% of the population papers were published in the 1980s, 1990s, 2000s, and 2010s, respectively (Figure 5.1). This finding suggests that the population papers have built upon the sample papers, primarily published in the 1990s, 2000s, and 2010s. Following this rising trend, particularly in the 2010s, it is expected that the number of papers will reach 50,000 by the end of the 2020s, more than doubling the current size.

The engagement of the institutions in this field at the global scale is significant as over 140 and 6,400 institutions contribute to the sample and population papers, respectively.

Ten top institutions publish 44% and 12.4% of the sample and population papers, respectively (Table 5.3). The top institution is the 'National Renewable Energy Laboratory' of the USA with seven sample papers and a 6% publication surplus. The other top institutions are the 'University of Massachusetts' of the USA, the 'Superior Council of Scientific Investigations' of Spain, and the 'University of Wisconsin' of the USA with six, five, and five sample papers, respectively. As in the case of the top authors, the most-prolific countries for these top institutions are the USA, China, and Spain. It is notable that some institutions with a heavy presence in the population papers are under-represented in the sample papers.

It is significant that only 38% and about 55% of the sample and population papers declare any funding, respectively. It is notable that the most-prolific country for these top funding bodies is the USA with four whilst China has two (Table 5.4). It is further notable that some top funding agencies for the population studies do not enter this top funding body list.

There are two Chinese bodies dominating this top funding table. This finding is in line with studies showing heavy research funding in China, where the NSFC is the primary funding agency (Wang et al., 2012).

The sample and population papers are published by 40 and over 1,380 journals, respectively. It is significant that the top 13 journals publish 67% and 7.6% of the sample and population papers, respectively (Table 5.5).

The top journal is 'Bioresource Technology', publishing 11 sample papers with a 5.2% publication surplus. The other top journals are 'Energy Conversion and Management', the 'Journal of Analytical and Applied Pyrolysis', and 'Energy Fuels' with seven, seven, and six sample papers, respectively.

The top categories are 'Energy Fuels', 'Engineering Chemical', and 'Thermodynamics' with nine, five, and three journals, respectively. It is notable that some journals with a heavy presence in the population papers are relatively under-represented in the sample papers.

In total, 24 and over 130 countries contribute to the sample and population papers, respectively. The top country is the USA, publishing 36% and 18.3% of the sample and population papers, respectively. China follows the USA with 14% and 21.4% of the sample and population papers, respectively (Table 5.6). This finding is in line with the studies arguing that the USA is not losing ground in science and technology (Leydesdorff and Wagner, 2009).

The other prolific countries are Canada, the UK, and Spain publishing ten, ten, and nine sample papers, respectively. These findings are in line with the studies showing that European countries have a superior publication performance in science and technology (Bordons et al., 2015; Youtie et al., 2008). Similarly, Canada also has a significant presence in science and technology (Fu and Ho, 2015).

On the other hand, the European and Asian countries represented in this table publish altogether 42% and 27% of the sample papers and 25% and 28.7% of the population papers, respectively.

It is notable that the publication surplus for the USA and these European and Asian countries is 17.7%, 17.0%, and –1.7%, respectively.

It is further notable that China has a significant publication deficit (–7.4%). This finding is in contrast with China's efforts to be a leading nation in science and technology (Zhou and Leydesdorff, 2006) but it is in line with the findings of Guan and Ma (2007) and Youtie et al. (2008) concerning China's performance in nanotechnology.

It is also notable that some countries do not have a presence in this top country table. Some of them are Japan, Italy, South Korea, Russia, Brazil, Malaysia, Sweden, Taiwan, Poland, Finland, Belgium, Thailand, Greece, and Portugal with at least a 1% presence in the population papers each (Bordons et al., 2015; Glanzel et al., 2006; Leydesdorff and Zhou, 2005; Oleinik, 2012).

The sample and population papers are indexed by 26 and 104 subject categories, respectively.

For the sample papers, the top subjects are 'Energy Fuels' with 54% and 46% of the sample and population papers, respectively (Table 5.7). This top subject category is followed by 'Engineering Chemical' with 32% and 41% of the sample and population papers, respectively. The other prolific subjects are 'Biotechnology Applied Microbiology', 'Chemistry Multidisciplinary', 'Agricultural Engineering', 'Environmental Sciences', and 'Thermodynamics' with at least 13 papers each.

It is notable that the publication surplus is most significant for 'Energy Fuels', 'Thermodynamics', 'Mechanics', and 'Biotechnology Applied Microbiology' with at least a 4.7% surplus each. On the other hand, the subjects with least impact are 'Chemistry Analytical', 'Engineering Chemical', 'Engineering Environmental', 'Chemistry Physical', and 'Chemistry Applied' with at least a –9.5% publication deficit each. These subject categories are under-represented in the sample papers.

These sample papers received about 60,000 citations as of January 2020. Thus, the average number of citations per paper is 600. Hence, the citation impact of these 100 top papers in this field has been significant.

Although a number of keywords are listed in the Appendix for the datasets related to this field, some of them are more significant for the sample papers. The most-prolific keyword is 'pyroly*' with 59 occurrences. This top keyword is followed by 'biomass' with 52 occurrences. The other prolific keywords are '*fuel*' and '*conversion' with 36 and 31 citations, respectively. 'Lignin', 'biochar', '*cellulos*', '*wood*', 'upgrading', 'waste', and 'torrefaction' are the other prolific keywords with 16, 10, 12, 8, 5, 5, and 5 citations, respectively. As expected, these keywords provide valuable information about the pathways of the research in this field.

Eight research fronts emerge from the examination of the sample papers: 'biomass pyrolysis' (Bridgwater and Peacocke, 2000; Mohan et al., 2006), 'biofuels from biomass pyrolysis' (Czernik and Bridgwater, 2004; Yaman, 2004), 'pyrolysis oil property and characterization' (Evans and Milne, 1987; Zhang et al., 2007), 'biomass conversion for biofuel production' (Alonso et al., 2010; Gallezot, 2012), 'biomass torrefaction' (Bridgeman et al., 2008; van der Stelt, 2011), 'biomass liquefaction' (Akhtar and Amin, 2011; Toor et al., 2011), 'biooil upgrading' (Saidi et al., 2014; Xiu and Shahbazi, 2012), and 'lignin conversion' (Ragauskas et al., 2014; Zakzeski et al., 2010) (Table 5.8).

The most-prolific research front is 'biomass conversion for biofuel production' with 46 sample papers. The other prolific research fronts are 'biomass pyrolysis', 'biofuels from biomass pyrolysis', 'lignin conversion', and 'pyrolysis oil property and characterization' with 19, 22, 16, and 12 sample papers, respectively.

The key emphasis in these research fronts is the exploration of the structure–processing–property relationships of biooils (Cheng and Ma, 2011; Konur and Matthews, 1989; Rogers and Hopfinger, 1994; Scherf and List, 2002).

5.5 CONCLUSION

This chapter has mapped the research on biooils using a scientometric method.

That there are over 20,000 population papers shows the public importance of this interdisciplinary research field. However, it is significant that the research has developed more in the technological aspects, rather than the social and humanitarian pathways.

Articles and reviews dominate the sample papers, primarily published in the 1990s and 2000s. The population papers, mainly published in the 2010s, build on these sample papers.

The data presented in the tables and in the one figure show that a small number of authors, institutions, funding bodies, journals, keywords, research fronts, subject categories, and countries have shaped the research in this field.

It is notable that the authors, institutions, and funding bodies in the USA, the UK, Spain, and China dominate the research in this field. Furthermore, China, Japan, Italy, South Korea, Russia, Brazil, Malaysia, Sweden, Taiwan, Poland, Finland, Belgium, Thailand, Greece, and Portugal are under-represented significantly in the sample papers.

These findings show the importance of the development of efficient incentive structures for the progression of the research in this field, as in others. It seems that the USA and European countries (such as the UK and Spain) have efficient incentive structures for the development of the research, contrary to China, Japan, Italy, South Korea, Russia, Brazil, Malaysia, Sweden, Taiwan, Poland, Finland, Belgium, Thailand, Greece, and Portugal.

It further seems that although research funding is a significant element of these incentive structures, it might not be a sole solution for increasing incentives in this field, as in the case of China, Japan, Italy, South Korea, Russia, Brazil, Malaysia, Sweden, Taiwan, Poland, Finland, Belgium, Thailand, Greece, and Portugal.

On the other hand, it seems there is more to do to reduce the significant gender deficit in this field, as in other fields of science and technology (Lariviere et al., 2013; Xie and Shauman, 1998).

The data on the research fronts, keywords, source titles, and subject categories provide valuable evidence for the interdisciplinary nature of the research in this field (Lariviere and Gingras, 2010; Morillo et al., 2010).

There is ample justification for the broad search strategy employed in this study due to the interdisciplinary nature of this research field, as evidenced by the top subject categories. The search strategy employed is in line with the strategies employed for related and other research fields (Konur, 2011, 2021a–n, 2015, 2016a–f, 2017a–f, 2018a, b, 2019a–b).

Eight research fronts emerge from the examination of the sample papers: 'biomass pyrolysis' (Bridgwater and Peacocke, 2000; Mohan et al., 2006), 'biofuels from biomass pyrolysis' (Czernik and Bridgwater, 2004; Yaman, 2004), 'pyrolysis oil property and characterization' (Evans and Milne, 1987; Zhang et al., 2007), 'biomass conversion for biofuel production' (Alonso et al., 2010; Gallezot, 2012), 'biomass torrefaction' (Bridgeman et al., 2008; van der Stelt, 2011), 'biomass liquefaction' (Akhtar and Amin, 2011; Toor et al., 2011), 'biooil upgrading' (Saidi et al., 2014; Xiu and Shahbazi, 2012), and 'lignin conversion' (Ragauskas et al., 2014; Zakzeski et al., 2010) (Table 5.8).

It is recommended that further scientometric studies are carried out for each of these research fronts, building on the pioneering studies in these fields.

ACKNOWLEDGMENTS

The contribution of the highly cited researchers in the field of biooils has been greatly acknowledged.

5.A APPENDIX

Biooils keyword set: TI=(*pyrolysis or torrefaction* or thermolysis or "hydro-thermal liquefaction" or "bio-oil*" or ((*liquefaction or "thermal decomposition" or "hydrothermal conversion" or "hydrothermal processing" or "thermochemical processing" or "thermo-chemical conversion" or "hydrothermal liquefaction" or "hydrothermal treatment*" or "thermochemical conversion" or torrefaction or *deoxygenation or hydrocracking or "catalytic cracking" or "zeolite cracking" or thermogravimet* or "thermal conversion" or liquefaction or "thermo-chemical conversion" or 'valori?ation' or biorefin* or hydroprocessing or hydrotreating or "catalytic conversion" or *pyroly* or densification) and (biomass or *wood* or *cellulos* or xylan or lignin or manure or sludge* or straw* or bagasse* or legume* or cake or palm or tree* or waste* or "bio-mass" or feedstock* or husk* or stalk* or *grass* or glycerol or tyre* or *plastic* or residue* or sawdust* or shell* or bunch* or tire* or pulp* or stone* or triglyceride* or *guaiac* or "model compound*" or furfural or sawdust* or *phenol* or bark or scrap* or glucose or crop* or "vegetable oil*" or furan* or litter* or anisole or cassava or willow* or corncob* or oxygenate* or rape* or sucrose or *seed* or kernel or hull* or cynara or stover* or cone or polysaccharide* or bunch* or corn* or pine* or "acetic acid*")) or hydrodeoxygenation or ((lignin* or xylan) and (valori?ation or depolymeri?ation or biorefin* or pyroly* or liquefaction or *conversion or upgrading)) or "pyroly*oil*" or (biomass and *conversion) or biocrude* or "bio-crude*" or biooil* or "py-gc/ms").

NOT (WC=(food* or nutr* or pharm* or meteor* or geo* or "agriculture dairy*" or materials* or polymer* or "engineering geol*" or biophysics or cell* or metall* or "engineering biomed*"or physiol* or physics* or "chemistry organic" or econ* or business* or "engineering petrol*" or radiol* or med*) OR TI=(phosphate or coal or cell* or nano* or "carbon dot*" or laser or spray or aerosol or shale* or humic or *alga* or "natural gas" or flame or shock or acetylene or "organic matter" or methane or epox* or asphalt* or lignite or fire or alcohol* or "heavy oil*" or petrol* or napht* or ethanol or "gas oil*" or "light cycle oil" or *arene* or urea or mercury or bacillus or kerogen or algorithm* or semiconductor* or mist or *butanol or *heptane or cuticle* or concrete* or doped or spectra or polyacrylonitrile or geo* or protein* or disilane or propane or ammonia or dibenzothiophene or dianion* or macrocycl* or azide* or soot* or diamond or explosive* or polyethylene or polypropylene or maceral or polystyrene or qrrk or nannochlorop* or ergosterol or chlorella or retracted or coke or spirulina or drug*)).

REFERENCES

Ahmadun, F. R., A. Pendashteh, L. C. Abdullah, et al. 2009. Review of technologies for oil and gas produced water treatment. *Journal of Hazardous Materials* 170: 530–551.

Akhtar, J. and Amin, N. A. S. 2011. A review on process conditions for optimum bio-oil yield in hydrothermal liquefaction of biomass. *Renewable & Sustainable Energy Reviews* 15: 1615–1624.

Alonso, D. M., J. Q. Bond and J. A. Dumesic. 2010. Catalytic conversion of biomass to biofuels. *Green Chemistry* 12: 1493–1513.

Atlas, R. M. 1981. Microbial degradation of petroleum hydrocarbons: An environmental perspective. *Microbiological Reviews* 45: 180–209.

Babich, I. V. and J. A. Moulijn. 2003. Science and technology of novel processes for deep desulfurization of oil refinery streams: A review. *Fuel* 82: 607–631.

Bordons, M., B. Gonzalez-Albo, J. Aparicio and L. Moreno. 2015. The influence of R & D intensity of countries on the impact of international collaborative research: Evidence from Spain. *Scientometrics* 102: 1385–1400.

Bridgeman, T. G., J. M. Jones, I. Shield and P. T. Williams. 2008. Torrefaction of reed canary grass, wheat straw and willow to enhance solid fuel qualities and combustion properties. *Fuel* 87: 844–856.

Bridgwater, A. V., and G. V. C. Peacocke. 2000. Fast pyrolysis processes for biomass. *Renewable & Sustainable Energy Reviews* 4: 1–73.

Cheng, Y. Q. and E. Ma. 2011. Atomic-level structure and structure–property relationship in metallic glasses. *Progress in Materials Science* 56: 379–473.

Chisti, Y. 2007. Biodiesel from microalgae. *Biotechnology Advances* 25: 294–306.

Chisti, Y. 2008. Biodiesel from microalgae beats bioethanol. *Trends in Biotechnology* 26: 126–131.

Clarivate Analytics. 2019. *Highly cited researchers: 2019 Recipients.* Philadelphia, PA: Clarivate Analytics. https://recognition. webofsciencegroup.com/awards/highly-cited/2019/ (accessed January, 3, 2020).

Czernik, S. and A. V. Bridgwater. 2004. Overview of applications of biomass fast pyrolysis oil. *Energy & Fuels* 18: 590–598.

Docampo, D. and L. Cram. 2019. Highly cited researchers: A moving target. *Scientometrics* 118: 1011–1025.

Evans, R. J. and T. A. Milne. 1987. Molecular characterization of the pyrolysis of biomass. 1. Fundamentals. *Energy & Fuels* 1: 123–137.

Fu, H. Z. and Y. S. Ho. 2015. Highly cited Canada articles in science citation index expanded: A bibliometric analysis. *Canadian Social Science* 11: 50.

Gallezot, P. 2012. Conversion of biomass to selected chemical products. *Chemical Society Reviews* 41: 1538–1558.

Garfield, E. 1955. Citation indexes for science. *Science* 122: 108–111.

Garfield, E. 1972. Citation analysis as a tool in journal evaluation. *Science* 178: 471–479.

Glanzel, W., J. Leta and B. Thijs. 2006. Science in Brazil. Part 1: A macro-level comparative study. *Scientometrics* 67: 67–86.

Guan, J. C. and N. Ma. 2007. China's emerging presence in nanoscience and nanotechnology: A comparative bibliometric study of several nanoscience 'giants'. *Research Policy* 36: 880–886.

Kilian, L. 2009. Not all oil price shocks are alike: Disentangling demand and supply shocks in the crude oil market. *American Economic Review* 99: 1053–1069.

Konur, O. 2000. Creating enforceable civil rights for disabled students in higher education: An institutional theory perspective. *Disability & Society* 15: 1041–1063.

Konur, O. 2002a. Access to Nursing Education by disabled students: Rights and duties of nursing programs. *Nurse Education Today* 22: 364–374.

Konur, O. 2002b. Assessment of disabled students in higher education: Current public policy issues. *Assessment and Evaluation in Higher Education* 27: 131–152.

Konur, O. 2002c. Access to employment by disabled people in the UK: Is the Disability Discrimination Act working? *International Journal of Discrimination and the Law* 5: 247–279.

Konur, O. 2006a. Participation of children with dyslexia in compulsory education: Current public policy issues. *Dyslexia* 12: 51–67.

Konur, O. 2006b. Teaching disabled students in Higher Education. *Teaching in Higher Education* 11: 351–363.

Konur, O. 2007a. A judicial outcome analysis of the Disability Discrimination Act: A windfall for the employers? *Disability & Society* 22: 187–204.

Konur, O. 2007b. Computer-assisted teaching and assessment of disabled students in higher education: The interface between academic standards and disability rights. *Journal of Computer Assisted Learning* 23: 207–219.

Konur, O. 2011. The scientometric evaluation of the research on the algae and bio-energy. *Applied Energy* 88: 3532–3540.

Konur, O. 2012a. Evaluation of the research on the social sciences in Turkey: A scientometric approach. *Energy Education Science and Technology Part B: Social and Educational Studies* 4: 1893–1908.

Konur, O. 2012b. Prof. Dr. Ayhan Demirbas' scientometric biography. *Energy Education Science and Technology Part A: Energy Science and Research* 28: 727–738.

Konur, O. 2012c. The evaluation of the biogas research: A scientometric approach. *Energy Education Science and Technology Part A: Energy Science and Research* 29: 1277–1292.

Konur, O. 2012d. The evaluation of the educational research: A scientometric approach. *Energy Education Science and Technology Part B: Social and Educational Studies* 4: 1935–1948.

Konur, O. 2012e. The evaluation of the global energy and fuels research: A scientometric approach. *Energy Education Science and Technology Part A: Energy Science and Research* 30: 613–628.

Konur, O. 2012f. The evaluation of the research on the Arts and Humanities in Turkey: A scientometric approach. *Energy Education Science and Technology Part B: Social and Educational Studies* 4: 1603–1618.

Konur, O. 2012g. The evaluation of the research on the biodiesel: A scientometric approach. *Energy Education Science and Technology Part A: Energy Science and Research* 28: 1003–1014.

Konur, O. 2012h. The evaluation of the research on the bioethanol: A scientometric approach. *Energy Education Science and Technology Part A: Energy Science and Research* 28: 1051–1064.

Konur, O. 2012i. The evaluation of the research on the biofuels: A scientometric approach. *Energy Education Science and Technology Part A: Energy Science and Research* 28: 903–916.

Konur, O. 2012j. The evaluation of the research on the biohydrogen: A scientometric approach. *Energy Education Science and Technology Part A: Energy Science and Research* 29: 323–338.

Konur, O. 2012k. The evaluation of the research on the microbial fuel cells: A scientometric approach. *Energy Education Science and Technology Part A: Energy Science and Research* 29: 309–322.

Konur, O. 2012l. The scientometric evaluation of the research on the production of bioenergy from biomass. *Biomass and Bioenergy* 47: 504–515.

Konur, O. 2012m. The scientometric evaluation of the research on the deaf students in higher education. *Energy Education Science and Technology Part B: Social and Educational Studies* 4: 1573–1588.

Konur, O. 2012n. The scientometric evaluation of the research on the students with ADHD in higher education. *Energy Education Science and Technology Part B: Social and Educational Studies* 4: 1547–1562.

Konur, O. 2015. Current state of research on algal biodiesel. In *Marine Bioenergy: Trends and Developments*, S. K. Kim, and C. G. Lee, ed., 487–512. Boca Raton, FL: CRC Press.

Konur, O. 2016a. Scientometric overview in nanobiodrugs. In *Nanoarchitectonics for Smart Delivery and Drug Targeting*, A. M. Holban and A.M. Grumezescu, ed., 405–428. Amsterdam: Elsevier.

Konur, O. 2016b. Scientometric overview regarding nanoemulsions used in the food industry. In *Emulsions: Nanotechnology in the Agri-Food Industry*, A. M. Grumezescu, ed., 689–711. Amsterdam: Elsevier.

Konur, O. 2016c. Scientometric overview regarding the nanobiomaterials in antimicrobial therapy. In *Nanobiomaterials in Antimicrobial Therapy,* A. M. Grumezescu, ed., 511–535. Amsterdam: Elsevier.

Konur, O. 2016d. Scientometric overview regarding the nanobiomaterials in dentistry. In *Nanobiomaterials in Dentistry,* A. M. Grumezescu, ed., 425–453. Amsterdam: Elsevier.

Konur, O. 2016e. Scientometric overview regarding the surface chemistry of nanobiomaterials. In *Surface Chemistry of Nanobiomaterials,* A. M. Grumezescu, ed., 463–486. Amsterdam: Elsevier.

Konur, O. 2016f. The scientometric overview in cancer targeting. In *Nanoarchitectonics for Smart Delivery and Drug Targeting,* A. M. Holban and A. Grumezescu, ed., 871–895. Amsterdam; Elsevier.

Konur, O. 2017a. Recent citation classics in antimicrobial nanobiomaterials. In *Nanostructures for Antimicrobial Therapy,* A. Ficai and A. M. Grumezescu, ed., 669–685. Amsterdam: Elsevier.

Konur, O. 2017b. Scientometric overview in nanopesticides. In *New Pesticides and Soil Sensors,* A. M. Grumezescu, ed. 719–744. Amsterdam: Elsevier.

Konur, O. 2017c. Scientometric overview regarding oral cancer nanomedicine. In *Nanostructures for Oral Medicine*, E. Andronescu, A. M. Grumezescu, ed., 939–962. Amsterdam: Elsevier.

Konur, O. 2017d. Scientometric overview regarding water nanopurification. In *Water Purification,* A. M. Grumezescu, ed., 693–716. Amsterdam: Elsevier.

Konur, O. 2017e. Scientometric overview in food nanopreservation. In *Food Preservation,* A. M. Grumezescu, ed., 703–729. Amsterdam: Elsevier.

Konur, O. 2017f. The top citation classics in alginates for biomedicine. In *Seaweed Polysaccharides: Isolation, Biological and Biomedical Applications*, J. Venkatesan, S. Anil, S. K. Kim, ed., 223–249. Amsterdam: Elsevier.

Konur, O. 2018a. Scientometric evaluation of the global research in spine: An update on the pioneering study by Wei et al. *European Spine Journal* 27: 525–529.

Konur, O. 2018b. Bioenergy and biofuels science and technology: Scientometric overview and citation classics. In *Bioenergy and Biofuels*, O. Konur, ed., 3–63. Boca Raton: CRC Press.

Konur, O. 2019a. Cyanobacterial bioenergy and biofuels science and technology: A scientometric overview. In *Cyanobacteria: From Basic Science to Applications*, ed. A. K. Mishra, D. N. Tiwari and A. N. Rai, 419–442. Amsterdam: Elsevier.

Konur, O. 2019b. Nanotechnology applications in food: A scientometric overview. In *Nanoscience for Sustainable Agriculture*, R. N. Pudake, N. Chauhan, and C. Kole, ed., 683–711. Cham: Springer.

Konur, O., ed. 2021a. *Handbook of Biodiesel and Petrodiesel Fuels: Science, Technology, Health, and Environment.* Boca Raton, FL: CRC Press.

Konur, O., ed. 2021b. *Handbook of Biodiesel and Petrodiesel Fuels: Science, Technology, Health, and Environment. Volume 1. Biodiesel Fuels: Science, Technology, Health, and Environment.* Boca Raton, FL: CRC Press.

Konur, O., ed. 2021c. *Handbook of Biodiesel and Petrodiesel Fuels: Science, Technology, Health, and Environment. Volume 2. Biodiesel Fuels based on the Edible and Nonedible Feedstocks, Wastes, and Algae: Science, Technology, Health, and Environment.* Boca Raton, FL: CRC Press.

Konur, O., ed. 2021d. *Handbook of Biodiesel and Petrodiesel Fuels: Science, Technology, Health, and Environment. Volume 3. Petrodiesel Fuels: Science, Technology, Health, and Environment.* Boca Raton, FL: CRC Press.

Konur, O. 2021e. Biodiesel and petrodiesel fuels: Science, technology, health, and environment. In *Handbook of Biodiesel and Petrodiesel Fuels: Science, Technology, Health, and Environment. Volume 1. Biodiesel Fuels: Science, Technology, Health, and Environment*, ed. O. Konur. Boca Raton, FL: CRC Press.

Konur, O. 2021f. Biodiesel and petrodiesel fuels: A scientometric review of the research. In *Handbook of Biodiesel and Petrodiesel Fuels: Science, Technology, Health, and*

Environment. Volume 1. Biodiesel Fuels: Science, Technology, Health, and Environment, ed. O. Konur. Boca Raton, FL: CRC Press.

Konur, O. 2021g. Biodiesel and petrodiesel fuels: A review of the research. In *Handbook of Biodiesel and Petrodiesel Fuels: Science, Technology, Health, and Environment. Volume 1. Biodiesel Fuels: Science, Technology, Health, and Environment*, ed. O. Konur. Boca Raton, FL: CRC Press.

Konur, O. 2021h Nanotechnology applications in the diesel fuels and the related research fields: A review of the research. In *Handbook of Biodiesel and Petrodiesel Fuels: Science, Technology, Health, and Environment. Volume 1. Biodiesel Fuels: Science, Technology, Health, and Environment*, ed. O. Konur. Boca Raton, FL: CRC Press.

Konur, O. 2021i. Biooils: A scientometric review of the research. In *Handbook of Biodiesel and Petrodiesel Fuels: Science, Technology, Health, and Environment. Volume 1. Biodiesel Fuels: Science, Technology, Health, and Environment*, ed. O. Konur. Boca Raton, FL: CRC Press.

Konur, O. 2021j. Characterization and properties of biooils: A review of the research. In *Handbook of Biodiesel and Petrodiesel Fuels: Science, Technology, Health, and Environment. Volume 1. Biodiesel Fuels: Science, Technology, Health, and Environment*, ed. O. Konur. Boca Raton, FL: CRC Press.

Konur, O. 2021k. Biomass pyrolysis and pyrolysis oils: A review of the research. In *Handbook of Biodiesel and Petrodiesel Fuels: Science, Technology, Health, and Environment. Volume 1. Biodiesel Fuels: Science, Technology, Health, and Environment*, ed. O. Konur. Boca Raton, FL: CRC Press.

Konur, O. 2021l. Biodiesel fuels: A scientometric review of the research. In *Handbook of Biodiesel and Petrodiesel Fuels: Science, Technology, Health, and Environment. Volume 1. Biodiesel Fuels: Science, Technology, Health, and Environment*, ed. O. Konur. Boca Raton, FL: CRC Press.

Konur, O. 2021m. Glycerol: A scientometric review of the research. In *Handbook of Biodiesel and Petrodiesel Fuels: Science, Technology, Health, and Environment. Volume 1. Biodiesel Fuels: Science, Technology, Health, and Environment*, ed. O. Konur. Boca Raton, FL: CRC Press.

Konur, O. 2021n. Propanediol production from glycerol: A review of the research. In *Handbook of Biodiesel and Petrodiesel Fuels: Science, Technology, Health, and Environment. Volume 1. Biodiesel Fuels: Science, Technology, Health, and Environment*, ed. O. Konur Boca Raton, FL: CRC Press. February 17, 2020.

Konur, O. 2021o. Edible oil-based biodiesel fuels: A scientometric review of the research. *In Handbook of Biodiesel and Petrodiesel Fuels: Science, Technology, Health, and Environment. Volume 2. Biodiesel Fuels based on the Edible and Nonedible Feedstocks, Wastes, and Algae: Science, Technology, Health, and Environment,* ed. O. Konur. Boca Raton, FL: CRC Press.

Konur, O. 2021p. Palm oil-based biodiesel fuels: A review of the research. In *Handbook of Biodiesel and Petrodiesel Fuels: Science, Technology, Health, and Environment. Volume 2. Biodiesel Fuels based on the Edible and Nonedible Feedstocks, Wastes, and Algae*, ed. O. Konur. Boca Raton, FL: CRC Press.

Konur, O. 2021q. Rapeseed oil-based biodiesel fuels: A review of the research. In *Handbook of Biodiesel and Petrodiesel Fuels: Science, Technology, Health, and Environment. Volume 2. Biodiesel Fuels based on the Edible and Nonedible Feedstocks, Wastes, and Algae,* ed. O. Konur. Boca Raton, FL: CRC Press.

Konur, O. 2021r. Nonedible oil-based biodiesel fuels: A scientometric review of the research. In *Handbook of Biodiesel and Petrodiesel Fuels: Science, Technology, Health, and Environment. Volume 2. Biodiesel Fuels based on the Edible and Nonedible Feedstocks, Wastes, and Algae: Science, Technology, Health, and Environment*, ed. O. Konur. Boca Raton, FL: CRC Press.

Konur, O. 2021s. Waste oil-based biodiesel fuels: A scientometric review of the research. In *Handbook of Biodiesel and Petrodiesel Fuels: Science, Technology, Health, and Environment. Volume 2. Biodiesel Fuels based on the Edible and Nonedible Feedstocks, Wastes, and Algae: Science, Technology, Health, and Environment*, ed. O. Konur. Boca Raton, FL: CRC Press.

Konur, O. 2021t. Algal biodiesel fuels: A scientometric review of the research. In *Handbook of Biodiesel and Petrodiesel Fuels: Science, Technology, Health, and Environment. Volume 2. Biodiesel Fuels based on the Edible and Nonedible Feedstocks, Wastes, and Algae: Science, Technology, Health, and Environment*, ed. O. Konur. Boca Raton, FL: CRC Press.

Konur, O. 2021u. Algal biomass production for biodiesel production: A review of the research. In *Handbook of Biodiesel and Petrodiesel Fuels: Science, Technology, Health, and Environment. Volume 2. Biodiesel Fuels based on the Edible and Nonedible Feedstocks, Wastes, and Algae*, Ed. O. Konur Boca Raton, FL: CRC Press. February 23, 2020.

Konur, O. 2021v. Algal biomass production in wastewaters for biodiesel production: A review of the research. In *Handbook of Biodiesel and Petrodiesel Fuels: Science, Technology, Health, and Environment. Volume 2. Biodiesel Fuels based on the Edible and Nonedible Feedstocks, Wastes, and Algae*, ed. O. Konur. Boca Raton, FL: CRC Press. February 23, 2020.

Konur, O. 2021x. Algal lipid production for biodiesel production: A review of the research. In *Handbook of Biodiesel and Petrodiesel Fuels: Science, Technology, Health, and Environment. Volume 2. Biodiesel Fuels based on the Edible and Nonedible Feedstocks, Wastes, and Algae*, Ed. O. Konur Boca Raton, FL: CRC Press.

Konur, O. 2021y. Crude oils: A scientometric review of the research. In *Handbook of Biodiesel and Petrodiesel Fuels: Science, Technology, Health, and Environment. Volume 3. Petrodiesel Fuels: Science, Technology, Health, and Environment*, ed. O. Konur. Boca Raton, FL: CRC Press.

Konur, O. 2021z. Petrodiesel fuels: A scientometric review of the research. In *Handbook of Biodiesel and Petrodiesel Fuels: Science, Technology, Health, and Environment. Volume 3. Petrodiesel Fuels: Science, Technology, Health, and Environment*, ed. O. Konur. Boca Raton, FL: CRC Press.

Konur, O. 2021aa. Bioremediation of petroleum hydrocarbons in the contaminated soils: A review of the research. In *Handbook of Biodiesel and Petrodiesel Fuels: Science, Technology, Health, and Environment. Volume 3. Petrodiesel Fuels: Science, Technology, Health, and Environment*, ed. O. Konur. Boca Raton, FL: CRC Press.

Konur, O. 2021ab. Desulfurization of diesel fuels: A review of the research. In *Handbook of Biodiesel and Petrodiesel Fuels: Science, Technology, Health, and Environment. Volume 3. Petrodiesel Fuels: Science, Technology, Health, and Environment*, ed. O. Konur. Boca Raton, FL: CRC Press.

Konur, O. 2021ac. Diesel fuel exhaust emissions: A scientometric review of the research. In *Handbook of Biodiesel and Petrodiesel Fuels: Science, Technology, Health, and Environment. Volume 3. Petrodiesel Fuels: Science, Technology, Health, and Environment*, ed. O. Konur. Boca Raton, FL: CRC Press.

Konur, O. 2021ad. The adverse health and safety impact of diesel fuels: A scientometric review of the research. In *Handbook of Biodiesel and Petrodiesel Fuels: Science, Technology, Health, and Environment. Volume 3. Petrodiesel Fuels: Science, Technology, Health, and Environment*, ed. O. Konur. Boca Raton, FL: CRC Press.

Konur, O. 2021ae. Respiratory illnesses caused by the diesel fuel exhaust emissions: A review of the research. In *Handbook of Biodiesel and Petrodiesel Fuels: Science, Technology, Health, and Environment. Volume 3. Petrodiesel Fuels: Science, Technology, Health, and Environment*, ed. O. Konur. Boca Raton, FL: CRC Press.

Konur, O. 2021af. Cancer caused by the diesel fuel exhaust emissions: A review of the research. In *Handbook of Biodiesel and Petrodiesel Fuels: Science, Technology, Health, and Environment. Volume 3. Petrodiesel Fuels: Science, Technology, Health, and Environment*, ed. O. Konur. Boca Raton, FL: CRC Press.

Konur, O. 2021ag. Cardiovascular and other illnesses caused by the diesel fuel exhaust emissions: A review of the research. In *Handbook of Biodiesel and Petrodiesel Fuels: Science, Technology, Health, and Environment. Volume 3. Petrodiesel Fuels: Science, Technology, Health, and Environment*, ed. O. Konur. Boca Raton, FL: CRC Press.

Konur, O. and F. L. Matthews. 1989. Effect of the properties of the constituents on the fatigue performance of composites: A review. *Composites* 20: 317–328.

Lapuerta, M., O. Armas and J. Rodriguez-Fernandez. 2008. Effect of biodiesel fuels on diesel engine emissions. *Progress in Energy and Combustion Science* 34: 198–223.

Lariviere, V. and Y. Gingras. 2010. On the relationship between interdisciplinarity and scientific impact. *Journal of the American Society for Information Science and Technology* 61: 126–131.

Lariviere, V., C. Ni, Y. Gingras, B. Cronin, B., and C.R. Sugimoto. 2013. Bibliometrics: Global gender disparities in science. *Nature News* 504: 211–213.

Leydesdorff, L. and Wagner, C. 2009. Is the United States losing ground in science? A global perspective on the world science system. *Scientometrics* 78: 23–36.

Leydesdorff, L. and P. Zhou. 2005. Are the contributions of China and Korea upsetting the world system of science? *Scientometrics* 63: 617–630.

Mohan, D., C. U. Pittman and P. H. Steele. 2006. Pyrolysis of wood/biomass for bio-oil: A critical review. *Energy & Fuels* 20: 848–889.

Moldowan, J. M., W. K. Seifert and E. J. Gallegos. 1985. Relationship between petroleum composition and depositional environment of petroleum source rocks. *AAPG Bulletin-American Association of Petroleum Geologists* 69: 1255–1268.

Morillo, F., M. Bordons and I. Gomez. 2001. An approach to interdisciplinarity through bibliometric indicators. *Scientometrics* 51: 203–222.

North, D. C. 1991a. *Institutions, Institutional Change and Economic Performance*. Cambridge, Mass.: Cambridge University Press.

North, D.C. 1991b. Institutions. *Journal of Economic Perspectives* 5: 97–112.

Oleinik, A. 2012. Publication patterns in Russia and the West compared. *Scientometrics* 93: 533–551.

Perron, P. 1989. The great crash, the oil price shock, and the unit root hypothesis. *Econometrica: Journal of the Econometric Society* 57: 1361–1401.

Ragauskas, AJ, G. T. Beckham, and M. J. Biddy, et al. 2014. Lignin valorization: Improving lignin processing in the biorefinery. *Science* 344: 1246843.

Rogers, D., and A. J. Hopfinger. 1994. Application of genetic function approximation to quantitative structure-activity relationships and quantitative structure-property relationships. *Journal of Chemical Information and Computer Sciences* 34: 854–866.

Saidi, M., F. Samimi, and D. Karimipourfard, et al. 2014. Upgrading of lignin-derived bio-oils by catalytic hydrodeoxygenation. *Energy & Environmental Science* 7: 103–129.

Scherf, U. and E. J. List. 2002. Semiconducting polyfluorenes-towards reliable structure–property relationships. *Advanced Materials* 14: 477–487.

Toor, S. S., L. Rosendahl and A. Rudolf. 2011. Hydrothermal liquefaction of biomass: A review of subcritical water technologies. *Energy* 36: 2328–2342.

van der Stelt, M.J.C., H. Gerhauser, J. H. A. Kiel and K. J. Ptasinski. 2011. Biomass upgrading by torrefaction for the production of biofuels: A review. *Biomass & Bioenergy* 35: 3748–3762.

Van Gerpen, J. 2005. Biodiesel processing and production. *Fuel Processing Technology* 86: 1097–1107.

Wang, X., D. Liu, K. Ding. and X. Wang. 2012. Science funding and research output: A study on 10 countries. *Scientometrics* 91: 591–599.

Xie, Y. and K. A. Shauman. 1998. Sex differences in research productivity: New evidence about an old puzzle. *American Sociological Review* 63: 847–870.

Xiu, S. N. and A. Shahbazi. 2012. Bio-oil production and upgrading research: A review. *Renewable & Sustainable Energy Reviews* 16: 4406–4414.

Yaman, S. 2004. Pyrolysis of biomass to produce fuels and chemical feedstocks. *Energy Conversion and Management* 45: 651–671.

Youtie, J. P., Shapira, and A. L. Porter. 2008. Nanotechnology publications and citations by leading countries and blocs. *Journal of Nanoparticle Research* 10: 981–986.

Zakzeski, J., P. C. A. Bruijnincx, A. L. Jongerius and B. M. Weckhuysen. 2010. The catalytic valorization of lignin for the production of renewable chemicals. *Chemical Reviews* 110: 3552–3599.

Zhang, Q., J. Chang, T. J. Wang and Y. Xu. 2007. Review of biomass pyrolysis oil properties and upgrading research. *Energy Conversion and Management* 48: 87–92.

Zhou, P. and L. Leydesdorff. 2006. The emergence of China as a leading nation in science. *Research Policy* 35: 83–104.

6 Characterization and Properties of Biooils

A Review of the Research

Ozcan Konur

CONTENTS

6.1 INTRODUCTION

Crude oils have been primary sources of energy and fuels, such as petrodiesel. However, significant public concerns about the sustainability, price fluctuations, and adverse environmental impact of crude oils have emerged since the 1970s (Ahmadun et al., 2009; Atlas, 1981; Babich and Moulijn, 2003; Kilian, 2009; Perron, 1989). Thus, biooils (Bridgwater et al., 1999; Bridgwater and Peacocke, 2000; Czernik and Bridgwater, 2004; Mohan et al., 2006; Zhang et al., 2007) and biooil-based biodiesel fuels (Chisti, 2007; Hill et al., 2006) have emerged as alternatives to crude oils and crude oil-based petrodiesel fuels in recent decades. Nowadays, although petrodiesel fuels are still used extensively, biodiesel fuels are being used increasingly in the transportation and power sectors (Konur, 2021a–ag). Therefore, there has been great public interest in the development of biooils, especially their properties and characterization (Bridgwater et al., 1999; Bridgwater and Peacocke, 2000; Czernik and Bridgwater, 2004; Mohan et al., 2006; Zhang et al., 2007).

However, for the efficient progression of the research in this field, it is necessary to develop efficient incentive structures for the primary stakeholders and to inform these stakeholders about the research (Konur, 2000, 2002a–c, 2006a–b, 2007a–b; North, 1991a–b).

Although there have been a number of reviews and book chapters in this field (Akhtar and Amin, 2011, 2012; Carpenter et al., 2014; Kan et al., 2016; Lehto et al., 2014; Lu et al., 2009; Oasmaa and Czernik, 1999; Zhang et al., 2007), there has been no review of the 25-most-cited articles in this field. This chapter reviews these articles by highlighting the key findings of these studies on the characterization and properties of biooils obtained from a variety of feedstocks. Then, it discusses these key findings.

6.2 MATERIALS AND METHODOLOGY

The search for the literature was carried out on the 'Web of Science' (WOS) database in February 2020. It contains the 'Science Citation Index-Expanded' (SCI-E), the Social Sciences Citation Index' (SSCI), the 'Book Citation Index-Science' (BCI-S), the 'Conference Proceedings Citation Index-Science' (CPCI-S), the 'Emerging Sources Citation Index' (ESCI), the 'Book Citation Index-Social Sciences and Humanities' (BCI-SSH), the 'Conference Proceedings Citation Index-Social Sciences and Humanities' (CPCI-SSH), the and 'Arts and Humanities Citation Index' (A&HCI).

The keywords for the search of the literature were collated from the screening of abstract pages for the first 1,000 highly cited papers on biooils. These keywords sets are provided in the Appendix of the related chapter (Konur, 2021i).

The 25-most-cited articles are selected for this review and the key findings are presented and discussed briefly.

6.3 RESULTS

6.3.1 Properties of Biooils

Biller and Ross (2011) study the yields and properties of biooil from the 'hydrothermal liquefaction' (HTL) of microalgae with different biochemical contents in a paper with 501 citations. They liquefy a range of model biochemical components, microalgae, and cyanobacteria with different biochemical contents under hydrothermal conditions at 350°C, at about 200 bar in water, with 1 M Na_2CO_3 and 1 M formic acid. The model compounds include albumin and a soya protein, starch and glucose, the triglyceride from sunflower oil, and two amino acids. They use *Chlorella vulgaris*, *Nannochloropsis occulata*, *Porphyridium cruentum*, and *Spirulina* and the yields and product distribution obtained for each model compound to predict the behavior of microalgae with different biochemical compositions and validate using microalgae and cyanobacteria. They observe broad agreement between predictive yields and actual yields for the microalgae based on their biochemical composition. The yields of algal biooil were 5–25 wt% higher than the lipid content of the algae, depending upon the biochemical composition. The yields of biooil follow the trend: lipids > proteins > carbohydrates.

Miao and Wu (2004) study the properties of algal pyrolysis oils in a paper with 334 citations. They develop an approach for increasing the yield of biooil production from fast pyrolysis after manipulating the metabolic pathway in microalgae through heterotrophic growth. They observe that the yield of biooil (57.9%) produced from heterotrophic *Chlorella protothecoides* cells was 3.4 times higher than from autotrophic cells by fast pyrolysis. The biooil had a much lower oxygen content, with a higher heating value (41 MJ kg^{-1}), a lower density (0.92 kgL^{-1}), and a lower viscosity (0.02 PAs) compared to those of biooil from autotrophic cells and wood. These properties were comparable to crude oil.

Miao et al. (2004) study the properties of algal pyrolysis oils in a paper with 328 citations. They perform fast pyrolysis tests of microalgae in a 'fluid bed reactor' at a temperature of 500°C with a heating rate of 600°Cs^{-1}, a sweep gas (N_2) flow rate of 0.4 m^3h^{-1}, and a vapor residence time of 2–3 s. In comparison with the studies on slow pyrolysis from microalgae in an autoclave, they produce a greater amount of high quality biooil from continuously processing microalgae feeds at a rate of 4 $gmin^{-1}$. They obtain the liquid product yields of 18 and 24% from the fast pyrolysis of *Chlorella protothecoides* and *Microcystis aeruginosa*, respectively. The saturated and polar fractions accounted for 1.14 and 31.17% of the biooils of microalgae on average, respectively, which were higher than those of biooil from wood. The H: C and O:C molar ratios of microalgae biooil were 1.7 and 0.3, respectively. The distribution of straight-chain alkanes of the saturated fractions from microalgal biooils were similar to diesel fuel. Biooil had low oxygen content with a higher heating value of 29 $MJkg^{-1}$, a density of 1.16 kg L^{-1}, and a viscosity of 0.10 PAs. These properties of microalgal biooil make it more suitable for fuel oil use than fast pyrolysis oils from lignocellulosic materials.

Fahmi et al. (2008) study the effect of lignin and inorganic species in a biomass on the yields, quality, and stability of pyrolysis oil in a paper with 319 citations. They use three low-lignin *Lolium Festuca* grasses. They observe that the mineral matter had a dominating effect on pyrolysis when compared to the lignin content, in terms of pyrolysis yields for organics, biochar, and biogases. However, the higher molecular weight compounds present in the pyrolysis oil were due to the lignin-derived compounds. The light organic fraction also increased in yield, but reduced in water content, as metals increased at the expense of the lignin content. They find that the fresh biooil and aged biooil had different compound intensities and concentrations, which was due to a large number of reactions occurring when the biooil was aged day by day. A large amount of repolymerization occurred as levoglucosan yields increased during the aging progress, while hydroxyacetaldehyde decreased. Washing of the biomass could improve biooil quality and stability for high ash feedstocks, but less so for the energy crops.

Ingram et al. (2008) study the pyrolysis of wood and bark in an 'Auger reactor' focusing on the physical properties and chemical analysis of the produced biooils in a paper with 296 citations. They produce biooil at 450°C by fast pyrolysis in a continuous Auger reactor using pine wood, pine bark, oak wood, and oak bark. They characterize the whole biooils and their pyrolytic lignin-rich ethyl acetate fractions by 'gas chromatograph mass spectrometer' GCMS, 'gel permeation chromatography' (GPC), calorific values, viscosity dependences on shear rates and temperatures, elemental analyses, ^1H and '^{13}C NMR spectroscopy', water analyses, and ash content. They show that these biooils were comparable to those produced by fast pyrolysis in fluidized bed and vacuum pyrolysis processes. Portable Auger reactors' might be used at locations in forests to generate biooil on-site for the transportation of less bulky biooil (versus raw biomass) to biorefineries or power generation units. The pyrolysis had lower heat transfer rates than those achieved in fluidized bed reactors.

Vardon et al. (2011) study the chemical properties of biooil from the HTL of *Spirulina*, swine manure, and digested anaerobic sludge in a paper with 293 citations. They explore the influence of wastewater feedstock composition on HTL biooil properties and the physicochemical characteristics. They convert these feedstocks under HTL conditions (300°C, 10–12 MPa, and 30 min reaction time). They observe that biooil yields ranged from 9.4% (digested sludge) to 32.6% (*Spirulina*). Although they estimate similar higher heating values (32.0–34.7 MJ/kg) for all product biooils, there were significant differences in biooil chemistry. Feedstock composition influenced the individual compounds identified as well as the biooil functional group chemistry. Molecular weights tracked with obdurate carbohydrate content followed the order: *Spirulina* < swine manure < digested sludge. They observe a similar trend in boiling point distributions and the long branched aliphatic contents. They show the importance of HTL feedstock composition and highlight the need for better understanding of biooil chemistries.

Oasmaa et al. (2003) describe the use of forestry residue pyrolysis liquids, their physicochemical properties, and the behavior of these biooils in a paper with 278 citations. They produce a 10–25 wt% top phase with a higher heating value from forestry residue due to the high content of extractives and low water content. However, it had high solid and ash contents. The main product, the bottom phase, was similar to bark-free wood pyrolysis biooil: volatile acids 8–10 wt%, aldehydes and ketones 10–15 wt%, water 25–30 wt%, sugar constituents 30–35 wt%, water-insoluble, mainly lignin-based constituents 15–20 wt %, and extractives 2–6 wt%. Its physical properties (water 28 wt %, pH 3.0, viscosity at 40°C 15 cSt, LHV 14 MJ/kg, solids less than 0.05 wt%) make it suitable for fuel oil use. The solids content was typically lower than in pine biooils. Needles and bark in forestry residue, especially in fresh green feedstock, yielded high alkali metal (400–1000 mg/kg), ash (0.1–0.2 wt%), and N (0.1–0.4 wt%) contents of the biooil compared to pine (50 mg/kg, 0.02–0.03 wt%, less than 0.1 wt%, respectively) biooils. This resulted in higher NO_x and particulate emissions in combustion.

Tsai et al. (2007) study the product yields and compositions of pyrolysis oils from rice husk in a paper with 265 citations. They prepare a series of pyrolysis oils and biochars from rice husk by the lab-scale fast pyrolysis system using induction heating. They examine the effect of process parameters, such as pyrolysis temperature, heating rate, holding time, N gas flow rate, condensation temperature, and particle size, on the pyrolysis product yields and their chemical compositions. They obtain the maximum biooil yield of over 40% at the proper pyrolysis conditions. The pyrolysis oils derived from the fast pyrolysis of rice husk contained considerable amounts of carbonyl groups and/or oxygen content, resulting in low pH and low heating values.

Jena et al. (2011) study optimum 'thermochemical liquefaction' (TCL) operating conditions for producing biooil from *Spirulina platensis* in a paper with 255 citations. They perform TCL experiments at various temperatures (200–380°C), holding times

(0–120 min), and solid concentrations (10–50%). They observe that TCL conversion at 350°C, a 60 min holding time, and 20% solids concentration produced the highest biooil yield of 39.9%, representing a 98.3% carbon conversion efficiency. Light fraction biooil appeared at 300°C or higher temperatures and represented 50–63% of the total biooil. Biooil obtained at 350–380°C had similar fuel properties to that of crude oil with an energy density of 34.7–39.9 MJ kg^{-1} compared to 42.9 MJ kg^{-1} for crude oil. Biooil from conversion at 300°C or above had 71–77% elemental carbon, 0.6–11.6% elemental oxygen, and viscosities in the range 40–68 cP. The GC/MS of biooil reported higher hydrocarbons (C_{16}–C_{17}), phenolics, carboxylic acids, esters, aldehydes, amines, and amides.

Diebold and Czernik (1997) study the additives that lower and stabilize the viscosity of pyrolysis oils during storage in a paper with 251 citations. The initial development of additives to stabilize the viscosity of biooil during long-term storage produced dramatic results. They use ethyl acetate, methyl isobutyl ketone and methanol, acetone, methanol, acetone and methanol, and ethanol as additives. These additives represent three chemical families, which all demonstrated the ability to drastically reduce the aging rate of biooil, as defined by the increase in viscosity with time. They perform accelerated aging tests at 90°C to screen the additives. The additives not only lowered the initial viscosity at 40°C by half but also reduced the aging rate of a hot gas filtered pyrolysis oil made from hybrid poplar (NREL run 175) by factors of 1–18 compared to the original pure biooil. With the best additive, methanol, at a 10 wt% level in the pyrolysis oil, the modified biooil was still a single-phase liquid and still met the ASTM No. 4 diesel fuel specification for viscosity, even after 96 h exposure at 90°C. Based on the aging rate at 90°C, recently determined for pure biooil without additives, the pure biooil tested would have exceeded the permitted ASTM No. 4 viscosity after only 2.6 h. In addition, the unmodified biooil formed a waxy precipitate that floated on top of the liquid phase after 8 h exposure at 90°C. Use of methanol with previously aged biooils greatly reduced the resultant viscosity, but not quite as effectively as the use of the methanol shortly after the pyrolysis oil was produced. The cost of the additive, e.g. methanol, may be offset by the heating value it adds to the pyrolysis oil, depending on the local cost of each.

Oasmaa and Kuoppala (2003) follow the storage stability of wood-based pyrolysis biooils by analysis of the changes in the physical properties and chemical composition during storage in a paper with 239 citations. They observe that the main physicochemical changes took place during the first six months. The high-molecular-mass (HMM) fraction of water-insolubles, which were originally lignin-derived material, increased, because of polymerization and condensation reactions of carbohydrate constituents, aldehydes, and ketones. Therefore, the average molecular mass of pyrolysis biooils increased, which was also observed as an increase in viscosity. There was a clear correlation of the average molecular mass with the viscosity, water-insolubles, and the HMM fraction of water-insolubles. The chemical changes in the aging were similar to those which occurred during the accelerated aging test. The decrease in volatile aldehydes and ketones increased the flash point of the bioliquids. The increase in viscosity increased the pour point. Water was formed as a by-product in various condensation reactions. Increases in the amount of water decreased the heating value. The density of the biooils increased, because the increase in the HMM lignin fraction was more significant than the increase in water.

Garcia-Perez et al. (2008) study the production of biooil, biochar, and pyrolytic biogases from the fast pyrolysis of mallee woody biomass, focusing on the effect of temperature on the yield and quality of pyrolysis products in a paper with 223 citations. The feedstock was ground, sieved to several narrow particle size ranges, and dried prior to pyrolysis in a novel laboratory-scale 'fluidized-bed reactor' at 350–600°C and a biomass particle size of 100–600 μm. They observe that the pyrolysis temperature had an important impact on the yield and composition of biooil, biochar, and biogases. Biomass particle size had a significant effect on the water content of biooil. The temperature for maximum biooil yield, between 450 and 475°C, resulted in a biooil with the highest content of oligomers and, consequently, with the highest viscosity. The increases in biooil yield with increasing temperature from 350 to 500°C were mainly due to the increases in the production of lignin-derived oligomers insoluble in water but soluble in CH_2Cl_2. They finally compare the yield and some fuel properties of the pyrolysis products with those obtained for pine as well as those reported in the literature for other lignocellulosic feedstocks, but using similar reactors.

Vardon et al. (2012) study the thermochemical conversion of raw and defatted algal biomass via HTL and slow pyrolysis in a paper with 206 citations. They use HTL (300°C and 10–12 MPa) and slow pyrolysis (heated to 450°C at a rate of 50°C/min) to produce biooils from *Scenedesmus* (raw and defatted) and *Spirulina* that were compared against Illinois shale oil. Although both thermochemical conversion routes produced energy dense biooil (35–37 MJ/kg) that approached shale oil (41 MJ/kg), biooil yields (24–45%) and physicochemical characteristics were highly influenced by conversion route and feedstock selection. They observe sharp differences in the mean biooil molecular weight (pyrolysis 280–360 Da; HTL 700–1330 Da) and the percentage of low boiling compounds (bp < 400°C) (pyrolysis 62–66%; HTL 45–54%). For a wet algal biomass there was an 80% moisture content. HTL was more favorable ('energy consumption ratio' (ECR) 0.44–0.63) than pyrolysis (ECR 0.92–1.24) due to required water volatilization in the latter technique.

6.3.2 Characterization of Biooils

Scholze and Meier (2001) characterize the water-insoluble fraction from pyrolysis oils from different fast pyrolysis processes in a paper with 358 citations. They obtain pyrolytic lignins from pyrolysis oil as a fine homogeneous powder by a novel precipitation method. Analysis methods comprise chromatography, spectroscopy, and wet chemical techniques. 'Fourier transform infrared spectroscopy' (FTIR) data indicate that a changing oxygen content mainly effects the intensity of carbonyl absorption bands. Therefore, FTIR analysis is valuable as a fast analytical method to elucidate the aging processes of pyrolysis oil. They observe that pyrolytic lignin was similar to technical lignins.

Garcia-Perez et al. (2007) describe an analytical approach to determine the chemical composition of biooils in terms of macrochemical families in a paper with 313 citations. They first fractionate biooils from the vacuum pyrolysis of softwood bark and hardwood using solvent extraction and characterize them using GCMS, 'thermogravimetric techniques', and 'gel permeation chromatography'. They interpret thermogravimetric and molar mass distribution curves of each fraction in terms of macrofamilies applying curve-fitting procedures. They find that the composition of the different macrofractions obtained was in agreement using both methods. This proposed procedure enabled a thorough description of biooil composition as a mixture of water, monolignols, polar compounds with moderate volatility, sugars, extractive-derived compounds, heavy polar and nonpolar compounds, methanol (MeOH)-toluene insolubles, and volatile organic compounds.

Sipila et al. (1998) develop an analytical scheme for the characterization of biomass-based flash pyrolysis oils in a paper with 297 citations. This scheme was based on fractionation of the biooils with water and on further extraction of the water-soluble fraction with diethylether. They analyze the chemical composition of the fractions by GCMS. They determine the physical and chemical nature of straw, pine, and hardwood pyrolysis oils and compare them with each other. They finally draw correlations between the physical properties and chemical compositions of the biooils.

Williams et al. (1990) study the impact of temperature and heating rate on product composition in a paper with 277 citations. They pyrolyze shredded automotive tire waste in a 200 cm^3 static batch reactor in an N_2 atmosphere. They determine the compositions and properties of the derived biogases, pyrolysis oils, and solid biochar in relation to pyrolysis temperatures up to 720°C and at heating rates between 5 and 80°C min^{-1}. As the pyrolysis temperature was increased, the percentage mass of solid biochar decreased, while biogas and biooil products increased until 600°C after which there was a minimal change to product yield, the scrap tires producing approximately 55% biooil, 10% biogas, and 35% biochar. There was a small effect of heating rate on the product yield. They identify the biogases as H_2, CO, CO_2, C_4H_6, CH_4, and C_2H_6, with lower concentrations of other hydrocarbon biogases. An increase in temperature produced a decrease in the proportion of aliphatic fractions and an increase in aromatic fractions for each heating rate. The molecular mass range of the biooils was up to 1,600 mass units with a peak in the 300–400 range. There was an increase in molecular mass range as the pyrolysis temperature was increased. They find the presence of alkanes, alkenes, ketones or aldehydes, aromatic, polyaromatic, and substituted aromatic groups, and there was a significant increase with increasing pyrolysis temperature and heating rate.

Horne and Williams (1996) study the impact of temperature on the products from the flash pyrolysis of biomass in a paper with 274 citations. They pyrolyze mixed wood waste in a fluidized bed reactor at 400, 450, 500, and 550°C. They analyze the biochar, biooil, and biogas products to determine their elemental composition and calorific value. The biogases evolved were CO_2, CO, and C_1–C_4 hydrocarbons. The biooils were homogeneous, of low viscosity, and highly oxygenated. The molecular weight range of the biooils was 50–1300 u. Only low quantities of hydrocarbons were present and the oxygenated and polar fractions were dominant. PAH up to MW 252 were present in the bioliquids. The concentration of PAH in the biooils increased with pyrolysis temperature, but even at the maximum pyrolysis temperature of 550°C the total concentration was less than 120 ppmw. The biooils contained significant quantities of phenolic compounds and the yield of phenol and its alkylated derivatives was highest at 500 and 550°C. Some of the oxygenated compounds identified were of high value.

Branca et al. (2003) characterize biooils generated from the low-temperature pyrolysis of wood in a paper with 265 citations. They perform the pyrolysis of beech wood for heating temperatures in the range 600–900°K, reproducing conditions of interest in countercurrent fixed-bed gasification. They observe that the yields of biooils (water and tars) increased with the heating temperature from about 40 to 55% of dry wood mass. Apart from qualitative identification of about 90 species, they apply GC/MS techniques to quantify 40–43% of tars (40 species). Decomposition of holocellulose led to the formation of furan derivatives and carbohydrates, with a temperature-dominated selectivity toward hydroxyacetaldehyde against levoglucosan. Syringols and guaiacols, originating from the primary degradation of lignin, presented a maximum for heating temperatures of about 750–800°K, whereas, owing to secondary degradation, phenols continuously increased.

Anastasakis and Ross (2011) study the hydrothermal liquefaction of *Laminaria saccharina* in a batch reactor focusing on the effect of reaction conditions on product distribution and composition in a paper with 263 citations. They assess the influence of reactor loading, residence time, temperature, and catalyst (KOH) loading. They obtain a maximum biooil yield of 19.3 wt% with a 1:10 biomass: water ratio at 350°C and a residence time of 15 min without the presence of the catalyst. The biooil had a higher 'heating value' (HHV) of 36.5 MJ/kg and was similar in nature to a heavy crude oil or bitumen. The solid residue had a high ash content and contained a large proportion of calcium and magnesium. The aqueous phase was rich in sugars and ammonium and contained a large proportion of potassium and sodium.

Cunliffe and Williams (1998) study the composition of biooils derived from the batch pyrolysis of tires in a paper with 251 citations. They use an N purged static-bed batch reactor to pyrolyze 3 kg batches of shredded scrap tires at temperatures between 450 and 600°C. They observe that the biooil yield decreased with increasing final pyrolysis temperature and that the yield of product biogases increased. They determine the fuel properties of the condensed biooil, including calorific values, ultimate analyses, flash point, moisture content, fluorine and chlorine content, as well as the concentration of polycyclic aromatic hydrocarbons (PAH) and lighter aromatic hydrocarbons. The derived tire biooils had fuel properties similar to those of a light crude oil. The influence of pyrolysis temperature showed an increase in the aromatic content of the biooils with increasing temperature, with a consequent decrease in aliphatic content. The total PAH content of the biooils increased from 1.5 to 3.5 wt% of the total biooil as the pyrolysis temperature was increased from 450 to 600°C. They identify biologically active compounds such as methylfluorenes, tri- and tetra-methylphenanthrenes, and chrysene in significant concentrations. The results of biogas analysis supported a Diels–Alder mechanism of alkane dehydrogenation to alkenes, followed by cyclization and aromatization. They identify limonene as a major component of the biooils, representing 3.1 wt% at 450°C, falling to 2.5 wt% total biooil at 600°C. They also find significant quantities of light aromatics such as benzene, toluene, xylene, and styrene.

Mullen and Boateng (2008) determine the chemical composition of biooils produced by fast pyrolysis of two energy crops in a paper with 244 citations. They analyze biooils from the fast pyrolysis of switchgrass forage and two sets of alfalfa stems by wet-chemical methods, GCMS, and 'high performance liquid chromatography' (HPLC). They perform pyrolysis experiments at 500°C under a nitrogen atmosphere in 2.5 kg/h fluidized bed reactor. They identify a total of 62 chemical species in the liquids and quantify 27 of them. While the compositions of the biooil from the two alfalfa stems were similar, there were numerous differences in the compositions of the alfalfa and switchgrass biooils. They note the higher levels of N, water, and aromatic hydrocarbons in biooils produced from alfalfa stems than from switchgrass and woody feedstocks that have been previously characterized. They also note a much lower concentration of levoglucosan and hydroxyacetaldehyde concentrations among biooils from alfalfa stems compared with biooil from switchgrass or woody biomass.

Zou et al. (2010) characterize biooil from the hydrothermal liquefaction of *Dunaliella tertiolecta* cake under various liquefaction temperatures, holding times, and catalyst dosages in a paper with 237 citations. They obtain a maximum biooil yield of 25.8% at a reaction temperature of 360°C and a holding time of 50 min using 5% Na_2CO_3 as a catalyst. They determine the various physical and chemical characteristics of biooil obtained under the most suitable conditions, and perform a detailed chemical compositional analysis of biooil using an elemental analyzer, FTIR, and GCMS. They observe that the biooil was composed of fatty acids, 'fatty acid methyl esters' (FAMEs), ketones, and aldehydes. Its empirical formula was $CH_{1.44}O_{0.29}N_{0.05}$ and its heating value was 30.74 MJ/kg. The biooil product was a possible eco-friendly green biofuel and chemical.

Demirbas (2007) studies the effect of temperature on the compounds in liquid products obtained from biomass pyrolysis in relation to the yield and composition of the product biooils in a paper with 219 citations. He finds that the biooils were composed of a range of cyclopentanone, methoxyphenol, acetic acid, methanol, acetone, furfural, phenol, formic acid, levoglucosan, guaiacol, and their alkylated phenol derivatives. The structural components of the biomass samples mainly affected the pyrolytic degradation products. He proposes a reaction mechanism which describes a possible reaction route for the formation of the characteristic compounds found in the biooils. The supercritical water extraction and liquefaction partial reactions also occurred during the pyrolysis. Acetic acid was formed in the thermal decomposition of all three main components of biomass. In the pyrolysis reactions of biomass: water was formed by dehydration; acetic acid comes from the elimination of acetyl groups originally linked to the xylose unit; furfural was formed by dehydration of the xylose unit; formic acid proceeded from carboxylic groups of uronic acid; and methanol arose from methoxyl groups of uronic acid.

Patwardhan et al. (2011) study the primary pyrolysis product distribution of hemicelluloses extracted and purified from switchgrass in a paper with 206 citations. They use a combination of several analytical techniques, including micro-pyrolyzer, GCMS, 'flame ionization detector' (FID), gas analysis, and capillary electrophoresis. They identify a total of 16 products and quantify which ones accounted for 85% of the overall mass balance. They observe that the pyrolysis behavior of hemicellulose was considerably different than cellulose and explain it on the basis of a proposed mechanism for glycosidic bond cleavage. Further, they examine the effect of minerals and temperature. They provide insight into the fast pyrolysis behavior of hemicellulose and provide a basis for developing models that can predict biooil composition resulting from overall biomass fast pyrolysis.

6.4 DISCUSSION

Table 6.1 provides information on the research fronts in this field. As this table shows the primary research fronts of 'properties of biooils' and 'characterization of biooils' comprise 52 and 48% of these papers, respectively.

6.4.1 Properties of Biooils

Biller and Ross (2011) study the yields and properties of biooil from the hydrothermal liquefaction (HTL) of microalgae with different biochemical content in a paper with 501 citations. Miao and Wu (2004) and Miao et al. (2004) study the properties of algal pyrolysis oils in papers with 334 and 328 citations, respectively. Fahmi et al. (2008) study the effect of lignin and inorganic species in biomass on yields, quality, and stability of pyrolysis oil in a paper with 319 citations.

Ingram et al. (2008) study the pyrolysis of wood and bark in an 'Auger reactor' focusing on the physical properties and chemical analysis of the produced biooils in a paper with 296 citations. Vardon et al. (2011) study the chemical properties of biooil from the HTL of *Spirulina*, swine manure, and digested anaerobic sludge in a paper with 293 citations. Oasmaa et al. (2003) describe the use of forestry residue pyrolysis oils, their physicochemical properties, and the behavior of these biooils in a paper with 278 citations.

Tsai et al. (2007) study the product yields and compositions of pyrolysis oils from rice husk in a paper with 265 citations. Jena et al. (2011) study the optimum thermochemical liquefaction operating conditions for producing biooil from *Spirulina platensis* in a paper with 255 citations. Diebold and Czernik (1997) study the additives to lower and stabilize the viscosity of pyrolysis oils during storage in a paper with 251 citations.

Oasmaa and Kuoppala (2003) follow the storage stability of wood-based pyrolysis bioliquids by analysis of the changes in their physical properties and chemical composition during storage in a paper with 239 citations. Garcia-Perez et al. (2008) study the production of biooil, biochar, and pyrolytic biogases from the fast pyrolysis of mallee woody biomass focusing on the effect of temperature on the yield and quality of pyrolysis products in a paper with 223 citations. Vardon et al. (2012) study the thermochemical conversion of raw and defatted algal biomass via HTL and slow pyrolysis in a paper with 206 citations.

TABLE 6.1
Research Fronts

	Research Front	No. of Papers (%)
1	Properties of biooils	52
2	Characterization of biooils	48

These prolific studies highlight the properties of biooils obtained from a variety of feedstock such as algae and lignocellulosic feedstocks for primarily biodiesel production.

6.4.2 Characterization of Biooils

Scholze and Meier (2001) characterize the water-insoluble fraction from pyrolysis oils from different fast pyrolysis processes in a paper with 358 citations. Garcia-Perez et al. (2007) describe an analytical approach to determine the chemical composition of biooils in terms of macrochemical families in a paper with 313 citations. Sipila et al. (1998) develop an analytical scheme for the characterization of biomass-based flash pyrolysis oils in a paper with 297 citations.

Williams et al. (1990) study the impact of temperature and heating rate on product composition in a paper with 277 citations. Horne and Williams (1996) study the impact of temperature on the products from the flash pyrolysis of biomass in a paper with 274 citations. Branca et al. (2003) characterize bioliquids generated from the low-temperature pyrolysis of wood in a paper with 265 citations.

Anastasakis and Ross (2011) study the hydrothermal liquefaction of *Laminaria saccharina* in a batch reactor focusing on the effect of reaction conditions on product distribution and composition in a paper with 263 citations. Cunliffe and Williams (1998) study the composition of biooils derived from the batch pyrolysis of tires in a paper with 251 citations. Mullen and Boateng (2008) determine the chemical composition of biooils produced by fast pyrolysis of two energy crops in a paper with 244 citations.

Zou et al. (2010) characterize biooil from hydrothermal liquefaction of *Dunaliella tertiolecta* cake under various liquefaction temperatures, holding times, and catalyst dosages in a paper with 237 citations. Demirbas (2007) studies the effect of temperature on the compounds in liquid products obtained from biomass pyrolysis in relation to the yield and composition of the product biooils in a paper with 219 citations. Patwardhan et al. (2011) study the primary pyrolysis product distribution of hemicelluloses extracted and purified from switchgrass in a paper with 206 citations.

These prolific studies highlight the characterization of biooils obtained from a variety of feedstock such as algae and lignocellulosic feedstocks using a variety of diagnostic and analytical tools for primarily biodiesel production.

6.5 CONCLUSION

This chapter presents the key findings of the 25-most-cited article papers in this field. Table 6.1 provides information on the research fronts. As this table shows the primary research fronts of 'characterization of biooils' and 'properties of biooils' comprise 48 and 52% of these papers, respectively.

These prolific studies on two complementary research fronts provide valuable evidence on the characterization and properties of biooils obtained from a variety of feedstock such as algae and wood.

It is recommended that similar studies are carried out for each research front as well.

ACKNOWLEDGMENTS

The contribution of the highly cited researchers in this field is greatly acknowledged.

REFERENCES

Ahmadun, F. R., A. Pendashteh, L. C. Abdullah, et al. 2009. Review of technologies for oil and gas produced water treatment. *Journal of Hazardous Materials* 170: 530–551.

Akhtar, J. and N. A. S. Amin. 2011. A review on process conditions for optimum bio-oil yield in hydrothermal liquefaction of biomass. *Renewable & Sustainable Energy Reviews* 15: 1615–1624.

Akhtar, J. and N. A. S. Amin. 2012. A review on operating parameters for optimum liquid oil yield in biomass pyrolysis. *Renewable & Sustainable Energy Reviews* 16: 5101–5109.

Anastasakis, K. and A. B. Ross. 2011. Hydrothermal liquefaction of the brown macro-alga *Laminaria Saccharina*: Effect of reaction conditions on product distribution and composition. *Bioresource Technology* 102: 4876–4883.

Atlas, R. M. 1981. Microbial degradation of petroleum hydrocarbons: An environmental perspective. *Microbiological Reviews* 45: 180–209.

Babich, I. V. and J. A. Moulijn. 2003. Science and technology of novel processes for deep desulfurization of oil refinery streams: A review. *Fuel* 82: 607–631.

Biller, P. and A. B. Ross. 2011. Potential yields and properties of oil from the hydrothermal liquefaction of microalgae with different biochemical content. *Bioresource Technology* 102: 215–225.

Branca, C., P. Giudicianni, and C. di Blasi. 2003. GC/MS characterization of liquids generated from low-temperature pyrolysis of wood. *Industrial & Engineering Chemistry Research* 42: 3190–3202.

Bridgwater, A. V. and G. V. C. Peacocke. 2000. Fast pyrolysis processes for biomass. *Renewable & Sustainable Energy Reviews* 4: 1–73.

Bridgwater, A. V., D. Meier, and D. Radlein. 1999. An overview of fast pyrolysis of biomass. *Organic Geochemistry* 30: 1479–1493.

Carpenter, D., T. L. Westover, S. Czernik, and W. Jablonski. 2014. Biomass feedstocks for renewable fuel production: A review of the impacts of feedstock and pretreatment on the yield and product distribution of fast pyrolysis bio-oils and vapors. *Green Chemistry* 16: 384–406.

Chisti, Y. 2007. Biodiesel from microalgae. *Biotechnology Advances* 25: 294–306.

Cunliffe, A. M. and P. T. Williams. 1998. Composition of oils derived from the batch pyrolysis of tyres. *Journal of Analytical and Applied Pyrolysis* 44: 131–152.

Czernik, S. and A. V. Bridgwater. 2004. Overview of applications of biomass fast pyrolysis oil. *Energy & Fuels* 18: 590–598.

Demirbas, A. 2007. The influence of temperature on the yields of compounds existing in bio-oils obtained from biomass samples via pyrolysis. *Fuel Processing Technology* 88: 591–597.

Diebold, J. P. and S. Czernik. 1997. Additives to lower and stabilize the viscosity of pyrolysis oils during storage. *Energy & Fuels* 11:1081–1091.

Fahmi, R., A. Bridgwater, I. Donnison, N. Yates, and J. M. Jones. 2008. The effect of lignin and inorganic species in biomass on pyrolysis oil yields, quality and stability. *Fuel* 87: 1230–1240.

Garcia-Perez, M., A. Chaala, H. Pakdel, D. Kretschmer, and C. Roy. 2007. Characterization of bio-oils in chemical families. *Biomass & Bioenergy* 31: 222–242.

Garcia-Perez, X. S. Wang, J. Shen, et al. 2008. Fast pyrolysis of oil mallee woody biomass: Effect of temperature on the yield and quality of pyrolysis products. *Industrial & Engineering Chemistry Research* 47: 1846–1854.

Hill, J., E. Nelson, D. Tilman, S. Polasky, and D. Tiffany. 2006. Environmental, economic, and energetic costs and benefits of biodiesel and ethanol biofuels. *Proceedings of the National Academy of Sciences of the United States of America* 103: 11206–11210.

Horne, P. A. and P. T. Williams. 1996. Influence of temperature on the products from the flash pyrolysis of biomass. *Fuel* 75: 1051–1059.

Ingram, L., D. Mohan, M. Bricka, et al. 2008. Pyrolysis of wood and bark in an Auger reactor: Physical properties and chemical analysis of the produced bio-oils. *Energy & Fuels* 22: 614–625.

Jena, U., K. C. Das, and J. R. Kastner. 2011. Effect of operating conditions of thermochemical liquefaction on biocrude production from *Spirulina platensis*. *Bioresource Technology* 102: 6221–6229.

Kan, T., V. Strezov, and T. J. Evans. 2016. Lignocellulosic biomass pyrolysis: A review of product properties and effects of pyrolysis parameters. *Renewable & Sustainable Energy Reviews* 57: 1126–1140.

Kilian, L. 2009. Not all oil price shocks are alike: Disentangling demand and supply shocks in the crude oil market. *American Economic Review* 99: 1053–1069.

Konur, O. 2000. Creating enforceable civil rights for disabled students in higher education: An institutional theory perspective. *Disability & Society* 15: 1041–1063.

Konur, O. 2002a. Access to Nursing Education by disabled students: Rights and duties of nursing programs. *Nurse Education Today* 22: 364–374.

Konur, O. 2002b. Assessment of disabled students in higher education: Current public policy issues. *Assessment and Evaluation in Higher Education* 27: 131–152.

Konur, O. 2002c. Access to employment by disabled people in the UK: Is the Disability Discrimination Act working? *International Journal of Discrimination and the Law* 5: 247–279.

Konur, O. 2006a. Participation of children with dyslexia in compulsory education: Current public policy issues. *Dyslexia* 12: 51–67.

Konur, O. 2006b. Teaching disabled students in Higher Education. *Teaching in Higher Education* 11: 351–363.

Konur, O. 2007a. A judicial outcome analysis of the Disability Discrimination Act: A windfall for the employers? *Disability & Society* 22: 187–204.

Konur, O. 2007b. Computer-assisted teaching and assessment of disabled students in higher education: The interface between academic standards and disability rights. *Journal of Computer Assisted Learning* 23: 207–219.

Konur, O., ed. 2021a. *Handbook of Biodiesel and Petrodiesel Fuels: Science, Technology, Health, and Environment*. Boca Raton, FL: CRC Press.

Konur, O., ed. 2021b. *Handbook of Biodiesel and Petrodiesel Fuels: Science, Technology, Health, and Environment. Volume 1. Biodiesel Fuels: Science, Technology, Health, and Environment.* Boca Raton, FL: CRC Press.

Konur, O., ed. 2021c. *Handbook of Biodiesel and Petrodiesel Fuels: Science, Technology, Health, and Environment. Volume 2. Biodiesel Fuels based on the Edible and Nonedible Feedstocks, Wastes, and Algae: Science, Technology, Health, and Environment.* Boca Raton, FL: CRC Press.

Konur, O., ed. 2021d. *Handbook of Biodiesel and Petrodiesel Fuels: Science, Technology, Health, and Environment. Volume 3. Petrodiesel Fuels: Science, Technology, Health, and Environment.* Boca Raton, FL: CRC Press.

Konur, O. 2021e. Biodiesel and petrodiesel fuels: Science, technology, health, and environment. In *Handbook of Biodiesel and Petrodiesel Fuels: Science, Technology, Health, and Environment. Volume 1. Biodiesel Fuels: Science, Technology, Health, and Environment*, ed. O. Konur. Boca Raton, FL: CRC Press.

Konur, O. 2021f. Biodiesel and petrodiesel fuels: A scientometric review of the research. In *Handbook of Biodiesel and Petrodiesel Fuels: Science, Technology, Health, and Environment. Volume 1. Biodiesel Fuels: Science, Technology, Health, and Environment*, ed. O. Konur. Boca Raton, FL: CRC Press.

Konur, O. 2021g. Biodiesel and petrodiesel fuels: A review of the research. In *Handbook of Biodiesel and Petrodiesel Fuels: Science, Technology, Health, and Environment. Volume 1. Biodiesel Fuels: Science, Technology, Health, and Environment*, ed. O. Konur. Boca Raton, FL: CRC Press.

Konur, O. 2021h. Nanotechnology applications in the diesel fuels and the related research fields: A review of the research. In *Handbook of Biodiesel and Petrodiesel Fuels: Science, Technology, Health, and Environment. Volume 1. Biodiesel Fuels: Science, Technology, Health, and Environment*, ed. O. Konur. Boca Raton, FL: CRC Press.

Konur, O. 2021i. Biooils: A scientometric review of the research. In *Handbook of Biodiesel and Petrodiesel Fuels: Science, Technology, Health, and Environment. Volume 1. Biodiesel Fuels: Science, Technology, Health, and Environment*, ed. O. Konur. Boca Raton, FL: CRC Press.

Konur, O. 2021j. Characterization and properties of biooils: A review of the research. In *Handbook of Biodiesel and Petrodiesel Fuels: Science, Technology, Health, and Environment. Volume 1. Biodiesel Fuels: Science, Technology, Health, and Environment*, ed. O. Konur. Boca Raton, FL: CRC Press.

Konur, O. 2021k. Biomass pyrolysis and pyrolysis oils: A review of the research. In *Handbook of Biodiesel and Petrodiesel Fuels: Science, Technology, Health, and Environment. Volume 1. Biodiesel Fuels: Science, Technology, Health, and Environment*, ed. O. Konur. Boca Raton, FL: CRC Press.

Konur, O. 2021l. Biodiesel fuels: A scientometric review of the research. In *Handbook of Biodiesel and Petrodiesel Fuels: Science, Technology, Health, and Environment. Volume 1. Biodiesel Fuels: Science, Technology, Health, and Environment*, ed. O. Konur. Boca Raton, FL: CRC Press.

Konur, O. 2021m. Glycerol: A scientometric review of the research. In *Handbook of Biodiesel and Petrodiesel Fuels: Science, Technology, Health, and Environment. Volume 1. Biodiesel Fuels: Science, Technology, Health, and Environment*, ed. O. Konur. Boca Raton, FL: CRC Press.

Konur, O. 2021n. Propanediol production from glycerol: A review of the research. In *Handbook of Biodiesel and Petrodiesel Fuels: Science, Technology, Health, and Environment. Volume 1. Biodiesel Fuels: Science, Technology, Health, and Environment*, ed. O. Konur Boca Raton, FL: CRC Press. February 7, 2020.

Konur, O. 2021o. Edible oil-based biodiesel fuels: A scientometric review of the research. *In Handbook of Biodiesel and Petrodiesel Fuels: Science, Technology, Health, and Environment. Volume 2. Biodiesel Fuels based on the Edible and Nonedible Feedstocks, Wastes, and Algae: Science, Technology, Health, and Environment,* ed. O. Konur. Boca Raton, FL: CRC Press.

Konur, O. 2021p. Palm oil-based biodiesel fuels: A review of the research. In *Handbook of Biodiesel and Petrodiesel Fuels: Science, Technology, Health, and Environment. Volume 2. Biodiesel Fuels based on the Edible and Nonedible Feedstocks, Wastes, and Algae*, ed. O. Konur. Boca Raton, FL: CRC Press.

Konur, O. 2021q. Rapeseed oil-based biodiesel fuels: A review of the research. In *Handbook of Biodiesel and Petrodiesel Fuels: Science, Technology, Health, and Environment. Volume 2. Biodiesel Fuels based on the Edible and Nonedible Feedstocks, Wastes, and Algae,* ed. O. Konur. Boca Raton, FL: CRC Press.

Konur, O. 2021r. Nonedible oil-based biodiesel fuels: A scientometric review of the research. In *Handbook of Biodiesel and Petrodiesel Fuels: Science, Technology, Health, and Environment. Volume 2. Biodiesel Fuels based on the Edible and Nonedible Feedstocks, Wastes, and Algae: Science, Technology, Health, and Environment*, ed. O. Konur. Boca Raton, FL: CRC Press.

Konur, O. 2021s. Waste oil-based biodiesel fuels: A scientometric review of the research. In *Handbook of Biodiesel and Petrodiesel Fuels: Science, Technology, Health, and*

Environment. Volume 2. Biodiesel Fuels based on the Edible and Nonedible Feedstocks, Wastes, and Algae: Science, Technology, Health, and Environment, ed. O. Konur. Boca Raton, FL: CRC Press.

Konur, O. 2021t. Algal biodiesel fuels: A scientometric review of the research. In *Handbook of Biodiesel and Petrodiesel Fuels: Science, Technology, Health, and Environment. Volume 2. Biodiesel Fuels based on the Edible and Nonedible Feedstocks, Wastes, and Algae: Science, Technology, Health, and Environment*, ed. O. Konur. Boca Raton, FL: CRC Press.

Konur, O. 2021u. Algal biomass production for biodiesel production: A review of the research. In *Handbook of Biodiesel and Petrodiesel Fuels: Science, Technology, Health, and Environment. Volume 2. Biodiesel Fuels based on the Edible and Nonedible Feedstocks, Wastes, and Algae*, Ed. O. Konur Boca Raton, FL: CRC Press. February 23, 2020.

Konur, O. 2021v. Algal biomass production in wastewaters for biodiesel production: A review of the research. In *Handbook of Biodiesel and Petrodiesel Fuels: Science, Technology, Health, and Environment. Volume 2. Biodiesel Fuels based on the Edible and Nonedible Feedstocks, Wastes, and Algae*, ed. O. Konur. Boca Raton, FL: CRC Press. February 23, 2020.

Konur, O. 2021x. Algal lipid production for biodiesel production: A review of the research. In *Handbook of Biodiesel and Petrodiesel Fuels: Science, Technology, Health, and Environment. Volume 2. Biodiesel Fuels based on the Edible and Nonedible Feedstocks, Wastes, and Algae*, Ed. O. Konur Boca Raton, FL: CRC Press.

Konur, O. 2021y. Crude oils: A scientometric review of the research. In *Handbook of Biodiesel and Petrodiesel Fuels: Science, Technology, Health, and Environment. Volume 3. Petrodiesel Fuels: Science, Technology, Health, and Environment*, ed. O. Konur. Boca Raton, FL: CRC Press.

Konur, O. 2021z. Petrodiesel fuels: A scientometric review of the research. In *Handbook of Biodiesel and Petrodiesel Fuels: Science, Technology, Health, and Environment. Volume 3. Petrodiesel Fuels: Science, Technology, Health, and Environment*, ed. O. Konur. Boca Raton, FL: CRC Press.

Konur, O. 2021aa. Bioremediation of petroleum hydrocarbons in the contaminated soils: A review of the research. In *Handbook of Biodiesel and Petrodiesel Fuels: Science, Technology, Health, and Environment. Volume 3. Petrodiesel Fuels: Science, Technology, Health, and Environment*, ed. O. Konur. Boca Raton, FL: CRC Press.

Konur, O. 2021ab. Desulfurization of diesel fuels: A review of the research. In *Handbook of Biodiesel and Petrodiesel Fuels: Science, Technology, Health, and Environment. Volume 3. Petrodiesel Fuels: Science, Technology, Health, and Environment*, ed. O. Konur. Boca Raton, FL: CRC Press.

Konur, O. 2021ac. Diesel fuel exhaust emissions: A scientometric review of the research. In *Handbook of Biodiesel and Petrodiesel Fuels: Science, Technology, Health, and Environment. Volume 3. Petrodiesel Fuels: Science, Technology, Health, and Environment*, ed. O. Konur. Boca Raton, FL: CRC Press.

Konur, O. 2021ad. The adverse health and safety impact of diesel fuels: A scientometric review of the research. In *Handbook of Biodiesel and Petrodiesel Fuels: Science, Technology, Health, and Environment. Volume 3. Petrodiesel Fuels: Science, Technology, Health, and Environment*, ed. O. Konur. Boca Raton, FL: CRC Press.

Konur, O. 2021ae. Respiratory illnesses caused by the diesel fuel exhaust emissions: A review of the research. In *Handbook of Biodiesel and Petrodiesel Fuels: Science, Technology, Health, and Environment. Volume 3. Petrodiesel Fuels: Science, Technology, Health, and Environment*, ed. O. Konur. Boca Raton, FL: CRC Press.

Konur, O. 2021af. Cancer caused by the diesel fuel exhaust emissions: A review of the research. In *Handbook of Biodiesel and Petrodiesel Fuels: Science, Technology, Health, and Environment. Volume 3. Petrodiesel Fuels: Science, Technology, Health, and Environment*, ed. O. Konur. Boca Raton, FL: CRC Press.

Konur, O. 2021ag. Cardiovascular and other illnesses caused by the diesel fuel exhaust emissions: A review of the research. In *Handbook of Biodiesel and Petrodiesel Fuels: Science, Technology, Health, and Environment. Volume 3. Petrodiesel Fuels: Science, Technology, Health, and Environment*, ed. O. Konur. Boca Raton, FL: CRC Press.

Lehto, J., A. Oasmaa, Y. Solantausta, M. Kyto, and D. Chiaramonti. 2014. Review of fuel oil quality and combustion of fast pyrolysis bio-oils from lignocellulosic biomass. *Applied Energy* 116: 178–190.

Lu, Q., W.Z. Li, and X. F. Zhu. 2009. Overview of fuel properties of biomass fast pyrolysis oils. *Energy Conversion and Management* 50: 1376–1383.

Miao, X. L. and Q. Y. Wu. 2004. High yield bio-oil production from fast pyrolysis by metabolic controlling of *Chlorella protothecoides*. *Journal of Biotechnology* 110: 85–93.

Miao, X. L., Q. Y. Wu, and C. Y. Yang. 2004. Fast pyrolysis of microalgae to produce renewable fuels. *Journal of Analytical and Applied Pyrolysis* 71: 855–863.

Mohan, D., C. U. Pittman, and P. H. Steele. 2006. Pyrolysis of wood/biomass for bio-oil: A critical review. *Energy & Fuels* 20: 848–889.

Mullen, C. A. and A. A. Boateng. 2008. Chemical composition of bio-oils produced by fast pyrolysis of two energy crops. *Energy & Fuels* 22: 2104–2109.

North, D. C. 1991a. *Institutions, Institutional Change and Economic Performance*. Cambridge, Mass.: Cambridge University Press.

North, D.C. 1991b. Institutions. *Journal of Economic Perspectives* 5: 97–112.

Oasmaa, A. and S. Czernik. 1999. Fuel oil quality of biomass pyrolysis oils: State of the art for the end user. *Energy & Fuels* 13: 914–921.

Oasmaa, A. and E. Kuoppala. 2003. Fast pyrolysis of forestry residue. 3. Storage stability of liquid fuel. *Energy & Fuels* 17: 1075–1084.

Oasmaa, A., E. Kuoppala, and Y. Solantausta. 2003. Fast pyrolysis of forestry residue. 2. Physicochemical composition of product liquid. *Energy & Fuels* 17: 433–443.

Patwardhan, P. R., R. C. Brown, and B. H. Shanks. 2011. Product distribution from the fast pyrolysis of hemicellulose. *Chemsuschem* 4: 636–643.

Perron, P. 1989. The great crash, the oil price shock, and the unit root hypothesis. *Econometrica: Journal of the Econometric Society* 57: 1361–1401.

Scholze, B. and D. Meier. 2001. Characterization of the water-insoluble fraction from pyrolysis oil (pyrolytic lignin). Part I. PY-GC/MS, FTIR, and functional groups. *Journal of Analytical and Applied Pyrolysis* 60: 41–54.

Sipila, K., E. Kuoppala, L. Fagernas, and A. Oasmaa. 1998. Characterization of biomass-based flash pyrolysis oils. *Biomass & Bioenergy* 14: 103–113.

Tsai, W. T., M. K. Lee, and Y. M. Chang. 2007. Fast pyrolysis of rice husk: Product yields and compositions. *Bioresource Technology* 98: 22–28.

Vardon, D. R., B. K. Sharma, J. Scott, et al. 2011. Chemical properties of biocrude oil from the hydrothermal liquefaction of *Spirulina* algae, swine manure, and digested anaerobic sludge. 2011. *Bioresource Technology* 102: 8295–8303.

Vardon, D. R., B. K. Sharma, G. V. Blazina, K. Rajagopalan, and T. J. Strathmann. 2012. Thermochemical conversion of raw and defatted algal biomass via hydrothermal liquefaction and slow pyrolysis. *Bioresource Technology* 109: 178–187.

Williams, P. T., S. Besler, and D. T. Taylor. 1990. The pyrolysis of scrap automotive tyres: The influence of temperature and heating rate on product composition. *Fuel* 69: 1474–1482.

Zhang, Q., J. Chang, T. J. Wang, and Y. Xu. 2007. Review of biomass pyrolysis oil properties and upgrading research. *Energy Conversion and Management* 48: 87–92.

Zou, S. P., Y. L. Wu, M. D. Yang, et al. 2010. Production and characterization of bio-oil from hydrothermal liquefaction of microalgae *Dunaliella tertiolecta* cake. *Energy* 35: 5406–5411.

7 Biomass Pyrolysis and Pyrolysis Oils

A Review of the Research

Ozcan Konur

CONTENTS

7.1 INTRODUCTION

Crude oils have been primary sources of energy and fuels, such as petrodiesel fuel. However, significant public concerns about the sustainability, price fluctuations, and adverse environmental impact of crude oils have emerged since the 1970s (Ahmadun et al., 2009; Atlas, 1981; Babich and Moulijn, 2003; Kilian, 2009; Perron, 1989). Thus, biooils (Bridgwater et al., 1999; Bridgwater and Peacocke, 2000; Czernik and Bridgwater, 2004; Mohan et al., 2006; Zhang et al., 2007) and biooil-based biodiesel fuels (Chisti, 2007; Hill et al., 2006) have emerged as alternatives to crude oils and crude oil-based petrodiesel fuels in recent decades. Nowadays, although petrodiesel fuels are still used extensively, biodiesel fuels are being used increasingly in the transportation and power sectors (Konur, 2021a–ag). Therefore, there has been great public interest in the development of biooils, especially pyrolysis oils and their upgrading (Bridgwater et al., 1999; Bridgwater and Peacocke, 2000; Czernik and Bridgwater, 2004; Mohan et al., 2006; Zhang et al., 2007).

However, for the efficient progression of the research in this field, it is necessary to develop efficient incentive structures for the primary stakeholders and to inform these stakeholders about the research (Konur, 2000, 2002a–c, 2006a–b, 2007a–b; North, 1991a–b).

Although there have been a number of reviews and book chapters in this field (Bridgwater et al., 1999; Bridgwater and Peacocke, 2000; Czernik and Bridgwater, 2004; Mohan et al., 2006; Zhang et al., 2007), there has been no review of the 25-most-cited articles. Thus, this chapter reviews these articles by highlighting the key findings of these studies on biomass pyrolysis and pyrolysis oils. Then, it discusses these key findings.

7.2 MATERIALS AND METHODOLOGY

The search for the literature was carried out in the 'Web of Science' (WOS) database in February 2020. It contains the 'Science Citation Index-Expanded' (SCI-E), the Social Sciences Citation Index' (SSCI), the 'Book Citation Index-Science' (BCI-S), the 'Conference Proceedings Citation Index-Science' (CPCI-S), the 'Emerging Sources Citation Index' (ESCI), the 'Book Citation Index-Social Sciences and Humanities' (BCI-SSH), the 'Conference Proceedings Citation Index-Social Sciences and Humanities' (CPCI-SSH), and the 'Arts and Humanities Citation Index' (A&HCI).

The keywords for the search of the literature are collated from the screening of abstract pages for the first 1,000 highly cited papers in biooils. These keyword sets are provided in the Appendix of the related book chapter (Konur, 2021i).

The 25-most-cited articles are selected for this review and the key findings are presented and discussed briefly.

7.3 RESULTS

7.3.1 Pyrolysis

7.3.1.1 Kinetic Studies

Yang et al. (2007) study the pyrolysis characteristics of biomass components using a 'thermogravimetric analyzer' (TGA) with a 'differential scanning calorimetry' (DSC) detector and a 'packed-bed pyrolyzer' in a paper with 2,891 citations. They find that, in thermal analysis, the pyrolysis of hemicellulose and cellulose occurred quickly, with the weight loss of hemicellulose mainly happening at 220–315°C and that of cellulose at 315–400°C. However, lignin was more difficult to decompose, as its weight loss happened over a wide temperature range (from 160 to 900°C) and the generated solid residue was very high (around 40 wt%). From the viewpoint of energy consumption in the course of pyrolysis, cellulose behaved differently from hemicellulose and lignin; the pyrolysis of the former was endothermic while that of the latter was exothermic. The main gas products from pyrolyzing the three components were similar, including CO_2, CO, CH_4, and some organics. Hemicellulose had a higher CO_2 yield, cellulose generated a higher CO yield, whilst lignin had higher H_2 and CH_4 yields.

Evans and Milne (1987) apply the technique of 'molecular beam mass spectrometry' (MBMS) sampling to the elucidation of the molecular pathways in the fast pyrolysis of wood and its principal isolated constituents in a paper with 633 citations. They optimize high-value biofuel products by thermal and catalytic means. They obtain the 'positive-ion mass spectra' shown from real-time, direct sampling of light gases, reactive intermediates, and condensable vapors simultaneously. They find that the cellulose, lignin, and hemicellulose components of wood pyrolyze largely to monomer and monomer-related fragments. Whole wood behaves as the sum of its constituents, with few if any vapor species derived from interaction of the main polymer constituents. An important interaction, however, is the influence of mineral matter in the wood on the carbohydrate pyrolysis pathways. Vapor phase cracking of the primary products proceeds through a stage of light hydrocarbons and oxygenates to the ultimate formation of aromatic tars and H_2, CO, CO_2, and H_2O. They propose a relatively simple pyrolysis reaction scheme.

Yang et al. (2006) study biomass pyrolysis based on three major components using a thermogravimetric analysis (TGA) in a paper with 559 citations. They first analyze the pyrolysis characteristics of these three components and divide the process of biomass pyrolysis into four ranges according to the temperatures specified by the individual components. Second, they develop synthesized biomass samples containing two or three of the biomass components on the basis of a simplex-lattice approach. The pyrolysis of the synthesized samples indicates negligible interaction among the three components and a linear relationship occurring between the weight loss and proportion of hemicellulose (or cellulose) and residues at the specified temperature ranges. Finally, they establish two sets of multiple linear-regression equations for predicting the component proportions in a biomass and the weight loss of a biomass during pyrolysis in TGA, respectively. They find that the results of the calculations for the synthesized samples are consistent with the experimental measurements. Furthermore, to validate the computational approach, they perform TGA experimental analysis of the three components of palm oil wastes.

Orfao et al. (1999) study the behavior of biomass components thermogravimetrically with linear temperature programming under N and air in a paper with 546 citations. They compare the results and determine the pyrolysis kinetics of cellulose, assuming a first-order kinetic function. They find that the thermal decomposition of xylan and lignin could not be modeled with acceptable errors by means of simple reactions. They determine thermograms for pine and eucalyptus woods and pine bark, under an inert (N) or oxidizing (air) atmosphere. They model the pyrolysis of these lignocellulosic materials with good approximation by three first-order independent reactions. One of these reactions is associated with the primary pyrolysis of cellulose, its parameters being previously determined and fixed in the model. The model parameters are the activation energies and pre-exponential factors for the pyrolysis of the remaining two pseudo-components and two additional parameters related to the biomass composition. They finally compare the results calculated in this way with data from the literature with good agreement.

Raveendran et al. (1995) study the influence of mineral matter on biomass pyrolysis characteristics in a paper with 465 citations. They find that deashing increased the volatile yield, initial decomposition temperature, and rate of pyrolysis in wood and 12 other types of biomass. However, coir pith, groundnut shell, and rice husk showed an increase in char yield on deashing due to their high lignin, potassium, and zinc contents. They support these results by studies on salt-impregnated, acid-soaked, and synthetic biomass. They develop a correlation to predict the influence of ash on the volatile yield. On deashing, liquid yield increased and gas yield decreased for all the biomasses studied. The active surface area increased on deashing; the heating value of the liquid increased, whereas the increase in the char heating value was only marginal.

Ralph and Hatfield (1991) obtain pyrolysis, 'gas chromatograph mass spectrometer' (GCMS) pyrograms from a series of alfalfa preparations, a grass, an angiosperm wood, a cellulose, and an arabinoxylan under pyrolytic conditions optimal for aromatic components of plant cell walls in a paper with 453 citations. They identify approximately 130 pyrolytic fragments by a combination of mass spectral interpretation, a comparison with the literature data, and, where possible, a confirmation with authentic compounds. They assign several new fragments including both guaiacyl- and syringylpropyne and a range of alcohols. They find diagnostic peaks from arabinose and xylose components of forages, along with markers for protein.

Liu et al. (2008) study the mechanism of wood lignin pyrolysis by using TGA-' Fourier Transform Infrared spectroscopy' (FTIR) analysis in a paper with 452 citations. They perform 'Van Soest's method' to extract lignin from different species of biomass. They carry out the experimental research on pyrolysis of lignins from fir and birch on a TGA coupled with FTIR spectrometry. They find that wood lignin undergoes three consecutive stages, corresponding to the evaporation of water, the formation of primary volatiles, and the subsequent release of small molecular gases. The main pyrolysis sections and the maximum weight loss rates are quite different for different wood species. Phenols are the main volatile products, in addition to alcohols, aldehydes, acids, and others. As the main gaseous products, CO, CO_2, and CH_4 are released greatly.

Raveendran et al. (1996) conduct biomass pyrolysis studies using both a TGA and a 'packed-bed pyrolyzer' in a paper with 425 citations. They find that each kind of biomass has a characteristic pyrolysis behavior which is explained according to its individual component characteristics. Based on the studies of isolated biomass components as well as synthetic biomass, they observe that the interactions among the components are not as significant as the composition of the biomass. Direct summative correlations based on biomass component pyrolysis adequately explain both the pyrolysis characteristics and product distribution of the biomass. They infer that there is no detectable interaction among the components during pyrolysis in either the TGA or the packed-bed pyrolyzer. However, ash present in the biomass has a strong influence on both the pyrolysis characteristics and the product distribution.

Patwardhan et al. (2010) study the effect of mineral salts including mixtures of salts in the form of switchgrass ash on the chemical speciation resulting from primary pyrolysis reactions of cellulose to gain an insight of the underlying mechanisms in a paper with 406 citations. They impregnate various concentrations of inorganic salts and switchgrass ash on pure cellulose. They pyrolyze these samples in a micro-pyrolyzer connected to a GCMS/FID system. They report the effects of minerals on the formation of low molecular weight species (formic acid, glycolaldehyde, and acetol), furan ring derivatives (2-furaldehyde and 5-hydroxy methyl furfural), and anhydro sugar (levoglucosan). Further, they report the effect of reaction temperatures ranging from 350 to 600^{0}C on the pyrolysis speciation of pure and ash-doped cellulose. The pyrolysis speciation revealed the competitive nature of the primary reactions. Mineral salts and higher temperatures accelerated the reactions that led to the formation of low molecular weight species from cellulose as compared to those leading to anhydro sugars.

Carlson et al. (2009) study the conversion of biomass compounds to aromatics by thermal decomposition in the presence of catalysts using a pyroprobe analytical pyrolyzer in a paper with 404 citations. The first step in this process is the thermal decomposition of the biomass to smaller oxygenates that then enter the catalysts' pores where they are converted to CO, CO_2, water, coke, and volatile aromatics. The desired reaction is the conversion of biomass into aromatics, CO_2, and water with the undesired products being coke and water. Both the reaction conditions and catalyst properties are critical in maximizing the desired product selectivity. High heating rates and a high catalyst: feed ratio favor aromatic production over coke formation. They obtain aromatics with carbon yields in excess of 30 molar carbon% from glucose, xylitol, cellobiose, and cellulose with ZSM-5 at the optimal reactor conditions. They find that the aromatic yield for all the products was similar, suggesting that all of these biomass-derived oxygenates go through a common intermediate. At lower catalyst: feed ratios, volatile oxygenates are formed including furan type compounds, acetic acid, and hydroxyacetaldehyde. The product selectivity is dependent on both the size of the catalyst pores and the nature of the active sites. They test five catalysts including ZSM-5, silicalite, beta, Y-zeolite, and silica-alumina. ZSM-5 had the highest aromatic yields (30% carbonW yield) and the least amount of coke.

Varhegyi et al. (1997) carry out kinetic modeling of biomass pyrolysis in a paper with 362 citations. They evaluate thermogravimetric and DSC curves at different T_t heating programs by the method of least squares. They employ pseudo-first-order models, parallel, successive, and competitive reaction schemes, and complex reaction networks in the modeling. They consider thermal decomposition of cellulose at low (2°C min^{-1}) and high (50–80°C min^{-1}) heating rates, low temperature phenomena, the validity of the 'Broido-Shafizadeh model', the effects of mineral catalysts, cellulose pyrolysis in closed sample holders, and the thermal decomposition kinetics of xylan, lignin, and lignocellulosic plant samples.

Lin et al. (2009) report the kinetics and chemistry of cellulose pyrolysis using both a 'pyroprobe reactor' and a 'TGA mass spectrometer' (TGA-MS). They identify more than 90% of the products from cellulose pyrolysis in a pyroprobe reactor with a liquid N trap. They observe that the first step in the cellulose pyrolysis is the depolymerization of solid cellulose to form levoglucosan (LGA). LGA can undergo dehydration and isomerization reactions to form other anhydrosugars including levoglucosenone (LGO) and '1,6-anhydro-beta-D-glucofuranose' (AGF). The anhydrosugars can react further to form furans, such as furfural ('furan-2-carbaldehyde') and hydroxymethylfurfural (HMF) by dehydration reactions or hydroxyacetone, glycolaldehyde, and glyceraldehyde by fragmentation and retroaldol condensation reactions. CO and CO_2 are formed from decarbonylation and decarboxylation reactions. Biochar is formed from the polymerization of the pyrolysis products. They fit the pyrolytic conversion of cellulose to two different reaction models. The first model (Model I) combined the first-order kinetic model with a thermal-lag model that assumed the temperature difference between the thermocouple and specimen in TGA to be directly proportional to the heating rate. The second model (Model II) combined the first-order kinetic model with an energy balance that took into account the heat transfer at the sample boundary, including the heat flow, by endothermic pyrolysis reaction. They observe that both models adequately fitted the empirical data. The kinetic parameters obtained from both models were similar. They note that Model I is computationally easier; however, Model II is physically more realistic. Importantly, they show that the intrinsic kinetics for cellulose pyrolysis are not a function of the heating rate. During the pyrolysis of cellulose, a thermal temperature gradient between the cellulose and heater can occur due to the endothermic pyrolysis reaction.

Ranzi et al. (2008) analyze the main kinetic features of biomass pyrolysis, devolatilization, and the gas phase reactions of the released species in a paper with 321 citations. They follow three complex steps in sequence: the characterization of biomasses, the description of the release of the species, and their chemical evolution in the gas phase. Biomass is characterized as a mixture of reference constituents: cellulose, hemicellulose, and lignin. They verify this assumption against the experimental data, mainly relating it to the thermal degradation of different biomasses. Devolatilization of biomasses is a complex process in which several chemical reactions take place in both the gas and the condensed phase alongside the mass and thermal resistances involved in the pyrolysis process. They include the successive gas phase reactions of released species into an existing detailed kinetic scheme of pyrolysis and oxidation of hydrocarbon fuels. Comparisons with experimental measurements in a drop tube reactor confirm the high potentials of the proposed modeling approach.

7.3.1.2 Pyrolysis Oils

Williams and Nugranad (2000) compare the products from the pyrolysis and catalytic pyrolysis of rice husks in a paper with 390 citations. They pyrolyze these rice husks in a 'fluidized bed reactor' at 400, 450, 500, 550, and 600°C. They next pyrolyze these rice husks at 550°C with a zeolite ZSM-5 catalyst upgrading of the pyrolysis vapors at catalyst temperatures of 400, 450, 500, 550, and 600°C. They collect the pyrolysis oils in a series of condensers and cold traps and analyze them to determine their yield and composition in relation to the process conditions. They find that the pyrolysis oils before catalysis were homogeneous, of low viscosity, and highly oxygenated. 'Polycyclic aromatic hydrocarbons' (PAHs) were present in the oils at low concentration and increased in concentration with the increasing temperature of pyrolysis. Oxygenated compounds in the oils consisted mainly of phenols: cresols, benzenediols, and guaiacol, and their alkylated derivatives. In the presence of the catalyst, the yield of oil was markedly reduced, although the oxygen content was reduced with the formation of coke on the catalyst. The influence of the catalyst was to convert the oxygen in the pyrolysis oil to largely H_2O at the lower catalyst temperatures and to largely CO and CO_2 at the higher catalyst temperatures. The molecular weight distribution of the oils was decreased after catalysis and further decreased with the increasing temperature of catalysis. The catalyzed oils were markedly increased in a single ring and PAH compared to uncatalyzed biomass pyrolysis oils. The concentration of aromatic and polycyclic aromatic species increased with increasing catalyst temperature.

Scholze and Meier (2001) characterize water-insoluble fraction from pyrolysis oil using pyrolysis GCMS, FTIR, and functional groups analysis in a paper with 358 citations. They obtain pyrolytic lignins from pyrolysis oil as a fine homogeneous powder by a novel precipitation method. They present results obtained for various pyrolytic lignins of pyrolysis oils from different fast pyrolysis processes. These were subjected to various physicochemical characterization methods. FTIR data indicate that a changing oxygen content mainly affects the intensity of carbonyl absorption bands. Therefore, FTIR analysis is valuable as a fast analytical method to elucidate the aging processes of pyrolysis oil. Analytical pyrolysis combined with GC/MS, FTIR data, and the results of functional group analysis show pyrolytic lignin was similar to technical lignins.

Mullen et al. (2010) study pyrolysis oil and biochar production from corncobs and corn stover by fast pyrolysis using a pilot scale fluidized bed reactor in a paper with 340 citations. They obtain yields of 60% (mass/mass) pyrolysis oil (high heating values are about 20 MJ kg^{-1}, and densities more than 1.0 Mg m^{-3}) from both corncobs and from corn stover. The high energy density of pyrolysis oil, about 20–32 times on a per unit volume basis over the raw corn residues, offers potentially significant savings in transportation costs, particularly for a distributed 'farm scale' bio-refinery system. Biochar yield was 18.9 and 17.0% (mass/mass) from corncobs and corn stover, respectively.

Miao and Wu (2004) report on an approach for increasing the yield of pyrolysis oil production from fast pyrolysis after manipulating the metabolic pathway in microalgae through heterotrophic growth in a paper with 334 citations. They find that the yield of pyrolysis oil (57.9%) produced from heterotrophic *Chlorella protothecoides* cells was 3.4 times higher than from autotrophic cells by fast pyrolysis. The pyrolysis oils had a much lower oxygen content, with a higher heating value (41 MJ kg^{-1}), a lower density (0.92 kg L^{-1}), and lower viscosity (0.02 Pa s) compared to those of pyrolysis oil from autotrophic cells and wood. These properties are comparable to crude oil.

Miao et al. (2004) perform fast pyrolysis tests of microalgae in a 'fluid bed reactor' in a paper with 328 citations. They complete the experiments at a temperature of 500°C with a heating rate of 600°C s^{-1}, a sweep gas (N_2) flow rate of 0.4 m^3 h^{-1}, and a vapor residence time of 2–3 s. In comparison with the previous studies on slow pyrolysis from microalgae in an autoclave, they find that a greater amount of high quality pyrolysis oil can be directly produced from continuously processing microalgae feeds at a rate of 4 g min^{-1}, which has the potential for commercial application for the large-scale production of liquid fuels. They obtain liquid product yields of 18 and 24% from the fast pyrolysis of *Chlorella protothecoides* and *Microcystis aeruginosa*. The saturated and polar fractions account for 1.14 and 31.17% of the pyrolysis oils of microalgae on average, which are higher than those of pyrolysis oil from wood. The H/C and O/C molar ratios of microalgal pyrolysis oil are 1.7 and 0.3, respectively. The distribution of straight-chain alkanes of the saturated fractions from these pyrolysis oils were similar to diesel fuel. Pyrolysis oils had low oxygen content with a higher heating value of 29 MJ/kg, a density of 1.16 kg/l, and a viscosity of 0.10 Pa s. They conclude that these properties of the pyrolysis oil of microalgae make it more suitable for fuel oil use than fast pyrolysis oils from lignocellulosic materials.

7.3.2 Pyrolysis Oil Upgrading

Vispute et al. (2010) show that pyrolysis oils can be converted into industrial commodity chemical feedstocks using an integrated catalytic approach that combines hydroprocessing with zeolite catalysis in a paper with 672 citations. They find that the hydroprocessing increases the intrinsic hydrogen content of the pyrolysis oil, producing polyols and alcohols. The zeolite catalyst then converts these hydrogenated products into light olefins and aromatic hydrocarbons in a yield as much as three times higher than that produced with the pure pyrolysis oil. The yield of aromatic hydrocarbons and light olefins from the biomass conversion over zeolite is proportional to the intrinsic amount of hydrogen added to the biomass feedstock during hydroprocessing. The total product yield can be adjusted depending on market values of the chemical feedstocks and the relative prices of the hydrogen and biomass.

French and Czernik (2010) evaluate a set of commercial and laboratory-synthesized catalysts for their hydrocarbon production performance via the pyrolysis/catalytic cracking route in a paper with 452 citations. They pyrolyze cellulose, lignin, and wood (batch experiments) in quartz boats in physical contact with the catalysts at temperatures ranging from 400 to 600°C and catalyst: biomass ratios of 510 by weight. They use 'molecular-beam mass spectrometry' (MBMS) to analyze the product vapor and gas composition. They obtain the highest yield of hydrocarbons (approximately 16 wt%, including 3.5 wt% of toluene) using nickel, cobalt, iron, and gallium-substituted ZSM-5. They observe the change in the composition of the volatiles produced by the pyrolysis/catalytic vapor cracking reactions as a function of the catalyst time-on-stream using a semi-continuous flow reactor. The deoxygenation activity decreased with time because of coke deposits formed on the catalyst.

Wang et al. (1997) study biohydrogen production through the fast pyrolysis of the biomass and catalytic steam reforming of the pyrolysis oil in a paper with 354 citations. They propose a regionalized system of biohydrogen production, where small- and medium-sized pyrolysis units (less than 500 Mg/day) provide pyrolysis oil to a central reforming unit to be catalytically converted to H_2 and CO_2. Thermodynamic modeling of the major constituents of the pyrolysis oil showed that reforming was possible within a wide range of temperatures and steam: carbon ratios.

In addition, screening tests aimed at catalytic reforming of model compounds to hydrogen using Ni-based catalysts achieved essentially complete conversion to H_2. They study the existing data on the catalytic reforming of oxygenates to guide catalyst selection. They finally discuss a process diagram for the pyrolysis and reforming operations and initial production cost estimates.

Wildschut et al. (2009) study the upgrading of fast pyrolysis oil by catalytic hydrotreatment using heterogeneous noble-metal catalysts in a paper with 346 citations. They test a variety of heterogeneous noble-metal catalysts for this purpose and compare the results to those obtained with typical hydrotreatment catalysts (sulfided $NiMo/Al_2O_3$ and $CoMo/Al_2O_3$). They carry out the reactions at temperatures of 250 and 350°C and pressures of 100 and 200 bar. They find that the Ru/C catalyst was superior to the classical hydrotreating catalysts with respect to oil yield (up to 60 wt%) and deoxygenation level (up to 90 wt%). The upgraded products were less acidic and contained less water than the original fast pyrolysis oil. The 'higher heating value' (HHV) was about 40 MJ/kg, which is about twice the value of pyrolysis oil. The upgraded pyrolysis oil had lower contents of organic acids, aldehydes, ketones, and ethers than the feed, whereas the amounts of phenolics, aromatics, and alkanes were considerably higher.

Zhao et al. (2011) study the 'hydrodeoxygenation' (HDO) of guaiacol as a model compound for pyrolysis oil on transition metal phosphide hydroprocessing catalysts in a paper with 337 citations. They find that the turnover frequency based on active sites titrated by the chemisorption of CO followed the order: $Ni_2P > Co_2P > Fe_2P$, WP, MoP. The major products from HDO of guaiacol for the most active phosphides were benzene and phenol, with a small amount of methoxybenzene being formed. Kinetic studies revealed the formation of reaction intermediates such as catechol and cresol at short contact times. A commercial catalyst 5% Pd/Al_2O_3 was more active than the metal phosphides at lower contact times but produced only catechol. A commercial $CoMoS/Al_2O_3$ deactivated quickly and showed little activity for the HDO of guaiacol at these conditions. Thus, transition metal phosphides are promising materials for the catalytic HDO of biofuels.

Wright et al. (2010) carry out a techno-economic study examining the fast pyrolysis of corn stover to pyrolysis oil with a subsequent upgrading of the pyrolysis oil to naphtha and diesel range biofuels in a paper with 331 citations. They develop two 2,000 dry tons per day scenarios: the first scenario separates a fraction of the pyrolysis oil to generate biohydrogen on-site for fuel upgrading, while the second scenario relies on commercial hydrogen. The modeling effort resulted in liquid fuel production rates of 134 and 220 million liters per year for the biohydrogen production and purchase scenarios, respectively. Capital costs for these plants are $287 and $200 million. Fuel product value estimates are $3.09 and $2.11 per gallon of gasoline equivalent ($0.82 and $0.56 per liter). Pioneer plant analysis estimates the capital costs at $911 and $585 million for the construction of a first-of-a-kind fast pyrolysis and upgrading biorefinery with product values of $6.55 and $3.41 per 'gasoline gallon equivalent' (GGE) ($1.73 and $0.90 per Liter).

Rioch et al. (2005) study biohydrogen production from pyrolysis oils in a paper with 323 citations. They focus on the use of noble metal-based catalysts for the steam reforming of a few model compounds and that of an actual pyrolysis oil. They perform the steam reforming of the model compounds in a temperature range of 650–950°C over Pt, Pd, and Rh supported on alumina and a ceria-zirconia sample. The model compounds used were acetic acid, phenol, acetone, and ethanol. They find that the nature of the support played a significant role in the activity of these catalysts. The use of ceria-zirconia, a redox mixed oxide, led to higher H_2 yields as compared to the case of the alumina-supported catalysts. The supported Rh and Pt catalysts were the most active for the steam reforming of these compounds, while Pd-based catalysts performed poorly. They also study the activity of the promising Pt and Rh catalysts for the steam reforming of a pyrolysis oil obtained from beech wood fast pyrolysis. Temperatures close to, or higher than, 800°C were required to achieve significant conversions to COx and H_2 (e.g. H_2 yields around 70%). The ceria-zirconia materials showed a higher activity than the corresponding alumina samples. A Pt/ceria-zirconia sample used for over 9 h showed essentially constant activity, while extensive carbonaceous deposits were observed on the quartz reactor walls early on. No benefit was observed by adding a small amount of O_2 to the steam/pyrolysis oil feed ('autothermal reforming'), probably partly due to the already high concentration of oxygen in the pyrolysis oil composition.

7.4 DISCUSSION

Table 7.1 provides information on the research fronts in this field. As this table shows the primary research fronts of 'pyrolysis' and 'upgrading of pyrolysis oils' comprise 72 and 28% of these papers, respectively. 'Kinetic studies' and 'pyrolysis oils' comprise 52 and 20% of the papers in the first group, respectively.

7.4.1 Pyrolysis

7.4.1.1 Kinetic Studies

Yang et al. (2007) study the pyrolysis characteristics of biomass components using a TGA with a DSC detector and a 'pack-bed pyrolyzer' in a paper with 2,891 citations. Evans and Milne (1987) apply the technique of MBMS sampling to the elucidation of the molecular pathways in the fast pyrolysis of wood and its principal isolated constituents in a paper with 633 citations. Yang et al. (2006) study biomass pyrolysis based on three major components using a TGA in a paper with 559 citations.

Orfao et al. (1999) study the behavior of biomass components thermogravimetrically with linear temperature programming under N and air in a paper with 546

TABLE 7.1
Research Fronts

	Research Front	No. of Papers (%)
1	Pyrolysis	72
1.1	Kinetic studies	52
1.2	Pyrolysis oils	20
2	Upgrading of pyrolysis oils	28

citations. Raveendran et al. (1995) study the influence of mineral matter on biomass pyrolysis characteristics in a paper with 465 citations. Ralph and Hatfield (1991) obtain pyrolysis-GCMS pyrograms from a series of alfalfa preparations, a grass, an angiosperm wood, a cellulose, and an arabinoxylan under pyrolytic conditions optimal for aromatic components of plant cell walls in a paper with 453 citations.

Liu et al. (2008) study the mechanism of wood lignin pyrolysis by using TG-FTIR analysis in a paper with 452 citations. Raveendran et al. (1996) conduct biomass pyrolysis studies using both a TGA and a 'packed-bed pyrolyzer' in a paper with 425 citations. Patwardhan et al. (2010) study the effect of mineral salts including mixtures of salts in the form of switchgrass ash on the chemical speciation resulting from primary pyrolysis reactions of cellulose to gain an insight of the underlying mechanisms in a paper with 406 citations.

Carlson et al. (2009) study the conversion of biomass compounds to aromatics by thermal decomposition in the presence of catalysts using a 'pyroprobe analytical pyrolyzer' in a paper with 404 citations. Varhegyi et al. (1997) carry out kinetic modeling of biomass pyrolysis in a paper with 362 citations. Lin et al. (2009) report the kinetics and chemistry of cellulose pyrolysis using both a 'pyroprobe reactor' and a TGA-MS. Ranzi et al. (2008) analyze the main kinetic features of biomass pyrolysis, devolatilization, and the gas phase reactions of the released species in a paper with 321 citations.

These prolific studies highlight the kinetics of biomass pyrolysis.

7.4.1.2 Pyrolysis Oils

Williams and Nugranad (2000) compare the products from the pyrolysis and catalytic pyrolysis of rice husks in a paper with 390 citations. Scholze and Meier (2001) characterize water-insoluble fraction from pyrolysis oil using pyrolysis GCMS, FTIR, and functional group analysis in a paper with 358 citations. Mullen et al. (2010) study pyrolysis oil and biochar production from corncobs and corn stover by fast pyrolysis using a pilot scale fluidized bed reactor in a paper with 340 citations.

Miao and Wu (2004) report on an approach for increasing the yield of pyrolysis oil production from fast pyrolysis after manipulating the metabolic pathway in microalgae through heterotrophic growth in a paper with 334 citations. Miao et al. (2004) perform fast pyrolysis tests of microalgae in a 'fluid bed reactor' in a paper with 328 citations.

These prolific studies highlight the production of pyrolysis oils.

7.4.2 Pyrolysis Oil Upgrading

Vispute et al. (2010) show that pyrolysis oils can be converted into industrial commodity chemical feedstocks using an integrated catalytic approach that combines hydroprocessing with zeolite catalysis in a paper with 672 citations. French and Czernik (2010) evaluate a set of commercial and laboratory-synthesized catalysts for their hydrocarbon production performance via the pyrolysis/catalytic cracking route in a paper with 452 citations. Wang et al. (1997) study the biohydrogen production through fast pyrolysis of the biomass and catalytic steam reforming of the pyrolysis oil in a paper with 354 citations.

Wildschut et al. (2009) study the upgrading of fast pyrolysis oil by catalytic hydrotreatment using heterogeneous noble-metal catalysts in a paper with 346 citations. Zhao et al. (2011) study the HDO of guaiacol as a model compound for pyrolysis oil on transition metal phosphide hydroprocessing catalysts in a paper with 337 citations. Wright et al. (2010) carry out a techno-economic study examining the fast pyrolysis of corn stover to pyrolysis oil with subsequent upgrading of the pyrolysis oil to naphtha and diesel range biofuels in a paper with 331 citations. Rioch et al. (2005) study biohydrogen production from pyrolysis oils in a paper with 323 citations.

These prolific studies highlight the upgrading of pyrolysis oils for biofuel production, including biohydrogen production.

7.5 CONCLUSION

This chapter has presented the key findings of the 25-most-cited article papers in this field. Table 7.1 provides information on the research fronts in this field. As this table shows the primary research fronts of 'pyrolysis' and 'upgrading of pyrolysis oils' comprise 72 and 28% of these papers, respectively. 'Kinetic studies' and 'pyrolysis oils' comprise 52 and 20% of the papers in the first group, respectively.

These prolific studies on two different research fronts provide valuable evidence on the biomass pyrolysis and upgrading of pyrolysis oils.

It is recommended that similar studies are carried out for each research front as well.

ACKNOWLEDGMENTS

The contribution of the highly cited researchers in this field is greatly acknowledged.

REFERENCES

Ahmadun, F. R., A. Pendashteh, L. C. Abdullah, et al. 2009. Review of technologies for oil and gas produced water treatment. *Journal of Hazardous Materials* 170: 530–551.

Atlas, R. M. 1981. Microbial degradation of petroleum hydrocarbons: An environmental perspective. *Microbiological Reviews* 45: 180–209.

Babich, I. V. and J. A. Moulijn. 2003. Science and technology of novel processes for deep desulfurization of oil refinery streams: A review. *Fuel* 82: 607–631.

Bridgwater, A. V. and G. V. C. Peacocke. 2000. Fast pyrolysis processes for biomass. *Renewable & Sustainable Energy Reviews* 4: 1–73.

Bridgwater, A. V., D. Meier, and D. Radlein. 1999. An overview of fast pyrolysis of biomass. *Organic Geochemistry* 30: 1479–1493.

Carlson, T. R., G. A. Tompsett, W. C. Conner, and G. W. Huber. 2009. Aromatic production from catalytic fast pyrolysis of biomass-derived feedstocks. *Topics in Catalysis* 52: 241–252.

Chisti, Y. 2007. Biodiesel from microalgae. *Biotechnology Advances* 25: 294–306.

Czernik, S. and A. V. Bridgwater. 2004. Overview of applications of biomass fast pyrolysis oil. *Energy & Fuels* 18: 590–598.

Evans, R. J. and T. A. Milne. 1987. Molecular characterization of the pyrolysis of biomass. 1. Fundamentals. *Energy & Fuels* 1: 123–137.

French, R. and S. Czernik. 2010. Catalytic pyrolysis of biomass for biofuels production. *Fuel Processing Technology* 91: 25–32.

Hill, J., E. Nelson, D. Tilman, S. Polasky, and D. Tiffany. 2006. Environmental, economic, and energetic costs and benefits of biodiesel and ethanol biofuels. *Proceedings of the National Academy of Sciences of the United States of America* 103: 11206–11210.

Kilian, L. 2009. Not all oil price shocks are alike: Disentangling demand and supply shocks in the crude oil market. *American Economic Review* 99: 1053–1069.

Konur, O. 2000. Creating enforceable civil rights for disabled students in higher education: An institutional theory perspective. *Disability & Society* 15: 1041–1063.

Konur, O. 2002a. Access to Nursing Education by disabled students: Rights and duties of nursing programs. *Nurse Education Today* 22: 364–374.

Konur, O. 2002b. Assessment of disabled students in higher education: Current public policy issues. *Assessment and Evaluation in Higher Education* 27: 131–152.

Konur, O. 2002c. Access to employment by disabled people in the UK: Is the Disability Discrimination Act working? *International Journal of Discrimination and the Law* 5: 247–279.

Konur, O. 2006a. Participation of children with dyslexia in compulsory education: Current public policy issues. *Dyslexia* 12: 51–67.

Konur, O. 2006b. Teaching disabled students in Higher Education. *Teaching in Higher Education* 11: 351–363.

Konur, O. 2007a. A judicial outcome analysis of the Disability Discrimination Act: A windfall for the employers? *Disability & Society* 22: 187–204.

Konur, O. 2007b. Computer-assisted teaching and assessment of disabled students in higher education: The interface between academic standards and disability rights. *Journal of Computer Assisted Learning* 23: 207–219.

Konur, O., ed. 2021a. *Handbook of Biodiesel and Petrodiesel Fuels: Science, Technology, Health, and Environment.* Boca Raton, FL: CRC Press.

Konur, O., ed. 2021b. *Handbook of Biodiesel and Petrodiesel Fuels: Science, Technology, Health, and Environment. Volume 1. Biodiesel Fuels: Science, Technology, Health, and Environment.* Boca Raton, FL: CRC Press.

Konur, O., ed. 2021c. *Handbook of Biodiesel and Petrodiesel Fuels: Science, Technology, Health, and Environment. Volume 2. Biodiesel Fuels based on the Edible and Nonedible Feedstocks, Wastes, and Algae: Science, Technology, Health, and Environment.* Boca Raton, FL: CRC Press.

Konur, O., ed. 2021d. *Handbook of Biodiesel and Petrodiesel Fuels: Science, Technology, Health, and Environment. Volume 3. Petrodiesel Fuels: Science, Technology, Health, and Environment.* Boca Raton, FL: CRC Press.

Konur, O. 2021e. Biodiesel and petrodiesel fuels: Science, technology, health, and environment. In *Handbook of Biodiesel and Petrodiesel Fuels: Science, Technology, Health, and Environment. Volume 1. Biodiesel Fuels: Science, Technology, Health, and Environment*, ed. O. Konur. Boca Raton, FL: CRC Press.

Konur, O. 2021f. Biodiesel and petrodiesel fuels: A scientometric review of the research. In *Handbook of Biodiesel and Petrodiesel Fuels: Science, Technology, Health, and Environment. Volume 1. Biodiesel Fuels: Science, Technology, Health, and Environment*, ed. O. Konur. Boca Raton, FL: CRC Press.

Konur, O. 2021g. Biodiesel and petrodiesel fuels: A review of the research. In *Handbook of Biodiesel and Petrodiesel Fuels: Science, Technology, Health, and Environment. Volume 1. Biodiesel Fuels: Science, Technology, Health, and Environment*, ed. O. Konur. Boca Raton, FL: CRC Press.

Konur, O. 2021h Nanotechnology applications in the diesel fuels and the related research fields: A review of the research. In *Handbook of Biodiesel and Petrodiesel Fuels: Science, Technology, Health, and Environment. Volume 1. Biodiesel Fuels: Science, Technology, Health, and Environment*, ed. O. Konur. Boca Raton, FL: CRC Press.

Konur, O. 2021i. Biooils: A scientometric review of the research. In *Handbook of Biodiesel and Petrodiesel Fuels: Science, Technology, Health, and Environment. Volume 1. Biodiesel Fuels: Science, Technology, Health, and Environment*, ed. O. Konur. Boca Raton, FL: CRC Press.

Konur, O. 2021j. Characterization and properties of biooils: A review of the research. In *Handbook of Biodiesel and Petrodiesel Fuels: Science, Technology, Health, and Environment. Volume 1. Biodiesel Fuels: Science, Technology, Health, and Environment*, ed. O. Konur. Boca Raton, FL: CRC Press.

Konur, O. 2021k. Biomass pyrolysis and pyrolysis oils: A review of the research. In *Handbook of Biodiesel and Petrodiesel Fuels: Science, Technology, Health, and Environment. Volume 1. Biodiesel Fuels: Science, Technology, Health, and Environment*, ed. O. Konur. Boca Raton, FL: CRC Press.

Konur, O. 2021l. Biodiesel fuels: A scientometric review of the research. In *Handbook of Biodiesel and Petrodiesel Fuels: Science, Technology, Health, and Environment. Volume 1. Biodiesel Fuels: Science, Technology, Health, and Environment*, ed. O. Konur. Boca Raton, FL: CRC Press.

Konur, O. 2021m. Glycerol: A scientometric review of the research. In *Handbook of Biodiesel and Petrodiesel Fuels: Science, Technology, Health, and Environment. Volume 1. Biodiesel Fuels: Science, Technology, Health, and Environment*, ed. O. Konur. Boca Raton, FL: CRC Press.

Konur, O. 2021n. Propanediol production from glycerol: A review of the research. In *Handbook of Biodiesel and Petrodiesel Fuels: Science, Technology, Health, and Environment. Volume 1. Biodiesel Fuels: Science, Technology, Health, and Environment*, ed. O. Konur Boca Raton, FL: CRC Press.

Konur, O. 2021o. Edible oil-based biodiesel fuels: A scientometric review of the research. *In Handbook of Biodiesel and Petrodiesel Fuels: Science, Technology, Health, and Environment. Volume 2. Biodiesel Fuels based on the Edible and Nonedible Feedstocks, Wastes, and Algae: Science, Technology, Health, and Environment,* ed. O. Konur. Boca Raton, FL: CRC Press.

Konur, O. 2021p. Palm oil-based biodiesel fuels: A review of the research. In *Handbook of Biodiesel and Petrodiesel Fuels: Science, Technology, Health, and Environment. Volume 2. Biodiesel Fuels based on the Edible and Nonedible Feedstocks, Wastes, and Algae*, ed. O. Konur. Boca Raton, FL: CRC Press.

Konur, O. 2021q. Rapeseed oil-based biodiesel fuels: A review of the research. In *Handbook of Biodiesel and Petrodiesel Fuels: Science, Technology, Health, and Environment. Volume 2. Biodiesel Fuels based on the Edible and Nonedible Feedstocks, Wastes, and Algae,* ed. O. Konur. Boca Raton, FL: CRC Press.

Konur, O. 2021r. Nonedible oil-based biodiesel fuels: A scientometric review of the research. In *Handbook of Biodiesel and Petrodiesel Fuels: Science, Technology, Health, and Environment. Volume 2. Biodiesel Fuels based on the Edible and Nonedible Feedstocks, Wastes, and Algae: Science, Technology, Health, and Environment*, ed. O. Konur. Boca Raton, FL: CRC Press.

Konur, O. 2021s. Waste oil-based biodiesel fuels: A scientometric review of the research. In *Handbook of Biodiesel and Petrodiesel Fuels: Science, Technology, Health, and Environment. Volume 2. Biodiesel Fuels based on the Edible and Nonedible Feedstocks, Wastes, and Algae: Science, Technology, Health, and Environment*, ed. O. Konur. Boca Raton, FL: CRC Press.

Konur, O. 2021t. Algal biodiesel fuels: A scientometric review of the research. In *Handbook of Biodiesel and Petrodiesel Fuels: Science, Technology, Health, and Environment. Volume 2. Biodiesel Fuels based on the Edible and Nonedible Feedstocks, Wastes, and Algae: Science, Technology, Health, and Environment*, ed. O. Konur. Boca Raton, FL: CRC Press.

Konur, O. 2021u. Algal biomass production for biodiesel production: A review of the research. In *Handbook of Biodiesel and Petrodiesel Fuels: Science, Technology, Health, and Environment. Volume 2. Biodiesel Fuels based on the Edible and Nonedible Feedstocks, Wastes, and Algae*, ed. O. Konur Boca Raton, FL: CRC Press. February 23, 2020.

Konur, O. 2021v. Algal biomass production in wastewaters for biodiesel production: A review of the research. In *Handbook of Biodiesel and Petrodiesel Fuels: Science, Technology, Health, and Environment. Volume 2. Biodiesel Fuels based on the Edible and Nonedible Feedstocks, Wastes, and Algae*, ed. O. Konur. Boca Raton, FL: CRC Press. February 23, 2020.

Konur, O. 2021x. Algal lipid production for biodiesel production: A review of the research. In *Handbook of Biodiesel and Petrodiesel Fuels: Science, Technology, Health, and Environment. Volume 2. Biodiesel Fuels based on the Edible and Nonedible Feedstocks, Wastes, and Algae*, Ed. O. Konur Boca Raton, FL: CRC Press.

Konur, O. 2021y. Crude oils: A scientometric review of the research. In *Handbook of Biodiesel and Petrodiesel Fuels: Science, Technology, Health, and Environment. Volume 3. Petrodiesel Fuels: Science, Technology, Health, and Environment*, ed. O. Konur. Boca Raton, FL: CRC Press.

Konur, O. 2021z. Petrodiesel fuels: A scientometric review of the research. In *Handbook of Biodiesel and Petrodiesel Fuels: Science, Technology, Health, and Environment. Volume 3. Petrodiesel Fuels: Science, Technology, Health, and Environment*, ed. O. Konur. Boca Raton, FL: CRC Press.

Konur, O. 2021aa. Bioremediation of petroleum hydrocarbons in the contaminated soils: A review of the research. In *Handbook of Biodiesel and Petrodiesel Fuels: Science, Technology, Health, and Environment. Volume 3. Petrodiesel Fuels: Science, Technology, Health, and Environment*, ed. O. Konur. Boca Raton, FL: CRC Press.

Konur, O. 2021ab. Desulfurization of diesel fuels: A review of the research. In *Handbook of Biodiesel and Petrodiesel Fuels: Science, Technology, Health, and Environment. Volume 3. Petrodiesel Fuels: Science, Technology, Health, and Environment*, ed. O. Konur. Boca Raton, FL: CRC Press.

Konur, O. 2021ac. Diesel fuel exhaust emissions: A scientometric review of the research. In *Handbook of Biodiesel and Petrodiesel Fuels: Science, Technology, Health, and Environment. Volume 3. Petrodiesel Fuels: Science, Technology, Health, and Environment*, ed. O. Konur. Boca Raton, FL: CRC Press.

Konur, O. 2021ad. The adverse health and safety impact of diesel fuels: A scientometric review of the research. In *Handbook of Biodiesel and Petrodiesel Fuels: Science, Technology, Health, and Environment. Volume 3. Petrodiesel Fuels: Science, Technology, Health, and Environment*, ed. O. Konur. Boca Raton, FL: CRC Press.

Konur, O. 2021ae. Respiratory illnesses caused by the diesel fuel exhaust emissions: A review of the research. In *Handbook of Biodiesel and Petrodiesel Fuels: Science, Technology, Health, and Environment. Volume 3. Petrodiesel Fuels: Science, Technology, Health, and Environment*, ed. O. Konur. Boca Raton, FL: CRC Press.

Konur, O. 2021af. Cancer caused by the diesel fuel exhaust emissions: A review of the research. In *Handbook of Biodiesel and Petrodiesel Fuels: Science, Technology, Health, and Environment. Volume 3. Petrodiesel Fuels: Science, Technology, Health, and Environment*, ed. O. Konur. Boca Raton, FL: CRC Press.

Konur, O. 2021ag. Cardiovascular and other illnesses caused by the diesel fuel exhaust emissions: A review of the research. In *Handbook of Biodiesel and Petrodiesel Fuels: Science, Technology, Health, and Environment. Volume 3. Petrodiesel Fuels: Science, Technology, Health, and Environment*, ed. O. Konur. Boca Raton, FL: CRC Press.

Lin, Y. C., J. Cho, G. A. Tompsett, P. R. Westmoreland, and G. W. Huber. 2009. Kinetics and mechanism of cellulose pyrolysis. *Journal of Physical Chemistry C* 113: 20097–20107.

Liu, Q., S. R. Wang, Y. Zheng, et al. 2008. Mechanism study of wood lignin pyrolysis by using TG-FTIR analysis. *Journal of Analytical and Applied Pyrolysis* 82: 170–177.

Miao, X. L. and Q. Y. Wu. 2004. High yield bio-oil production from fast pyrolysis by metabolic controlling of *Chlorella protothecoides*. *Journal of Biotechnology* 110: 85–93.

Miao, X. L., Q. Y. Wu, and C. Y. Yang. 2004. Fast pyrolysis of microalgae to produce renewable fuels. *Journal of Analytical and Applied Pyrolysis* 71: 855–863.

Mohan, D., C. U. Pittman, and P. H. Steele. 2006. Pyrolysis of wood/biomass for bio-oil: A critical review. *Energy & Fuels* 20: 848–889.

Mullen, C. A., A. A. Boateng, and N. M. Goldberg, et al. 2010. Bio-oil and bio-char production from corn cobs and stover by fast pyrolysis. *Biomass & Bioenergy* 34: 67–74.

North, D. C. 1991a. *Institutions, Institutional Change and Economic Performance*. Cambridge, Mass.: Cambridge University Press.

North, D.C. 1991b. Institutions. *Journal of Economic Perspectives* 5: 97–112.

Orfao, J. J. M., F. J. A. Antunes, and J. L. Figueiredo. 1999. Pyrolysis kinetics of lignocellulosic materials: Three independent reactions model. *Fuel* 78: 349–358.

Patwardhan, P. R., J. A. Satrio, R. C. Brown, and B. H. Shanks. 2010. Influence of inorganic salts on the primary pyrolysis products of cellulose. *Bioresource Technology* 101: 4646–4655.

Perron, P. 1989. The great crash, the oil price shock, and the unit root hypothesis. *Econometrica: Journal of the Econometric Society* 57: 1361–1401.

Ralph, J. and R. D. Hatfield. 1991. Pyrolysis-GC-MS characterization of forage materials. *Journal of Agricultural and Food Chemistry* 39: 1426–1437.

Ranzi, E., A. Cuoci, and T. Faravelli, et al. 2008. Chemical kinetics of biomass pyrolysis. *Energy & Fuels* 22: 4292–4300.

Raveendran, K., A. Ganesh, and K. C. Khilar. 1995. Influence of mineral matter on biomass pyrolysis characteristics. *Fuel* 74.1812–1822.

Raveendran, K., A. Ganesh, and K. C. Khilar. 1996. Pyrolysis characteristics of biomass and biomass components. *Fuel* 75: 987–998.

Rioche, C., S. Kulkarni, F. C. Meunier, J. P. Breen, and R. Burch. 2005. Steam reforming of model compounds and fast pyrolysis bio-oil on supported noble metal catalysts. *Applied Catalysis B-Environmental* 61: 130–139.

Scholze, B. and D. Meier. 2001. Characterization of the water-insoluble fraction from pyrolysis oil (pyrolytic lignin). Part I. PY-GC/MS, FTIR, and functional groups. *Journal of Analytical and Applied Pyrolysis* 60: 41–54.

Varhegyi, G., M. J. Antal, E. Jakab, and P. Szabo. 1997. Kinetic modeling of biomass pyrolysis. *Journal of Analytical and Applied Pyrolysis* 42: 73–87.

Vispute, T. P., H. Y. Zhang, and A. Sanna, et al. 2010. Renewable chemical commodity feedstocks from integrated catalytic processing of pyrolysis oils. *Science* 330: 1222–1227.

Wang, D., S. Czernik, D. Montane, M. Mann, and E. Chornet. 1997. Biomass to hydrogen via fast pyrolysis and catalytic steam reforming of the pyrolysis oil or its fractions. *Industrial & Engineering Chemistry Research* 36: 1507–1518.

Wildschut, J., F. H. Mahfud, R. H. Venderbosch, and H. J. Heeres. 2009. Hydrotreatment of fast pyrolysis oil using heterogeneous noble-metal catalysts. *Industrial & Engineering Chemistry Research* 48: 10324–10334.

Williams, P. T. and N. Nugranad. 2000. Comparison of products from the pyrolysis and catalytic pyrolysis of rice husks. *Energy* 25: 493–513.

Wright, M. M., D. E. Daugaard, J. A. Satrio, and R. C. Brown. 2010. Techno-economic analysis of biomass fast pyrolysis to transportation fuels. *Fuel* 89: S11–S19.

Yang, H. P., R. Yan, and H. P. Chen, et al. 2006. In-depth investigation of biomass pyrolysis based on three major components: Hemicellulose, cellulose and lignin. *Energy & Fuels* 20: 388–393.

Yang, H. P., R. Yan, H. P. Chen, D. H. Lee, and C. G. Zheng. 2007. Characteristics of hemicellulose, cellulose and lignin pyrolysis. *Fuel* 86: 1781–1788.

Zhang, Q., J. Chang, T. J. Wang, and Y. Xu. 2007. Review of biomass pyrolysis oil properties and upgrading research. *Energy Conversion and Management* 48: 87–92.

Zhao, H. Y., D. Li, P. Bui, and S. T. Oyama. 2011. Hydrodeoxygenation of guaiacol as model compound for pyrolysis oil on transition metal phosphide hydroprocessing catalysts. *Applied Catalysis A-General* 391: 305–310.

8 An Overview of Catalytic Bio-oil Upgrading, Part 1: *Processing Aqueous-Phase Compounds*

Jianghao Zhang
Junming Sun
Yong Wang

CONTENTS

8.1 INTRODUCTION

The concerns of dwindling fossil resources have motivated the exploration of alternative sustainable energies, such as wind power, solar energy, controlled nuclear fusion, hydroelectricity, and biomass. Among these alternatives, biomass, the only renewable organic carbon resource in nature (Li et al., 2015), has been recognized as the most readily available energy source to produce hydrocarbon fuels (Ruddy et al., 2014), considering our reliance on liquid fuels for transportation vehicles (trucks, ships, airplanes, etc.). Among all forms of biomass, due to its abundancy and inedibility, lignocellulosic biomass (woods, grass, energy crop, agricultural waste, etc.) has attracted the greatest interest in both industry and academia for the production of biofuels without competing with human food and animal feed (Liu et al., 2014; Wang, et al., 2013). After decades of effort, several promising approaches have been developed, including gasification followed by 'Fischer–Tropsch synthesis', a fermentative process, pyrolysis or 'hydrothermal liquefaction' (HTL) to bio-oil, followed by catalytic upgrading (Ayodele et al., 2019; Gollakota et al., 2018; Wang et al., 2013).

Comparing with other approaches in lignocellulosic biomass conversion, pyrolysis and hydrothermal liquefaction integrated with upgrading have been demonstrated to be a cost-effective and feasible one in techno-economic analysis (Anex et al., 2010; de Jong et al., 2015). The liquid product derived from these processes has been considered as a low-cost liquid bio-oil (Chiaramonti et al., 2007) for the efficient and large-scale production of fuels and valuable chemicals (Xiu and Shahbazi, 2012).

Crude bio-oil is instable and corrosive, and it has high viscosity, water content, and low heating value. Bio-oil upgrading is indispensable and attained through oxygen removal and molecular weight reduction (Elliott, 2007). The organic compounds in bio-oil can be separated into two phases: an aqueous phase and an oil phase (Garcia-Perez et al., 2007a). In this chapter, we focus on an overview of the progresses in catalytically upgrading the aqueous-phase compounds in bio-oil. In particular, a fundamental understanding of the surface chemistry of catalysts in the modeling reaction, such as reforming, dehydration, and C–C coupling, will be reviewed. The promising catalysts and proposed reaction mechanism are carefully discussed in an attempt to correlate catalyst properties (e.g. acid-base pairs) with catalytic performance which may guide the future rational design of catalysts in bio-oil upgrading.

8.2 OBJECTIVES AND CHALLENGES IN THE BIO-OIL UPGRADING PROCESS

Bio-oil usually has a dark brown color and a distinctive smoky odor. It contains 10–30 wt% water and hundreds of organic oxygenates (Mullen and Boateng, 2008), including carboxylic acids and esters, aliphatic aldehydes and ketones, furans and pyrans, aliphatic alcohol and sugars, and aromatics and oxygenates. The relative content of oxygenates is highly dependent on the feedstocks and the process conditions employed. For example, Bertero et al. (2012) reported that pyrolysis of pine sawdust produced ~26% aromatic oxygenates and ~24% aliphatic aldehydes/ketones in bio-oil. In contrast, these numbers in mesquite sawdust pyrolysis oil are ~49 and ~16%, respectively. Wang et al. (2015) pyrolyzed four different types of lignin and found significant changes of the relative contents of each phenolic compound with a pyrolytic temperature. However, the elemental components in bio-oils are usually similar (Azeez et al., 2010; Elliott et al., 2012).

Table 8.1 compares the chemical composition and physical properties of bio-oil with crude oil and diesel fuel that meets the EN590 standard published by European Committee for Standardization (Dickerson and Soria, 2013; Lehto et al., 2014; Liaquat et al., 2013; Munoz et al., 2016). Several major differences can be perceived, i.e. high oxygen and water content, high viscosity (comparing with the diesel), and a low heating value. In addition, bio-oil contains a certain amount of organic acids that cause corrosion problems (Aziz et al., 2013; Elliott, 2007). The sulfur and nitrogen

TABLE 8.1
A Comparison of Components and Properties of Bio-oil, Crude Oil, and Diesel Fuel that Meets the EN590 Standard

	Typical Bio-oil	Heavy Crude Oil	Diesel
C (wt%)	50–65	83–86	—
H (wt%)	5–7	11–14	—
N (wt%)	< 0.4	< 1	0.02
S (wt%)	< 0.05	< 4	< 0.001
O (wt%)	28–40	< 1	~0
Ash (wt%)	0.01–0.1	0.1	< 0.01
Water (wt%)	10–30	0.1	~0
Viscosity (cSt)	15–35 at 40°C	40–175 at 50°C	2.0–4.5 at 40°C
HHV (MJ/kg) [a]	15–19	44	45.5

[a] HHV: higher heating values.
Source: Lehto et al., (2014), Dickerson and Soria (2013), Munoz et al. (2016), Liaquat et al. (2013).

contents of bio-oil, however, are lower than those of crude oil, which mitigates the workload for the desulfurization and denitrogenation of bio-oil.

Given the above issues, bio-oil must be upgraded to be compatible with conventional infrastructures (e.g. the combustion engine). Essentially, the issues are caused by the high oxygen content. Besides the low heating value, the presence of large amounts of aldehydes and ketones make the oil hydrophilic and highly hydrated, leading to the difficulty of removing water from the bio-oil (Zhang et al., 2007). The viscosity can also be significantly reduced if the oxygen content is decreased to a certain level (Elliott, 2007). Therefore, upgrading will be ultimately achieved by extensive oxygen elimination facilitated by catalysts (Wang et al., 2013), the presence of which can also promote the catalytic cracking of large oligomers in the bio-oil.

In the U.S. Department of Energy's 'Multi-Year Program Plan 2016' for the Bioenergy Technologies Office, developing oxygen removal strategies and hydrotreating catalysts that are highly selective to desired end products were counted as the main challenges and barriers in biomass valorization (Department of Energy US, 2016). Other than the oxygen removal, high selectivity is proposed to retain the carbon in the product, as well as to preserve the unsaturated compounds in the bio-oil (aromatics, olefins, etc.) in order to minimize the consumption of valuable hydrogen (Hicks, 2011; Resasco, 2011). Moreover, the presence of impurities such as alkalis, S- and N-containing compounds, and coking in the upgrading process tend to severely deactivate the catalysts, which is another challenge for catalytic bio-oil upgrading.

8.3 THE PROCESSING OF AQUEOUS-PHASE COMPOUNDS

With water addition, bio-oils can be separated into two phases: the oil phase that is hydrophobic and the aqueous phase that contains light organic fraction and a large amount of water (Garcia-Perez et al., 2007b; Imam and Capareda, 2012; Su-Ping, 2003; Wang et al., 2013). In the oil phase, the main oxygen-containing compounds are aromatic oxygenates, sugars, furans, cyclic ketones, and larger oligomers (Garcia-Perez et al., 2007a–b; Imam and Capareda, 2012). It may also contain a certain amount of fatty acids (Garcia-Perez et al., 2007a; Imam and Capareda, 2012). Light oxygenates such as carboxylic acids, alcohols, aldehydes, and ketones are the major organic components in the aqueous phase (Bae et al., 2011; Imam and Capareda, 2012). Depending on the biomass feedstocks, the content of carboxylic acid, alcohols, and aldehydes/ketones could reach to 18 wt%, 6–10 wt%, and 13–24 wt%, respectively, for all the organic compounds of bio-oil (Bertero et al., 2012). Therefore, upgrading the aqueous-phase fraction of bio-oil focuses on processing those oxygenates. It should be noted that the discussed reactions in this section can also be used in upgrading the oil-phase compounds that share the same functional groups.

8.3.1 Reforming to Produce Hydrogen

The acetic acid, alcohols, and ketones/aldehydes in the aqueous phase of bio-oil are good candidates for the 'steam reforming' (SR) reaction (Equation 8.1) (Kechagiopoulos et al., 2006). Since these compounds contain C-C chains, besides the activation of the C–H bond and water, the catalysts should also bear the functionality cracking the C–C bond for efficient reforming as well as inhibited coking reactions (Chen et al., 2017a–b; Davidson et al., 2014b). Moreover, the catalyst needs also to be active in the water-gas shift reaction to remove CO from metal surfaces (Hu and Lu, 2010b). Therefore, SR using the compounds in the aqueous phase can only be achieved with the catalysts that possess the above multifunction. One alternative is the tandem combination of pre-reformer and reformer, i.e. the pre-reformer converts these oxygenates to CH_4, CO, CO_2, and H_2, and the produced CH_4 is further reformed with the conventional SR setup (Trane et al., 2012).

$$C_nH_mO_k + (2n-k)H_2O \rightarrow nCO_2 + (2n+m/2-k)H_2 \tag{8.1}$$

From the thermodynamic point of view, the SR of oxygenates is favorable at high temperatures and low pressures (Trane et al., 2012). However, the high temperature also favors the formation of CO due to the reverse water-gas shift reaction, which necessitates a lower-temperature shift downstream (Basagiannis and Verykios, 2007). Another way to decrease the CO selectivity and increase the H_2 selectivity is by increasing the steam: carbon ratio (Hu and Lu, 2010a), which shifts the equilibrium towards CO_2 and H_2. The higher steam content will demand more energy for water evaporation and thus is less energy efficient.

In terms of catalysts for the SR of oxygenates in the aqueous phase of bio-oil, Ni-based, Rh-based, and Co-based ones have displayed high activity and H_2 yield (Sun et al., 2013; Trane et al., 2012; Yu et al., 2016), as shown in Table 8.2. Hu et al. (2010b) compared the performances of Al_2O_3-supported Ni, Co, Fe, and Cu in an SR of acetic acid. Comparing with two other catalysts, Ni/Al_2O_3 and Co/Al_2O_3 were active for the cracking of both C–C and C–H bonds, and thus displayed much higher reactivity. Kechagiopoulos et al. (2006) reported that the commercial C11-NK catalyst (major component Ni-K/$CaAl_2O_4$) is effective in the SR of acetic acid, acetone, ethylene glycol, and a mixture of the three compounds.

TABLE 8.2
Representative Catalysts for the Steam Reforming of Oxygenates in the Aqueous Phase of Bio-oil

Catalyst	Reactant	Reaction Condition				Conv. (%)	S_{H2} (%)	Reference
		T (°C)	P (bar)	S/C[a] Ratio	Space Velocity[b]			
Ni/Al_2O_3	Acetic acid	550	1	7.5	LHSV = 8.3 h^{-1}	100	92	Hu and Lu (2010b)
Co/Al_2O_3	Acetic acid	550	1	7.5	LHSV = 8.3 h^{-1}	100	90	Hu and Lu (2010b)
Rh/Al_2O_3	Acetic acid	800	1	3	WHSV ≈ 14 h^{-1}	78	96	Basagiannis and Verykios 2007)
Pd/Ni-Co-Mg-AlO_x	Acetic acid	575	1	3	WHSV = 0.893 h^{-1}	87	>99	Gil et al. (2016)
Co-La/Al_2O_3	Acetic acid	450	1	7.5	LHSV = 10.1 h^{-1}	100	96	Hu and Lu (2010a)
Ni-La/Al_2O_3	Acetaldehyde	600	1	5	LHSV = 12.7 h^{-1}	55	82	Zhang et al. (2020)
Ni-La/Al_2O_3	Acetone	600	1	5	LHSV = 12.7 h^{-1}	60	93	Zhang et al. (2020)
Ni-K/$CaAl_2O_4$	Acetone	650	1	6	LHSV = 8.9 h^{-1}	100	~90	Kechagiopoulos et al. (2006)
Ni-K/$CaAl_2O_4$	Ethylene glycol	700	1	6	LHSV = 8.5 h^{-1}	100	~90	Kechagiopoulos et al. (2006)
Rh-Ce/Al_2O_3	Ethylene glycol	680	1	4.5	Contact time ~ 10 ms	100	92	Dauenhauer et al. (2006)

[a] S/C: steam to carbon molar ratio.
[b] LHSV: liquid hourly space velocity; WHSV: weight hourly space velocity.

Since the high steam: carbon ratio benefits the reforming reaction (Chen et al. 2017b), the researchers also directly fed the aqueous phase fraction of bio-oil containing a large amount of water, with about a 60% H_2 yield being obtained. In addition, support also influences the performance, since it is reported to activate the steam and change the reaction pathways (Sun et al., 2015). An appropriate support and control of the metal/support interface area could benefit the SR reaction (Chen et al., 2017b, 2019).

Though certain catalysts show good initial performance, coking brings the main challenge in SR that gradually deactivates the catalysts during the time-on-stream operations. This problem was proposed to be solved in two directions: modifying the catalysts to change the surface chemistry or changing the reactor type and operating conditions (Trane et al., 2012). Coking usually happens on the acidic site of support (Wang et al., 1998) and metals with imbalanced water activation and the gasification of formed carbons during the SR (Davidson et al., 2014a; Sun et al., 2013). Mitigation of coking may be achieved by choosing the appropriate support (Sun et al., 2013) or adding the basic additives (Davidson et al., 2016; Kechagiopoulos et al., 2006), as well as controlling the reaction conditions. The second direction involves the co-feeding of O_2 during the SR which can gasify part of the coke and provide the energy for this endothermic reaction (Cavallaro et al., 2003) and catalytic partial oxidation using O_2 supplied below stoichiometric for combustion, prereforming, etc. (Trane et al., 2012). Readers are referred to the literature (Chen and He, 2011; Chen et al., 2017a–b; Nabgan et al., 2017; Trane et al., 2012) for detailed discussion.

8.3.2 C–C Coupling of Small Oxygenates

The most oxygen-abundant functional group in bio-oil is the carboxylic one and directly hydrotreating carboxylic acids will consume a large amount of valuable hydrogen (Deng et al., 2009). On the other hand, C–C coupling reactions of the small oxygenates, such as ketonization, are able to remove oxygen from carboxylic acids while retaining the carbon in the product (Resasco, 2011). Deng et al. (2009) has proposed a potential route to convert these small oxygenates: ketonization (Equation 8.2) of two carboxylic acids to remove the carboxyl group and form a larger ketone, which can then react with other ketones or aldehydes via (cross) aldol condensation (Equation 8.3) to further increase the carbon number in the carbon backbone. The produced enone compound is then processed by hydrogenation and deoxygenation for the production of heavier hydrocarbons. Also note that the ketonization and aldol condensation do not consume H_2 and only form carbon dioxide and water as the by-products, which is another advantage of these processes.

$$R'COOH + R''COOH \rightarrow R'RCO)R'' + H_2O + CO_2 \tag{8.2}$$

$$R')CO)R'' + R'''\left(CO\right)R''''H_2 \rightarrow R'''COR'''' = CR'R'' + H_2O \tag{8.3}$$

Most ketonization studies and reports focus on carboxylic acids as substrates. Whereas there is still a considerable debate about the reaction mechanism, the 'coordinatively unsaturated site' (CUS) with intermediate acid-base strength is generally agreed to play a crucial role in the ketonization reaction (Pham et al., 2012; Wang and Iglesia, 2017). Table 8.3 shows several representative catalysts applied in the ketonization of carboxylic acids. Glinski et al. (1995) screened 20 oxide catalysts supported on SiO_2, TiO_2, and Al_2O_3, among which the Al_2O_3-supported CeO_2 and MnO_2 exhibited the highest performances for the ketonization of acetic acid. The Ce-based and Mn-based catalysts were also active for ketonization of other investigated (C3, C6, C7) carboxylic acids. However, the leaching of Mn oxide into the acid environment means that it may not be appropriate for the long-term reaction. The research shows that Ce-based, Ti-based, and Zr-based catalysts are among the most effective for this reaction (Davidson et al., 2019b; Kumar et al., 2018; Randery et al., 2002). Gaertner et al. (2009) conducted a kinetic study of hexanoic acid ketonization over $CeZrO_x$ in the presence of pentanol and butanone.

The results showed the activation energy is 132 kJ/mol, meaning that ketonization usually needs to be processed at a relatively high temperature. However, Pham et al. (2012) reported that $Ru/TiO_2/C$ pretreated *in situ* with an H_2 reduction exhibited high reactivity and 100% selectivity to acetone in the conversion of acetic acid at 180°C which is much lower than the typical reaction temperature. A coordinative unsaturated Ti^{3+} species was proposed to benefit ketonization; hydrophobic carbon support mitigated the inhibition to ketonization from water. The ketonization catalyst could be deactivated by contaminants in bio-oil, such as Ca and Na (Davidson et al., 2019b). To solve this problem, before ketonization over La_2O_3/ZrO_2, Davidson et al. (2019b) conducted a cleanup process involving active carbon treatment, extraction, and dilution. Comparing with the reaction using unpretreated HTL-derived aqueous-phase feed showing rapid deactivation, ~100 h of stability was achieved with feed that had been subjected to a cleanup process. Overall, the decarboxylative coupling of carboxylic acid with the currently developed catalysts can produce a ketone containing a longer chain with high selectivity.

As mentioned above, the ketone species can be further upgraded via aldol condensation. Aldol condensation, generally performed at mild temperatures (0–200°C), is another widely utilized reaction to upgrade carbonyl-containing compounds to larger products (Wu et al., 2016). In the upgrading of biomass-derived compounds using heterogeneous catalysts, it has been recognized that the catalyst with a combination of moderate strength acidic and basic sites is ideal to conduct aldol condensation with high performance (Cueto et al., 2017; Shen et al., 2012; Wang et al., 2016). It is also concluded that, in vapor phase aldol condensation, if the catalyst only preserves either strong acidic or basic sites, the severe deactivation seems usually to happen by the occupation of carbonaceous by-products of these sites (Shen et al., 2012). The investigated catalysts are various and can be categorized into (mixed) metal oxides (Liang et al., 2016; Shen et al., 2012), zeolites (Sharma et al., 2007), hydroxyapatite (Young et al., 2016), hydrotalcite (Sharma et al., 2007), resins (Bui et al., 2017), etc. Though the proposed mechanisms differ depending on the utilized catalysts, several reaction steps have been agreed to complete the reaction, including α-C–H activation/cleavage, C–C coupling, and dehydration.

The aldol condensation of ketones/aldehydes form a larger ketone that may act as a reactant for the secondary condensation (Salvapati et al., 1989), and there may be more than one site for α-C–H cleavage (site for the C–C bond formation) generating

TABLE 8.3
Representative Catalysts for the Ketonization of Carboxylic Acids

Catalyst	Reactant	Reaction Condition			Conv. (%)	S_{ketone} (%)	Reference
		T (°C)	P (bar)	Space Velocity[a]			
MnO_2/Al_2O_3	Acetic acid	350	1	LHSV = 2 ml g^{-1} h^{-1}	100	97	Glinski et al. (1995)
CeO_2/Al_2O_3	Acetic acid	350	1	LHSV = 2 ml g^{-1} h^{-1}	100	97	Glinski et al. (1995)
CeO_2-K_2O/TiO_2	Acetic acid	375	1	LHSV = 4 ml g^{-1} h^{-1}	99	99	Deng et al. (2009)
$CeZrO_x$	Propionic acid	350	1	WHSV = 20 h^{-1}	32	>95	Ding et al. (2018)
$Zn_xZr_yO_z$	Acetic acid/Propionic acid	415	1	WHSV = 0.3 h^{-1}	100	[b]	Baylon et al. (2016)
$LaZrO_x$	Acetic acid	295	98	WHSV = 3.8 h^{-1}	28	100	Lopez-Ruiz et al. (2017)
$Ru/TiO_2/C$	Acetic acid	180	28–55	Batch m_{acid}/m_{cat} = 14.5, 5h	54.2	100	Pham et al. (2012)
SiO_2-Al_2O_3	Acetic acid	300	1	WHSV = 10 h^{-1}	~10	~100	Gumidyala et al. (2016)
$CeTiO_x$	Acetic acid	350	1	LHSV = 6 ml g^{-1} h^{-1}	100	>98	Lu et al. (2017)

[a] LHSV: liquid hourly space velocity; WHSV: weight hourly space velocity.
[b] ~90% theoretical carbon yield to C3–C6 olefins

several isomers in the products (Liang et al., 2016). Therefore, the products from aldol condensation are complex. After this condensation, other processes such as dehydration/hydrogenation and self-deoxygenation can further convert the heavier ketones to fuels. For example, Huber et al. (2005) converted the products from aldol condensation of 5-hydroxymethylfurfural, furfural, and acetone species to C9–C15 alkanes with dehydration/hydrogenation over the bifunctional catalysts containing both acid and metal sites. Our recent studies (Baylon et al. 2016, 2018; Sun et al. 2011, 2016) valorized ethanol, acetone, and acetic acid over $Zn_xZr_yO_z$ where the ZnO addition selectively passivated the strong Lewis acidic sites on ZrO_2 and introduced the basicity such that the balanced surface Lewis acid–base pairs were obtained to enhance the condensation reaction and, most importantly, the following self-deoxygenation to produce light olefins in the absence of H_2.

Other C–C bond formation reactions, such as alkylation and hydroalkylation, also build up larger compounds. Alkylation is the introduction of an alkyl group into an organic compound by substitution or addition. Similar to alkylation, hydroalkylation reactions also increase the carbon number of furans and phenols. Corma et al. (2011) converted 2-methylfuran with a consecutive hydroxyalkylation, alkylation, and hydrodeoxygenation. By hydroalkylation reaction, 2-methylfuran firstly reacted with aldehyde forming the furanic alcohol. Further alkylation with the second 2-methyl furan followed by HDO formed 6-alkyl undecane. Readers are referred to the literature for detailed discussion (Corma et al., 2011; Li et al., 2014; Wu et al., 2016; Zhao et al., 2012).

8.3.3 Other Processes

Besides the above, other approaches were also studied for bio-oil upgrading. Bio-oil contains both carboxylic acids and alcohols that can react via esterification to form larger molecules. Esterification is a route to reduce the acid concentration and to improve stability. In this reaction, the oxygen content is also partially reduced in the form of water prior to hydrotreating (Milina et al., 2014). Wang et al. (2010) studied the esterification of bio-oil catalyzed over ion-exchange resins at 50°C. After the upgrading, the amount of the carboxyl group was reduced by ~90% and the heating values increased by ~30%. The stability was studied by determining the viscosity of bio-oil aging at 80°C. Comparing with the drastic increase of viscosity of the original bio-oil, the upgraded one did not show significant changes within 120 h. The recently reported esterification catalysts applied in bio-oil upgrading includes resins (Fu et al., 2015; Wang et al., 2010), zeolites (Li et al., 2017; Milina et al., 2014; Peng et al., 2009), and other solid acid/base catalysts (Peng et al. 2008; Zhang et al. 2006).

The dehydration of alcohols followed by other processes has also been widely studied to produce alkanes or other value-added chemicals (Huber et al., 2004; Sun and Wang, 2014). Different mechanisms have been proposed for dehydration that involve carbocation intermediate with C–O cleavage, or carbanion intermediate with C–H cleavage, or a concerted elimination of –OH and α–H (Sun and Wang, 2014). Other parallel reaction pathways may also exist. For example, ethanol has been revealed to process intermolecular dehydration to diethyl ether, followed by dehydration of diethyl ether to ethylene. The produced small olefins can undergo polymerization to build hydrocarbons with a larger carbon backbone (another reaction of C–C coupling). For example, the polymerization of ethylene derived from ethanol dehydration produces various polyethylene products (Sun and Wang, 2014).

The ethylene can also be epoxidized to obtain ethylene oxide as another important chemical. Bio-oil contains sugar alcohols that can be converted with dehydration combined with hydrogenation. Huber et al. (2004) employed the bifunctional reaction pathway over Pt/$SiAlO_x$ catalyst where $SiAlO_x$ catalyzed the sorbitol dehydration and Pt catalyzed the reforming of sorbitol to generate H_2 and a following hydrogenation of the dehydrated intermediates. The sorbitol was eventually converted to alkanes plus CO_2 and water. The carbon selectivity of alkanes could be controlled by co-feeding H_2 so that only alkanes without CO_2 were produced. The dehydration also commonly involved in the HDO reaction to convert the alcohol to alkene (Zhao et al., 2009) or arenes (Nie et al., 2014).

8.4 OUTLOOK

Given that the biomass is hydrogen deficient, the reforming of the small oxygenates in the aqueous phase minimized the external H_2 supply that is used in the hydrotreating process of bio-oil (Wang et al., 2013). It could also be coupled with other reactions so that the *in situ* generated hydrogen is directly used for the production of alkanes via hydrogenation/dehydration in the aqueous-phase processing (Huber et al., 2004). Several of the organic compounds (e.g. carboxylic acid) in the aqueous phase have high oxygen content, thus the direct hydrotreating (if applied) will consume costly H_2. Alternatively, the ketonization and aldol condensation remove the oxygen-rich carboxyl and keto group in the form of H_2O and CO_2, which may be used before the hydrotreating/hydrodeoxygenation to reduce the H_2 consumption.

Recent results have shown $Zn_xZr_yO_z$ with balanced surface Lewis acid–base pairs are versatile in directly converting carboxylic acids (Baylon et al., 2016; Crisci et al., 2014), ketone (Sun et al., 2016), and alcohol (Sun et al., 2011) to olefins via ketonization, aldol condensation, and self-deoxygenation reactions. The viability of using $Zn_xZr_yO_z$ for upgrading the aqueous-phase compounds in real bio-oil has also been demonstrated by Davidson et al. (2019a) for the production of olefins. These developed approaches to upgrading the aqueous phase will be beneficial in the utilization of bio-oil as a whole.

REFERENCES

Anex, R. P., A. Aden, and F. K. Kazi, et al.. 2010 Techno-economic comparison of biomass-to-transportation fuels via pyrolysis, gasification, and biochemical pathways. *Fuel* 89: S29–S35.

Ayodele, B. V., S. I. Mustapa, T. A. R. B. T. Abdullah, and S. F. Salleh. 2019. A mini-review on hydrogen-rich syngas production by thermo-catalytic and bioconversion of biomass and its environmental implications. *Frontiers in Energy Research* 7: 00118.

Azeez, A. M., D. Meier, J. Odermatt, and T. Willner. 2010. Fast pyrolysis of African and European lignocellulosic biomasses using Py-GC/MS and fluidized bed reactor. *Energy & Fuels* 24: 2078–2085.

Aziz, S. M. A., R. Wahi, Z. Ngaini, and S. Hamdan. 2013. Bio-oils from microwave pyrolysis of agricultural wastes. *Fuel Processing Technology* 106: 744–750.

Bae, Y. J., C. Ryu, J. K. Jeon, et al. 2011. The characteristics of bio-oil produced from the pyrolysis of three marine macroalgae. *Bioresource Technology* 102: 3512–3520.

Basagiannis, A. C. and X. E. Verykios. 2007. Catalytic steam reforming of acetic acid for hydrogen production. *International Journal of Hydrogen Energy* 32: 3343–3355.

Baylon, R. A. L., J. M. Sun, K. J. Martin, P. Venkitasubramanian, and Y. Wang. 2016. Beyond ketonization: Selective conversion of carboxylic acids to olefins over balanced Lewis acid-base pairs. *Chemical Communications* 52: 4975–4978.

Baylon, R. A. L., J. M. Sun, and L. Kovarik, et al. 2018. Structural identification of $Zn_x Zr_yO_z$ catalysts for cascade aldolization and self-deoxygenation reactions. *Applied Catalysis B-Environmental* 234: 337-346.

Bertero, M., G. de la Puente, and U. Sedran. 2012. Fuels from bio-oils: Bio-oil production from different residual sources, characterization and thermal conditioning. *Fuel* 95: 263-271.

Bui, T. V., T. Sooknoi, and D. E. Resasco. 2017. Simultaneous upgrading of furanics and phenolics through hydroxyalkylation/aldol condensation reactions. *ChemSusChem* 10: 1631–1639.

Cavallaro, S., V. Chiodo, A. Vita, and S. Freni. 2003. Hydrogen production by auto-thermal reforming of ethanol on Rh/Al_2O_3 catalyst. *Journal of Power Sources* 123: 10–16.

Chen, D. and L. He. 2011. Towards an efficient hydrogen production from biomass: A review of processes and materials. *Chem CatChem* 3: 490–511.

Chen, G., J. Tao, and C. Liu, et al. 2017a. Hydrogen production via acetic acid steam reforming: A critical review on catalysts. *Renewable and Sustainable Energy Reviews* 79: 1091–1098.

Chen, J., J. Sun, and Y. Wang. 2017b. Catalysts for steam reforming of bio-oil: A Review. *Industrial & Engineering Chemistry Research* 56: 4627–4637.

Chen, S., C. Pei, and J. Gong. 2019. Insights into interface engineering in steam reforming reactions for hydrogen production. *Energy & Environmental Science* 12: 3473–3495.

Chiaramonti, D., A. Oasmaa, and Y. Solantausta. 2007. Power generation using fast pyrolysis liquids from biomass. *Renewable and Sustainable Energy Reviews* 11: 1056–1086.

Corma, A., O. de la Torre, M. Renz, and N. Villandier. 2011. Production of high-quality diesel from biomass waste products. *Angewandte Chemie International Edition* 50: 2375–2378.

Crisci, A. J., H. Dou, T. Prasomsri, and Y. Roman-Leshkov. 2014. Cascade reactions for the continuous and selective production of isobutene from bioderived acetic acid over zinc-zirconia catalysts. *ACS Catalysis* 4: 4196–4200.

Cueto, J., L. Faba, E. Diaz, and S. Ordonez. 2017. Performance of basic mixed oxides for aqueous-phase 5-hydroxymethylfurfural-acetone aldol condensation. *Applied Catalysis B: Environmental* 201: 221–231.

Dauenhauer, P., J. Salge, and L. Schmidt. 2006. Renewable hydrogen by autothermal steam reforming of volatile carbohydrates. *Journal of Catalysis* 244: 238–247.

Davidson, S. D., H. Zhang, J. M. Sun, and Y. Wang. 2014a. Supported metal catalysts for alcohol/sugar alcohol steam reforming. *Dalton Transactions* 43: 11782–11802.

Davidson, S. D., H. Zhang, J. Sun, and Y. Wang. 2014b. Supported metal catalysts for alcohol/sugar alcohol steam reforming. *Dalton Transactions* 43: 11782.

Davidson, S. D., J. Sun, and Y. Wang. 2016. The effect of ZnO addition on H_2O activation over Co/ZrO_2 catalysts. *Catalysis Today* 269: 140–147.

Davidson, S. D., J. A. Lopez-Ruiz, M. Flake, et al. 2019a. Cleanup and conversion of biomass liquefaction aqueous phase to C_3–C_5 olefins over $Zn_xZr_yO_z$ catalyst. *Catalysts* 9: 923.

Davidson, S. D., J. A. Lopez-Ruiz, and Y. Zhu, et al. 2019b. Strategies to valorize the hydrothermal liquefaction-derived aqueous phase into fuels and chemicals. *ACS Sustainable Chemistry & Engineering* 7: 19889–19901.

de Jong, S., R. Hoefnagels, A. Faaij, et al. 2015. The feasibility of short-term production strategies for renewable jet fuels: A comprehensive techno-economic comparison. *Biofuels, Bioproducts and Biorefining* 9 (6): 778–800.

Deng, L., Y. Fu, and Q. X. Guo. 2009. Upgraded acidic components of bio-oil through catalytic ketonic condensation. *Energy & Fuels* 23: 564–568.

Department of Energy, US. 2016. *Bioenergy Technologies Office Multi-Year Program Plan: March 2016*. Washington, DC: Department of Energy.

Dickerson, T. and J. Soria. 2013. Catalytic fast pyrolysis: A review. *Energies* 6 (1): 514–538.

Ding, S., H. Wang, J. Han, X. Zhu, and Q. Ge. 2018. Ketonization of propionic acid to 3-pentanone over $Ce_xZr_{1-x}O_2$ catalysts: The importance of acid–base balance. *Industrial & Engineering Chemistry Research* 57: 17086–17096.

Elliott, D. C. 2007. Historical developments in hydroprocessing bio-oils. *Energy & Fuels* 21: 1792–1815.

Elliott, D. C., T. R. Hart, G. Gary, et al. 2012. Catalytic hydroprocessing of fast pyrolysis bio-oil from pine sawdust. *Energy & Fuels* 26: 3891–3896.

Fu, J., L. Chen, P. Lv, L. Yang, and Z. Yuan. 2015. Free fatty acids esterification for biodiesel production using self-synthesized macroporous cation exchange resin as solid acid catalyst. *Fuel* 154: 1–8.

Gaertner, C. A., J. C. Serrano-Ruiz, D. J. Braden, and J. A. Dumesic. 2009. Catalytic coupling of carboxylic acids by ketonization as a processing step in biomass conversion. *Journal of Catalysis* 266: 71–78.

Garcia-Perez, M., A. Chaala, H. Pakdel, D. Kretschmer, and C. Roy. 2007a. Vacuum pyrolysis of softwood and hardwood biomass. *Journal of Analytical and Applied Pyrolysis* 78: 104–116.

Garcia-Perez, M., T. T. Adams, J. W. Goodrum, D. P. Geller, and K. C. Das. 2007b. Production and fuel properties of pine chip bio-oil/biodiesel blends. *Energy & Fuels* 21: 2363–2372.

Gil, M. V., J. Fermoso, C. Pevida, D. Chen, and F. Rubiera. 2016. Production of fuel-cell grade H2 by sorption enhanced steam reforming of acetic acid as a model compound of biomass-derived bio-oil. *Applied Catalysis B: Environmental* 184: 64–76.

Glinski, M., J. Kijenski, and A. Jakubowski. 1995. Ketones from monocarboxylic acids: Catalytic ketonization over oxide systems. *Applied Catalysis A: General* 128: 209–217.

Gollakota, A. R. K., N. Kishore, and S. Gu. 2018. A review on hydrothermal liquefaction of biomass. *Renewable and Sustainable Energy Reviews* 81: 1378–1392.

Gumidyala, A., T. Sooknoi, and S. Crossley. 2016. Selective ketonization of acetic acid over HZSM-5: The importance of acyl species and the influence of water. *Journal of Catalysis* 340: 76–84.

Hicks, J. C. 2011. Advances in C–O bond transformations in lignin-derived compounds for biofuels production. *Journal of Physical Chemistry Letters* 2: 2280–2287.

Hu, X. and G. Lu. 2010a. Acetic acid steam reforming to hydrogen over Co–Ce/Al_2O_3 and Co–La/Al_2O_3 catalysts: The promotion effect of Ce and La addition. *Catalysis Communications* 12: 50–53.

Hu, X. and G. Lu. 2010b. Comparative study of alumina-supported transition metal catalysts for hydrogen generation by steam reforming of acetic acid. *Applied Catalysis B: Environmental* 99: 289–297.

Huber, G. W., R. D. Cortright, and J. A. Dumesic. 2004. Renewable alkanes by aqueous-phase reforming of biomass-derived oxygenates. *Angewandte Chemie International Edition* 43: 1549–1551.

Huber, G. W., J. N. Chheda, C. J. Barrett, and J. A. Dumesic. 2005. Production of liquid alkanes by aqueous-phase processing of biomass-derived carbohydrates. *Science* 308: 1446–1450.

Imam, T. and S. Capareda. 2012. Characterization of bio-oil, syn-gas and bio-char from switchgrass pyrolysis at various temperatures. *Journal of Analytical and Applied Pyrolysis* 93: 170–177.

Kechagiopoulos, P. N., S. S. Voutetakis, A. A. Lemonidou, and I. A. Vasalos. 2006. Hydrogen production via steam reforming of the aqueous phase of bio-oil in a fixed bed reactor. *Energy & Fuels* 20: 2155–2163.

Kumar, R., N. Enjamuri, S. Shah, et al. 2018. Ketonization of oxygenated hydrocarbons on metal oxide based catalysts. *Catalysis Today* 302: 16–49.

Lehto, J., A. Oasmaa, Y. Solantausta, M. Kyto, and D. Chiaramonti. 2014. Review of fuel oil quality and combustion of fast pyrolysis bio-oils from lignocellulosic biomass. *Applied Energy* 116: 178–190.

Li, G., N. Li, J. Yang, et al. 2014. Synthesis of renewable diesel range alkanes by hydrodeoxygenation of furans over Ni/Hβ under mild conditions. *Green Chemistry* 16: 594–599.

Li, C., X. Zhao, A. Wang, G. W. Huber, and T. Zhang. 2015. Catalytic transformation of lignin for the production of chemicals and fuels. *Chemical Reviews* 115: 11559–11624.

Li, J., H. Liu, T. An, Y. Yue, and X. Bao. 2017. Carboxylic acids to butyl esters over dealuminated-realuminated beta zeolites for removing organic acids from bio-oils. *RSC Advances* 7: 33714–33725.

Liang, G., A. Wang, X. Zhao, N. Lei, and T. Zhang. 2016. Selective aldol condensation of biomass-derived levulinic acid and furfural in aqueous-phase over MgO and ZnO. *Green Chemistry* 18: 3430–3438.

Liaquat, A. M., H. H. Masjuki, M. A. Kalam, et al. 2013. Effect of coconut biodiesel blended fuels on engine performance and emission characteristics. *Procedia Engineering* 56: 583–590.

Liu, C., H. Wang, A. M. Karim, J. Sun, and Y. Wang. 2014. Catalytic fast pyrolysis of lignocellulosic biomass. *Chemical Society Reviews* 43: 7594–7623.

Lopez-Ruiz, J. A., A. R. Cooper, G. Li, and K. O. Albrecht. 2017. Enhanced hydrothermal stability and catalytic activity of $La_x Zr_y O_z$ mixed oxides for the ketonization of acetic acid in the aqueous condensed phase. *ACS Catalysis* 7: 6400–6412.

Lu, F., B. Jiang, J. Wang, et al. 2017. Promotional effect of Ti doping on the ketonization of acetic acid over a CeO_2 catalyst. *RSC Advances* 7: 22017–22026.

Milina, M., S. Mitchell, and J. Perez-Ramirez. 2014. Prospectives for bio-oil upgrading via esterification over zeolite catalysts. *Catalysis Today* 235: 176–183.

Mullen, C. A. and A. A. Boateng. 2008. Chemical composition of bio-oils produced by fast pyrolysis of two energy crops. *Energy & Fuels* 22: 2104–2109.

Munoz, J. A. D., J. Ancheyta, and L. C. Castaneda. 2016. Required viscosity values to ensure proper transportation of crude oil by pipeline. *Energy & Fuels* 30: 8850–8854.

Nabgan, W., T. Abdullah, R. Mat, et al. 2017. Renewable hydrogen production from bio-oil derivative via catalytic steam reforming: An overview. *Renewable and Sustainable Energy Reviews* 79: 347–357.

Nie, L., P. M. de Souza, F. B. Noronha, et al. 2014. Selective conversion of m-cresol to toluene over bimetallic Ni-Fe catalysts. *Journal of Molecular Catalysis A: Chemical* 388–9: 47–55.

Peng, J., P. Chen, H. Lou, and X. Zheng. 2008. Upgrading of bio-oil over aluminum silicate in supercritical ethanol. *Energy & Fuels* 22: 3489–3492.

Peng, J., P. Chen, H. Lou, and X. Zheng. 2009. Catalytic upgrading of bio-oil by HZSM-5 in sub- and super-critical ethanol. *Bioresource Technology* 100: 3415–3418.

Pham, T. N., D. Shi, T. Sooknoi, and D. E. Resasco. 2012. Aqueous-phase ketonization of acetic acid over Ru/TiO_2/carbon catalysts. *Journal of Catalysis* 295: 169–178.

Randery, S., J. Warren, and K. Dooley. 2002. Cerium oxide-based catalysts for production of ketones by acid condensation. *Applied Catalysis A: General* 226: 265–280.

Resasco, D. E. 2011. What should we demand from the catalysts responsible for upgrading biomass pyrolysis oil? *Journal of Physical Chemistry Letters* 2: 2294–2295.

Ruddy, D. A., J. A. Schaidle, J. R. F. Iii, et al. 2014. Recent advances in heterogeneous catalysts for bio-oil upgrading via *ex situ* catalytic fast pyrolysis: Catalyst development through the study of model compounds. *Green Chemistry* 16: 454–490.

Salvapati, G. S., K. V. Ramanamurty, and M. Janardanarao. 1989. Selective catalytic self-condensation of acetone. *Journal of Molecular Catalysis* 54: 9–30.

Sharma, S. K., S. A. Parikh, and R. V. Jasra. 2007. Solvent free aldol condensation of propanal to 2-methylpentenal using solid base catalysts. *Journal of Molecular Catalysis A: Chemical* 278: 135–144.

Shen, W., G. A. Tompsett, R. Xing, W. C. Conner, and G. W. Huber. 2012. Vapor phase butanal self-condensation over unsupported and supported alkaline earth metal oxides. *Journal of Catalysis* 286: 248–259.

Sun, J., K. Zhu, F. Gao, et al. 2011. Direct conversion of bio-ethanol to isobutene on nanosized $Zn_xZr_yO_z$ mixed oxides with balanced acid-base sites. *Journal of the American Chemical Society* 133 (29): 11096–11099.

Sun, J. M., D. H. Mei, A. M. Karim, A. K. Datye, and Y. Wang. 2013. Minimizing the formation of coke and methane on Co Nanoparticles in steam reforming of biomass-derived oxygenates. *Chem CatChem* 5: 1299–1303.

Sun, J. and Y. Wang. 2014. Recent advances in catalytic conversion of ethanol to chemicals. *ACS Catalysis* 4: 1078–1090.

Sun, J. M., A. M. Karim, D. H. Mei, et al. 2015. New insights into reaction mechanisms of ethanol steam reforming on Co-ZrO_2. *Applied Catalysis B-Environmental* 162: 141–148.

Sun, J., R. A. Baylon, C. Liu, et al. 2016. Key roles of Lewis acid-base pairs on Xn_x Zr_YO_z in direct ethanol/acetone to isobutene conversion. *Journal of the American Chemical Society* 138: 507–517.

Su-Ping, Z. 2003. Study of hydrodeoxygenation of bio-oil from the fast pyrolysis of biomass. *Energy Sources* 25: 57–65.

Trane, R., S. Dahl, M. S. Skjoth-Rasmussen, and A. D. Jensen. 2012. Catalytic steam reforming of bio-oil. *International Journal of Hydrogen Energy* 37: 6447–6472.

Wang, D., S. Czernik, and E. Chornet. 1998. Production of hydrogen from biomass by catalytic steam reforming of fast pyrolysis oils. *Energy & Fuels* 12: 19–24.

Wang, J. J., J. Chang, and J. Fan. 2010. Upgrading of bio-oil by catalytic esterification and determination of acid number for evaluating esterification degree. *Energy & Fuels* 24: 3251–3255.

Wang, H., J. Male, and Y. Wang. 2013. Recent advances in hydrotreating of pyrolysis bio-oil and its oxygen-containing model compounds. *ACS Catalysis* 3: 1047–1070.

Wang, S., B. Ru, H. Lin, W. Sun, and Z. Luo. 2015. Pyrolysis behaviors of four lignin polymers isolated from the same pine wood. *Bioresource Technology* 182: 120–127.

Wang, S., K. Goulas, and E. Iglesia. 2016. Condensation and esterification reactions of alkanals, alkanones, and alkanols on TiO_2: Elementary steps, site requirements, and synergistic effects of bifunctional strategies. *Journal of Catalysis* 340: 302–320.

Wang, S. and E. Iglesia. 2017. Experimental and theoretical assessment of the mechanism and site requirements for ketonization of carboxylic acids on oxides. *Journal of Catalysis* 345: 183–206.

Wu, L., T. Moteki, A. A. Gokhale, D. W. Flaherty, and F. D. Toste. 2016. Production of fuels and chemicals from biomass: Condensation reactions and beyond. *Chem* 1: 32–58.

Xiu, S. and A. Shahbazi. 2012. Bio-oil production and upgrading research: A review. *Renewable and Sustainable Energy Reviews* 16: 4406–4414.

Young, Z. D., S. Hanspal, and R. J. Davis. 2016. Aldol condensation of acetaldehyde over titania, hydroxyapatite, and magnesia. *ACS Catalysis* 6: 3193–3202.

Yu, N., H. Zhang, S. D. Davidson, J. Sun, and Y. Wang. 2016. Effect of ZnO facet on ethanol steam reforming over Co/ZnO. *Catalysis Communications* 73: 93–97.

Zhang, Q., J. Chang, Wang, and Y. Xu. 2006. Upgrading bio-oil over different solid catalysts. *Energy & Fuels* 20: 2717–2720.

Zhang, Q., J. Chang, T. Wang, and Y. Xu. 2007. Review of biomass pyrolysis oil properties and upgrading research. *Energy Conversion and Management* 48: 87–92.

Zhang, L., Z. Yu, J. Li, et al. 2020. Steam reforming of typical small organics derived from bio-oil: Correlation of their reaction behaviors with molecular structures. *Fuel* 259: 116214.

Zhao, C., Y. Kou, A. A. Lemonidou, X. Li, and J. A. Lercher. 2009. Highly selective catalytic conversion of phenolic bio-oil to alkanes. *Angewandte Chemie International Edition* 48: 3987–3990.

Zhao, C., W. Song, and J. A. Lercher. 2012. Aqueous phase hydroalkylation and hydrodeoxygenation of phenol by dual functional catalysts comprised of Pd/C and H/La-BEA. *ACS Catalysis* 2: 2714–2723.

9 An Overview of Catalytic Bio-oil Upgrading, Part 2: *Processing Oil-Phase Compounds and Real Bio-oil*

Jianghao Zhang
Junming Sun
Yong Wang

CONTENTS

9.1 INTRODUCTION

Due to its higher viscosity, water content, and lower heating value, bio-oil upgrading is an essential process through oxygen removal and molecular weight reduction (Elliott, 2007). The bio-oil may be divided into two phases: the oil phase that is hydrophobic and the aqueous phase that contains a light organic fraction and a large amount of water (Garcia-Perez et al., 2007a, 2007b; Imam and Capareda 2012; Su-Ping, 2003). In the previous chapter (Zhang et al., 2021), we summarized the developed approaches, including the promising catalysts and proposed reaction mechanism for the catalytic conversions of the small oxygenates in the aqueous phase of bio-oil to produce valuable fuels/chemicals.

In this part, we mainly present the progress with respect to the catalytic bio-oil upgrading of oil-phase compounds and real bio-oil. We first review the fundamental understanding of the surface chemistry of catalysts for upgrading model molecules that are representative in the oil phase. Based on the fundamental understandings derived from upgrading compounds in the two phases (the aqueous phase is shown in Zhang et al. 2021 and the oil phase), the processing of real bio-oil is then discussed with representative examples including hydrotreating and zeolite cracking. Finally, we present our perspective on future catalyst design and potential processes in bio-oil upgrading.

9.2 THE PROCESSING OF OIL-PHASE COMPOUNDS

In the oil phase, the main oxygen-containing compounds are aromatic oxygenates, sugars, furans, cyclic ketones, and larger oligomers (Garcia-Perez et al. 2007a, 2007b; Imam and Capareda 2012). It may also contain a certain amount of fatty acids (Garcia-Perez et al. 2007b; Imam and Capareda, 2012). Though other approaches have also been developed, e.g. steam reforming of the oil-phase oxygenates (Chen et al., 2017; Zhang et al., 2020c) and C-C coupling of substituted furans with other oxygenates (Liang et al., 2016), the studies for upgrading the compounds in the oil phase mainly focuses on hydrotreating/hydrodeoxygenation (HDO). Moreover, the products from C-C coupling still need further deoxygenation, mainly by HDO, to complete the upgrading. In the HDO process, the oxygenates are exposed to hydrogen in the presence of a catalyst at a temperature usually between 200 and 500°C. The oxygen in the substrates is removed in the form of water and/or other small oxygenates (CO_x, methanol, etc.) (Goncalves et al., 2017a, 2017b; Hong et al., 2016; Zhao et al., 2010). This section focuses on the HDO of two types of compounds: phenolic and furanic oxygenates, which involve the elimination of hydroxyl group bonding to aromatic and aliphatic carbon, carbonyl, aldehyde, and ether that can represent most of the oxygen-containing functional groups. Other oxygenates in the oil phase such as cyclic ketones are not discussed.

9.2.1 Catalytic Hydrodeoxygenation of Lignin-Derived Aromatic Oxygenates

The dominant oxygen-containing functional groups in aromatic oxygenates could be classified as ether ($C_{aromatic}$–O–C bond) and hydroxyl ($C_{aromatic}$–OH bond). The 'bonding dissociation energies' (BDE) follow the order of $C_{aromatic}$–OH > $C_{aromatic}$–OC > $C_{aromatic}$O–$C_{aliphatic}$. It has been shown that the activation energy of hydrogenolysis of the ether bond is highly dependent on the BDE (He et al., 2012). Therefore, most of the catalysts primarily cleave to the weakest $C_{aromatic}$O–$C_{aliphatic}$ bond, forming the strongest $C_{aromatic}$–OH bond in the HDO process (Guvenatam et al., 2014; Yang et al., 2014). Moreover, our recent study has shown that, regarding the aromatic oxygenate containing two hydroxyls, such as catechol, one of the hydroxyls is readily eliminated to form phenol even over a carbon material (Sun et al., 2013). As a result, the HDO of each type of aromatic oxygenates eventually converges to the removal of oxygen in the aromatics possessing a single $C_{aromatic}$–OH group, with phenol as representative.

From the thermodynamic point of view, low pressure and high temperature favor the formation of arene, and vice versa for the ring-saturated hydrocarbons. To minimize hydrogen consumption, producing aromatic hydrocarbons via a selective HDO process is always desirable (Hicks, 2011; Resasco, 2011). In terms of the reaction mechanism, three primary reaction pathways have been proposed in the HDO of phenols, i.e. direct aromatic ring hydrogenation, direct C–O bond cleavage, and keto-enol tautomerization (isomerization of phenol to form cyclohexadienone followed by hydrogenation of C=C and C=O bonds) (Gu et al., 2016; Zhang et al., 2020a–b), as shown in Figure 9.1 for the case of phenol. Though each pathway can lead to both aromatic and ring-saturated hydrocarbons, the product distribution is highly dependent on the reaction pathway, catalyst, and reaction conditions.

For the ring saturation pathway (i.e. the direct ring hydrogenation and the tautomerization reaction pathways in Figure 9.1), where high H_2 pressure can enhance the activity for hydrogenation of C=C in the reactant/intermediates together with the thermodynamically inhibited dehydrogenation of cyclohexene intermediate, benzene production is unfavorable. Under low H_2 pressure, however, dehydrogenation of cyclohexane is no longer a thermodynamic bottleneck for the decreased hydrogenation activity: both direct ring hydrogenation and the tautomerization pathway are able to produce benzene with relatively high selectivity, as demonstrated by previous studies (Nie and Resasco, 2014; Nie et al., 2014). For a direct

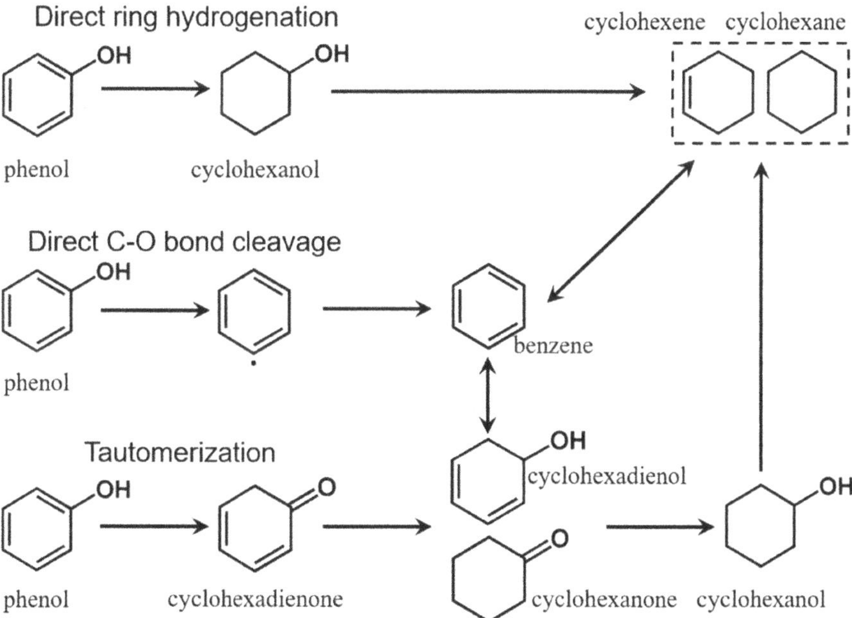

FIGURE 9.1 The proposed reaction pathways for HDO of phenol.

C–O bond cleavage pathway, benzene is produced as a primary product. Though low hydrogen pressure reduces the secondary hydrogenation reaction of benzene to ring saturation products, an optimum pressure would balance the selectivity and activity since high hydrogen pressure improves hydrogen supply and enhances the selective HDO reaction rate.

Other than thermodynamic considerations, the dominant reaction pathway and thus product distribution is also highly dependent on the surface properties of a catalyst (e.g. reaction kinetics). Table 9.1 lists several representative catalysts in the HDO of phenolics. Several factors, including the adsorption configuration of the reactant, the 'coordinatively unsaturated site' (CUS), and oxophilicity, have been used to interpret the structure–function relationship that may be used as guidance for future catalyst design. The adsorption configuration has been demonstrated to affect the reaction pathway (Badawi et al., 2011; Shi et al., 2016; Wang et al. 2018), i.e. planar adsorption favors ring hydrogenation whereas the perpendicular one favors direct C–O cleavage.

For example, Shi et al. (2016) showed that the strong interaction between the phenyl ring of anisole and the surface Pt(111) facilitated ring saturation, as suggested

TABLE 9.1
Representative Catalysts in the HDO of Aromatic Oxygenates

	Reaction Condition						
Catalyst	**T (°C)**	**P (bar)**	**Solvent/ reactor[a]**	**Substrate**	**Conv. (%)**	**S_{arene} (%)**	**Reference**
MoS_2	325	41	Decalin/[b]	Cresol	24	73	Whiffen and Smith (2010)
Co-$^{S}MoS_2$	300	30	Decalin/[b]	Cresol	83.6	99.2	Liu et al. (2017)
$NiMoWS_x$	300	30	Decalin/[b]	Cresol	100	94.4	Chen et al. (2013)
Fe	300	1	~/[c]	Cresol	21	90	Hong et al. (2014)
MoO_3	325	1	~/[c]	Cresol	48.9	99.4	Prasomsri et al. (2014)
Ni/SiO_2	275	100	~/[b]	Phenol	80	0	Mortensen et al. (2013)
Co/SBA-15	300	5	~/[c]	Anisole	>80	42	Yang et al. (2017b)
Ru/C	275	50	Hexadecane	Eugenol	100	0	Bjelic et al. (2018)
Pd/C	275	100	~/[b]	Phenol	100	0	Mortensen et al. (2013)
Ru/Nb_2O_5	250	5	Water/[b]	Cresol	100	80	Shao et al. (2017)
Ru-WO_x/SiAl	270	20	Water/[b]	Butylphenol	100	81	Huang et al. (2015)
Pd-Fe/C	450	1	~/[c]	Guaiacol	100	83	Sun et al. (2013)
Pt-WO_x/C	300	36	Dodecane/[b]	Cresol	61	98	Wang et al. (2018)
MoC-SiO_2	320	60	Hexadecane/[b]	Anisole	~65	~70	Smirnov et al. (2017)
Ni_2P/SiO_2	340	40	dodecane/[c]	Cresol	18.7	8.2	Goncalves et al. (2017b)
Mo_2N/TiO_2	350	25	Decane/[c]	Phenol	~9	~91	Boullosa-Eiras et al. (2014)

[a,b] batch reactor.
[c] continuous flow reactor.

by temperature programmed desorption and electron energy loss spectroscopy. In contrast, over the Zn-modified Pt(111), the anisole bond to the surface by oxygen at Zn or adjacent Pt sites and the phenyl ring was tilted away from the surface. This change of adsorption geometry made the PtZn/C catalyst display ~70% selectivity to phenol from CH_3–O bond cleavage, in comparison with ~100% selectivity to ring saturation over Pt/C (Shi et al., 2016). A CUS is proposed to be responsible for the turnover of the HDO reaction, i.e. the catalytic loop consists of an abstraction of O in phenolics by CUS and the elimination of O by H_2 to regenerate the CUS. Therefore, controlling the density of CUS influences both the direct deoxygenation reaction rate and arene selectivity. Goncalves et al. (2017a) correlated the number of CUS (oxygen vacancies) of MoO_x on different supports measured by oxygen chemisorption with the reaction rates of cresol deoxygenation, further suggesting CUS plays an important role in the deoxygenation reaction.

Oxophilicity describes the tendency of a metal to form oxide. Tan et al. (2017) investigated the HDO of anisole over Fe and noble metals (Pt and Ru) by a combined experimental and theoretical study where oxophilicity was expressed as the adsorption energies of atomic oxygen. With the increase of the oxophilicity of metal, the energy barrier for the direct C–O bond cleavage decreases, whereas the barrier for hydrogenation increases, which in return, impedes the further elimination of the oxygen-containing functional group on the metal surface, resulting in the oxidation and thus deactivation of the catalyst (Hong et al., 2017). To maintain the catalytic turnover, a metal with the appropriate oxyphilicity, together with the modification of a second metal (e.g. noble metal), was shown to be promising (Hong et al., 2014).

9.2.2 Catalytic Hydrodeoxygenation of Carbohydrate-Derived Furanic Oxygenates

Furanic oxygenates (e.g. 5-hydroxymethylfurfural (HMF) and furfural) are derived from the dehydration of carbohydrate in the biomass (i.e. C6 sugar units in cellulose and C5 sugar units in hemicellulose). Furanic oxygenates and their derivatives such as cyclic ketones constitute a large proportion of bio-oil (Bertero et al., 2012; Nakagawa et al., 2015). Several of these oxygenates can be initially processed with C–C coupling to increase the carbon number in the carbon backbone and meanwhile obtain a partial deoxygenation. Complete oxygen removal by HDO is still critical to the production of biofuels (Nakagawa et al., 2015). This section presents a fundamental understanding of the catalysts in the HDO of modeling furanic oxygenates.

The desired product with high selectivity can be gained by controlling the reaction pathway using catalysts. Similar to the HDO of aromatic oxygenate, oxophilicity is also correlated with the performance of a catalyst. Readers are referred to Kepp (2016) for the oxophilicity index of a specific metal. Goulas et al. (2019) conducted kinetic and extensive characterization studies on the C–O bond cleavage of furfuryl alcohol. The reaction rate for C–O bond cleavage in furfuryl alcohol displayed a volcano-type dependence on the 'Gibbs free energy' of metal oxide formation (another way to evaluate the oxophilicity of metal). This correlation (Goulas et al.,

2019) for oxide catalysts originated from the reverse 'Mars-van Krevelen C–O bond activation mechanism', which requires the balancing of the activation of the O abstraction from furanic oxygenates and its further elimination from the vacancy site. Therefore, the optimum catalyst is the one with moderate oxophilicity (Xiong et al., 2014) or a combination of both low and high oxophilic metals (Huang et al., 2014) to achieve high activity and selectivity.

Another factor that has been widely reported to influence the selectivity is the adsorption configuration of the functional groups in furanic oxygenates with the catalyst surface. For example, depending on the nature of the catalyst surface, the carbonyl group may adsorb on the surface via η^1-(O) or η^2-(C, O) configuration (Singh, 2018). The former facilitates the deoxygenation reaction pathway while the latter favors decarbonylation or hydrogenation. Yang et al. (2017a) modified the Pd/SiO_2 with FeOx species to restrain the decarbonylation by hindering the η^2-(C, O) adsorption on the Pd surface. Moreover, the Pd enhanced the reducibility of Fe species which might promote deoxygenation. As a result, the catalyst produced 87–94% jet fuel range alkanes under the tested conditions. This aligns with reports regarding the enhanced reactivity of phenolic HDO over a Pd-Fe bimetallic catalyst (Hong et al., 2014; Sun et al., 2013), suggesting the underlying mechanism for HDO of furanic oxygenates may be similar to that of aromatic oxygenates. Table 9.2 displays the representative catalysts applied in the HDO of furanic oxygenates. These catalysts are similar to those reported in the HDO of aromatic oxygenates, though the sulfide catalysts are not widely reported in this type of reaction.

9.3 THE PROCESSING OF REAL BIO-OILS

Though there have been tremendous advances in understanding the fundamentals of each type of reaction, the complexity of bio-oil creates a barrier in applying this understanding to the upgrading of real bio-oil. When real bio-oil is fed, it is still challenging to predict how each family of compounds will behave in the presence of others (Resasco and Crossley, 2015). Therefore, benefiting from knowledge obtained in modeling reaction, processing of real bio-oil has been extensively investigated.

9.3.1 HYDROTREATING

Hydrotreating is the most common route for upgrading bio-oil that usually takes place between 100 and 500°C (Wang et al., 2013). The goal of hydrotreating is to eliminate the reactive functional groups (e.g. –C=O and –COO) in bio-oil that may also be accompanied with the hydrogenation of the C=C bond (Patel and Kumar, 2016). Jensen et al. (2016) hydrotreated bio-oil in a batch reactor with sulfided a commercial NiMo/Al_2O_3 catalyst under different operating conditions. The optimized condition (350°C and 337 L H_2/L bio-crude) generated a product with 0.3 wt% O, an HHV of 43.9 MJ/kg, and a density of 894 kg/m^3.

These chemical and thermophysical properties are similar to North Sea fossil crude oil, making the hydrotreated product amenable to further upgrading in a refinery. On the other hand, a continuous-flow reactor is more desirable in the industrial processing of bio-oil (Wang et al., 2013). It was found that, however, hydrotreating

TABLE 9.2
Representative Catalysts in the HDO of Furanic Oxygenates

Catalyst	Reactant	Reaction Condition			Conv.	Selectivity	Reference
		T (°C)	P (bar)	Space Velocity	(%)		
Pd-FeO_x/SiO_2	Furanic C10, C11, C14	300	60	1.3 ml g^{-1} h^{-1}	100	87%–94% to jet fuel range alkanes	Yang et al. (2017a)
Pd/C+$ZnCl_2$	HMF	150	22	batch, sub/cat=10, 8h	99	85% to dimethyl furan	Saha et al. (2014)
$PdNb_2O_5/SiO_2$	4-(2-furyl)-3-buten-2-one	170	25	batch, sub/cat=1, 16h	100	95% to C8 alkane	Shao et al. (2015)
Ni/Hβ	5,5′-(butane-1,1-diyl)bis(2-methylfuran)	230	60	1.3 ml g^{-1} h^{-1}	100	87% to C9–C15 range alkanes	Li et al. (2014)
Fe/C	HMF	240	40	batch, sub/cat~0.6, 12h	100	86% to dimethyl furan	Li et al. (2017)
Mo_2C	Furfural	150	1	~0.1 h^{-1}	2–70	~70% to methyl furan	Xiong et al. (2014)
Ni-W_2C/AC	HMF	180	40	batch, sub/cat~1, 4h	100	96% to dimethyl furan	Huang et al. (2014)
Fe_2P/Al_2O_3	2-furyl methyl ketone	400	1	~1.3 h^{-1}	93	~93% to alkanes	Ly et al. (2017)

bio-oil at a high temperature in a continuous-flow reactor produced chars plugged reactor and led to the quick deactivation of the catalyst via coking (Elliott, 2007), due to the presence of unsaturated compounds (e.g. olefins, aldehydes, ketones) that undergo condensation reactions to form heavy oligomers. To solve this problem, a two-step process was developed: a hydrotreating step at temperatures below 300°C to remove the reactive compounds and stabilize the bio-oils, followed by another hydrotreating step under more severe conditions to achieve deep deoxygenation.

Besides using bio-oil as feedstock, another option is directly hydrotreating biomass in the presence of a catalyst. Shao et al. (2017) reported on a Ru/Nb_2O_5 catalyst for the direct HDO of organosolv lignin with a 64 wt% arene selectivity and a total mass yield of 35.5 wt% in a 20 h reaction. This catalyst for the HDO of cresol displayed an 80% arene selectivity at 250°C and 5 bar H_2. A combined inelastic neutron scattering and DFT calculation confirmed that Nb_2O_5 strongly adsorbed phenol and significantly reduced the energy barrier for $C_{aromatic}$–O bond cleavage while Ru contributed to dissociation/activation of H_2. In addition, the generated arene desorbed readily which further mitigated the hydrogenation of the aromatic ring.

9.3.2 Zeolite Cracking

Similar to the catalytic cracking in a petroleum refinery, using zeolite catalysts allows the reduction of the oxygen content as well as the cracking of heavy molecules and thus improves the thermal stability of the obtained bio-oil. In the cracking process, oxygen in the compounds of bio-oils is mainly removed in the form of CO, CO_2, and H_2O (Wang et al., 2013). The obtained products are mostly aromatic and aliphatic hydrocarbons following a series of reactions (e.g. cracking, dehydration, decarbonylation/decarboxylation, and aromatization) (Wang et al., 2013).

Wang et al. (2014) studied the cracking of a simulated bio-oil mixture (i.e. hydroxypropanone, cyclopentanone, and acetic acid) in ethanol. An increased H: C ratio and inhibited coking were observed on the Ga_2O_3/HZSM-5 at 300–450°C and 20 bar. Acid sites on HZSM-5 were investigated to catalyze the deoxygenation reaction, and Ga_2O_3 was used to promote the aromatization of light olefin intermediates. As a result, the yield of oil-phase fuel reached up to 39.2% (containing ~95% aromatic hydrocarbons), and the formation of gaseous hydrocarbons was inhibited. Another type of catalytic cracking used in the upgrading process is the one found in the presence of hydrogen, namely hydrocracking (Yang et al., 2015). Remon et al. (2017) hydrocracked bio-oil in a batch microbomb reactor using a co-precipitated Ni-Co/Al-Mg catalyst. Statistical analysis showed that temperature and pressure are the operating parameters with the greatest effect on the C content and O removal. After the hydrocracking at 450°C and 150 bar for 1 h, the O content in the liquid oil was decreased from ~41 to 9 wt% and the HHV increased from 19 to 37 MJ/kg.

The integration of hydrotreating with zeolite cracking has been reported as an effective way to upgrade bio-oil into industrial commodity chemical feedstocks (Vispute et al., 2010). The bio-oil was firstly hydrotreated over Ru/C at 100°C and 100 bar to form polyols and alcohols, which were then converted over HZSM-5 at 600°C to produce light olefins and aromatic hydrocarbons. The hydrotreated bio-oil yielded 32.6% olefins and aromatics, much higher than that without hydrotreating (a 20% yield). In addition, the arene: olefins ratio can be tuned, depending on the reaction conditions, and the olefins can also be converted to arenes by recycling back to the zeolite reactor.

Co-processing bio-oil in refinery 'fluid catalytic cracking' (FCC) is a potentially economic approach since it reduces the cost by directly utilizing the existing infrastructure and distribution systems, and avoids the risk of establishing a new parallel system for bio-oil (Bui et al., 2009). The cracking of a blend of bio-oil (20 wt%) and vacuum gasoil (VGO, 80 wt%) was studied by Ibarra et al. (2016) at 500–560°C within a riser simulator reactor. In comparison with individual oil processing, the co-processing promoted the production of C_3–C_4 and C_5–C_{12} (gasoline range), and meanwhile inhibited the formation of CO, CO_2, and coke. This promotion was attributed to the water in the feed which attenuated the cracking reactions of the gasoline components. In addition, the competitive adsorption of water with coke precursors on the acid sites, together with the enhanced hydrogen transfer from hydrocarbons in VGO to phenolic intermediates, were also proposed to alleviate coke formation.

9.3.3 Other Processes

By adding water into bio-oil, the relatively polar oxygenates are extracted into the separated aqueous fraction. Processes to upgrade this aqueous phase fraction have been investigated, such as an aqueous-phase reforming process to produce hydrogen (Wei et al., 2012) or a dehydration/hydrogenation process to produce alkanes (Huber et al., 2004; Vispute and Huber, 2009). For example, Vispute and Huber (2009) upgraded the aqueous fraction with a tandem process. The first step was hydrotreating the unsaturated functional groups to hydroxyl over Ru/C at 125–175°C. The produced alcohols could undergo either aqueous-phase reforming to produce hydrogen (60% selectivity) over Pt/Al_2O_3 at 265°C and 55.1 bar or dehydration/hydrogenation to produce alkane (45% selectivity) over Pt/SiO_2-Al_2O_3 at 260°C and 51.7 bar. Prior to the dehydration/hydrogenation, one alternative route is a C–C coupling reaction that elongates the carbon backbone to obtain heavier alkanes, as reported by Huber et al. (2005).

Not only the aqueous fraction, but all the bio-oil can be reformed to produce hydrogen (Chen et al., 2017). Due to the complexity of bio-oil, the current fundamental knowledge of bio-oil steam reforming (SR) is mainly based on single compound models, as discussed in Section 3.2.1. The proposed strategies to improve catalytic performance include mitigation of coking, inhibition of methanation, and enhancement of the water gas shift (Chen et al., 2017). In terms of catalysts, Ni-based ones, widely applied in the SR of fossil feedstocks, is also very active for the C–C and C–H bond cleavage of bio-oil components (Fu et al., 2014). Fu et al. (2014) investigated the SR of bio-oil over the nickel/alumina supported catalysts modified by cerium. By optimizing the amount of loaded nickel and the reaction conditions (i.e. reaction temperature, steam: carbon ratio, and space velocity), the hydrogen yield reached up to 71.4%. Ochoa et al. (2017) studied the SR of bio-oil over supported Ni (i.e. Ni/La_2O_3-Al_2O_3) in a two-stage system.

The first stage was the thermal treatment of bio-oil at 500°C to retain the pyrolytic lignin, while the second stage was the in-line SR of the remaining oxygenates in fluidized bed reactor. The main reason for the deactivation was demonstrated to be coke deposition by highly oxygenated compounds. It was found that the phenols and alcohols significantly contributed to coking, while acids, ketones, and aldehydes showed lower impact. Coking could be significantly mitigated by adjusting the reacting temperature and steam:carbon ration (i.e. 700°C and a ratio of 6).

9.4 OUTLOOK

The conversion of biomass into fuels and chemicals is one prospective approach for sustainable energy. The main challenge in this technology is the catalytic upgrading of bio-oil due to its complex composition (Resasco and Crossley, 2015). In the upgrading process, besides the productivity and lifetime of catalysts, maximizing carbon retention and minimizing the external hydrogen supply should be emphasized to gain a high yield of liquid fuels and chemicals at reduced cost. A multistage process may be employed, as suggested by Resasco (2011), i.e. a C–C coupling reaction of small oxygenates may be conducted prior to hydrotreating. For example, ketonization forms a larger carbon backbone and removes the carboxyl groups in the form of CO_2 and H_2O, which not only reduces the acidity, but also alleviates the H_2 consumption in hydrotreating. Moreover, it is desirable to develop a compatible upgrading process that can be dropped into the existing petroleum refinery infrastructure and fuel distribution system to reduce the investment and the risk.

Highly selective and durable catalysts are the key in successful upgrading technology (Hicks, 2011). The numerous studies conducted on the model compound reactions have made great contributions to the fundamental understanding of active sites which guide the catalyst design for the upgrading process. Taking the hydrotreating/HDO process as an example, the oxophilicity of the metal is proposed as a descriptor for the selective HDO of aromatic/furancic oxygenates (Goulas et al., 2019; Zhang et al., 2020a). In the HDO of aromatic oxygenates, low oxophilic metals are not active enough for the abstraction of O to achieve direct C–O bond cleavage, but they do display high activity in hydrogenating the aromatic ring; on the other hand, high oxophilic metals strongly interact with the abstracted O, making it difficult to release the O, even in the presence of H_2, such that the redox loop cannot be closed. Only the metals with moderate oxophilicity or an appropriate combination of low and high oxyphilic metal (e.g. PdFe; Hong et al., 2014; Sun et al., 2013) are able to not only abstract the O from the aromatic oxygenates but readily eliminate it from the catalyst surface.

Catalyst design for other families of reactions may also be guided by several references, such as the acid-base pairs for aldol condensation (Sun et al., 2011) and acid sites for hydration (Huber et al., 2004). Understanding the deactivation mechanism will guide not only the design of catalysts but also that of upgrading processes to elongate the catalysts' lifetimes. Given the complex composition of the products of bio-oil, the deactivation mechanisms may vary depending on impurities, reacting conditions, and catalysts. The common cause of deactivation is the blocking of the active site by coke or strongly adsorbed species, surface phase transformation of the catalyst during the upgrading process, poisoning of the active site by impurities such as an alkali metal, sintering that decreases the amount of active sites, etc. Future study needs to focus on understanding the deactivation mechanism and developing efficient methods to mitigate deactivation.

Though the studies on the upgrading process are plenty, the investigation is still lacking knowledge in terms of how the upgrading of one family of compounds will be influenced by the presence of other families (Dwiatmoko et al., 2014). Obtaining this knowledge will be critical for guiding the whole upgrading process, e.g. the sequence in multistage processes when the real bio-oil mixture is used as the feed. Another alternative is the separation of each family of compounds, followed by individual upgrading, which requires multiple step hydrolysis treatment (Jae et al. 2010) and the pyrolysis of solid lignin residue. These separated compounds could be upgraded with the catalytic reactions discussed above.

REFERENCES

Badawi, M., J. F. Paul, S. Cristol, and E. Payen. 2011. Guaiacol derivatives and inhibiting species adsorption over MoS_2 and CoMoS catalysts under HDO conditions: A DFT study. *Catalysis Communications* 12: 901–905.

Bertero, M., G. de la Puente, and U. Sedran. 2012. Fuels from bio-oils: Bio-oil production from different residual sources, characterization and thermal conditioning. *Fuel* 95: 263–271.

Bjelic, A., M. Grilc, and B. Likozar. 2018. Catalytic hydrogenation and hydrodeoxygenation of lignin-derived model compound eugenol over Ru/C: Intrinsic microkinetics and transport phenomena. *Chemical Engineering Journal* 333: 240–259.

Boullosa-Eiras, S., R. Lodeng, H. Bergem, et al. 2014. Catalytic hydrodeoxygenation (HDO) of phenol over supported molybdenum carbide, nitride, phosphide and oxide catalysts. *Catalysis Today* 223: 44–53.

Bui, van N., G. Toussaint, D. Laurenti, C. Mirodatos, and C. Geantet. 2009. Co-processing of pyrolysis bio oils and gas oil for new generation of bio-fuels: Hydrodeoxygenation of guaiacol and SRGO mixed feed. *Catalysis Today* 143: 172–178.

Chen, D., Z. Qu, Y. Sun, K. Gao, and Y. Wang. 2013. Identification of reaction intermediates and mechanism responsible for highly active HCHO oxidation on Ag/MCM-41 catalysts. *Applied Catalysis B: Environmental* 142–143: 838–848.

Chen, J., J. Sun, and Y. Wang. 2017. Catalysts for steam reforming of bio-oil: A review. *Industrial & Engineering Chemistry Research* 56: 4627–4637.

Dwiatmoko, A. A., S. Lee, H. C. Ham, et al. 2014. Effects of carbohydrates on the hydrodeoxygenation of lignin-derived phenolic compounds. *ACS Catalysis* 5: 433–437.

Elliott, D. C. 2007. Historical developments in hydroprocessing bio-oils. *Energy & Fuels* 21: 1792–1815.

Fu, P., W. Yi, Z. Li, et al. 2014. Investigation on hydrogen production by catalytic steam reforming of maize stalk fast pyrolysis bio-oil. *International Journal of Hydrogen Energy* 39: 13962–13971.

Garcia-Perez, M., A. Chaala, H. Pakdel, D. Kretschmer, and C. Roy. 2007a. Vacuum pyrolysis of softwood and hardwood biomass. *Journal of Analytical and Applied Pyrolysis* 78: 104–116.

Garcia-Perez, M., T. T. Adams, J. W. Goodrum, D. P. Geller, and K. C. Das. 2007b. Production and fuel properties of pine chip bio-oil/biodiesel blends. *Energy & Fuels* 21: 2363–2372.

Goncalves, V. O. O., C. Ciotonea, S. Arrii-Clacens, et al. 2017a. Effect of the support on the hydrodeoxygenation of *m*-cresol over molybdenum oxide based catalysts. *Applied Catalysis B: Environmental* 214: 57–66.

Goncalves, V. O. O., P. M. de Souza, V. T. da Silva, F. B. Noronha, and F. Richard. 2017b. Kinetics of the hydrodeoxygenation of cresol isomers over Ni_2P/SiO_2: Proposals of nature of deoxygenation active sites based on an experimental study. *Applied Catalysis B: Environmental* 205: 357–367.

Goulas, K. A., A. V. Mironenko, G. R. Jenness, T. Mazal, and D. G. Vlachos. 2019. Fundamentals of C–O bond activation on metal oxide catalysts. *Nature Catalysis* 2: 269–276.

Gu, G. H., C. A. Mullen, A. A. Boateng, and D. G. Vlachos. 2016. Mechanism of dehydration of phenols on noble metals via first-principles microkinetic modeling. *ACS Catalysis* 6: 3047–3055.

Guvenatam, B., O. Kursun, E. H. J. Heeres, E. A. Pidko, and E. J. M. Hensen. 2014. Hydrodeoxygenation of mono- and dimeric lignin model compounds on noble metal catalysts. *Catalysis Today* 233: 83–91.

He, J., C. Zhao, and J. A. Lercher. 2012. Ni-catalyzed cleavage of aryl ethers in the aqueous phase. *Journal of the American Chemical Society* 134: 20768–20775.

Hicks, J. C. 2011. Advances in C–O bond transformations in lignin-derived compounds for biofuels production. *Journal of Physical Chemistry Letters* 2: 2280–2287.

Hong, Y., H. Zhang, J. Sun, et al. 2014. Synergistic catalysis between Pd and Fe in gas phase hydrodeoxygenation of *m*-Cresol. *ACS Catalysis* 4: 3335–3345.

Hong, Y., A. Hensley, J.-S. McEwen, and Y. Wang. 2016. Perspective on catalytic hydrodeoxygenation of biomass pyrolysis oils: Essential roles of Fe-based catalysts. *Catalysis Letters* 146: 1621–1633.

Hong, Y., S. Zhang, F. F. Tao, and Y. Wang. 2017. Stabilization of iron-based catalysts against oxidation: An *in situ* ambient-pressure X-ray photoelectron spectroscopy (AP-XPS) study. *ACS Catalysis* 7: 3639–3643.

Huang, Y. B., M. Y. Chen, L. Yan, Q. X. Guo, and Y. Fu. 2014. Nickel-tungsten carbide catalysts for the production of 2, 5-dimethylfuran from biomass-derived molecules. *ChemSusChem* 7: 1068–1072.

Huang, Y.-B., L. Yan, M.-Y. Chen, Q.-X. Guo, and Y. Fu. 2015. Selective hydrogenolysis of phenols and phenyl ethers to arenes through direct C–O cleavage over ruthenium–tungsten bifunctional catalysts. *Green Chemistry* 17: 3010–3017.

Huber, G. W., J. N. Chheda, C. J. Barrett, and J. A. Dumesic. 2005. Production of liquid alkanes by aqueous-phase processing of biomass-derived carbohydrates. *Science* 308: 1446–1450.

Huber, G. W., R. D. Cortright, and J. A. Dumesic. 2004. Renewable alkanes by aqueous-phase reforming of biomass-derived oxygenates. *Angewandte Chemie International Edition* 43: 1549–1551.

Ibarra, A., E. Rodriguez, U. Sedran, J. M. Arandes, and J. Bilbao. 2016. Synergy in the cracking of a blend of bio-oil and vacuum gasoil under fluid catalytic cracking conditions. *Industrial & Engineering Chemistry Research* 55: 1872–1880.

Imam, T., and S. Capareda. 2012. Characterization of bio-oil, syn-gas and bio-char from switchgrass pyrolysis at various temperatures. *Journal of Analytical and Applied Pyrolysis* 93: 170–177.

Jae, J., G. A. Tompsett, Y.-C. Lin, et al. 2010. Depolymerization of lignocellulosic biomass to fuel precursors: Maximizing carbon efficiency by combining hydrolysis with pyrolysis. *Energy & Environmental Science* 3: 358–365.

Jensen, C. U., J. Hoffmann, and L. A. Rosendahl. 2016. Co-processing potential of HTL bio-crude at petroleum refineries. Part 2: A parametric hydrotreating study. *Fuel* 165: 536–543.

Kepp, K. P. 2016. A quantitative scale of oxophilicity and thiophilicity. *Inorganic Chemistry* 55: 9461–9470.

Li, G., N. Li, J. Yang, et al. 2014. Synthesis of renewable diesel range alkanes by hydrodeoxygenation of furans over Ni/Hβ under mild conditions. *Green Chemistry* 16: 594–599.

Li, J., J. L. Liu, H. Y. Liu, et al. 2017. Selective hydrodeoxygenation of 5-hydroxymethylfurfural to 2, 5-dimethylfuran over heterogeneous iron catalysts. *ChemSusChem* 10: 1436–1447.

Liang, G., A. Wang, X. Zhao, N. Lei, and T. Zhang. 2016. Selective aldol condensation of biomass-derived levulinic acid and furfural in aqueous-phase over MgO and ZnO. *Green Chemistry* 18:3430–3438.

Liu, G., A. W. Robertson, M. M.-J. Li, et al. 2017. MoS_2 monolayer catalyst doped with isolated Co atoms for the hydrodeoxygenation reaction. *Nature Chemistry* 9:810–816.

Ly, H. V., E. Galiwango, S.-S. Kim, et al. 2017. Hydrodeoxygenation of 2-furyl methyl ketone as a model compound of algal *Saccharina Japonica* bio-oil using iron phosphide catalyst. *Chemical Engineering Journal* 317: 302–308.

Mortensen, P. M., J.-D. Grunwaldt, P. A. Jensen, and A. D. Jensen. 2013. Screening of catalysts for hydrodeoxygenation of phenol as a model compound for bio-oil. *ACS Catalysis* 3: 1774–1785.

Nakagawa, Y., S. Liu, M. Tamura, and K. Tomishige. 2015. Catalytic total hydrodeoxygenation of biomass-derived polyfunctionalized substrates to alkanes. *ChemSusChem* 8: 1114–1132.

Nie, L. and D. E. Resasco. 2014. Kinetics and mechanism of m-cresol hydrodeoxygenation on a Pt/SiO_2 catalyst. *Journal of Catalysis* 317: 22–29.

Nie, L., P. M. de Souza, F. B. Noronha, et al. 2014. Selective conversion of *m*-cresol to toluene over bimetallic Ni–Fe catalysts. *Journal of Molecular Catalysis A: Chemical* 388-389: 47–55.

Ochoa, A., B. Aramburu, B. Valle, et al. 2017. Role of oxygenates and effect of operating conditions in the deactivation of a Ni supported catalyst during the steam reforming of bio-oil. *Green Chemistry* 19: 4315–4333.

Patel, M., and A. Kumar. 2016. Production of renewable diesel through the hydroprocessing of lignocellulosic biomass-derived bio-oil: A review. *Renewable and Sustainable Energy Reviews* 58: 1293–1307.

Prasomsri, T., M. Shetty, K. Murugappan, and Y. Roman-Leshkov. 2014. Insights into the catalytic activity and surface modification of MoO_3 during the hydrodeoxygenation of lignin-derived model compounds into aromatic hydrocarbons under low hydrogen pressures. *Energy & Environmental Science* 7: 2660–2669.

Remon, J., P. Arcelus-Arrillaga, J. Arauzo, L. Garcia, and M. Millan-Agorio. 2017. Pyrolysis bio-oil upgrading to renewable liquid fuels by catalytic hydrocracking: effect of operating conditions on the process. In *Mediterranean Green Buildings & Renewable Energy*, ed. A. Sayigh, 491–500. Cham: Springer.

Resasco, D. E. 2011. What should we demand from the catalysts responsible for upgrading biomass pyrolysis oil? *Journal of Physical Chemistry Letters* 2: 2294–2295.

Resasco, D. E., and S. P. Crossley. 2015. Implementation of concepts derived from model compound studies in the separation and conversion of bio-oil to fuel. *Catalysis Today* 257: 185–199.

Saha, B., C. M. Bohn, and M. M. Abu-Omar. 2014. Zinc-assisted hydrodeoxygenation of biomass-derived 5-hydroxymethylfurfural to 2, 5-dimethylfuran. *ChemSusChem* 7: 3095–3101.

Shao, Y., Q. Xia, X. Liu, G. Lu, and Y. Wang. 2015. $Pd/Nb_2O_5/SiO_2$ catalyst for the direct hydrodeoxygenation of biomass-related compounds to liquid alkanes under mild conditions. *ChemSusChem* 8: 1761–1767.

Shao, Y., Q. Xia, L. Dong, et al. 2017. Selective production of arenes via direct lignin upgrading over a niobium-based catalyst. *Nature Communications* 8: 16104.

Shi, D., L. Arroyo-Ramirez, and J. M. Vohs. 2016. The use of bimetallics to control the selectivity for the upgrading of lignin-derived oxygenates: Reaction of anisole on Pt and PtZn catalysts. *Journal of Catalysis* 340: 219–226.

Singh, S. K. 2018. Heterogeneous bimetallic catalysts for upgrading biomass-derived furans. *Asian Journal of Organic Chemistry* 7: 1901–1923.

Smirnov, A. A., Zh Geng, S. A. Khromova, et al. 2017. Nickel molybdenum carbides: Synthesis, characterization, and catalytic activity in hydrodeoxygenation of anisole and ethyl caprate. *Journal of Catalysis* 354: 61–77.

Sun, J., A. M. Karim, H. Zhang, et al. 2013. Carbon-supported bimetallic Pd-Fe catalysts for vapor-phase hydrodeoxygenation of guaiacol. *Journal of Catalysis* 306: 47–57.

Sun, J., K. Zhu, F. Gao, et al. 2011. Direct conversion of bio-ethanol to isobutene on nanosized $Zn_xZr_yO_z$ mixed oxides with balanced acid-base sites. *Journal of the American Chemical Society* 133: 11096–11099.

Su-Ping, Z. 2003. Study of hydrodeoxygenation of bio-oil from the fast pyrolysis of biomass. *Energy Sources* 25: 57–65.

Tan, Q., G. Wang, A. Long, et al. 2017. Mechanistic analysis of the role of metal oxophilicity in the hydrodeoxygenation of anisole. *Journal of Catalysis* 347: 102–115.

Vispute, T. P. and G. W. Huber. 2009. Production of hydrogen, alkanes and polyols by aqueous phase processing of wood-derived pyrolysis oils. *Green Chemistry* 11: 1433–1445.

Vispute, T. P., H. Zhang, A. Sanna, R. Xiao, and G. W. Huber. 2010. Renewable chemical commodity feedstocks from integrated catalytic processing of pyrolysis oils. *Science* 330: 1222–1227.

Wang, H., J. Male, and Y. Wang. 2013. Recent advances in hydrotreating of pyrolysis bio-oil and its oxygen-containing model compounds. *ACS Catalysis* 3: 1047–1070.

Wang, S., Q. Cai, J. Chen, et al. 2014. Green aromatic hydrocarbon production from cocracking of a bio-oil model compound mixture and ethanol over Ga_2O_3/HZSM-5. *Industrial & Engineering Chemistry Research* 53: 13935–13944.

Wang, C., A. V. Mironenko, A. Raizada, et al. 2018. Mechanistic study of the direct hydrodeoxygenation of *m*-cresol over WO_x-decorated Pt/C Catalysts. *ACS Catalysis* 8: 7749–7759.

Wei, Z., J. Sun, Y. Li, A. K. Datye, and Y. Wang. 2012. Bimetallic catalysts for hydrogen generation. *Chemical Society Reviews* 41: 7994–8008.

Whiffen, V. M. L., and K. J. Smith. 2010. Hydrodeoxygenation of 4-methylphenol over unsupported MoP, MoS_2, and MoO_x catalysts. *Energy & Fuels* 24: 4728–4737.

Xiong, K., W. S. Lee, A. Bhan, and J. G. Chen. 2014. Molybdenum carbide as a highly selective deoxygenation catalyst for converting furfural to 2-methylfuran. *ChemSusChem* 7: 2146–2149.

Yang, J., S. Li, L. Zhang, et al. 2017a. Hydrodeoxygenation of furans over Pd-FeO_x/SiO_2 catalyst under atmospheric pressure. *Applied Catalysis B: Environmental* 201: 266–277.

Yang, Y., C. Ochoa-Hernandez, V. A. de la Pena O'Shea, et al. 2014. Effect of metal–support interaction on the selective hydrodeoxygenation of anisole to aromatics over Ni-based catalysts. *Applied Catalysis B: Environmental* 145: 91–100.

Yang, Z., A. Kumar, and R. L. Huhnke. 2015. Review of recent developments to improve storage and transportation stability of bio-oil. *Renewable and Sustainable Energy Reviews* 50: 859–870.

Yang, Y., G. Lv, L. Deng, et al. 2017b. Renewable aromatic production through hydrodeoxygenation of model bio-oil over mesoporous Ni/SBA-15 and Co/SBA-15. *Microporous and Mesoporous Materials* 250: 47–54.

Zhang, J., J. Sun, B. Sudduth, X. P. Hernandez, and Y. Wang. 2020a. Liquid-phase hydrodeoxygenation of lignin-derived phenolics on Pd/Fe: A mechanistic study. *Catalysis Today* 339: 305–311.

Zhang, J., J. Sun, and Y. Wang. 2020b. Recent advances in selectively catalytic hydrodeoxygenation of lignin-derived oxygenates to arenes. *Green Chemistry* 22: 1072–1098.

Zhang, J., J. Sun, and Y. Wang. 2021. An overview of catalytic bio-oil upgrading. Part I: Processing aqueous-phase compounds. In *Handbook of Biodiesel and Petrodiesel Fuels: Science, Technology, Health, and Environment. Volume 1. Biodiesel Fuels: Science, Technology, Health, and Environment*, ed. O. Konur. Boca Raton, FL: CRC Press.

Zhang, L., Z. Yu, J. Li, et al. 2020c. Steam reforming of typical small organics derived from bio-oil: Correlation of their reaction behaviors with molecular structures. *Fuel* 259: 116214.

Zhao, C., Y. Kou, A. A. Lemonidou, X. Li, and J. A. Lercher. 2010. Hydrodeoxygenation of bio-derived phenols to hydrocarbons using RANEY[R] Ni and Nafion/SiO_2 catalysts. *Chemical Communications* 46: 412–414.

10 Bio-oil Production through Hydrothermal Liquefaction (HTL) of Biomass

Recent Developments and Future Prospects

Leichang Cao
Daniel C. W. Tsang
Shicheng Zhang

CONTENTS

10.1 INTRODUCTION

Since the Industrial Revolution of the 18th century, fossil fuels have become the means of accessing energy for human society, which currently account for nearly 85% of the total energy consumed worldwide (Jahangiri et al., 2018). The high level of development and utilization of fossil fuels has promoted technological progress and social development. However, traditional fossil fuels damage the environment and cause a large amount of greenhouse gas emissions in the process of their production and use.

In addition, the reserves of these fuels, such as coal, oil, and natural gas, are very limited and cannot be used sustainably by everyone (Posmanik et al., 2018). Faced with the increasing exhaustion of fossil energy and the deterioration of the ecological environment caused by its combustion, the development of renewable and sustainable new energy sources has become an urgent need for human survival and development.

New energy sources mainly include wind energy, water energy, nuclear energy, solar energy, and biomass energy (Duongbia et al., 2019; Si et al., 2019). Amongst these, biomass has advantages, such as the diversity of its raw material, its high energy reserves, and its lack of additional carbon emissions. Biomass is considered as the new energy with the highest potential to replace traditional fossil fuels. In addition, it is the only resource of renewable energy that can be converted into liquid fuel (Bi et al., 2017; Yang et al., 2018). Compared with fossil energy, biomass energy has a wide range of sources, and is clean, storable, and can realize the carbon cycle, so it has received widespread attention (Han et al., 2020; Sunphorka et al., 2015).

Biomass mainly includes agricultural and forestry waste, energy crops, algae, and organic waste. According to statistics, the total biomass produced by green plants worldwide through photosynthesis is as much as 220 billion tons, but less than 1% of it is used as energy (Li et al., 2019). Generally, biomass energy conversion and utilization technologies can be roughly divided into three categories (Figure 10.1) (Chang et al., 2015; Mishra and Mohanty, 2020): first, biomass is directly combusted to obtain heat or is used for power generation; second, biomass is subjected to HTL, direct pyrolysis, gasification, and transesterification (from extracted oil) to obtain bio-based liquid fuels, chemicals, and combustible gases; third, biomass is subjected to anaerobic fermentation to obtain biogas, or by combining enzymolysis and fermentation to obtain bioethanol.

HTL has broad research prospects in the production of bio-oil from biomass. It is favored by scientific researchers and has become one of the hot topics of research (Lin et al., 2017; Mishra and Mohanty, 2020). The HTL of biomass uses water or green organic solvent to liquefy biomass to produce a value-added product under a certain temperature (200–400°C), pressure (5–25 MPa), and catalyst. The products are mainly bio-oils, with co-products such as coke, water-soluble substances, and gases (Wang et al., 2018). HTL has advantages unmatched by other biomass treatment technologies, but there are still many problems that need to be solved urgently. At present, research on the preparation of bio-oil from the HTL of biomass is mainly at the laboratory stage (Hayward et al., 2015; Wang et al., 2019b).

The main problems are: (1) the temperature and pressure are high during the reaction and the choice of parameters basically depends on experience with a certain blindness (Sun et al., 2020); (2) the composition of bio-oil is complex and the quality is low as a drop-in liquid fuel, the N content of some bio-oils (e.g. from algae) is high, and it is difficult to upgrade and refine bio-oils (Prajitno et al., 2018); (3) the solubility of coke, tar, and solid residues formed in the reaction process in water is very small, which can easily cause deposition and block the equipment (Hirano et al., 2020).

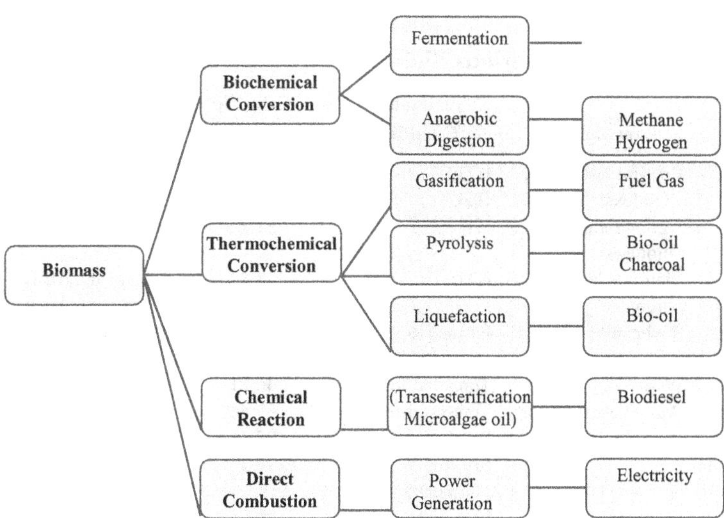

FIGURE 10.1 Conversion routes for biomass and its main products.

In response to the above issues, this chapter first briefly introduces the hydrothermal degradation pathway of different components of various biomasses and the main factors influencing bio-oil yield from the HTL of biomass. The catalytic upgrading of bio-oil is comprehensively reviewed and analyzed in a following section.

10.2 COMPARISON OF HTL AND OTHER TECHNOLOGIES FOR BIOFUEL PRODUCTION

10.2.1 Biofuels from Various Biomasses

Among the renewable energy sources, biofuels derived from biomass are the most favorable ones for sustainable development. The reasons are: (1) biomass has the advantages of huge reserves and extensive distribution which ensure a supply for biofuel production (Bi et al., 2018); (2) biofuel has a high energy density and is comparable to petroleum (Alma et al., 2016; Bunt et al., 2018) (Table 10.1); (3) biofuel can realize zero carbon emissions since it is derived from biomass (Eboibi et al., 2015); (4) some high value-added chemicals can be generated from biofuels as an alternative for petroleum biorefinery (Wang et al., 2019a). In the past decade, biofuels such as ethanol and biodiesel have been successfully developed and used. As an alternative fuel for fossil fuels, it can be blended with gasoline or diesel for engines. Because of the great advantages of biofuels, they have been

TABLE 10.1
The Calorific Values of Various Biofuels

Biofuel	Raw Material	Technology Condition	Calorific Value (MJ/kg)	References
Bio-oil	Low-fat microalgae	HTL	37.4	Li et al. (2014b)
Bio-oil	Cellulosic biomass	HTL	25.3	Patil et al. (2014))
Bio-oil	Agricultural waste biomass	HTL	21.1	Huang et al. (2013)
Bio-oil	Marginal land biomass	HTL	28.3	Feng et al. (2018)
Biodiesel	Soybean	Extractive transesterification	39.77	Fassinou et al. (2012)
Biodiesel	Waste cooking oil	Transesterification	40.11	Fassinou et al. (2012)
Biodiesel	Waste vegetable oil	Transesterification	39.44	Fassinou et al. (2012)
Bio-oil	Seaweed	Pyrolysis	25.4	Yanik et al. (2013)
Bioethanol	Corn	Enzymolysis fermentation	29.7	Lopez-Gonzalez et al. (2015)
Bioethanol	Sugarcane	Enzymolysis fermentation	30	Feng et al. (2018)
Petroleum crude oil	—	Exploitation	42.9	Caspeta et al. (2014)

TABLE 10.2
Four Typical Generations of Biofuels

Biofuel	Fuel Representation	Technique	Merits	Disadvantages
1st generation	Corn ethanol	Fermentation	Concentrated feedstock, industrially feasible	Competing food with human beings
	Sugarcane ethanol	Fermentation	Concentrated feedstock, industrially feasible	Competing food with human beings
	Soybean biodiesel	Transesterification	Concentrated feedstock, industrially feasible	Competing food with human beings
2nd generation	Cellulose ethanol	Enzymatic fermentation	Concentrated feedstock, industrially feasible	Competing food with human beings
	Algae ethanol	Enzymatic fermentation	No occupation of cultivated land	Low production efficiency
3rd generation	Microalgae biodiesel	Transesterification	No occupation of cultivated land	Serious energy waste
	Microalgae bio-oil	Thermochemistry	No occupation of cultivated land	Oil needs to be upgraded
4th generation	Microalgae bio-oil	Thermochemistry	Short feedstock cultivation cycle	Oil needs to be upgraded

rapidly developed and utilized in recent years, and their development has gone through three or four generations (Gu et al., 2020; Kumar and Pant, 2015; Subagyono et al., 2015) (Table 10.2).

10.2.2 Merits of HTL for Biomass Conversion

There are various methods of biomass energy utilization: direct combustion, biochemical conversion, and thermochemical conversion (Gu et al., 2020; Remon et al., 2019). The technologies have different features (Table 10.3). At present, the main methods are biological and thermochemical conversion. The biological method is to use the activity of certain specific enzymes or bacteria to destroy polymers in biomasses and transform cellulose, hemicellulose, lignin, sugars, lipids, and proteins into bioethanol or biogas (Tsegaye et al., 2020). Thermochemical methods mainly include direct pyrolysis and HTL to produce bio-oil (Kumar and Pant, 2015; Remon et al., 2019).

Compared with other biomass conversion technologies, HTL has the following obvious advantages. First, it does not require drying the raw materials in advance, which significantly reduces operating costs for the waste with high moisture content. Second, the reaction conditions are relatively mild. A pressurized hot solvent is used

TABLE 10.3
Features of Different Biomass Utilization Methods

Method	Principle	Product	Features
Direct combustion	Stove combustion	Thermal energy	Low thermal efficiency and gradually eliminated
	Waste incineration	Electrical energy/ thermal energy	Broad prospects and environmentally friendly
	Compression molding fuel combustion	Electrical energy/ thermal energy	High thermal efficiency and simple process
	Combined combustion	Electrical energy/ thermal energy	Widely used, energy saving, and emission reduction
Biochemical transformation	Anaerobic digestion	Biogas	Widely used and slow conversion rate
	Alcohol fermentation	Fuel ethanol	Slow conversion rate
Thermochemical conversion	Dry distillation	Combustible gas	Isolate air and supply collective gas
	Gasification	Combustible gas	High energy consumption and difficult product storage
	Pyrolysis	Gas/oil/charcoal	Requires dry materials and high energy consumption
	HTL	Bio-oil	Fit for wet biomass but bio-oil needs upgrading

as the reaction medium and reaction reagent, and bio-oil can even be produced without adding other chemicals. The whole process is environmentally friendly. Third, the HTL process is less corrosive to equipment.

10.3 HTL OF BIOMASS

10.3.1 Major Components of Biomass and Their Decomposition Routes in the HTL Process

10.3.1.1 Cellulose

The temperature ranges of the main components in lignocellulose are: hemicellulose (0'–200°C), cellulose (200'–300°C), and lignin (250'–340 °C) (Cao et al., 2016). Cellulose is made up of glucose through β-1,4 glycosides. When the temperature rises, β-1,4 glycoside bonds will gradually break and glucose monomers will be formed. Glucose monomers will further transform some oligomers under hydrothermal conditions (Li et al., 2015). The main components of the degradation products are furfural, ketones, and phenol compounds (Cao et al., 2017; Usman et al., 2019).

10.3.1.2 Hemicellulose

Hemicellulose is the second most abundant polysaccharide in nature after cellulose, accounting for 14–50% of the dry weight of lignocellulosic biomass (Baloch et al., 2018). Hemicellulose includes a variety of structural units, including five carbons (xylose and arabinose) and six monosaccharides (mannose, galactose, and glucose).

The most abundant hemicellulose in hardwood and many agricultural residues is 'xylan' (mainly consisting of xylose) (Cao et al., 2018b). HTL is one of the ideal methods for treating hemicellulose. Under appropriate conditions, it can produce soluble products such as monosaccharides and degradation products of monosaccharides, such as furfural or hydroxymethylfurfural, and carboxylic acid.

10.3.1.3 Lignin

Lignin is the most abundant aromatic hydrocarbon polymer, which is polymerized from phenolic compounds (Cao et al., 2018a–b; Younas et al., 2017). The precursors of these phenolic compounds are three aromatic alcohols, namely coumarol, coniferyl alcohol, and glucosinolate. More than two-thirds of these monomers are connected through ether bonds (C–O–C), and the rest are connected through C–C bonds. The three aromatic components in the polymer are called *p*-hydroxyphenyl (H), guaiacyl (G), and syringal (S). The structure of lignin indicates that it can be used as a source of valuable chemicals, especially phenolic compounds (Cao et al., 2018c). Lignin is usually combined with hemicellulose, which is known as the skeleton structure of plant cell walls. It is not only physically bonded, but also covalently bonded. In hydrothermal treatment, the cross-linked structures of lignin and lignin-hemicellulose can be degraded, partially depolymerized, and severely degraded so that their structures are rearranged (Cao et al., 2017). The hydrothermal products are monocyclic phenolic compounds, including phenol, 2-methoxy-phenol, 4-methyl-1,2-benzenediol, 1,2-benzenediol, 3-methyl-1, and 2-benzenediol (Lyckeskog et al., 2017).

10.3.1.4 Protein

Protein is the main source of nitrogen heterocyclic compounds in bio-oils (Gu et al., 2020; Teri et al., 2014). At temperatures of 0–100°C, proteins undergo hydrolysis reactions to form various amino acids. As the temperature rises, various reactions occur in the generated amino acids. The decarboxylation and the deamination are the two main reactions and, between 100 and 200°C, carboxyl functional groups in some amino acids are decarboxylated to form amine compounds, in which carboxyl groups are released as gas-phase products, such as CO_2 (Teri et al., 2014). Another part of the amino acid is formed into organic acid through a deamination reaction, in which the amino group is released in the form of NH_3. In addition, the amino acid may undergo a Maillard reaction' with a carbohydrate hydrolysate such as reducing sugars to form a variety of heterocyclic nitrogen oxides, including pyrrole, pyrrolidone, pyridine, and imidazole, and finally to form a nitrogen-rich polymerized substance which is called melanin (Feng et al., 2018).

10.3.1.5 Lipids

A lipid is the component closest to bio-oil in biomass, and the conversion efficiency of it into bio-oil is also the highest (Chen et al., 2015; Gu et al., 2020). These make it an ideal component for preparing bio-oil. The HTL reaction process of lipids has been deduced and lipids have been hydrolyzed to produce glycerol and long-chain fatty acids in the temperature range 0–100°C (Qiu et al., 2019). As the temperature rises, some fatty acids are converted into long-chain hydrocarbons. In addition, the hydrolysis intermediates of fatty acids also undergo decarboxylation reactions and polymerization reactions to produce alkane and olefin hydrocarbons.

10.3.1.6 Carbohydrates

Although a carbohydrate is ideal feedstock for fine chemical production through acidic hydrolysis with various liquid or solid catalysts, it is one of the least easily ones to be converted into bio-oil among the major components of biomass (Cao et al., 2019). Within the temperatures 0–100°C, carbohydrates are hydrolyzed to produce reducing and non-reducing sugars. As the temperature rises, the reducing sugar and non-reducing sugar molecular bonds are broken and repolymerized to form epoxy compounds. During the hydrothermal process, some sugars are converted into the oil phase, water phase, and gas phase, but most of the sugar compounds are converted into the solid phase. Therefore, it is wise to use the two-step liquefaction method that first extracts the sugar compounds in the biomass (mainly algae) and then prepares the bio-oil without degrading its quality.

10.3.2 Effect of HTL Parameters

10.3.2.1 Biomass Feedstock Composition

The specific compositions of different biomasses are quite different, and even under the same reaction conditions, the yield and composition of the obtained liquid products will vary greatly. It is found that biomass with a higher cellulose or carbohydrate content is easier to liquefy, and lignin has the greatest effect on the composition and quality of liquefied bio-oils.

Huang et al. (2013) studied the liquefaction reaction process of three different biomasses (straw, microalgae, and sludge), and found that sludge liquefaction had the highest bio-oil yield, reaching 39.5%, which was higher than those of straw (21.1%) and microalgae (34.5%). In addition, the sludge bio-oil had the highest calorific value (36.14 MJ/kg). The conversion rate of microalgae was the highest and reached 79.7%. Feng et al. (2014) liquefied the bark of white pine, spruce, and birch at 300°C and an initial N_2 pressure of 2.0 MPa for 15 mins using ethanol and water as the reaction solvent. They investigated the effect of ash content in biomass on bio-oil yield and composition. The results showed that when the three kinds of biomass bark (white pine, spruce, white birch) were pre-deashed, their conversion rate and bio-oil yield decreased, while when the basic catalysts of K_2CO_3 and $Ca(OH)_2$ were added, the yield of the liquefied bio-oils increased and the composition changed, indicating that the ash substances (compounds containing K and Ca) play a catalytic role in the liquefaction process (Feng et al., 2014).

10.3.2.2 Reaction Temperature

The reaction temperature is the most important factor in biomass HTL, which affects its conversion, the yield of liquefied products, the distribution of products, and the calorific value of products (Lin et al., 2017; Wang et al., 2019b). As the reaction temperature rises, the chemical bonds in the macromolecular compounds of the biomass will be broken, and the depolymerization reaction will occur with the free radical concentration increase. The probability of repolymerization of the decomposed small fragments will also increase, which will be beneficial to the production of bio-oil.

When the temperature approaches or exceeds the critical point of water, the higher the temperature, the more severe the secondary reaction will be, resulting in the decomposition reaction of the intermediate products to produce gas, and the polycondensation reaction to produce solid residues, leading to a decrease in the yield of bio-oil. Zhu et al. (2015) studied the HTL process of barley straw in the temperature range 280–400°C. The results showed that low temperature was beneficial to the production of bio-oil and that its yield was the highest (34.9%) at 300°C, though when the reaction temperature further increased to 400°C, the yield was reduced to 19.9%. This decrease in bio-oil yield was due to the polycondensation reaction of intermediates at high temperature to produce solid residues (Zhu et al., 2015).

10.3.2.3 Reaction Time

The reaction time refers to a period after the liquefaction reaction rises to a set reaction temperature and is maintained at that temperature, and which does not include heating and cooling time. Many researchers have studied the effect of the reaction time on the HTL of biomass and found that the yield of bio-oil is closely related to the length of the reaction time (Pourkarimi et al., 2019; Xu and Savage, 2015). In the process of biomass HTL, usually a short reaction time is conducive to the production of a large amount of bio-oil, but a too short period of time will make the reaction incomplete (Chan et al., 2015). In addition, if the reaction time is too long, it will cause the polymerization of the intermediate products and generate solid residues, which will reduce the yield of bio-oil. Therefore, it is very important to choose a suitable reaction time. Generally, the favorable reaction time ranges from 30 to 120 min for bio-oil production.

10.3.2.4 Reaction Solvents

The reaction solvent is an important factor that affects the liquefaction reaction process. It dissolves the biomass raw materials and stabilizes the free radicals generated by the reaction. It can inhibit the polymerization of intermediate products and provide active hydrogen. During the HTL of biomass, the solvents used in biomass liquefaction are mainly methanol (Han et al., 2020), ethanol (Jogi et al., 2019), water (Posmanik et al., 2017), isopropanol (Wang et al., 2019a), glycerol (Kosmela et al., 2017), ethyl acetate (Tsubaki et al., 2019), oxygenates (Pedersen and Rosendahl, 2015), hydrocarbons (Kaur et al., 2019), and other organic solvents (Wang et al., 2020). Among these solvents, water is the cheapest reaction solvent and is non-toxic and does not cause any pollution to the environment. But the use of water as a solvent often requires severer conditions, and compared to bio-oil produced in ethanol and other organic solvents the yield in water is low (Jogi et al., 2019).

Methanol is a toxic organic solvent that can be harmful to humans, though it is rarely used. As we all know, ethanol is the most commonly used organic solvent in the liquefaction of biomass. It is non-toxic and can be recycled. Compared with water, ethanol dissolves more easily high molecular weight cellulose, hemicellulose, and lignin derivatives due to its low dielectric constant. In recent years, many researchers have found that there is a synergy between ethanol and water in the HTL of biomass, which not only affects the biomass conversion rate, the bio-oil yield, but also the composition of bio-oil (Han et al., 2020; Yang et al., 2018).

10.3.2.5 Reaction Pressure

Reaction pressure is another factor in the liquefaction of biomass. By keeping the reaction pressure above the critical pressure of the solvent medium, the rate of biomass hydrolysis and dissolution can be controlled. At the same time, high pressure will increase the density of the solvent, and high-density media can effectively penetrate the molecules of biomass components, thereby strengthening the decomposition of biomass molecules (Gu et al., 2020; Subagyono et al., 2015). However, once supercritical liquefaction conditions are reached, the effect of pressure on liquid oil yield and gas yield is negligible, because in the supercritical region, pressure has very little effect on the performance of the solvent medium.

10.4 CATALYTIC UPGRADING OF BIOCRUDE OIL

The physical and chemical properties of bio-oil determine that it cannot be used directly as a fuel (Prajitno et al., 2018). It must be modified and refined before it can be burned in the existing internal combustion engine, individually or mixed with other commercial fuels. In recent years, researchers have extensively studied the refining of bio-oil by drawing on the upgrading and refining methods of petroleum oil, and developed modification technologies for different biomass liquefied oils. Based on the characteristics of liquefied bio-oil, the refining methods include in general catalytic hydrogenation', cracking, and esterification.

10.4.1 Catalytic Hydrogenation

The bio-oil hydrogenation process is generally carried out under conditions of high temperature (200–400°C) and high pressure (7–14 MPa) with a catalyst (or catalyst-free) to realize the hydrogenation and deoxygenation of biocrude oil (Hansen et al., 2020; Weber and Holladay, 2018). Hydrogenation treatment is performed under initially introduced H_2 at a certain pressure or in the presence of a hydrogen-supplying solvent. The carbon-oxygen bonds are broken and unsaturated bonds become saturated bonds. Oxygen in bio-oil can be removed by being transferred into reaction-generated H_2O and CO_2 (Hansen et al., 2020). After hydrogenation, the hydrogen content in the bio-oil is significantly increased, while the oxygen content is greatly reduced. The physical and chemical characteristics of the bio-oil are improved to realize its direct use as a drop-in fuel (or mixed with fossil fuels). Catalytic hydrogenation is by far the most mature and widely used technology in bio-oil upgrading and refining.

Hydrogen is indispensable to the catalytic hydrogenation process. The addition of hydrogen or a hydrogen donor helps reduce the oxygen content of bio-oil. Catalytic hydrogenation of bio-oil can be divided into direct hydrogen supply (H_2) and *in situ* hydrogen supply, according to the source of the hydrogen.

10.4.1.1 Direct Hydrogen Donor

Direct hydrogen supply means that a certain pressure of industrial hydrogen is introduced directly into the reactor. Catalytic hydrogenation not only can significantly reduce the oxygen content in bio-oil and increase its energy density, it can also remove hydrophilic functional groups in the oil and stabilize free radicals (Hansen et al., 2020). Li et al. (2014a) investigated bio-oil production through the HTL of *Nannochloropsis salina* and larvae-vermicompost under both non-hydrogenation and hydrogenation conditions under an H_2 atmosphere with Ni–Mo/Al_2O_3 as a catalyst.

This indicates that hydrogenation-promoted bio-oil yields and high heating values (HHVs) from 55.6% and 36.30 MJ/kg to 78.5% and 37.53 MJ/kg (*Nannochloropsis salina*), whereas for vermicompost the yields and HHVs increased from 33.2% and 32.89 MJ/kg to 43.5% and 34.24 MJ/kg. Compared with the non-hydrogenation HTL process, the contents of hydrocarbons increased along with the decrease of acids, amides, phenols, and alcohols. Molecular weights and polydispersity of the hydrogenated biocrudes also decreased. Results implied that hydrogenation was able to enhance the stabilization of radicals and the removal of hydrophilic functional groups, thereby inhibiting the mass loss toward liquid and gaseous products; the oil quality was upgraded (Li et al., 2014a).

Liu et al. (2015) used NiMoCe/Al_2O_3 as a catalyst for the hydrogenation of Jatropha oil. It was found that the yield of C15–C18 hydrocarbon was 80% and that the conversion rate reached 89%. The reaction temperature obviously affected the concentration of small molecules in bio-oils. The doping of Ce on the NiMo/Al_2O_3 catalyst increased the dispersion of NiMo, forming more active centers, and significantly increased the absorption of H_2 (Liu et al., 2015).

Murnieks et al. (2014) studied the effects of toluene and ethanol as solvents and Ni/SiO_2-Al_2O_3 as catalysts on the product distribution of hydrogenated (H_2 atmosphere) liquefaction of wheat straw. The results indicate that the yield and composition of bio-oil were both strongly impacted by hydrogenation. In addition ethanol was helpful in promoting the decomposition of the macromolecules into smaller molecules, such as 1,2-ethylene glycol, while toluene was more helpful in the formation of furans and cyclic substances (Murnieks et al., 2014).

10.4.1.2 Indirect Hydrogen Supply (*in situ* Hydrogen Supply)

'*In situ* hydrogen supply' means the use of inorganic hydrogen donors mainly composed of salts or organic hydrogen donors such as tetralin and small molecular liquid alcohols (Li et al., 2014a; Ma et al., 2017). It is believed that as long as one compound contains a mobile C-H bond, it can be used as a hydrogen supply solvent, such as tetralin, ethanol, or methanol, which diversifies the supply of hydrogen (Haghighat et al., 2019; Li et al., 2018). *In situ* hydrogenation during the preparation of bio-oil is an efficient internal hydrogenation method. Hydrogen is generated by the hydrolysis and thermochemical reaction of the hydrogen supply agent, which is directly hydrodeoxygenated in the liquefaction process. Compared with a direct hydrogen supply, *in situ* hydrogen supply does not require a large amount of energy, and can also avoid the issues of hydrogen storage, transportation, and safety, so it has received widespread attention (Cole et al., 2016; Feng et al., 2017). The hydrogen supply agent mainly includes the following four types.

1. *Inorganic hydrogen donors.* The hydrothermal conversion of some salts can produce hydrogen, such as the conversion of carbonate and formate or the hydrolysis reaction of formate. However, hydrogen production using these additives is inefficient and the separation and recovery of them are difficult. Coke and corrosion are easily generated in the reactor (Taghipour et al., 2019).
2. *Organic hydrogen donors.* Generally, tetralin and small molecule liquid alcohols are the frequently used organic hydrogen donors. Among them, tetralin is the most widely used one (Isa et al., 2018; Koriakin et al., 2017). Although the organic hydrogen supply agent has a better hydrogenation effect, it has a higher cost and is generally toxic. In particular, it easily pollutes aquatic organisms and the environment.
3. *Hydrogen production from metal hydrolysis.* Generating *in situ* hydrogen through hydrolysis reactions of metals is a promising way to store and transport renewable energy. Active metals that can react with water include Be, Al, Zn, Mg, Ca, Li, Na, and K (Cheng et al., 2018). Among them, aluminum is the most abundant metal on the earth, and an aluminum hydrolysis reaction can not only generate hydrogen, but also release a large amount of energy. At room temperature, aluminum can readily react with water and generate hydrogen. For example, aluminum-water and zinc-water reactions have been used for the hydrogenation of coal, asphalt, and model compounds under supercritical water (Yang et al., 2017; Yang et al., 2019). During the process, zinc and aluminum are oxidized and the hydrogen produced is used for the hydrogenation of organic matter (Yang et al., 2017). In addition, after the reaction the generated nanosized zinc oxide and alumina also possess catalytic activity as a candidate for other applications.
4. *Plastic waste.* As the main components of plastic, resin polymer is used in large amounts and is difficult to degrade. Researchers have found that bio-oil products can be obtained through liquefaction. The basic structure of C–H type resin polymer (e.g. polyethylene) is a hydrocarbon chain which will crack under high temperature (Raikova et al., 2019). During the cracking process, a suitable catalyst can break the C–H bonds and generate a large amount of H_2 and hydrogen ions. Therefore, plastic can be used as a stable source of hydrogen during liquefaction. Researchers have extensively studied the co-liquefaction reaction of plastic and biomass, and found that there is a synergistic effect in the co-liquefaction process. The addition of plastic has significantly improved the properties of bio-oil. In recent years, supercritical technology has become increasingly mature.

Scholars have focused on the co-liquefaction research of plastics and other substances (Dimitriadis and Bezergianni, 2017). At present, co-liquefaction research on plastics and biomass is mainly concentrated in supercritical water conditions. Raikova et al. (2019) studied the hydrothermal co-liquefaction experiments of biomass and plastic under subcritical and supercritical conditions. It was found that the HHV of co-liquefied oil was higher than that of biomass alone, which was mainly composed of saturated and unsaturated straight-chain hydrocarbons, and less oxygenated compounds were formed. During the co-liquefaction, the interaction between the biomass and the fragments generated by the thermal decomposition of plastic occurred, and the plastic and biomass co-liquefaction process had a significant synergistic effect (Raikova et al., 2019). Wang et al. (2014, 2015) studied the co-liquefaction of lignite, straw, and plastic in supercritical water and found that the three have synergistic effects. When the mixing ratio was 5:4:1, the oil yield and gas yield were the highest ones. Yang et al. (2016) found a synergistic effect during co-liquefaction experiments on wood chips and plastics. The oil production rate reached a maximum of 24% at 380°C.

10.4.1.3 Catalytic Cracking

The high-molecular substances in bio-oil can undergo a cracking reaction under the condition of high temperature and the presence of a catalyst. The process of generating small-molecule substances and removing the oxygen-containing groups therein is called catalytic cracking (Hirano et al., 2020). The catalytic cracking process mainly breaks the C–C bond and C–O bond in the bio-oil, and the oxygen atoms are removed in the form of H_2O, CO_2, and CO during the cracking process, thereby improving the physical and chemical properties of the bio-oil to achieve the applicable purpose. A selective molecular sieve catalyst such as HZSM-5 is generally used in the bio-oil cracking reaction (Kong et al., 2016). Catalytic cracking faces the issues of lower refined bio-oil yield, a high coking rate, and a short catalyst life. However, it does not require hydrogen during the reaction, and the reaction conditions are mild; thus with low requirements for reaction equipment, it has good industrial development potential.

Yuan et al. (2019) used an HZSM-5 catalyst to crack the macroalgal bio-oil, resulting in a yield of up to 46.75 wt%. The C and H content increased significantly, and the viscosity was also significantly reduced. The calorific value of the component aromatic compounds was as high as 44 MJ/kg, and heteroatoms such as O, S, and N were removed to varying degrees (Yuan et al., 2019). Alper et al. (2019) used zeolite to catalytically crack the liquefied woody biomass oil. After modification, a large number of long chain compounds were successfully cracked into alkanes and alkenes and their isomers. The oxygen content of the bio-oil was reduced, its viscosity was significantly reduced, and the calorific value was increased to 27.11 MJ/kg (Alper et al., 2019).

In addition, Yigezu and Muthukumar (2014) have successfully cracked vegetable oils to produce light oils such as gasoline and biodiesel. The use of HZSM-5 can significantly increase the content of phenols in bio-oils, and can reduce the generation of coke during bio-catalytic cracking. When HZSM-5 and zeolite are used as catalysts, the aromatic hydrocarbon content was significantly greater than the aliphatic hydrocarbon content, while the effect of the other three catalysts was the opposite. It was also found that HZSM-5, H-Y, and zeolite catalysts have stronger activity on the cracking of bio-oil, and acidic catalysts are favorable for its cracking reaction (Barreiro et al., 2016).

The catalytic cracking process can crack the macromolecules in bio-oil into small-molecule substances. It can also remove some oxygen, sulfur, and other heteroatoms in the bio-oil. The reaction does not require the use of hydrogen and thus reduces the operating cost. However, during the catalytic cracking process, the catalyst surface tends to be coked and deactivates the catalyst, and the reaction temperature is higher than that of catalytic hydrogenation (Santillan-Jimenez et al., 2019).

10.4.2 Catalytic Esterification

Catalytic esterification is a commonly used method for upgrading biocrude oils. Generally, short chain alcohols are used as additives to the oil, under the action of the catalyst, and the carboxyl groups in the oil react with the alcohol to convert into esters, thereby increasing the pH of the bio-oil, reducing its acidity and corrosivity, at the same time increasing its calorific value and stability (Yang et al., 2018). This method has gradually attracted attention due to its low energy consumption and mild reaction conditions. However, the addition of alcohol in the catalytic esterification process increases the cost. The high viscosity of the biocrude oil will also reduce the selectivity of the catalyst (e.g. molecular sieve), and other cross-reactions will occur. The moisture contained in the bio-oil will also inhibit the esterification reaction, and the problems of catalyst selection and recovery also affect the large-scale application of the method.

Solid acids and base catalysts have been used for catalytic esterification in biocrude oil upgrading with the development of green chemistry. Kassargy et al. (2016) studied the simultaneous transesterification and liquefaction of *Pistacia atlantica* using ethanol as an additive. It was found that the friction properties and other physical and chemical properties of the modified bio-oil were greatly improved. The H/C value increased significantly and the O/C value decreased significantly, so the calorific value was also greatly enhanced. In addition, the density, water content, and viscosity of bio-oil all decreased (Kassargy et al., 2016). Xu et al. (2014) used ethanol as an additive using solid acid catalysts to catalytically esterify bio-oils. The refined oils had reduced density, increased calorific value, increased pH, and improved stability (Xu et al., 2014). In recent years, ion exchange resins have received widespread attention as catalysts in bio-oil esterification. Common strong acidic cation-exchange resins, including D001, NKC-9, D061, and 732, have shown good catalytic performance (Oh et al., 2019).

10.5 CONCLUSIONS AND PROSPECTS

The efficient hydrothermal conversion of biomass to produce bio-oil is in line with the current concepts of environmental conservation and sustainable development. The cheap and ready availability of biomass demonstrates great development potential and market prospects. However, the research on catalytic hydrothermal conversion of biomass is still at the lab-scale or pilot stage, and future research should focus on (1) the development of highly active, hydrothermally stable, green, and easily recoverable new catalysts to realize the production of high-quality bio-oils through hydrothermal conversion; (2) the cost-effective separation of the main biomass as the route to achieve value-added products with high yield and selectivity; (3) green and environmentally friendly biphasic or multiphasic solvents should gain more emphasis in terms of both efficient conversion and product purification.

ACKNOWLEDGMENTS

This study was supported by the International Cooperation Project of Shanghai Municipal Science and Technology Commission (No. 18230710700), the National Key Research and Development Program of China (No. 2017YFC0212205), and the National Natural Science Foundation of China (No. 21876030).

REFERENCES

Alma, M. H., T. Salan, and A. Temiz. 2016. A novel approach for the liquefaction of wood powder: Usage of pyrolytic bio-oil as a reaction medium. *International Journal of Energy Research* 40: 1986–2001.

Alper, K., K. Tekin, and S. Karagoz. 2019. Hydrothermal and supercritical ethanol processing of woody biomass with a high-silica zeolite catalyst. *Biomass Conversion and Biorefinery* 9: 669–680.

Baloch, H. A., S. Nizamuddin, M. T. H. Siddiqui, et al. 2018. Sub-supercritical liquefaction of sugarcane bagasse for production of bio-oil and char: Effect of two solvents. *Journal of Environmental Chemical Engineering* 6: 6589–6601.

Barreiro, D. L., B. R. Gomez, F. Ronsse, et al. 2016. Heterogeneous catalytic upgrading of biocrude oil produced by hydrothermal liquefaction of microalgae: State of the art and own experiments. *Fuel Processing Technology* 148: 117–127.

Bi, Z. T., J. Zhang, E. Peterson, et al. 2017. Biocrude from pretreated sorghum bagasse through catalytic hydrothermal liquefaction. *Fuel* 188: 112–120.

Bi, Z. T., J. Zhang, Z. Y. Zhu, Y. N. Liang, and T. Wiltowski. 2018. Generating biocrude from partially defatted *Cryptococcus curvatus* yeast residues through catalytic hydrothermal liquefaction. *Applied Energy* 209: 435–444.

Bunt, J. R., S. Marx, F. B. Waanders, and N. T. Leokaoke. 2018. Green coal development for application in fixed-bed catalytic gasification. *Journal of the Southern African Institute of Mining and Metallurgy* 118: 419–429.

Cao, L. C., H. H. Chen, D. C. W. Tsang, et al. 2018b. Optimizing xylose production from pinewood sawdust through dilute-phosphoric-acid hydrolysis by response surface methodology. *Journal of Cleaner Production* 178: 572–579.

Cao, L. C., K. M. Iris, D.-W. Cho, et al. 2019. Microwave-assisted low-temperature hydrothermal treatment of red seaweed (*Gracilaria lemaneiformis*) for production of levulinic acid and algae hydrochar. *Bioresource Technology* 273: 251–258.

Cao, L. C., G. Luo, D. C. W. Tsang, et al. 2018a. A novel process for obtaining high quality cellulose acetate from green landscaping waste. *Journal of Cleaner Production* 176: 338–347.

Cao, L. C., I. K. M. Yu, Y. Y. Liu, et al. 2018c. Lignin valorization for the production of renewable chemicals: State-of-the-art review and future prospects. *Bioresource Technology* 269: 465–475.

Cao, L. C., C. Zhang, H. H. Chen, et al. 2017. Hydrothermal liquefaction of agricultural and forestry wastes: state-of-the-art review and future prospects. *Bioresource Technology* 245: 1184–1193.

Cao, L. C., C. Zhang, S. L. Hao, et al. 2016. Effect of glycerol as co-solvent on yields of bio-oil from rice straw through hydrothermal liquefaction. *Bioresource Technology* 220: 471–478.

Caspeta, L., Y. Chen, P. Ghiaci, et al. 2014. Altered sterol composition renders yeast thermotolerant. *Science* 346: 75–78.

Chan, Y. H., S. Yusup, A. T. Quitain, et al. 2015. Effect of process parameters on hydrothermal liquefaction of oil palm biomass for bio-oil production and its life cycle assessment. *Energy Conversion and Management* 104: 180–188.

Chang, Z. F., P. G. Duan, Y. P. Xu. 2015. Catalytic hydropyrolysis of microalgae: Influence of operating variables on the formation and composition of bio-oil. *Bioresource Technology* 184: 349–354.

Chen, H. H., D. Zhou, G. Luo, S. C. Zhang, and J. M. Chen. 2015. Macroalgae for biofuels production: Progress and perspectives. *Renewable and Sustainable Energy Reviews* 47: 427–437.

Cheng, S. Y., L. Wei, and M. Rabnawaz. 2018. Catalytic liquefaction of pine sawdust and in-situ hydrogenation of biocrude over bifunctional Co-Zn/HZSM-5 catalysts. *Fuel* 223: 252–260.

Cole, A., Y. Dinburg, B. S. Haynes, et al. 2016. From macroalgae to liquid fuel *via* waste-water remediation, hydrothermal upgrading, carbon dioxide hydrogenation and hydrotreating. *Energy & Environmental Science* 9: 1828–1840.

Dimitriadis, A. and S. Bezergianni. 2017. Hydrothermal liquefaction of various biomass and waste feedstocks for biocrude production: A state of the art review. *Renewable and Sustainable Energy Reviews* 68: 113–125.

Duongbia, N., S. Chaiwongsar, C. Chaichana, and S. Chaiklangmuang. 2019. Acidic hydrolysis performance and hydrolyzed lipid characterizations of wet *Spirulina platensis. Biomass Conversion and Biorefinery* 9: 305–319.

Eboibi, B. E., D. M. Lewis, P. J. Ashman, and S. Chinnasamy. 2015. Influence of process conditions on pretreatment of microalgae for protein extraction and production of biocrude during hydrothermal liquefaction of pretreated *Tetraselmis* sp. *RSC Advances* 5: 20193–20207.

Fassinou, W. F., A. Sako, A. Fofana, K. B. Koua, and S. Toure. 2012. Fatty acids composition as a means to estimate the high heating value (HHV) of vegetable oils and biodiesel fuels. *Energy* 35:4949–4954.

Feng, H., B. Zhang, Z. X. He, et al. 2018. Study on co-liquefaction of *Spirulina* and *Spartina alterniflora* in ethanol-water co-solvent for bio-oil. *Energy* 155: 1093–1101.

Feng, J. F., Z. Z. Yang, C. Y. Hse, et al. 2017. *In situ* catalytic hydrogenation of model compounds and biomass-derived phenolic compounds for bio-oil upgrading. *Renewable Energy* 105: 140–148.

Feng, S. H., Z. S. Yuan, M. Leitch, and C. C. Xu. 2014. Hydrothermal liquefaction of barks into bio-crude: Effects of species and ash content/composition. *Fuel* 116: 214–220.

Gu, X., J. S. Martinez-Fernandez, N. Pang, X. Fu, and S. Chen. 2020. Recent development of hydrothermal liquefaction for algal biorefinery. *Renewable and Sustainable Energy Reviews* 121: 109707.

Haghighat, P., A. Montanez, G. R. Aguilera, et al. 2019. Hydrotreating of Hydrofaction™ biocrude in the presence of presulfided commercial catalysts. *Sustainable Energy and Fuels* 3: 744–759.

Han, F. Y., M. Komiyama, Y. Uemura, and N. E. Rabat. 2020. Catalytic alcohothermal liquefaction of wet microalgae with supercritical methanol. *Journal of Supercritical Fluids* 157: 104704.

Hansen, S., A. Mirkouei, and L. A. Diaz. 2020. A comprehensive state-of-technology review for upgrading bio-oil to renewable or blended hydrocarbon fuels. *Renewable and Sustainable Energy Reviews* 118: 109548.

Hayward, J. A., D. A. O'Connell, R. J. Raison, et al. 2015. The economics of producing sustainable aviation fuel: A regional case study in Queensland, Australia. *Global Change Biology Bioenergy* 7: 497–511.

Hirano, Y., Y. Miyata, M. Taniguchi, et al. 2020. Fe-assisted hydrothermal liquefaction of cellulose: Effects of hydrogenation catalyst addition on properties of water-soluble fraction. *Journal of Analytical and Applied Pyrolysis* 145: 104719.

Huang, H. J., X. Z. Yuan, H. N. Zhu, et al. 2013. Comparative studies of thermochemical liquefaction characteristics of microalgae, lignocellulosic biomass and sewage sludge. *Energy* 56: 52–60.

Isa, K. M., T. A. T. Abdullah, and U. F. M. Ali. 2018. Hydrogen donor solvents in liquefaction of biomass: A review. *Renewable and Sustainable Energy Reviews* 81: 1259–1268.

Jahangiri, H., A. Osatiashtiani, J. A. Bennett, et al. 2018. Zirconia catalysed acetic acid ketonisation for pre-treatment of biomass fast pyrolysis vapours. *Catalysis Science & Technology* 8: 1134–1141.

Jogi, R., P. Maki-Arvela, P. Virtanen, et al. 2019. Biocrude production through hydro-liquefaction of wood biomass in supercritical ethanol using iron silica and iron beta zeolite catalysts. *Journal of Chemical Technology and Biotechnology* 94: 3736–3744.

Kassargy, C., S. Awad, K. Kahine, et al. 2016. Study on the simultaneous lipids transesterification and cellulosic matter liquefaction of oleaginous seeds of Pistacia atlantica. *Energy Conversion and Management* 124: 369–376.

Kaur, R., P. Gera, M. K. Jha, and T. Bhaskar. 2019. Reaction parameters effect on hydrothermal liquefaction of castor (*Ricinus Communis*) residue for energy and valuable hydrocarbons recovery. *Renewable Energy* 141: 1026–1041.

Kong, X. J., X. L. Li, S. X. Wu, X. Zhang, and J. H. Liu. 2016. Efficient conversion of cotton stalks over a Fe modified HZSM-5 catalyst under microwave irradiation. *RSC Advances* 6: 28532–28537.

Koriakin, A., S. Moon, D. W. Kim, and C. H. Lee. 2017. Liquefaction of oil palm empty fruit bunch using sub- and supercritical tetralin, n-dodecane, and their mixture. *Fuel* 208: 184–192.

Kosmela, P., P. Kazimierski, K. Formela, J. Haponiuk, and L. Piszczyk. 2017. Liquefaction of macroalgae *Enteromorpha* biomass for the preparation of biopolyols by using crude glycerol. *Journal of Industrial and Engineering Chemistry* 56: 399–406.

Kumar, D. and K. K. Pant. 2015. Production and characterization of biocrude and biochar obtained from non-edible de-oiled seed cakes hydrothermal conversion. *Journal of Analytical and Applied Pyrolysis* 115: 77–86.

Li, H. Y., J. Hu, Z. J. Zhang, et al. 2014a. Insight into the effect of hydrogenation on efficiency of hydrothermal liquefaction and physico-chemical properties of biocrude oil. *Bioresource Technology* 163: 143–151.

Li, H. Y., Z. D. Liu, Y. H. Zhang, et al. 2014b. Conversion efficiency and oil quality of low-lipid high-protein and high-lipid low-protein microalgae via hydrothermal liquefaction. *Bioresource Technology* 154: 322–329.

Li, R. D., B. S. Li, T. H. Yang, et al. 2015. Sub-supercritical liquefaction of rice stalk for the production of bio-oil: Effect of solvents. *Bioresource Technology* 198: 94–100.

Li, R. D., B. S. Li, T. H. Yang, X. P. Kai, and W. Q. Zhang. 2018. Hydrogenation of rice stalk in situ in supercritical ethanol-water co-solvent via catalytic ethanol steam reforming. *Journal of Supercritical Fluids* 133: 309–317.

Li, Z. Y., Z. P. Zhong, B. Zhang, W. Wang, and W. T. Wu. 2019. Catalytic fast pyrolysis of rice husk over hierarchical micro-mesoporous composite molecular sieve: Analytical Py-GC/MS study. *Journal of Analytical and Applied Pyrolysis* 138: 103–113.

Lin, Qs. S., Y. Chen, Y. Tang, et al. 2017. Catalytic hydrothermal liquefaction of *D. tertiolecta* over multifunctional mesoporous silica-based catalysts with high stability. *Microporous and Mesoporous Materials* 250: 120–127.

Liu, J., J. D. Lei, J. He, et al. 2015. Hydroprocessing of Jatropha oil for production of green diesel over non-sulfided Ni-PTA/Al_2O_3 Catalyst. *Scientific Reports* 5: 11327.

Lopez-Gonzalez, D., M. Puig-Gamero, F. G. Acien, et al. 2015. Energetic, economic and environmental assessment of the pyrolysis and combustion of microalgae and their oils. *Renewable and Sustainable Energy Reviews* 51: 1752–1770.

Lyckeskog, H. N., C. Mattsson, L. Olausson, et al. 2017. Thermal stability of low and high Mw fractions of bio-oil derived from lignin conversion in subcritical water. *Biomass Conversion and Biorefinery* 7: 401–414.

Ma, Q. H., D. D. Chen, L. F. Wei, et al. 2017. Bio-oil production from hydrogenation liquefaction of rice straw over metal (Ni, Co, Cu)-modified CeO_2 catalysts. *Energy Sources, Part A: Recovery, Utilization, and Environmental Effects* 40: 200–206.

Mishra, S. and K. Mohanty. 2020. Co-HTL of domestic sewage sludge and wastewater treatment derived microalgal biomass: An integrated biorefinery approach for sustainable biocrude production. *Energy Conversion and Management* 204: 112312.

Murnieks, R., V. Kampars, K. Malins, and L. Apseniece. 2014. Hydrotreating of wheat straw in toluene and ethanol. *Bioresource Technology* 163: 106–111.

Oh, S., I. G. Choi, and J. W. Choi. 2019. Pretreatment of bio-oil with ion exchange resin to improve fuel quality and reduce char during hydrodeoxygenation upgrading with Pt/C. *Environmental Technology* 2019: 1658810.

Patil, P. T., U. Armbruster, and A. Martin. 2014. Hydrothermal liquefaction of wheat straw in hot compressed water and subcritical water-alcohol mixtures. *Journal of Supercritical Fluids* 93: 121–129.

Pedersen, T. H. and L. A. Rosendahl. 2015. Production of fuel range oxygenates by supercritical hydrothermal liquefaction of lignocellulosic model systems. *Biomass & Bioenergy* 83: 206–215.

Posmanik, R., D. A. Cantero, A. Malkani, D. L. Sills, and J. W. Tester. 2017. Biomass conversion to bio-oil using sub-critical water: Study of model compounds for food processing waste. *Journal of Supercritical Fluids* 119: 26–35.

Posmanik, R., C. M. Martinez, B. Cantero-Tubilla, et al. 2018. Acid and alkali catalyzed hydrothermal liquefaction of dairy manure digestate and food waste. *ACS Sustainable Chemistry & Engineering* 6: 2724–2732.

Pourkarimi, S., A. Hallajisani, A. Alizadehdakhel, and A. Nouralishahi. 2019. Biofuel production through micro- and macroalgae pyrolysis: A review of pyrolysis methods and process parameters. *Journal of Analytical and Applied Pyrolysis* 142: 104599.

Prajitno, H., J. Park, C. Ryu, et al. 2018. Effects of solvent participation and controlled product separation on biomass liquefaction: A case study of sewage sludge. *Applied Energy* 218: 402-416.

Qiu, Y., J. Cheng, H. Guo, et al. 2019. Mild hydrothermal treatment on microalgal biomass in batch reactors for lipids hydrolysis and solvent-free extraction to produce biodiesel. *Energy* 189: 116308.

Raikova, S., T. D. J. Knowles, M. J. Allen, and C. J. Chuck. 2019. Co-liquefaction of macroalgae with common marine plastic pollutants. *ACS Sustainable Chemistry & Engineering* 7: 6769–6781.

Remon, J., J. Randall, V. L. Budarin, and J. H. Clark. 2019. Production of bio-fuels and chemicals by microwave-assisted, catalytic, hydrothermal liquefaction (MAC-HTL) of a mixture of pine and spruce biomass. *Green Chemistry* 21: 284–299.

Santillan-Jimenez, E., R. Pace, T. Morgan, et al. 2019. Co-processing of hydrothermal liquefaction algal bio-oil and petroleum feedstock to fuel-like hydrocarbons via fluid catalytic cracking. *Fuel Processing Technology* 188: 164–171.

Si, B. C., L. B. Yang, X. F. Zhou, et al. 2019. Anaerobic conversion of the hydrothermal liquefaction aqueous phase: Fate of organics and intensification with granule activated carbon/ozone pretreatment. *Green Chemistry* 21: 1305–1318.

Subagyono, D. J. N., M. Marshall, W. R. Jackson, and A. L. Chaffee. 2015. Pressurized thermal and hydrothermal decomposition of algae, wood chip residue, and grape marc: A comparative study. *Biomass and Bioenergy* 76: 141–157.

Sun, J., X. A. Xie, D. Fan, X. Wang, and W. T. Liao. 2020. Effect of TEMPO and characterization of bio-oil from cellulose liquefaction in supercritical ethanol. *Renewable Energy* 145: 1949–1956.

Sunphorka, S., K. Prapaiwatcharapan, N. Hinchiranan, K. Kangvansaichol, and P. Kuchonthara. 2015. Biocrude oil production and nutrient recovery from algae by two-step hydrothermal liquefaction using a semi-continuous reactor. *Korean Journal of Chemical Engineering* 32: 79–87.

Taghipour, A., J. A. Ramirez, R. J. Brown, and T. J. Rainey. 2019. A review of fractional distillation to improve hydrothermal liquefaction biocrude characteristics: Future outlook and prospects. *Renewable and Sustainable Energy Reviews* 115: 109355.

Teri, G., L. G. Luo, and P. E. Savage. 2014. Hydrothermal treatment of protein, polysaccharide, and lipids alone and in mixtures. *Energy & Fuels* 28: 7501–7509.

Tsegaye, B., C. Balomajumder, and P. Roy. 2020. Organosolv pretreatments of rice straw followed by microbial hydrolysis for efficient biofuel production. *Renewable Energy* 148: 923–934.

Tsubaki, S., K. Oono, A. Onda, et al. 2019. Microwave-assisted solubilization of microalgae in high-temperature ethylene glycol. *Biomass and Bioenergy* 130: 105360.

Usman, M., S. L. Hao, H. H. Chen, et al. 2019. Molecular and microbial insights towards understanding the anaerobic digestion of the wastewater from hydrothermal liquefaction of sewage sludge facilitated by granular activated carbon (GAC). *Environment International* 133: 105257.

Wang, B. F., Y. R. Huang, and J. J. Zhang. 2014. Hydrothermal liquefaction of lignite, wheat straw and plastic waste in sub-critical water for oil: Product distribution. *Journal of Analytical and Applied Pyrolysis* 110: 382–389.

Wang, B. F., Y. R. Huang, and J. J. Zhang. 2015. Sulfur distribution during hydrothermal liquefaction of lignite, wheat straw and plastic waste in sub-critical water. *China Petroleum Processing & Petrochemical Technology* 17: 24–30.

Wang, J. P., B. Su, Y. L. Xie, et al. 2018. Comparative research on deoxy-liquefaction of marine and terrestrial biomasses. *Journal of Analytical and Applied Pyrolysis* 131: 28–34.

Wang, Y. H., Y. H. Han, W. Y. Hu, D. L. Fu, and G. Wang. 2020. Analytical strategies for chemical characterization of bio-oil. *Journal of Separation Science* 43: 360–371.

Wang, Y. L., X. J. Pan, Y. Y. Ye, et al. 2019a. Process optimization of biomass liquefaction in isopropanol/water with Raney nickel and sodium hydroxide as combined catalysts. *Biomass and Bioenergy* 122: 305–312.

Wang, Y. X., Y. H. Zhang, and Z. D. Liu. 2019b. Effect of aging in nitrogen and air on the properties of biocrude produced by hydrothermal liquefaction of *Spirulina*. *Energy & Fuels* 33: 9870–9878.

Weber, R. S. and J. E. Holladay. 2018. Modularized production of value-added products and fuels from distributed waste carbon-rich feedstocks. *Engineering* 4: 330–335.

Xu, D. H. and P. E. Savage. 2015. Effect of reaction time and algae loading on water-soluble and insoluble biocrude fractions from hydrothermal liquefaction of algae. *Algal Research-Biomass Biofuels and Bioproducts* 12: 60–67.

Xu, Y. F., X. J. Zheng, X. G. Hu, K. D. Dearn, and H. M. Xu. 2014. Effect of catalytic esterification on the friction and wear performance of bio-oil. *Wear* 311: 93–100.

Yang, T. H., Y. F. Jie, B. S. Li, et al. 2016. Catalytic hydrodeoxygenation of crude bio-oil over an unsupported bimetallic dispersed catalyst in supercritical ethanol. *Fuel Processing Technology* 148: 19–27.

Yang, T. H., L. P. Shi, R. D. Li, B. S. Li, and X. P. Kai. 2019. Hydrodeoxygenation of crude bio-oil in situ in the bio-oil aqueous phase with addition of zero-valent aluminum. *Fuel Processing Technology* 184: 65–72.

Yang, T. H., J. Wang, B. S. Li, et al. 2018. Behaviors of rice straw two-step liquefaction with sub/supercritical ethanol in carbon dioxide atmosphere. *Bioresource Technology* 258: 287–294.

Yang, T. H., W. Q. Zhang, R. D. Li, et al. 2017. Deoxy-liquefaction of corn stalk in subcritical water with hydrogen generated in situ via aluminum-water reaction. *Energy & Fuels* 31: 9605–9612.

Yanik, J., R. Stahl, N. Troeger, and A. Sinag. 2013. Pyrolysis of algal biomass. *Journal of Analytical and Applied Pyrolysis* 103: 134–141.

Yigezu, Z. D. and K. Muthukumar. 2014. Catalytic cracking of vegetable oil with metal oxides for biofuel production. *Energy Conversion and Management* 84: 326–333.

Younas, R., S. L. Hao, L. W. Zhang, and S. C. Zhang. 2017. Hydrothermal liquefaction of rice straw with NiO nanocatalyst for bio-oil production. *Renewable Energy* 113: 532–545.

Yuan, C., S. Wang, L. L. Qian, est al. 2019. Effect of cosolvent and addition of catalyst (HZSM-5) on hydrothermal liquefaction of macroalgae. *International Journal of Energy Research* 43: 8841–8851.

Zhu, Z., L. Rosendahl, S. S. Toor, D. H. Yu, and G. Y. Chen. 2015. Hydrothermal liquefaction of barley straw to bio-crude oil: Effects of reaction temperature and aqueous phase recirculation. *Applied Energy* 137: 183–192.

Part III

Biodiesel Fuels in General

11 Biodiesel Fuels
A Scientometric Review of the Research

Ozcan Konur

CONTENTS

11.1 INTRODUCTION

Crude oils have been primary sources of energy and fuels, such as petrodiesel fuel. However, significant public concerns about the sustainability, price fluctuations, and adverse environmental impact of crude oils have emerged since the 1970s (Ahmadun et al., 2009; Atlas, 1981; Babich and Moulijn, 2003; Kilian, 2009; Perron, 1989). Thus, biooils (Bridgwater and Peacocke, 2000; Czernik and Bridgwater, 2004; Gallezot, 2012; Mohan et al., 2006) and biooil-based biodiesel fuels (Chisti, 2007;

Hill et al., 2006; Lapuerta et al., 2008; Mata et al., 2010; Zhang et al., 2003) have emerged as alternatives to crude oils and crude oil-based petrodiesel fuels in recent decades. Nowadays, both petrodiesel and biodiesel fuels are being used extensively (Konur, 2021a–ag).

However, for the efficient progression of the research in this field, it is necessary to develop efficient incentive structures for the primary stakeholders and to inform these stakeholders about the research (Konur, 2000, 2002a–c, 2006a–b, 2007a–b; North, 1991a–b).

Scientometric analysis of the research offers ways to evaluate the research in a respective field (Garfield, 1955, 1972). This method has been used to evaluate research in a number of other fields (Konur, 2011, 2012a–n, 2015, 2016a–f, 2017a–f, 2018a–b, 2019a–b). However, there has been no current scientometric study of the research on biodiesel fuels in general.

This chapter presents a study on the scientometric evaluation of the research on biodiesel fuels in general using two datasets. The first dataset includes the 100-most-cited papers (n = 100 sample papers) whilst the second set includes population papers (n = over 6,500 population papers) published between 1980 and 2019.

The data on the indices, document types, authors, institutions, funding bodies, source titles, 'Web of Science' subject categories, keywords, research fronts, and citation impact are presented and discussed.

11.2 MATERIALS AND METHODOLOGY

The search for the literature was carried out in the 'Web of Science' (WOS) database in February 2020. It contains the 'Science Citation Index Expanded' (SCI-E), the Social Sciences Citation Index' (SSCI), the 'Book Citation Index-Science' (BCI-S), the 'Conference Proceedings Citation Index-Science' (CPCI-S), the 'Emerging Sources Citation Index' (ESCI), the 'Book Citation Index-Social Sciences and Humanities' (BCI-SSH), the 'Conference Proceedings Citation Index-Social Sciences and Humanities' (CPCI-SSH), and the 'Arts and Humanities Citation Index' (A&HCI).

The keywords for the search of the literature were collated from the screening of abstract pages for the first 1,000 highly cited papers. This keyword set is provided in the Appendix.

Two datasets are used for this study. The highly cited 100 papers comprise the first dataset (sample dataset, n = 100 papers) whilst all the papers form the second dataset (population dataset, n = over 6,500 papers).

The data on the indices, document types, publication years, institutions, funding bodies, source titles, countries, 'Web of Science' subject categories, citation impact, keywords, and research fronts are collated from these datasets. The key findings are provided in the relevant tables and a figure, supplemented with explanatory notes in the text. The findings are discussed, a number of conclusions are drawn, and a number of recommendations for further study are made.

11.3 RESULTS

11.3.1 Indices and Documents

There are over 8,000 papers in this field in the 'Web of Science' as of February 2020. This original population dataset is refined for the document type (article, review, book chapter, book, editorial material, note, and letter) and language (English), resulting in over 6,500 papers comprising over 78.8% of the original population dataset.

The primary index is the SCI-E for both the sample and population papers. About 92.7% of the population papers are indexed by the SCI-E database. Additionally 3.7, 3.9, and 3.0% of these papers are indexed by the CPCI-S, ESCI, and BCI-S databases, respectively. The papers on the social and humanitarian aspects of this field are relatively negligible with only 2.2 and 0.1% of the population papers indexed by the SSCI and A&HCI, respectively.

Brief information on the document types for both datasets is provided in Table 11.1. The key finding is that article types of documents are the primary documents for both sample and population papers, whilst reviews form 59% of the sample papers.

11.3.2 Authors

Brief information about the 31-most-prolific authors with at least two sample papers each is provided in Table 11.2. Around 290 and 14,800 authors contribute to the sample and population papers, respectively.

The most-prolific author is 'Gerhard Knothe' with nine sample papers primarily on 'biodiesel properties'. The other prolific authors are 'Mustafa Canakci', 'Ayhan Demirbas', and 'Jon H. van Gerpen' with four, three, and three sample papers,

TABLE 11.1
Document Types

	Document Type	Sample Dataset (%)	Population Dataset (%)	Difference (%)
1	Article	38	91.1	–53.1
2	Review	59	7.3	51.7
3	Book chapter	0	2.8	–2.8
4	Proceeding paper	2	3.8	–1.8
5	Editorial material	1	0.9	0.1
6	Letter	0	0.6	–0.6
7	Book	0	0.2	–0.2
8	Note	0	0.1	–0.1

Note: Originally there were 53 articles and 46 reviews as classified by the database.

TABLE 11.2
Authors

	Author	No. of Sample Papers (%)	No. of Population Papers (%)	Surplus (%)	Institution	Country	Research Fronts
1	Knothe, Gerhard	9	0.3	8.7	Dept. Agric.	USA	Properties
2	Canakci. Mustafa	4	0.2	3.8	Kocaeli Univ.	Turkey	Production, properties
3	Demirbas, Ayhan	3	0.3	2.7	Sila Sci.	Turkey	Production
4	Van Gerpen, Jon H.	3	0.3	2.7	Iowa State Univ.	USA	Production, properties, emissions
5	Agarwal, Avinash K.	2	0.5	1.5	Univ. Wisconsin	USA	Properties
6	Wilson, Karen F	2	0.4	1.6	York Univ.	UK	Production
7	Lee, Adam F.	2	0.3	1.7	Bioenerg. Res. Inst.	UK	Production
8	Sharma, Yogesh C.	2	0.3	1.7	Banaras Hindu Univ.	India	Production, properties
9	Aracil, Jose	2	0.2	1.8	Univ. Complutense	Spain	Production
10	Aroua, Mohamed K. *	2	0.2	1.8	Univ. Malaya	Malaysia	Production, properties
11	Boehman, Andre L.	2	0.2	1.8	Penn. State Univ.	USA	Properties, emissions
12	Ellis, Naoko F	2	0.2	1.8	Univ. British Columbia	Canada	Production
13	Martinez, Mercedes F	2	0.2	1.8	Univ. Complutense	Spain	Production
14	Singh, Bhaskar	2	0.2	1.8	Banaras Hindu Univ.	India	Production, properties
15	Alam, Mahabuhul	2	0.1	1.9	Penn. State Univ.	USA	Properties
16	Balat, Havva F	2	0.1	1.9	Sila Sci.	Turkey	Production
17	Balat, Mustafa	2	0.1	1.9	Sila Sci.	Turkey	Production
18	De Andrade, Jailson B.	2	0.1	1.9	Univ. Fed. Bahia	Brazil	Production, properties
19	Fogli, Thomas A.	2	0.1	1.9	Dept. Agric.	USA	Production

20	Guarieiro, Lilian L. N. F	2	0.1	1.9	Univ. Fed. Rio de Janeiro	Brazil	Production, properties
21	Hoekman, S. Kent	2	0.1	1.9	Desert Res. Inst.	USA	Emissions
22	Meher, Lekha C.	2	0.1	1.9	Univ. Saskatchewan	Canada	Production
23	Pinto, Angelo C.	2	0.1	1.9	Univ. Fed. Bahia	Brazil	Production, properties
24	Rezende, Michelle J. C. F	2	0.1	1.9	Univ. Fed. Rio de Janeiro	Brazil	Production, properties
25	Ribeiro, Nubia M. F	2	0.1	1.9	Univ. Fed. Rio de Janeiro	Brazil	Production, properties
26	Robbins, Curtis	2	0.1	1.9	Desert Res. Inst.	USA	Emissions, properties
27	Song, Juhun	2	0.1	1.9	Penn. State Univ.	USA	Properties, emissions
28	Steidley, Kevin R.	2	0.1	1.9	Dept. Agric.	USA	Properties
29	Torres, Ednildo A.	2	0.1	1.9	Univ. Fed. Bahia	Brazil	Production, properties
30	Vicente, Gemma F	2	0.1	1.9	Univ. Complutense	Spain	Production
31	Westbrook, Charles K. *	2	0.1	1.9	Lawrence Livermore Natl. Lab.	USA	Properties

*'Highly Cited Researchers' in 2019 (Clarivate Analytics. 2019); F: Female.

respectively, mostly on the 'properties and production of biodiesel fuels'. The other prolific authors have two papers each.

On the other hand, a number of authors have a significant presence in the population papers: 'Hassan H. Masjuki', 'M. Abul Kalam', 'Dehua Liu', 'Wei Du', 'Chang Sik Lee', 'Wenming Yang', 'Barat Ghobadian', 'Hu Li', 'Yunjun Yan', 'Joao A. P. Coutinho', 'Gholamhassan Najafi', 'Georgios Karavalakis', 'Shiro Saka', 'Chun Shun Cheung', 'Breda Kegl', 'Rizalman Mamat', 'Yuvarajan Devarajan', 'Akihiko Kondo', 'Stamoulis Stournas', and 'Meisam Tabatabaei' with at least 0.31% of the population papers each.

The most-prolific institutions for these top authors are the US 'Department of Agriculture', 'Pennsylvania State University', 'Sila Science' of Turkey, the 'University of Complutense' of Spain, the 'Federal University of Bahia', and the 'Federal University of Rio de Janeiro' of Brazil, with three sample papers each. The other prolific institutions with two authors are 'Banaras Hindu University' of India and the US 'Desert Research Institute'. Thus, in total, 17 institutions house these top authors.

It is notable that two of these top researchers are listed in the 'Highly Cited Researchers' (HCR) in 2019 (Clarivate Analytics, 2019; Docampo and Cram, 2019).

The most-prolific country for these top authors is the USA with 11. This top country is followed by Brazil and Turkey with six and four sample papers, respectively. The other prolific countries are Spain with three authors, and Canada, India, and the UK with two authors each. Thus, in total, eight countries contribute to these top papers.

There are three key topical research fronts for these top researchers: 'production of biodiesel fuels', 'properties of biodiesel fuels', and 'emissions from biodiesel fuels' with 22, 19, and 5 sample papers, respectively.

It is further notable that there is a significant gender deficit among these top authors as only six of them are female (Lariviere et al., 2013; Xie and Shauman, 1998).

The author with the most impact is 'Gerhard Knothe' with an 8.7% publication surplus. The other authors with the most impact are 'Mustafa Canakci', 'Ayhan Demirbas', and 'Jon H. van Gerpen' with 3.8, 2.7, and 2.7% publication surpluses, respectively. On the other hand, the authors with the least impact are 'Avinash K. Agarwal', 'Karen Wilson', 'Yogesh C. Sharma', and 'Adam F. Lee' with at least a 1.5% publication surplus each.

11.3.3 Publication Years

The information about the publication years for both datasets is provided in Figure 11.1. This figure shows that 0, 5, 71, and 24% of the sample papers and 0, 1, 14.1, and 82% of the population papers were published in the 1980s, 1990s, 2000s, and 2010s, respectively.

Similarly, the most-prolific publication years for the sample dataset are 2008, 2009, 2010, 2006, and 2007 with 15, 14, 11, 11, and 10 papers, respectively. On the other hand, the most-prolific publication years for the population dataset are 2019, 2017, 2018, and 2016 with 11.1, 10.2, 9.9, and 9.1% of the population papers,

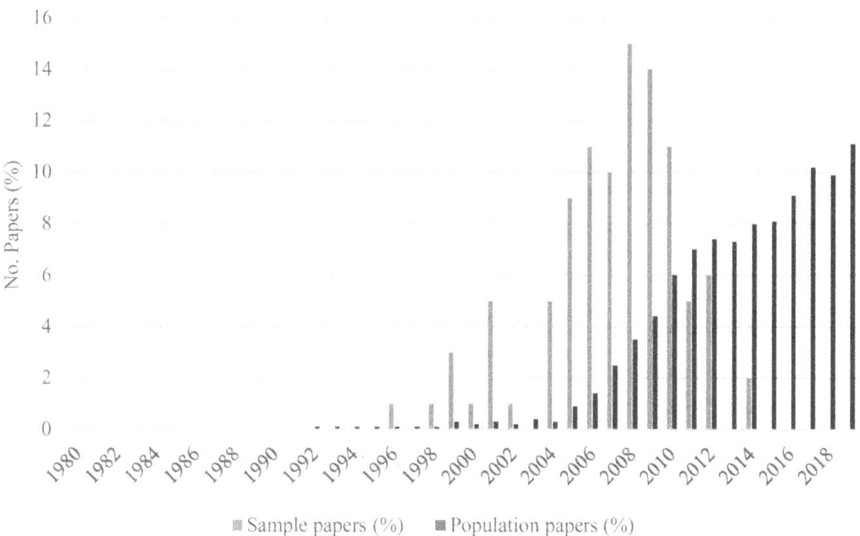

FIGURE 11.1 Research output between 1980 and 2019.

respectively. It is notable that there is a sharply rising trend for the population papers in the 2000s and 2010s.

11.3.4 Institutions

Brief information on the top 19 institutions with at least 2% of the sample papers each is provided in Table 11.3. In total, around 120 and 3,400 institutions contribute to the sample and population papers, respectively. Additionally, 1.4% of the population papers have no institutional information on their abstract pages.

These top institutions publish 58 and 13.5% of the sample and population papers, respectively. The top institution is the US 'Department of Agriculture' with 12 and 1.2% of the sample and population papers, respectively. This top institution is followed by the 'Indian Institute of Technology', the US 'Department of Energy', and 'Sila Science' of Turkey with five, four, and four sample papers, respectively. The 'Chinese Academy of Sciences', the 'University of Malaya', and 'Kocaeli University' of Turkey are the other prolific institutions with three sample papers each.

The most-prolific country for these top institutions is the USA with seven. This top country is followed by Turkey with three institutions. The other prolific countries are Brazil, India, and Spain with two institutions each.

The institution with the most impact is the US 'Department of Agriculture' with a 10.8% publication surplus. This top institution is followed by 'Sila Science' and 'Kocaeli University' of Turkey and the US 'Department of Energy' with at least a 2.8% publication surplus each.

On the other hand, the institutions with the least impact are 'Rio de Janeiro Federal University', the 'Chinese Academy of Sciences', and the 'University of Malaya' with at least a 0.8% publication surplus each.

TABLE 11.3
Institutions

	Institution	Country	No. of Sample Papers (%)	No. of Population Papers (%)	Difference (%)
1	Dept. Agric.	USA	12	1.2	10.8
2	Indian Inst. Technol.	India	5	2.6	2.4
3	Dept. Ener.	USA	4	1.1	2.9
4	Sila Sci.	Turkey	4	0.2	3.8
5	Chinese Acad. Sci.	China	3	1.9	1.1
6	Univ. Malaya	Malaysia	3	1.9	1.1
7	Kocaeli Univ.	Turkey	3	0.2	2.8
8	Rio de Janeiro Fed. Univ.	Brazil	2	1.2	0.8
9	Castilla la Mancha Univ.	Spain	2	0.6	1.4
10	Bahia Fed. Univ.	Brazil	2	0.6	1.4
11	Iowa State Univ.	USA	2	0.4	1.6
12	Banaras Hindu Univ.	India	2	0.2	1.8
13	Complutense Univ.	Spain	2	0.2	1.8
14	Desert Res. Inst.	USA	2	0.2	1.8
15	Lawrence Livermore Natl. Lab.	USA	2	0.2	1.8
16	Natl. Renew. Ener. Lab.	USA	2	0.2	1.8
17	Pennsylvania State Univ.	USA	2	0.2	1.8
18	Selcuk Univ.	Turkey	2	0.2	1.8
19	Univ. British Columbia	Canada	2	0.2	1.8

It is notable that some institutions have a heavy presence in the population papers: the 'State University of Campinas' and the 'University of Sao Paulo' of Brazil, the 'National Council of Scientific and Technical Research' of Argentina, 'Anna University' of India, 'Tsinghua University' of China, the 'National Scientific Research Center' of France, the 'Federal University of Rio Grande do Sul' of Brazil, the 'Council of Scientific Industrial Research' of India, 'Hanyang University' of South Korea, the 'National University of Singapore', 'Paulista State University', 'Parana Federal University' of Brazil, and 'Shanghai Jiao Tong University' of China have at least a 0.7% presence in the population papers each.

11.3.5 Funding Bodies

Brief information about the top three funding bodies with at least 2% of the sample papers each is provided in Table 11.4. It is significant that only 21 and 57% of the sample and population papers declare any funding, respectively. Around 40 and 3,500 bodies fund the research for the sample and population papers, respectively.

The most prolific funding bodies are the US 'Coordinating Research Council', the 'Ministry of Education' of Malaysia, and the 'National Natural Science Foundation' of China with two sample papers each.

It is notable that some top funding agencies have a heavy presence in the population studies. Some of them are the 'National Council for Scientific and Technological Development', the 'CAPES Foundation', the 'Sao Paulo State Research Funding

TABLE 11.4
Funding Bodies

	Funding Body	Country	No. of Sample Papers (%)	No. of Population Papers (%)	Difference (%)
1	Coordinating Res. Counc.	USA	2	0.1	1.9
2	Ministry Educ.	Malaysia	2	0.5	1.5
3	Natl. Natr. Sci. Found.	China	2	7.6	–4.6

Foundation', 'Science Technology and Innovation' of Brazil, the US 'Department of Energy', the 'European Union', the 'Natural Sciences and Engineering Research Council' of Canada, the 'National High Technology Research and Development Program' of China, the 'National Science Foundation' of the USA, the 'University of Malaya', the 'National Basic Research Program' of China, the 'National Council for Science and Technology' of Mexico, the 'Carlos Chagas Filho Foundation for Research Support of the State of Rio de Janeiro' of Brazil, the 'Department of Science Technology' of India, the 'Minas Gerais State Research Foundation' of Brazil, the 'Engineering Physical Sciences Research Council' of the UK, and the 'National Science Council' of Taiwan with at least 0.7% of the population papers each. These funding bodies are from Brazil, Mexico, China, Canada, Europe, the UK, and India.

11.3.6 Source Titles

Brief information about the top 12 source titles with at least three sample papers each is provided in Table 11.5. In total, around 40 and 1,000 source titles publish the sample and population papers, respectively. On the other hand, these top 12 journals publish 69 and 32% of the sample and population papers, respectively.

The top journals are 'Renewable Sustainable Energy Reviews' and 'Bioresource Technology', publishing 14 and 13 sample papers, respectively. These top journals are followed by 'Fuel', 'Fuel Processing Technology', 'Energy Fuels', and 'Energy Conversion and Management' with seven, seven, five, and five sample papers, respectively.

Although these journals are indexed by 12 subject categories, the top category is 'Energy Fuels' with ten journals. The other most-prolific subject categories are 'Engineering Chemical', 'Green Sustainable Science Technology', 'Agricultural Engineering', and 'Biotechnology and Applied Microbiology' with four, three, two, and two journals, respectively.

The journals with the most impact are 'Renewable Sustainable Energy Reviews', 'Bioresource Technology', and 'Fuel Processing Technology' with 11.3, 9.0, and 5.0% publication surpluses, respectively. On the other hand, the journals with the least impact are 'Fuel', 'Renewable Energy', and 'Energy Fuels' with –3.3, 0.2, and 0.9% publication surpluses/deficits, respectively.

TABLE 11.5
Source Titles

	Source Title	WOS Subject Categories	No. of Sample Papers (%)	No. of Population Papers (%)	Difference (%)
1	Renewable Sustainable Energy Reviews	Green Sust. Sci. Technol., Ener. Fuels	14	2.7	11.3
2	Bioresource Technology	Agr. Eng., Biot. Appl. Microb., Ener. Fuels	13	4.0	9.0
3	Fuel	Ener. Fuels, Eng. Chem.	7	10.3	–3.3
4	Fuel Processing Technology	Chem. Appl., Ener. Fuels, Eng. Chem.	7	2.0	5.0
5	Energy Fuels	Ener. Fuels, Eng. Chem.	5	4.1	0.9
6	Energy Conversion and Management	Therm., Ener. Fuels. Mechs.	5	2.2	2.8
7	Renewable Energy	Green Sust. Sci. Technol., Ener. Fuels	3	2.8	0.2
8	Applied Energy	Ener. Fuels, Eng. Chem	3	1.8	1.2
9	Biomass Bioenergy	Agr. Eng., Biot. Appl. Microb., Ener. Fuels	3	0.7	2.3
10	Green Chemistry	Chem. Mult., Green Sust. Sci. Technol.	3	0.6	2.4
11	Journal of Molecular Catalysis B Enzymatic	Bioch. Mol. Biol., Chem. Phys.	3	0.5	2.5
12	Progress in Energy and Combustion Science	Therm., Ener. Fuels, Eng. Chem., Eng. Mech.	3	0.3	2.7

It is notable that some journals have a heavy presence in the population papers. Some of them are 'Energy Sources Part A Recovery Utilization and Environmental Effects', 'Energy', 'Fuel Processing Technology', 'Industrial Engineering Chemistry Research', 'Journal of the American Oil Chemists Society', 'RSC Advances', 'Journal of Cleaner Production', 'Environmental Science and Pollution Research', 'Biofuels UK', 'Energies', and 'Applied Thermal Engineering' with at least a 0.7% presence in the population papers each.

11.3.7 Countries

Brief information about the top 14 countries with at least two sample papers each is provided in Table 11.6. In total, around 25 and 100 countries contribute to the sample and population papers, respectively.

TABLE 11.6
Countries

	Country	No. of Sample Papers (%)	No. of Population Papers (%)	Difference (%)
1	USA	31	13.0	18.0
2	India	14	13.4	0.6
3	China	10	14.4	–4.4
4	Turkey	10	3.2	6.8
5	Japan	8	2.8	5.2
6	Malaysia	6	5.2	0.8
7	Spain	6	4.1	1.9
8	Canada	4	3.0	1.0
9	UK	3	3.5	–0.5
10	Brazil	2	12.2	–10.2
11	Italy	2	2.6	–0.6
12	Thailand	2	1.7	0.3
13	Indonesia	2	1.2	0.8
14	Denmark	2	0.9	1.1
	Europe-4	13	11.1	1.9
	Asia-5	32	24.7	7.7

The top country is the USA, publishing 31 and 13% of the sample and population papers, respectively. India, China, Turkey, and Japan follow the USA with 14, 10, 10, and 8 sample papers, respectively.

On the other hand, the European and Asian countries represented in this table publish altogether 13 and 32% of the sample papers and 11.1 and 24.7% of the population papers, respectively.

It is notable that the publication surplus for the USA and these European and Asian countries is 18.0, 1.9, and 7.7%, respectively. On the other hand, the countries with the most impact are the USA, Turkey, and Japan with 18.9, 6.8, and 5.2% publication surpluses, respectively. Furthermore, the countries with the least impact are the UK, Italy, Thailand, and India with –0.5, –0.6, 0.3, and 0.6% publication surpluses/deficits, respectively.

It is also notable that some countries have a heavy presence in the population papers. The major producers of the population papers are Iran, South Korea, Taiwan, Australia, Germany, France, Portugal, Mexico, Greece, Argentina, Netherlands, and Saudi Arabia with at least 1.0% of the population papers each.

11.3.8 'Web of Science' Subject Categories

Brief information about the top 17 'Web of Science' subject categories with at least two sample papers each is provided in Table 11.7. The sample and population papers are indexed by around 25 and 125 subject categories, respectively.

For the sample papers, the top subject is 'Energy Fuels' with 68 and 48.5% of the sample and population papers, respectively. This top subject category is followed by 'Engineering Chemical', 'Biotechnology Applied Microbiology', 'Green Sustainable

TABLE 11.7
Web of Science Subject Categories

	Subject	No. of Sample Papers (%)	No. of Population Papers (%)	Difference (%)
1	Energy Fuels	68	48.5	19.5
2	Engineering Chemical	31	35.9	–4.9
3	Biotechnology Applied Microbiology	23	9.5	13.5
4	Green Sustainable Science Technology	20	10.7	9.3
5	Agricultural Engineering	18	5.1	12.9
6	Thermodynamics	11	9.4	1.6
7	Chemistry Applied	11	5.7	5.3
8	Chemistry Multidisciplinary	7	7.5	–0.5
9	Chemistry Physical	6	7.6	–1.6
10	Engineering Mechanical	6	4.8	1.2
11	Environmental Sciences	5	12.2	–7.2
12	Mechanics	5	3.2	1.8
13	Food Science Technology	3	2.8	0.2
14	Biochemistry Molecular Biology	3	1.7	1.3
15	Engineering Environmental	2	5.6	–3.6
16	Engineering Multidisciplinary	2	1.8	0.2
17	Multidisciplinary Sciences	2	1.3	0.7

Science Technology', and 'Agricultural Engineering' with 31, 23, 20, and 18% of the sample papers, respectively. The other prolific subject categories are 'Thermodynamics', 'Chemistry Applied', 'Chemistry Multidisciplinary', 'Chemistry Physical', and 'Engineering Mechanical' with at least six sample papers each.

It is notable that the publication surplus is most significant for 'Energy Fuels', 'Biotechnology Applied Microbiology', 'Agricultural Engineering', and 'Green Sustainable Science Technology' with 19.5, 13.5, 12.9, and 9.3% publication surpluses, respectively. On the other hand, the subjects with least impact are 'Environmental Sciences', 'Engineering Chemical', 'Engineering Environmental', and 'Chemistry Physical' with –7.2, –4.9, –3.6, and –1.6% publication deficits, respectively. This latter group of subject categories are under-represented in the sample papers.

Additionally, some subject categories have also a heavy presence in the population papers: 'Chemistry Analytical', 'Materials Science Multidisciplinary', 'Transportation Science Technology', 'Engineering Petroleum', 'Environmental Studies', 'Meteorology Atmospheric Sciences', 'Electrochemistry', 'Agronomy', 'Biochemical Research Methods', 'Economics', and 'Spectroscopy' with at least a 0.7% presence in the population papers each.

11.3.9 Citation Impact

These sample and population papers received about 51,000 and 190,000 citations, respectively, as of February 2020. Thus, the average number of citations per paper for these papers is about 510 and 135, respectively.

11.3.10 Keywords

Although a number of keywords are listed in the Appendix for the datasets related to this field, some of them are more significant for the sample papers.

The most-prolific keyword for the keyword set related to biodiesel fuels is 'biodiesel*' with 100 occurrences. There are two and one papers for 'bio-diesel' and 'renewable diesel', respectively.

The other prolific keywords are 'production' (50), '*cataly*' (32), 'engine' (14), 'emission*' (13), 'processing' (10), '*esterification' (11), 'propert*' (9), 'character*' (8), 'enzymatic' (7), 'combustion' (6), 'synthesis' (5), 'blend*' (5), 'feedstock*' (4), 'lipase*' (4), 'performance' (4), and 'viscosity' (4).

11.3.11 Research Fronts

Brief information about the key research fronts is provided in Table 11.8. There are three major topical research fronts for these sample papers: 'production of biodiesel fuels' (Fukuda et al., 2001; Leung et al., 2010; Lotero et al., 2005; Ma and Hanna, 1999; Marchetti et al., 2007; Meher et al., 2006; van Gerpen, 2005, Vicente et al., 2004), 'properties of biodiesel fuels' (Agarwal, 2007; Atabani et al., 2012; Hoekman et al., 2012; Knothe, 2005, 2008; Ramos et al., 2009; Sharma et al., 2008), and 'emissions from biodiesel fuels' (Knothe ====et al., 2006; Lapuerta et al., 2008; Monyem and van Gerpen, 2001; McCormick et al., 2001; Xue et al., 2011) with 77, 46, and 13 sample papers, respectively.

11.4 DISCUSSION

The size of the research in this field has increased to over 6,500 papers as of February 2020. It is expected that the number of the population papers in this field will exceed 15,000 papers by the end of the 2020s.

The research has developed more in the technological aspects of this field, rather than the social and humanitarian pathways, as evidenced by the negligible number of population papers in the indices of the 'Web of Science', SSCI, and A&HCI.

The article types of documents are the primary ones for both datasets and reviews are over-represented by 51.7% in the sample papers, whilst articles are under-represented by 53.1% (Table 11.1). Thus, the contribution of reviews by 59% of the sample papers in this field is highly exceptional (cf. Konur, 2011, 2012a–n, 2015, 2016a–f, 2017a–f, 2018a–b, 2019a–b).

TABLE 11.8
Research Fronts

	Research Front	No. of Sample Papers (%)
1	Production of biodiesel fuels	77
2	Properties of biodiesel fuels	46
3	Emissions from biodiesel fuels	13

Thirty-one authors from 17 institutions have at least two sample papers each (Table 11.2). Eleven, six, and four of these authors are from the USA, Brazil, and Turkey, respectively. The other prolific countries are Spain, Canada, India, and the UK.

There are three key topical research fronts for these top researchers: 'production of biodiesel fuels', 'properties of biodiesel fuels', and 'emissions from biodiesel fuels' with 22, 19, and 5 sample papers, respectively.

It is significant that there is ample 'gender deficit' among these top authors as only six of them are female (Lariviere et al., 2013; Xie and Shauman, 1998).

The population papers have built on the sample papers, primarily published in the 2000s and to a lesser extent in the 2010s (Figure 11.1). Following this rising trend, particularly in the 2000s and 2010s, it is expected that the number of papers will reach 15,000 by the end of the 2020s, more than doubling the current size.

The engagement of the institutions in this field at the global scale is significant as around 120 and 3,400 institutions contribute to the sample and population papers, respectively.

Nineteen top institutions publish 58 and 13.5% of the sample and population papers, respectively (Table 11.3). The top institutions are the US 'Department of Agriculture', the 'Indian Institute of Technology', the US 'Department of Energy', and 'Sila Science' of Turkey. The 'Chinese Academy of Sciences', the 'University of Malaya', and 'Kocaeli University' of Turkey are the other prolific institutions.

The most-prolific countries for these top institutions are the USA and Turkey. The other prolific countries are Brazil, India, and Spain. It is notable that some institutions with a heavy presence in the population papers are under-represented in the sample papers.

It is significant that only 21 and about 57% of the sample and population papers declare any funding, respectively. These funding bodies are from the USA, China, and Malaysia (Table 11.4). It is further notable that some top funding agencies for the population studies do not enter this top funding body list.

However, the substantial lack of Chinese bodies in this top funding body table is notable. This finding is in contrast with the studies showing that the heavy research funding in China and the NSFC is the primary funding agency in that country (Wang et al., 2012).

The sample and population papers are published by around 40 and 1,000 journals, respectively. It is significant that the top 12 journals publish 69 and 32% of the sample and population papers, respectively (Table 11.5).

The top journals are 'Renewable Sustainable Energy Reviews' and 'Bioresource Technology'. These top journals are followed by 'Fuel', 'Fuel Processing Technology', 'Energy Fuels', and 'Energy Conversion and Management'.

The top categories for these journals are 'Energy Fuels', 'Engineering Chemical', 'Green Sustainable Science Technology', 'Agricultural Engineering', and 'Biotechnology and Applied Microbiology'. It is notable that some journals with a heavy presence in the population papers are relatively under-represented in the sample papers.

In total, around 25 and 100 countries contribute to the sample and population papers, respectively. The top country is the USA (Table 11.6). This finding is in line with studies arguing that the USA is not losing ground in science and technology (Leydesdorff and Wagner, 2009).

The other prolific countries are India, China, Turkey, and Japan. These findings are in line with studies showing heavy research activity in these countries in recent decades (Kumar and Jan, 2014; Negishi et al., 2004; Prathap, 2017; Zhou and Leydesdorff, 2006).

On the other hand, the European and Asian countries represented in this table publish altogether 13 and 32% of the sample papers and 11.1 and 24.7% of the population papers, respectively. These findings are in line with studies showing that both European and Asian countries have a superior publication performance in science and technology (Bordons et al., 2015; Glanzel and Schlemmer, 2007; Okubo et al., 1998; Youtie et al., 2008).

It is notable that the publication surplus for the USA and these European and Asian countries is 18.0, 1.9, and 7.7%, respectively. On the other hand, the countries with the most impact are the USA, Turkey, and Japan. Furthermore, the countries with the least impact are the UK, Italy, Thailand, and India.

China's presence in this top table is notable. This finding is in line with China's efforts to be a leading nation in science and technology (Guan and Ma, 2007; Youtie et al., 2008; Zhou and Leydesdorff, 2006).

It is also notable that some countries have a heavy presence in the population papers. The major producers of the population papers are Iran, South Korea, Taiwan, Australia, Germany, France, Portugal, Mexico, Greece, Argentina, the Netherlands, and Saudi Arabia with at least 1.0% of the population papers each (Hassan et al., 2012; Huang et al., 2006; Rinia, 2000).

The sample and population papers are indexed by around 25 and 125 subject categories, respectively. For the sample papers, the top subject is 'Energy Fuels' with 68 and 48.5% of the sample and population papers, respectively (Table 11.7). This top subject category is followed by 'Engineering Chemical', 'Biotechnology Applied Microbiology', 'Green Sustainable Science Technology', and 'Agricultural Engineering'. The other prolific subject categories are 'Thermodynamics', 'Chemistry Applied', 'Chemistry Multidisciplinary', 'Chemistry Physical', and 'Engineering Mechanical'.

It is notable that the publication surplus is most significant for 'Energy Fuels', 'Biotechnology Applied Microbiology', 'Agricultural Engineering', and 'Green Sustainable Science Technology'. On the other hand, the subjects with least impact are 'Environmental Sciences', 'Engineering Chemical', 'Engineering Environmental', and 'Chemistry Physical'. This latter group of subject categories are under-represented in the sample papers.

These sample and population papers received about 51,000 and 190,000 citations, respectively, as of February 2020. Thus, the average number of citations per paper for these papers is about 510 and 135, respectively. Hence, the citation impact of the top 100 papers in this field has been significant.

Although a number of keywords are listed in the Appendix for the datasets related to this field, some of them are more significant for the sample papers.

The most-prolific keyword for the keyword set related to biodiesel fuels is 'biodiesel*' with 100 occurrences. The other prolific keywords are 'production', '*cataly*', 'engine', 'emission*', 'processing', '*esterification', 'propert*', 'character*', 'enzymatic', 'combustion', 'synthesis', 'blend*', 'feedstock*', 'lipase*', 'performance', and 'viscosity'. As expected, these keywords provide valuable information about the pathways of research in this field.

There are three major topical research fronts for these sample papers: 'production of biodiesel fuels', 'properties of biodiesel fuels', and 'emissions from biodiesel fuels'.

The key emphasis in these research fronts is the exploration of the structure–processing–property relationships of biodiesel fuels (Cheng and Ma, 2011; Konur and Matthews, 1989; Rogers and Hopfinger, 1994; Scherf and List, 2002).

11.5 CONCLUSION

This chapter has mapped the research on biodiesel fuels in general using a scientometric method.

The size of over 6,500 population papers shows the public importance of this interdisciplinary research field. However, it is significant that the research has developed more in the technological aspects in this field, rather than the social and humanitarian pathways.

Articles dominate both the sample and population papers. However, there are 59 review papers. The population papers, primarily published in the 2010s, build on these sample papers, mainly published in the 2000s and to a lesser extent in the 2010s.

The data presented in the tables and in the figure show that a small number of authors, institutions, funding bodies, journals, keywords, research fronts, subject categories, and countries have shaped the research in this field.

It is notable that the authors, institutions, and funding bodies in the USA, Brazil, Turkey, Spain, Canada, India, China, and the UK dominate the research in this field. Furthermore, it is also notable that some countries have a heavy presence in the population papers. The major producers of these papers are Iran, South Korea, Taiwan, Australia, Germany, France, Portugal, Mexico, Greece, Argentina, the Netherlands, and Saudi Arabia. Additionally, China and Brazil are under-represented significantly in the sample papers.

These findings show the importance of the development of efficient incentive structures for the development of the research in this field as in other fields. It seems that some countries (such as the USA, Brazil, Turkey, Spain, Canada, India, China, and the UK) have efficient incentive structures for the development of the research in this field, contrary to Iran, South Korea, Taiwan, Australia, Germany, France, Portugal, Mexico, Greece, Argentina, the Netherlands, and Saudi Arabia.

It further seems that although the research funding is a significant element of these incentive structures, it might not be a sole solution for increasing the incentives for the research in this field, as in the case of Iran, South Korea, Taiwan, Australia, Germany, France, Portugal, Mexico, Greece, Argentina, the Netherlands, and Saudi Arabia.

On the other hand, it seems there is more to do to reduce the significant gender deficit in this field as in others of science and technology (Lariviere et al., 2013; Xie and Shauman, 1998).

The data on the research fronts, keywords, source titles, and subject categories provide valuable evidence for the interdisciplinary nature of the research in this field (Lariviere and Gingras, 2010; Morillo et al., 2001).

There is ample justification for the broad search strategy employed in this study due to the interdisciplinary nature of this research field as evidenced by the top subject categories. The search strategy employed in this study is in line with the search strategies employed for related and other research fields (Konur, 2011, 2012a–n, 2015, 2016a–f, 2017a–f, 2018a–b, 2019a–b). It is particularly noted that only 68 and 48.5% of the sample and population papers are indexed by the 'Energy Fuels' subject category, respectively.

There are three major topical research fronts for these sample papers: 'production of biodiesel fuels', 'properties of biodiesel fuels', and 'emissions from biodiesel fuels'.

It is recommended that further scientometric studies are carried out for each of these research fronts building on the pioneering studies in these fields.

ACKNOWLEDGMENTS

The contribution of the highly cited researchers in the field of biodiesel fuels is greatly acknowledged.

11.A APPENDIX

The keyword set for biodiesel fuels

TI=(*biodiesel* or "*bio-diesel" or "renewable diesel" or "green diesel*" or "myco-diesel") NOT TI=(*alga* or vegetable* or fats or fat or *edible* or waste* or coffee or cooking or palm* or jatropha or sunflower* or rape* or *seed* or castor* or fish or soy* or karanj* or polanga or calophyllum or neem or rubber or frying or mahua or croton or sterculia or tallow or canola or moringa or *nut or cotton or rice or cynara or brassica or sesame or safflower or olive or camelina or horn or eucalyptus or corn or tall or mango or pistacia or almond or date or linseed or rubber or tobacco or jojoba or pennycress or poon or tung or cuphea or acrocomia or orange or *melon or sapium or grease or soapstock* or sludge or nannochloropsis or chlorella or scenedesmus or cyanobacter* or haematococcus or dunaliella or arginine or human* or rna or pongamia or glycerol or glycerin* or "plant oil*" or chlamydomonas or "co-product*" or "by-product*" or madhuca or salmon or manure or neochloris or ceiba or "soldier fly" or pine or hevea or monoraphidium or insect* or guizotia or kusum or oreochromis or laurel or babassu or terminalia or fruit*).

REFERENCES

Agarwal, A. K. 2007. Biofuels (alcohols and biodiesel) applications as fuels for internal combustion engines. *Progress in Energy and Combustion Science* 33: 233–271.

Ahmadun, F. R., A. Pendashteh, L. C. Abdullah, et al. 2009. Review of technologies for oil and gas produced water treatment. *Journal of Hazardous Materials* 170: 530–551.

Atabani, A. E., A. S. Silitonga, I. A. Badruddin, et al. 2012. A comprehensive review on biodiesel as an alternative energy resource and its characteristics. *Renewable & Sustainable Energy Reviews* 16: 2070–2093.

Atlas, R. M. 1981. Microbial degradation of petroleum hydrocarbons: An environmental perspective. *Microbiological Reviews* 45: 180–209.

Babich, I. V. and J. A. Moulijn. 2003. Science and technology of novel processes for deep desulfurization of oil refinery streams: A review. *Fuel* 82: 607–631.

Bordons, M., B. Gonzalez-Albo, J. Aparicio, and L. Moreno. 2015. The influence of R & D intensity of countries on the impact of international collaborative research: Evidence from Spain. *Scientometrics* 102: 1385–1400.

Bridgwater, A. V. and G. V. C. Peacocke. 2000. Fast pyrolysis processes for biomass. *Renewable & Sustainable Energy Reviews* 4: 1–73.

Cheng, Y. Q. and E. Ma. 2011. Atomic-level structure and structure–property relationship in metallic glasses. *Progress in Materials Science* 56: 379–473.

Chisti, Y. 2007. Biodiesel from microalgae. *Biotechnology Advances* 25: 294–306.

Clarivate Analytics. 2019. *Highly cited researchers: 2019 Recipients*. Philadelphia, PA: Clarivate Analytics. https://recognition.webofsciencegroup.com/awards/highly-cited/2019/ (accessed January, 3, 2020).

Czernik, S. and A. V. Bridgwater. 2004. Overview of applications of biomass fast pyrolysis oil. *Energy & Fuels* 18: 590–598.

Docampo, D. and L. Cram. 2019. Highly cited researchers: A moving target. *Scientometrics* 118: 1011–1025.

Fukuda, H., A. Kondo, and H. Noda. 2001. Biodiesel fuel production by transesterification of oils. *Journal of Bioscience and Bioengineering* 92: 405–416.

Gallezot, P. 2012. Conversion of biomass to selected chemical products. *Chemical Society Reviews* 41: 1538–1558.

Garfield, E. 1955. Citation indexes for science. *Science* 122: 108–111.

Garfield, E. 1972. Citation analysis as a tool in journal evaluation. *Science* 178: 471–479.

Glanzel, W. and B. Schlemmer. 2007. National research profiles in a changing Europe (1983–2003): An exploratory study of sectoral characteristics in the Triple Helix. *Scientometrics* 70: 267–275.

Guan, J. C. and N. Ma. 2007. China's emerging presence in nanoscience and nanotechnology: A comparative bibliometric study of several nanoscience 'giant'. *Research Policy* 36: 880–886.

Hassan, S. U., P. Haddawy, P. Kuinkel, A. Degelsegger, and C. Blasy. 2012. A bibliometric study of research activity in ASEAN related to the EU in FP7 priority areas. *Scientometrics* 91: 1035–1051.

Hill, J., E. Nelson, D. Tilman, S. Polasky, and D. Tiffany. 2006. Environmental, economic, and energetic costs and benefits of biodiesel and ethanol biofuels. *Proceedings of the National Academy of Sciences of the United States of America* 103: 11206–11210.

Hoekman, S. K., A. Broch, C. Robbins, E. Ceniceros, and M. Natarajan. 2012. Review of biodiesel composition, properties, and specifications. *Renewable & Sustainable Energy Reviews* 16: 143–169.

Huang, M. H., H. W. Chang, and D. Z. Chen. 2006. Research evaluation of research-oriented universities in Taiwan from 1993 to 2003. *Scientometrics* 67: 419–435.

Kilian, L. 2009. Not all oil price shocks are alike: Disentangling demand and supply shocks in the crude oil market. *American Economic Review* 99: 1053–1069.

Knothe, G. 2005. Dependence of biodiesel fuel properties on the structure of fatty acid alkyl esters. *Fuel Processing Technology* 86: 1059–1070.

Knothe, G. 2008. Designer biodiesel: Optimizing fatty ester (composition to improve fuel properties). *Energy & Fuels* 22: 1358–1364.

Knothe, G., C. A. Sharp, and T. W. Ryan. 2006. Exhaust emissions of biodiesel, petrodiesel, neat methyl esters, and alkanes in a new technology engine. *Energy & Fuels* 20: 403–408.

Konur, O. 2000. Creating enforceable civil rights for disabled students in higher education: An institutional theory perspective. *Disability & Society* 15: 1041–1063.
Konur, O. 2002a. Access to Nursing Education by disabled students: Rights and duties of nursing programs. *Nurse Education Today* 22: 364–374.
Konur, O. 2002b. Assessment of disabled students in higher education: Current public policy issues. *Assessment and Evaluation in Higher Education* 27: 131–152.
Konur, O. 2002c. Access to employment by disabled people in the UK: Is the Disability Discrimination Act working? *International Journal of Discrimination and the Law* 5: 247–279.
Konur, O. 2006a. Participation of children with dyslexia in compulsory education: Current public policy issues. *Dyslexia* 12: 51–67.
Konur, O. 2006b. Teaching disabled students in Higher Education. *Teaching in Higher Education* 11: 351–363.
Konur, O. 2007a. A judicial outcome analysis of the Disability Discrimination Act: A windfall for the employers? *Disability & Society* 22: 187–204.
Konur, O. 2007b. Computer-assisted teaching and assessment of disabled students in higher education: The interface between academic standards and disability rights. *Journal of Computer Assisted Learning* 23: 207–219.
Konur, O. 2011. The scientometric evaluation of the research on the algae and bio-energy. *Applied Energy* 88: 3532–3540.
Konur, O. 2012a. Evaluation of the research on the social sciences in Turkey: A scientometric approach. *Energy Education Science and Technology Part B: Social and Educational Studies* 4: 1893–1908.
Konur, O. 2012b. Prof. Dr. Ayhan Demirbas' scientometric biography. *Energy Education Science and Technology Part A: Energy Science and Research* 28: 727–738.
Konur, O. 2012c. The evaluation of the biogas research: A scientometric approach. *Energy Education Science and Technology Part A: Energy Science and Research* 29: 1277–1292.
Konur, O. 2012d. The evaluation of the educational research: A scientometric approach. *Energy Education Science and Technology Part B: Social and Educational Studies* 4: 1935–1948.
Konur, O. 2012e. The evaluation of the global energy and fuels research: A scientometric approach. *Energy Education Science and Technology Part A: Energy Science and Research* 30: 613–628.
Konur, O. 2012f. The evaluation of the research on the Arts and Humanities in Turkey: A scientometric approach. *Energy Education Science and Technology Part B: Social and Educational Studies* 4: 1603–1618.
Konur, O. 2012g. The evaluation of the research on the biodiesel: A scientometric approach. *Energy Education Science and Technology Part A: Energy Science and Research* 28: 1003–1014.
Konur, O. 2012h. The evaluation of the research on the bioethanol: A scientometric approach. *Energy Education Science and Technology Part A: Energy Science and Research* 28: 1051–1064.
Konur, O. 2012i. The evaluation of the research on the biofuels: A scientometric approach. *Energy Education Science and Technology Part A: Energy Science and Research* 28: 903–916.
Konur, O. 2012j. The evaluation of the research on the biohydrogen: A scientometric approach. *Energy Education Science and Technology Part A: Energy Science and Research* 29: 323–338.
Konur, O. 2012k. The evaluation of the research on the microbial fuel cells: A scientometric approach. *Energy Education Science and Technology Part A: Energy Science and Research* 29: 309–322.

Konur, O. 2012l. The scientometric evaluation of the research on the production of bioenergy from biomass. *Biomass and Bioenergy* 47: 504–515.

Konur, O. 2012m. The scientometric evaluation of the research on the deaf students in higher education. *Energy Education Science and Technology Part B: Social and Educational Studies* 4: 1573–1588.

Konur, O. 2012n. The scientometric evaluation of the research on the students with ADHD in higher education. *Energy Education Science and Technology Part B: Social and Educational Studies* 4: 1547–1562.

Konur, O. 2015. Current state of research on algal biodiesel. In *Marine Bioenergy: Trends and Developments*, S. K. Kim, and C. G. Lee, ed., 487–512. Boca Raton, FL: CRC Press.

Konur, O. 2016a. Scientometric overview in nanobiodrugs. In *Nanoarchitectonics for Smart Delivery and Drug Targeting*, A. M. Holban and A.M. Grumezescu, ed., 405–428. Amsterdam: Elsevier.

Konur, O. 2016b. Scientometric overview regarding nanoemulsions used in the food industry. In *Emulsions: Nanotechnology in the Agri-Food Industry*, A. M. Grumezescu, ed., 689–711. Amsterdam: Elsevier.

Konur, O. 2016c. Scientometric overview regarding the nanobiomaterials in antimicrobial therapy. In *Nanobiomaterials in Antimicrobial Therapy,* A. M. Grumezescu, ed., 511–535. Amsterdam: Elsevier.

Konur, O. 2016d. Scientometric overview regarding the nanobiomaterials in dentistry. In *Nanobiomaterials in Dentistry,* A. M. Grumezescu, ed., 425–453. Amsterdam: Elsevier.

Konur, O. 2016e. Scientometric overview regarding the surface chemistry of nanobiomaterials. In *Surface Chemistry of Nanobiomaterials,* A. M. Grumezescu, ed., 463–486. Amsterdam: Elsevier.

Konur, O. 2016f. The scientometric overview in cancer targeting. In *Nanoarchitectonics for Smart Delivery and Drug Targeting,* A. M. Holban and A. Grumezescu, ed., 871–895. Amsterdam; Elsevier.

Konur, O. 2017a. Recent citation classics in antimicrobial nanobiomaterials. In *Nanostructures for Antimicrobial Therapy,* A. Ficai and A. M. Grumezescu, ed., 669–685. Amsterdam: Elsevier.

Konur, O. 2017b. Scientometric overview in nanopesticides. In *New Pesticides and Soil Sensors,* A. M. Grumezescu, ed. 719–744. Amsterdam: Elsevier.

Konur, O. 2017c. Scientometric overview regarding oral cancer nanomedicine. In *Nanostructures for Oral Medicine*, E. Andronescu, A. M. Grumezescu, ed., 939–962. Amsterdam: Elsevier.

Konur, O. 2017d. Scientometric overview regarding water nanopurification. In *Water Purification,* A. M. Grumezescu, ed., 693–716. Amsterdam: Elsevier.

Konur, O. 2017e. Scientometric overview in food nanopreservation. In *Food Preservation,* A. M. Grumezescu, ed., 703–729. Amsterdam: Elsevier.

Konur, O. 2017f. The top citation classics in alginates for biomedicine. In *Seaweed Polysaccharides: Isolation, Biological and Biomedical Applications*, J. Venkatesan, S. Anil, S. K. Kim, ed., 22–249. Amsterdam: Elsevier.

Konur, O. 2018a. Scientometric evaluation of the global research in spine: An update on the pioneering study by Wei et al. *European Spine Journal* 27: 525–529.

Konur, O. 2018b. Bioenergy and biofuels science and technology: Scientometric overview and citation classics. In *Bioenergy and Biofuels*, O. Konur, ed., 3–63. Boca Raton: CRC Press.

Konur, O. 2019a. Cyanobacterial bioenergy and biofuels science and technology: A scientometric overview. In *Cyanobacteria: From Basic Science to Applications*, ed. A. K. Mishra, D. N. Tiwari and A. N. Rai, 419–442. Amsterdam: Elsevier.

Konur, O. 2019b. Nanotechnology applications in food: A scientometric overview. In *Nanoscience for Sustainable Agriculture*, R. N., Pudake, N. Chauhan, and C. Kole, ed., 683–711. Cham: Springer.

Konur, O., ed. 2021a. *Handbook of Biodiesel and Petrodiesel Fuels: Science, Technology, Health, and Environment.* Boca Raton, FL: CRC Press.

Konur, O., ed. 2021b. *Handbook of Biodiesel and Petrodiesel Fuels: Science, Technology, Health, and Environment. Volume 1. Biodiesel Fuels: Science, Technology, Health, and Environment.* Boca Raton, FL: CRC Press.

Konur, O., ed. 2021c. *Handbook of Biodiesel and Petrodiesel Fuels: Science, Technology, Health, and Environment. Volume 2. Biodiesel Fuels based on the Edible and Nonedible Feedstocks, Wastes, and Algae: Science, Technology, Health, and Environment.* Boca Raton, FL: CRC Press.

Konur, O., ed. 2021d. *Handbook of Biodiesel and Petrodiesel Fuels: Science, Technology, Health, and Environment. Volume 3. Petrodiesel Fuels: Science, Technology, Health, and Environment.* Boca Raton, FL: CRC Press.

Konur, O. 2021e. Biodiesel and petrodiesel fuels: Science, technology, health, and environment. In *Handbook of Biodiesel and Petrodiesel Fuels: Science, Technology, Health, and Environment. Volume 1. Biodiesel Fuels: Science, Technology, Health, and Environment*, ed. O. Konur. Boca Raton, FL: CRC Press.

Konur, O. 2021f. Biodiesel and petrodiesel fuels: A scientometric review of the research. In *Handbook of Biodiesel and Petrodiesel Fuels: Science, Technology, Health, and Environment. Volume 1. Biodiesel Fuels: Science, Technology, Health, and Environment*, ed. O. Konur. Boca Raton, FL: CRC Press.

Konur, O. 2021g. Biodiesel and petrodiesel fuels: A review of the research. In *Handbook of Biodiesel and Petrodiesel Fuels: Science, Technology, Health, and Environment. Volume 1. Biodiesel Fuels: Science, Technology, Health, and Environment*, ed. O. Konur. Boca Raton, FL: CRC Press.

Konur, O. 2021h Nanotechnology applications in the diesel fuels and the related research fields: A review of the research. In *Handbook of Biodiesel and Petrodiesel Fuels: Science, Technology, Health, and Environment. Volume 1. Biodiesel Fuels: Science, Technology, Health, and Environment*, ed. O. Konur. Boca Raton, FL: CRC Press.

Konur, O. 2021i. Biooils: A scientometric review of the research. In *Handbook of Biodiesel and Petrodiesel Fuels: Science, Technology, Health, and Environment. Volume 1. Biodiesel Fuels: Science, Technology, Health, and Environment*, ed. O. Konur. Boca Raton, FL: CRC Press.

Konur, O. 2021j. Characterization and properties of biooils: A review of the research. In *Handbook of Biodiesel and Petrodiesel Fuels: Science, Technology, Health, and Environment. Volume 1. Biodiesel Fuels: Science, Technology, Health, and Environment*, ed. O. Konur. Boca Raton, FL: CRC Press.

Konur, O. 2021k. Biomass pyrolysis and pyrolysis oils: A review of the research. In *Handbook of Biodiesel and Petrodiesel Fuels: Science, Technology, Health, and Environment. Volume 1. Biodiesel Fuels: Science, Technology, Health, and Environment*, ed. O. Konur. Boca Raton, FL: CRC Press.

Konur, O. 2021l. Biodiesel fuels: A scientometric review of the research. In *Handbook of Biodiesel and Petrodiesel Fuels: Science, Technology, Health, and Environment. Volume 1. Biodiesel Fuels: Science, Technology, Health, and Environment*, ed. O. Konur. Boca Raton, FL: CRC Press.

Konur, O. 2021m. Glycerol: A scientometric review of the research. In *Handbook of Biodiesel and Petrodiesel Fuels: Science, Technology, Health, and Environment. Volume 1. Biodiesel Fuels: Science, Technology, Health, and Environment*, ed. O. Konur. Boca Raton, FL: CRC Press.

Konur, O. 2021n. Propanediol production from glycerol: A review of the research. In *Handbook of Biodiesel and Petrodiesel Fuels: Science, Technology, Health, and Environment. Volume 1. Biodiesel Fuels: Science, Technology, Health, and Environment*, ed. O. Konur Boca Raton, FL: CRC Press.

Konur, O. 2021o. Edible oil-based biodiesel fuels: A scientometric review of the research. *In Handbook of Biodiesel and Petrodiesel Fuels: Science, Technology, Health, and Environment. Volume 2. Biodiesel Fuels based on the Edible and Nonedible Feedstocks, Wastes, and Algae: Science, Technology, Health, and Environment,* ed. O. Konur. Boca Raton, FL: CRC Press.

Konur, O. 2021p. Palm oil-based biodiesel fuels: A review of the research. In *Handbook of Biodiesel and Petrodiesel Fuels: Science, Technology, Health, and Environment. Volume 2. Biodiesel Fuels based on the Edible and Nonedible Feedstocks, Wastes, and Algae*, ed. O. Konur. Boca Raton, FL: CRC Press.

Konur, O. 2021q. Rapeseed oil-based biodiesel fuels: A review of the research. In *Handbook of Biodiesel and Petrodiesel Fuels: Science, Technology, Health, and Environment. Volume 2. Biodiesel Fuels based on the Edible and Nonedible Feedstocks, Wastes, and Algae,* ed. O. Konur. Boca Raton, FL: CRC Press.

Konur, O. 2021r. Nonedible oil-based biodiesel fuels: A scientometric review of the research. In *Handbook of Biodiesel and Petrodiesel Fuels: Science, Technology, Health, and Environment. Volume 2. Biodiesel Fuels based on the Edible and Nonedible Feedstocks, Wastes, and Algae: Science, Technology, Health, and Environment*, ed. O. Konur. Boca Raton, FL: CRC Press.

Konur, O. 2021s. Waste oil-based biodiesel fuels: A scientometric review of the research. In *Handbook of Biodiesel and Petrodiesel Fuels: Science, Technology, Health, and Environment. Volume 2. Biodiesel Fuels based on the Edible and Nonedible Feedstocks, Wastes, and Algae: Science, Technology, Health, and Environment*, ed. O. Konur. Boca Raton, FL: CRC Press.

Konur, O. 2021t. Algal biodiesel fuels: A scientometric review of the research. In *Handbook of Biodiesel and Petrodiesel Fuels: Science, Technology, Health, and Environment. Volume 2. Biodiesel Fuels based on the Edible and Nonedible Feedstocks, Wastes, and Algae: Science, Technology, Health, and Environment*, ed. O. Konur. Boca Raton, FL: CRC Press.

Konur, O. 2021u. Algal biomass production for biodiesel production: A review of the research. In *Handbook of Biodiesel and Petrodiesel Fuels: Science, Technology, Health, and Environment. Volume 2. Biodiesel Fuels based on the Edible and Nonedible Feedstocks, Wastes, and Algae*, ed. O. Konur Boca Raton, FL: CRC Press.

Konur, O. 2021v. Algal biomass production in wastewaters for biodiesel production: A review of the research. In *Handbook of Biodiesel and Petrodiesel Fuels: Science, Technology, Health, and Environment. Volume 2. Biodiesel Fuels based on the Edible and Nonedible Feedstocks, Wastes, and Algae*, ed. O. Konur. Boca Raton, FL: CRC Press.

Konur, O. 2021x. Algal lipid production for biodiesel production: A review of the research. In *Handbook of Biodiesel and Petrodiesel Fuels: Science, Technology, Health, and Environment. Volume 2. Biodiesel Fuels based on the Edible and Nonedible Feedstocks, Wastes, and Algae*, ed. O. Konur Boca Raton, FL: CRC Press.

Konur, O. 2021y. Crude oils: A scientometric review of the research. In *Handbook of Biodiesel and Petrodiesel Fuels: Science, Technology, Health, and Environment. Volume 3. Petrodiesel Fuels: Science, Technology, Health, and Environment*, ed. O. Konur. Boca Raton, FL: CRC Press.

Konur, O. 2021z. Petrodiesel fuels: A scientometric review of the research. In *Handbook of Biodiesel and Petrodiesel Fuels: Science, Technology, Health, and Environment. Volume 3. Petrodiesel Fuels: Science, Technology, Health, and Environment*, ed. O. Konur. Boca Raton, FL: CRC Press.

Konur, O. 2021aa. Bioremediation of petroleum hydrocarbons in the contaminated soils: A review of the research. In *Handbook of Biodiesel and Petrodiesel Fuels: Science, Technology, Health, and Environment. Volume 3. Petrodiesel Fuels: Science, Technology, Health, and Environment*, ed. O. Konur. Boca Raton, FL: CRC Press.

Konur, O. 2021ab. Desulfurization of diesel fuels: A review of the research. In *Handbook of Biodiesel and Petrodiesel Fuels: Science, Technology, Health, and Environment. Volume 3. Petrodiesel Fuels: Science, Technology, Health, and Environment*, ed. O. Konur. Boca Raton, FL: CRC Press.

Konur, O. 2021ac. Diesel fuel exhaust emissions: A scientometric review of the research. In *Handbook of Biodiesel and Petrodiesel Fuels: Science, Technology, Health, and Environment. Volume 3. Petrodiesel Fuels: Science, Technology, Health, and Environment*, ed. O. Konur. Boca Raton, FL: CRC Press.

Konur, O. 2021ad. The adverse health and safety impact of diesel fuels: A scientometric review of the research. In *Handbook of Biodiesel and Petrodiesel Fuels: Science, Technology, Health, and Environment. Volume 3. Petrodiesel Fuels: Science, Technology, Health, and Environment*, ed. O. Konur. Boca Raton, FL: CRC Press.

Konur, O. 2021ae. Respiratory illnesses caused by the diesel fuel exhaust emissions: A review of the research. In *Handbook of Biodiesel and Petrodiesel Fuels: Science, Technology, Health, and Environment. Volume 3. Petrodiesel Fuels: Science, Technology, Health, and Environment*, ed. O. Konur. Boca Raton, FL: CRC Press.

Konur, O. 2021af. Cancer caused by the diesel fuel exhaust emissions: A review of the research. In *Handbook of Biodiesel and Petrodiesel Fuels: Science, Technology, Health, and Environment. Volume 3. Petrodiesel Fuels: Science, Technology, Health, and Environment*, ed. O. Konur. Boca Raton, FL: CRC Press.

Konur, O. 2021ag. Cardiovascular and other illnesses caused by the diesel fuel exhaust emissions: A review of the research. In *Handbook of Biodiesel and Petrodiesel Fuels: Science, Technology, Health, and Environment. Volume 3. Petrodiesel Fuels: Science, Technology, Health, and Environment*, ed. O. Konur. Boca Raton, FL: CRC Press.

Konur, O. and F. L. Matthews. 1989. Effect of the properties of the constituents on the fatigue performance of composites: A review. *Composites* 20: 317–328.

Kumar, S. and J. M. Jan. 2014. Research collaboration networks of two OIC nations: Comparative study between Turkey and Malaysia in the field of 'Energy Fuels', 2009–2011. *Scientometrics* 98: 387–414.

Lapuerta, M., O. Armas, and J. Rodriguez-Fernandez. 2008. Effect of biodiesel fuels on diesel engine emissions. *Progress in Energy and Combustion Science* 34: 198–223.

Lariviere, V. and Y. Gingras. 2010. On the relationship between interdisciplinarity and scientific impact. *Journal of the American Society for Information Science and Technology* 61: 126–131.

Lariviere, V., C. Ni, Y. Gingras, B. Cronin, and C. R. Sugimoto. 2013. Bibliometrics: Global gender disparities in science. *Nature News* 504: 211–213.

Leung, D. Y. C., X. Wu, and M. K. H. Leung. 2010. A review on biodiesel production using catalyzed transesterification. *Applied Energy* 87: 1083–1095.

Leydesdorff, L. and C. Wagner. 2009. Is the United States losing ground in science? A global perspective on the world science system. *Scientometrics* 78: 23–36.

Lotero, E., Y. J. Liu, D. E. Lopez, et al. 2005. Synthesis of biodiesel via acid catalysis. *Industrial & Engineering Chemistry Research* 44: 5353–5363.

Ma, F. R. and M. A. Hanna. 1999. Biodiesel production: A review. *Bioresource Technology* 70: 1–15.

Marchetti, J. M., V. U. Miguel, and A. F. Errazu. 2007. Possible methods for biodiesel production. *Renewable & Sustainable Energy Reviews* 11: 1300–1311.

Mata, T. M., A. A. Martins, and N. S. Caetano. 2010. Microalgae for biodiesel production and other applications: A review. *Renewable & Sustainable Energy Reviews* 14: 217–232.

McCormick, RL, M. S. Graboski, T. L. Alleman, and A. M. Herring. 2001. Impact of biodiesel source material and chemical structure on emissions of criteria pollutants from a heavy-duty engine. *Environmental Science & Technology* 35: 1742–1747.

Meher, L. C., D. V. Sagar, and S. N. Naik. 2006. Technical aspects of biodiesel production by transesterification: A review. *Renewable & Sustainable Energy Reviews* 10: 248–268.

Mohan, D., C. U. Pittman, and P. H. Steele. 2006. Pyrolysis of wood/biomass for bio-oil: A critical review. *Energy & Fuels* 20: 848–889.

Monyem, A. and J. H. Van Gerpen. 2001. The effect of biodiesel oxidation on engine performance and emissions. *Biomass & Bioenergy* 20: 317–325.

Morillo, F., M. Bordons, and I. Gomez. 2001. An approach to interdisciplinarity through bibliometric indicators. *Scientometrics* 51: 203–222.

Negishi, M., Y. Sun, and K. Shigi. 2004. Citation database for Japanese papers: A new bibliometric tool for Japanese academic society. *Scientometrics* 60: 333–351.

North, D. C. 1991a. *Institutions, Institutional Change and Economic Performance*. Cambridge, Mass.: Cambridge University Press.

North, D.C. 1991b. Institutions. *Journal of Economic Perspectives* 5: 97–112.

Okubo, Y., J. C. Dore, T. Ojasoo, and J. F. Miquel. 1998. A multivariate analysis of publication trends in the 1980s with special reference to South-East Asia. *Scientometrics* 41: 273.

Perron, P. 1989. The great crash, the oil price shock, and the unit root hypothesis. *Econometrica: Journal of the Econometric Society* 57: 1361–1401.

Prathap, G. 2017. A three-dimensional bibliometric evaluation of recent research in India. *Scientometrics* 110: 1085–1097.

Ramos, M. J., C. M. Fernandez, A. Casas, L. Rodriguez, and A. Perez. 2009. Influence of fatty acid composition of raw materials on biodiesel properties. *Bioresource Technology* 100: 261–268.

Rinia, E. J. 2000. Scientometric studies and their role in research policy of two research councils in the Netherlands. *Scientometrics* 47: 363–378.

Rogers, D. and A. J. Hopfinger. 1994. Application of genetic function approximation to quantitative structure-activity relationships and quantitative structure-property relationships. *Journal of Chemical Information and Computer Sciences* 34: 854–866.

Scherf, U. and E. J. List. 2002. Semiconducting polyfluorenes-towards reliable structure–property relationships. *Advanced Materials* 14: 477–487.

Sharma, Y. C., B. Singh, and S. N. Upadhyay. 2008. Advancements in development and characterization of biodiesel: A review. *Fuel* 87: 2355–2373.

Van Gerpen, J. 2005. Biodiesel processing and production. *Fuel Processing Technology* 86: 1097–1107.

Vicente, G., M. Martinez, and J. Aracil. 2004. Integrated biodiesel production: a comparison of different homogeneous catalysts systems. *Bioresource Technology* 92: 297–305.

Wang, X., D. Liu, K. Ding, and X. Wang. 2012. Science funding and research output: a study on 10 countries. *Scientometrics* 91: 591–599.

Xie, Y. and K. A. Shauman. 1998. Sex differences in research productivity: New evidence about an old puzzle. *American Sociological Review* 63: 847–870.

Xue, J. L., T. E. Grift, and A. C. Hansen. 2011. Effect of biodiesel on engine performances and emissions. *Renewable & Sustainable Energy Reviews* 15: 1098–1116.

Youtie, J, P. Shapira, and A. L. Porter. 2008. Nanotechnology publications and citations by leading countries and blocs. *Journal of Nanoparticle Research* 10: 981–986.

Zhang, Y., M. A. Dube, D. D. McLean, and M. Kates. 2003. Biodiesel production from waste cooking oil: 1. Process design and technological assessment. *Bioresource Technology* 89: 1–16.

Zhou, P. and L. Leydesdorff. 2006. The emergence of China as a leading nation in science. *Research Policy* 35: 83–104.

12 Biomass-based Catalyst-Assisted Biodiesel Production

Celine Ming Hui Goh
Yie Hua Tan
Jibrail Kansedo
N. M. Mubarak
Mohd Lokman Ibrahim

CONTENTS

12.1 INTRODUCTION

Although the concerns may seem to outweigh the benefits by a small margin today, it is also vital to be reminded that the main concern of energy generation is its dependency upon non-renewable sources that are depleting at an exponential rate. While the disadvantages apparent today will undoubtedly be solved by greater minds and the further advancement of technology in the future, the depletion of fossil fuels is still inevitable at the present-day consumption rate. As a result of the unceasing decline of the fossil fuel inventory and the growing demand for it due to industrialization, biofuel has become a promising alternative to replace conventional petrodiesel (Ma and Hanna, 1999). Biodiesel has been one of the most promising biofuels in the world for the past two decades (Mittelbach, 2012). Compared to petrodiesel, biodiesel is non-toxic, biodegradable (Ajala et al., 2015), and clean burning as it does not contain any aromatic group (McCormick et al., 2001). Also, it has a lower sulfur content, 5 ppm lower than ultra-low sulfur diesel (ULSD) (Janaun and Ellis, 2011). Moreover, air pollutants such as carbon dioxide, carbon monoxide, sulfur oxides, particulate matter, and unburned hydrocarbons emitted from biodiesel combustion are less than for petrodiesel (Gryglewicz, 1999).

According to Konwar et al. (2014), the refined oil cost as the raw oil contributes to nearly 80% of the overall biodiesel production cost. Therefore, an acid homogeneous catalyst such as sulfuric acid (H_2SO_4) or phosphoric acid (H_3PO_4) is employed as a pretreatment in biodiesel production to reduce the 'free fatty acids' (FFA) content in waste oil. The use of a homogeneous catalyst is still plagued with issues such as its unrecoverability and the huge generation of wastewater, which requires a process to separate the homogeneous phase product during the purification stage and which incurs a high cost (Tang et al., 2018). Therefore, the need for research on alternative production pathways.

There are two other types of catalyst, namely the heterogeneous and the enzymatic. The enzymatic catalyst, i.e. lipase, was extensively researched to produce biodiesel via spontaneous reactions in hydrolysis and transesterification in mild conditions. However, the high synthesis cost of the catalyst itself and the long incubation time discourages the potential of the enzymatic catalyst to be used on an industrial scale (Narowska et al., 2019). On the other hand, a heterogeneous catalyst provides comparable catalytic performance to a homogeneous catalyst, while being easily separable and reusable. It has been shown to be a promising alternative for the current industrial scale of a base homogeneous catalyst.

The environmentally friendly heterogeneous catalyst can be derived from various biomass resources, for example mud/rocks, animal bones, seashells, and biochar. These can be found abundantly in any countries based on their natural and waste resources. The massive daily waste generation from developing countries is a pertinent concern as it consumes a large landfill area during disposal. In order to protect the environment, wide-ranging waste can be reused for a heterogeneous biomass-based catalyst for biodiesel production. The proposed strategies are not only important for achieving low catalyst cost, they are important for reducing waste intensiveness. Implementation of the biomass-based catalyst would drive production of a waste catalyst for use within the biodiesel industry. In this chapter, a brief summary of a biomass-based catalyst for biodiesel production is described. In addition, various experimental data for a transesterification process is summarized, and the future challenges for a biomass-based catalyst for biodiesel production is discussed.

12.2 SYNTHESIS OF A BIOMASS-BASED CATALYST

A biomass-based catalyst is typically synthesized by the thermal conversion of biomass using the following technologies: pyrolysis, 'hydrothermal carbonization' (HTC), and template-directed production. High temperature treatment, which is also known as calcination or direct carbonization, is the direct application of heat at elevated temperatures on biomass without the addition of other compounds (Lee et al., 2019). On the other hand, HTC offers an eco-friendly technique by heating a mixture of biomass and water at high temperatures to form a biomass-based catalyst. Finally, a biomass-based catalyst can also be produced from the method of template-directed synthesis by first impregnating the hard template with the biomass before being subjected to calcination. Figure 12.1 provides a clear flow chart of the various methods mentioned to produce a biomass-based catalyst using biomass waste as the precursor. As observed, an extra procedure of treatment with a specific activating agent is needed to produce an activated biomass-based catalyst before charring begins.

12.2.1 HYDROTHERMAL CARBONIZATION

HTC is the thermochemical treatment of biomass to be converted into a biomass-based catalyst. It is typically applied on biomass in the presence of water and involves steps of hydrolysis, condensation, decarboxylation, aromatization, and dehydration

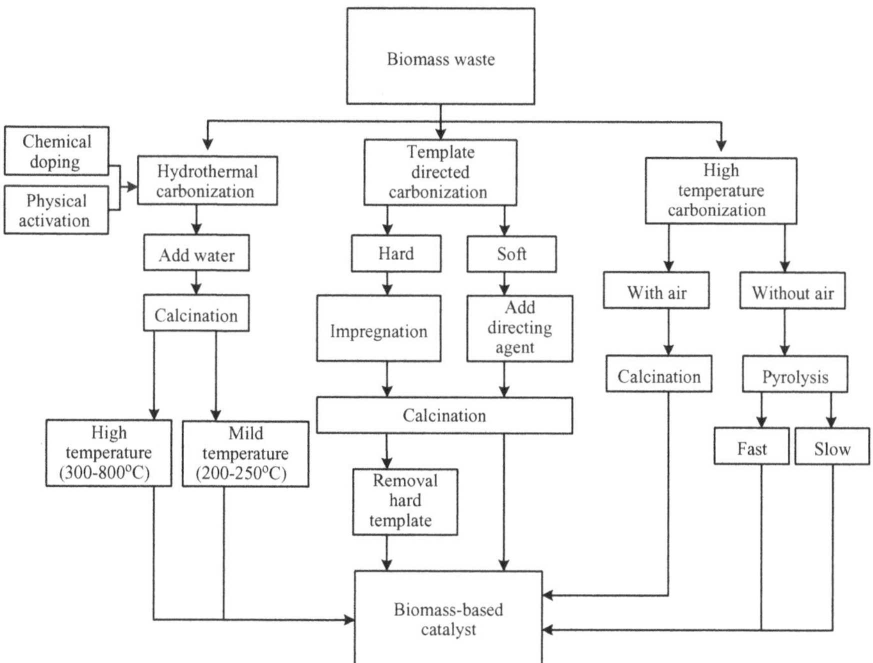

FIGURE 12.1 The production pathways of a biomass-based catalyst.

reactions (Reza et al., 2014). The reason for the compulsory presence of water in this method is for the hydrolysis of glycosidic groups that are found in abundance in cellulose and hemicellulose, which are the main compositions of plant-based biomass. Thus, water is imperative for a higher yield of porous hydrochar that comprises approximately 45–75% carbon. Apart from that, the process is usually conducted under high pressure conditions to inhibit water molecules from escaping when subjected to a suppressive force throughout the reaction (Liu et al., 2013). Generally, HTC liberates solid biochar, gaseous compounds, and liquid biooil as its main products.

12.2.2 Template-Directed Carbonization

A biomass-based catalyst is synthesized from biomass waste via the template-directed procedure with the desired pore diameter according to the template arrangement that is applied. The method is divided into two categories: soft and hard template-directed production. The hard template technique requires the utilization of artificial templates made of silica, such as 'halloysite nanotubes' (HNT), throughout production (Janaun and Ellis, 2011). This is carried out by the integration of a solution of biomass into a hard template prior to calcination of the mixture. Chemical solvents like hydrogen fluoride (HF) are then used for the washing of the product and removal of the template from the biomass-based catalyst produced from carbonization. On the other hand, in the process of soft template-directed biomass-based catalyst formation, the templates as well as the operating conditions that are required are highly specific (Xie et al., 2016). This factor is the main inhibitor of the large-scale production of this process. Organic compounds, surfactants, and block copolymers are prime examples of typical soft templates used. The application of a soft template with biomass as the material to be converted proves challenging, as the scarce interaction between the template and biomass, caused by the operating temperature for the soft template being significantly lower than that of a hard template, renders it inefficient (Tang et al., 2018).

12.2.3 High Temperature Carbonization

Due to its simplicity, direct calcination and pyrolysis are the most common synthesis pathways adopted for the formation of a biomass-based catalyst. Direct calcination refers to heating to high temperatures in air or oxygen, whereas pyrolysis showcases the direct application of high heat in the absence of oxygen without the need for chemical or physical enhancements on the precursor (Shan et al., 2018). Calcination is performed by heating the biomass to a higher temperature to remove volatile substances, render the biomass friable, and oxidize some components in the biomass to turn it into a biomass-based catalyst. Prior to pyrolysis, carbonization is first done on lignocellulosic biomass for the removal of all volatile matter via thermal disintegration at high heat to obtain porous carbon. Following this step, pyrolysis, which is operated under inert conditions, functions to strip off the carbon material of non-carbon components that include nitrogen, oxygen, hydrogen, and remaining volatile material in their gaseous form (Abdullah et al., 2017). The final product formed will be a rigid aromatic product of structures resembling that of a skeleton. Pyrolysis is divided into two categories, fast pyrolysis and slow pyrolysis. Fast pyrolysis is conducted in an environment with a high heating rate and short duration, while slow pyrolysis is carried out with a low heating rate over longer time (Chua et al., 2020).

12.3 ACTIVATION OF A BIOMASS-BASED CATALYST

The production of biomass-based catalysts by the above-mentioned synthesis pathways is insufficient. Harmas et al. (2016) reported that the biochar materials produced from those methods possessed specific surface areas as low as 100 m^2/g. Therefore, activation methods are employed to overcome this issue. Activation methods induce the enhancement of a biomass-based catalyst in terms of pore diameter, pore volume, specific surface area, and porosity (Vakros, 2018) as they are prone to surface functionalization when their precursors are reacted with acidic or basic solutions (Narowska et al., 2019). It is also reported that biomass-based catalysts present excellent stability under heat, radiation, acidic, as well as alkaline conditions, and are mechanically durable, which implies that it also offers high reusability. Moreover, they can provide high catalytic activity due to their surface properties and the presence of metal oxides and an acidic or alkaline functional group on the surface. For these reasons, they make good candidates for a low-cost biomass-derived catalyst to produce biodiesel.

Chemical activation is favored rather than its physical counterpart as it results in a final product, which reflects a relatively higher yield of activated carbon, an increased surface area, enhanced porosity, and reduced time of reaction. In a chemical activation process, impregnation of the biomass waste will first be done with a chemical activating agent of an acidic, basic, or oxidative nature (Jain et al., 2015). This step is then succeeded by a calcination process at high temperatures to yield the final activated carbon catalyst, while for physical activation it is completed via thermal activation using a furnace. NaOH and 'potassium hydroxide' (KOH) are the common basic alkalines, whereas H_2SO_4 and H_3PO_4 are the common acids used for the chemical activation of biomass (Tang et al., 2018).

The process is normally initiated by the pre-calcination of biomass to form biochar. Next, the formed biomass is mixed with the activating agents in a KOH: carbon ratio of a range between 2 and 5 for impregnation and is subjected to an extra dehydration step before further carbonization at temperatures in the range of 400–900°C to yield the final product (Wang and Kaskel, 2012). This process was reported to have generated porous carbons of a high specific surface area exceeding 2000 m^2/g and pore volumes that consist of micropores and mesopores.

However, it is worth noting that the elemental composition and microstructure of the activated carbon produced are greatly dependent upon the carbon used as a precursor as well as the parameters adopted during the activation process. Vakros (2018) also concluded that the contents of inorganic elements that remain in biomass as well as the temperature of calcination are of the utmost importance in terms of basicity. The inorganic elements present in the biomass allow its conversion to metal oxides when undergoing carbonization; and these metal oxides formed are the main

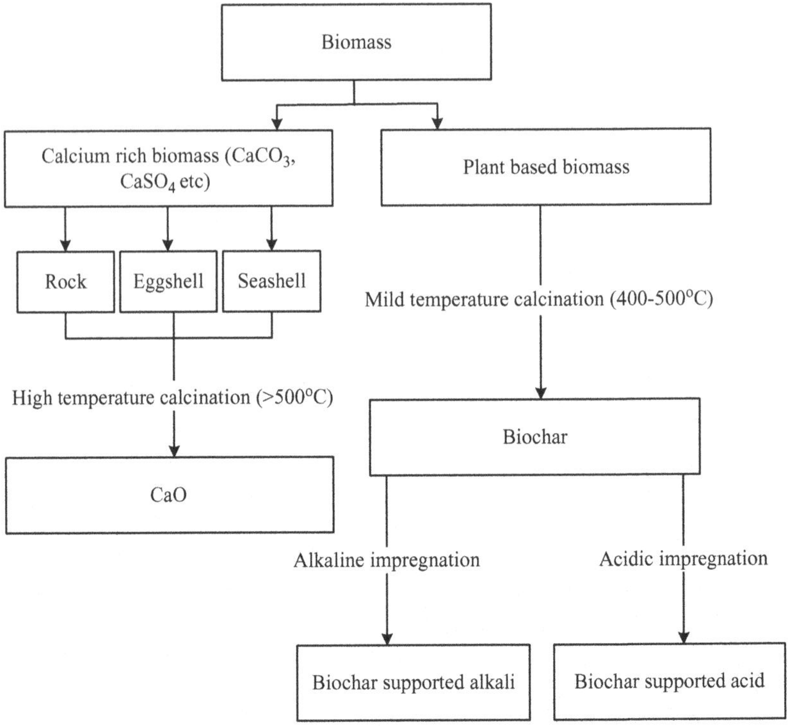

FIGURE 12.2 Calcination pathways for various biomass-derived base catalysts.

contributors to the basicity of the catalyst. Apart from that, higher temperatures of carbonization will result in increased ash content of the catalyst and thus increase its basicity.

The activation method for basic biomass-derived catalysts is clearly explained in the form a flow chart shown in Figure 12.2. The following sections summarize the various sources of biomass mentioned in the literature. However, they are not limited to these and there exist countless more sources which have recently surfaced.

12.4 CALCIUM RICH BIOMASS

Organic waste with a high composition of calcium carbonate ($CaCO_3$) can be utilized for the synthesis of green catalysts. Seashells and animal bones are prime examples of this sort (Verziu et al., 2011). $CaCO_3$ will undergo conversion to become calcium oxide (CaO) when subjected under combustion to elevated temperatures. This component is the base catalyst that provides high activity for biodiesel synthesis. CaO is typically acquired from $CaCO_3$ found in limestone, but it comes from ecologically harmful and unsustainable limestone mining, thus rendering it economically unfeasible. Examples of biomass sources that are high in $CaCO_3$ include waste animal bones, waste eggshells, waste coral fragments, waste animal shells, and fish scales. These are made up of between 96 and 98% $CaCO_3$ with water, organic substances, calcium phosphate $Ca_3(PO_4)_2$, magnesium carbonate ($MgCO_3$), and strontium carbonate ($SrCO_3$), which are present in small traces (Abdullah et al., 2017). Evidently, the high $CaCO_3$ content of these biomass wastes makes them promising renewable sources to produce CaO-based catalysts. To top it off, the production pathway is cheap, short, and eco-friendly.

12.4.1 Mud or Rocks

Limestone is also a non-renewable source of energy which contradicts the initial goal of achieving a sustainable solution for the depletion of non-renewable sources (Correia et al., 2014). Thus, biomass sources which are enriched with calcium can serve as a sustainable substitute for heterogeneous biomass-based catalysts. Limestone is predominantly made up of $CaCO_3$; cement, mud, and rock are the sources of $CaSO_4$. The existing internal mineral compounds in them, such as quartz, portlandite, albite, and calcite, are reported to further enhance the overall heterogeneous catalytic activity (Gimbun et al., 2013). Each year, around 180 million tonnes of shattered concrete are disposed of from construction sites. Hence, this waste can be fully utilized to form a biomass-based catalyst as an alternative in biodiesel production (Ling et al., 2019). Gimbun et al. (2013) have reported that the highest conversion of rubber seed oil at 96.9% was achieved using an activated cement clinker catalyst, with 5 wt% catalyst loading, a 4:1 methanol: oil molar ratio at 65°C, and 4 hours of reaction time. The biodiesel obtained met the ASTM D6751 specification.

12.4.2 Animal Bones

About 60% of slaughtered farm animals is made of bones, tendons, blood, plasma, and offal (Ghanei et al., 2016). Most of this has to be safely disposed or transformed in a proper way. One of the options is to recycle animal bones as a calcium resource in synthesizing biomass-based catalysts. The environmental solution is to transform the animal bones into a biomass-based catalyst, which contains hydroxyapatite that can reduce the negative environmental effect of the waste. There are various types of animal bones, for example fish, chicken, pig, bovine, turkey, and sheep bones. It is economical to transform these bones to maximize profit and turn it into a valuable product such as a heterogeneous catalyst in biodiesel production. Smith et al. (2013) reported the use of bovine bones as a precursor for solid catalyst synthesis for biodiesel production. The major composition of bones is hydroxyapatite and calcium carbonate. These components are converted into CaO following a calcination step at temperatures of 650–950°C for 6 hours. The biodiesel yield using the resulting catalyst was reported to be 97% at a methanol:oil ratio of 6: 1 and a catalyst loading of 8% at 65°C within 3 hours of reaction.

12.4.3 Seashells

Seashells, which are disposed as unwanted waste, contain rather high amounts of calcium. Seashells consist of over 95% $CaCO_3$, which is used in many agricultural and engineering applications (Rezaei et al., 2013). Thus, crushed seashells are qualified as a calcium rich resource which can be transformed into a value-added product, as a biomass-based catalyst for biodiesel production. Fragments of seashells can be recovered and subjected to elevated heat treatment to convert the $CaCO_3$ rich compound into a CaO-based catalyst, which can be employed in the biodiesel industry. Reusing seashell waste can improve the sustainability of the aquaculture industry and provide secondary economic benefits to shellfish growers and processors. Boey et al. (2011) also conducted a study on biodiesel production using waste cockle shell from the species *Anadara granosa*. The catalyst reported had a 71% composition of Ca following calcination at 900°C for 2 hours. A biodiesel yield of 97.48% was obtained from the transesterification reaction of palm olein oil with a methanol:oil ratio of 0.54:1 and a catalyst loading of 4.9 wt% at a reaction time of 3 hours. It was also reported that the catalyst could be reused for at least three cycles with a biodiesel yield of 96.5%.

12.4.4 Eggshells

Due to the high consumption of eggs for daily nutritional requirement, there are about 8 million tonnes of eggshells discarded worldwide each year. The disposal of a massive number of eggshells can be challenging, in terms of landfill size and management cost. Thus, this has raised an environmental concern and grabbed the attention of researchers to cater for this issue by reusing this waste as a valuable product. Eggshell is made up of 85–95% $CaCO_3$ and 5–15% of other compounds, such as calcium phosphate ($Ca_3(PO_4)_2$), magnesium carbonate ($MgCO_3$), sodium (Na), potassium (K), zinc (Zn), manganese (Mn), iron (Fe), copper (Cu), and proteins (Tan et al., 2015). This has shown that eggshell is a promising valuable biomaterial that can be transformed into a CaO catalyst. Correia et al. (2014) conducted a study on the transesterification reaction of sunflower oil, comparing the catalytic properties between waste crab shells and eggshells. Both waste shells were converted into CaO by calcination at 900°C for 2 hours. The biodiesel yield obtained in the presence of the crab shell was 83.1% at optimal conditions of a 6:1 methanol:oil ratio and a 3 wt% catalyst loading at 60°C for a reaction time of 4 hours. On the other hand, the yield of biodiesel was 97.75% when reaction took place in the presence of eggshells. The optimum conditions of the reaction were a methanol:oil ratio of 9:1, a catalyst loading of 3 wt%, a reaction time of 3 hours, and a temperature of 60°C.

12.4.5 Plant-based Biomass

Alternatively, a biochar-based catalyst can also be derived from biomass waste containing high compositions of hydrogen and carbon. Biomass ashes have also been the topic of investigation in various studies of catalysts for biodiesel synthesis due to the high content of basic elements found in their combusted counterparts. An organic compound is naturally enriched with carbon (C), oxygen (O), and metal salts which encompass calcium (Ca), sodium (Na), and magnesium (Mg) (Luque et al., 2012). During the combustion process, which occurs at elevated temperatures, the amount of C and O initially present in the compound will instantaneously decrease. As a result, the major active component that remains in the ash will include a variety of alkali metal oxides such as magnesium oxide (MgO), potassium oxide (K_2O), and CaO (Abdullah et al., 2017). Consequently, the catalytic activity of the ash produced is significantly increased due to its high content of basic oxides (Chouhan and Sarma, 2013). Additionally, it is also worth noting that the effects of biomass-based potash on the synthesis of biodiesel were investigated, which confirmed that it also provides high catalytic activity as a potassium-bearing basic ash catalyst (Ofori-Boateng and Lee, 2013).

Previously, tars and alkali ash were used as catalysts for the transesterification reaction by Luque et al. (2012). The catalyst was synthesized by calcinating the by-products obtained from a syngas production system at temperatures between 500 and 800°C for 4 hours. Due to the mass loss from the removal of volatile components following calcination, the catalyst produced is mainly composed of metal oxides, with 36 wt% of the composition made up of Ca. This was used to catalyze biodiesel production from sunflower oil to obtain a yield of 75% in a 12-hour reaction. Vadery et al. (2014) also reported the use of ashes produced from the calcination of coconut husks at 250–500°C for an hour.

The resulting catalyst mainly comprised K species which include K_2SO_4, KCl, and $K_2Si_2O_5$. A biodiesel yield of 97% was recorded from the transesterification reaction using Jatropha oil in the presence of the catalyst within 30 minutes of the reaction. The highest yield was determined to be 99.8% under an optimal methanol:oil ratio recorded at 12:1 and a catalyst loading of 7 wt% at 45°C for 45 minutes. In a study conducted by Deka and Basumatary (2011), a solid base catalyst was derived from the ash of a banana (*M. balbisiana Colla*) trunk for the conversion of *T. peruvinia* seed oil into biodiesel. Pieces of the trunk were burned and their ashes were analyzed to show they contained mainly CO_3, Cl, K, and Na with various other metals in small amounts. The biodiesel yield in the presence of this catalyst was 96% under the conditions of a 20:1 methanol:oil ratio and a catalyst loading of 20 wt% at 32°C for 3 hours.

Furthermore, an activation method may be applied on biochar, which is typically produced from pyrolyzed biomass, or directly onto lignocellulosic biomass to generate activated carbon of an amorphous nature and high porosity (Konwar et al., 2014). Biomasses used for this category of catalyst synthesis are usually those with a large amount of carbon content, such as coconut shells, wood, oil palm biomass, and rice husk. Due to the formation of pores in large numbers, activated carbons are able to cater to the requirement of a good solid base catalyst by possessing a relatively higher specific surface area. This allows for attachment and adsorption of active metals onto the catalyst for increased reactivity (Zabeti et al., 2009). Harmas et al. (2016) reported that activated carbon displays an expansion of its specific surface area by up to 50 times more than that of non-activated carbon.

This form of catalyst can be synthesized from the combination of activation and carbonization by a single- or double-step procedure based on the method of selection. On the other hand, the activation process can be done in two ways: by chemical or physical activation (Tang et al., 2018). Examples of activation agents that may be applied to synthesize activated carbon include steam, air, CO_2, KOH, sodium hydroxide (NaOH), and phosphoric acid (H_3PO_4). The latter three are more commonly utilized than the former three as they result in a higher percentage yield, a greater porosity of catalyst, a mild temperature for activation, and a short activation period.

Chakraborty et al. (2010) synthesized a CaO catalyst using fly ash, which is the by-product of coal combustion, as the supporting material. The catalyst was first loaded with 30 wt% CaO via a wet impregnation method before being subjected to calcination at a temperature of 1000°C for 2 hours as the activation step. Biodiesel was then produced from soybean oil as the feedstock with the aid of the activated fly-ash-based catalyst. A 'fatty acid methyl ester' (FAME) yield of 96.9% was reported using reaction conditions of a 7:1 methanol:oil ratio and a 1 wt% catalyst loading at 70°C for 5 hours. In another study, Ofori-Boateng and Lee (2013) utilized waste cocoa pod husks to successfully synthesize an activated carbon-supported catalyst.

Potash (K_2CO_3) was produced from the calcination of the husk at 650°C for 4 hours. This component was then doped with MgO before it was used to catalyze the transesterification reaction of soybean oil at a catalyst loading of 7 wt% and a methanol:oil ratio of 6:1 at a temperature of 40°C for 30 minutes. The final biodiesel yield was 98.7%. Similarly, Riadi et al. (2014) produced a solid base catalyst using empty fruit bunches as the vitamin A, C support. The precursor was activated with 1.5 wt% of KOH by a method of wet impregnation. The resulting catalyst was able to yield high amounts of both long-chain and short-chain methyl esters at 655.29 and 85.72 mg/L, respectively, with a catalyst loading of 17.3 wt%. To compare the types of biomass typically utilized in the synthesis of biodiesel, a summary of the various biomass-sourced base catalysts corresponding to their optimum conditions for catalyst preparation and transesterification reaction is shown in Table 12.1 for reference.

12.5 CHALLENGES OF BIODIESEL PRODUCTION VIA BIOMASS-BASED CATALYSTS

Condensed CaO nanograins reacted rapidly to form calcium hydroxide ($Ca(OH)_2$) when exposed to moisture or the atmosphere (Tan et al., 2017). This would be caused by the binding of a water molecule with CaO to form $Ca(OH)_2$. Thus, the obtained Ca rich biomass-based catalyst needs to be sealed tightly to avoid any contact with the atmosphere. Also, heat reactivation to the elevated temperature of the catalyst is suggested after usage to restore the Ca rich biomass-based catalytic performance in biodiesel production (Liu et al., 2018). Wang et al. (2018) have reported that their recyclable biomass-based catalyst showed a reduction in catalytic performance due to catalyst deactivation.

Clogging of active sites on the surface of a catalyst by crude glycerol, soap, and other impurities during the transesterification process would lead to catalyst deactivation. This could be solved by solvent washing, such as with n-hexane, to remove these impurities. Another possible challenge for a biomass-based catalyst would be the leaching of chemically active species, such as acidic or basic groups, from the catalyst solid framework and the slow reduction of its acidic or basicity density, thus the reduction in overall catalytic activity. Anyhow, a protective outer shell or coating can be established to improve catalytic performance substantially (Wang et al., 2018).

TABLE 12.1
Biodiesel Yield at Optimum Conditions for Transesterification Using Catalysts Derived from Various Types of Biomass

		Catalyst Synthesis Condition			Transesterification Reaction Condition				FAME	
Biomass	**Feedstock**	**CI**	**Ct (h)**	**CT (°C)**	**M:O ratio**	**t (h)**	**T (°C)**	**CL (wt%)**	**(%)**	**Reference**
Mud/rocks										
Limestone ($CaCO_3$)	Soybean oil	–	1.5	900	12:1	1	64.7	/	93.0	Kouzu et al. (2008)
Cement	Soybean oil	–	3	450	24:1	3	65	4	98.5	Wang et al. (2012)
Activated cement clinker	Rubber seed oil	–	7	700	4:1	4	65	5	96.9	Gimbun et al. (2013)
Lime mud	Refined peanut oil	–	–	900	15:1	3	64	6	94.4	Li et al. (2014)
Red mud	Soybean oil	–	5	200	24:1	3	65	4	94.0	Konwar et al. (2015)
Dolomite rock	Palm kernel oil	–	2	800	30:1	3	60	6	99.9	Ngamcharussrivichai et al. (2007)
Animal bones										
Waste fish scale of *Labeo rohita*	Soybean oil	–	2	900	6.27:1	5	70	1.01	97.7	Chakraborty et al. (2011)
Pig bone waste	Refined palm oil	K_2CO_3	4	600	9:1	1.5	65	8	>90	Chen et al. (2015)
Bovine bone waste	Soybean oil	–	6	650–950	6:1	3	65	8	97.0	Smith et al. (2013)
Sheep bone waste	Canola oil	$Ca(NO_3)_2$	8	600	12:1	5	60	5	95.2	Ghanei et al. (2016)
Turkey bone waste	Indian mustard oil	–	4	909.4	9.9:1	3	70	4.97	91.2	Chakraborty et al. (2015)
Chicken and fish bones	Used cooking oil	–	4	1000	10:1	1.54	65	1.98	89.5	Tan et al. (2019)
Seashells										
Freshwater mussel shell	Chinese tallow oil	–	3	900	12:1	1.5	70	5	90.0	Hu et al. (2011)
Mussel waste shell	Soybean	–	2	1050	24:1	8	60	12	94.1	Rezaei et al. (2013)
Crab shell	Sunflower oil	–	2	900	6:1	4	60	3	83.1	Correia et al. (2014)
Mud crab waste shell	Palm olein	–	2	900	0.5:1	3	65	2.5	98.8	Hamid et al. (2009)
Waste cockle shell of *Anadara granosa*	Palm olein oil	–	2	900	0.54:1	3	–	4.9	97.4	Boey et al. (2011)
Waste clam shell	Palm olein	–	4	900	9:1	2	65	1	98.0	Asikin-Mijan et al. (2015)
Waste capiz shell of *Amusium cristatum*	Palm oil	–	2	900	8:1	6	60	3	93.0	Suryaputra et al. (2013)

(Continued)

TABLE 12.1 (Continued)

Biomass	Feedstock	Catalyst Synthesis Condition			Transesterification Reaction Condition				FAME	Reference
		CI	Ct (h)	CT (°C)	M:O ratio	t (h)	T (°C)	CL (wt%)	(%)	
Waste oyster shell	Soybean oil	KI	4	1000	10:1	4	50	1 mmol/g	85.0	Jairam et al. (2012)
Waste oyster shell	Soybean oil	–	3	700	6:1	5	65	25	98.4	Nakatani et al. (2009)
Waste shell of *Turbonilla striatula*	Mustard oil	–	4	600–900	9:1	3	65	3	93.3	Boro et al. (2011)
Waste shell of golden apple snail	Palm olein oil	–	2–4	800	18:1	2	60	10	93.2	Viriya-Empirikul et al. (2012)
Scallop waste shell	Palm oil	–	4	1000	9:1	3	65	10	95.4	Buasri et al. (2014)
Waste shell of *Meretrix venus*	Palm olein oil	–	2–4	800	18:1	2	60	10	92.3	Viriya-Empirikul et al. (2012)
Eggshells										
Chicken eggshell	Palm oil	–	4	900	9:1	4	60	20	94.4	Buasri et al. (2013)
Duck eggshell	Palm oil	–	4	900	9:1	4	60	20	92.9	Buasri et al. (2013)
Eggshell	Sunflower oil	–	2	900	9:1	3	60	3	97.8	Correia et al. (2014)
Waste quail eggshell	Palm oil	–	2	800	12:1	2	65	1	98.0	Cho and Seo (2010)
Waste eggshell	Palm olein oil	–	2–4	800	18:1	2	60	10	94.1	Viriya-Empirikul et al. (2012)
Waste ostrich eggshell	Waste cooking oil	–	4	1000	10:1	2	65	1.5	98	Tan et al. (2017)
Biomass ashes										
Tar (alkali) ashes	Sunflower oil	–	4	600–800	–	12	–	–	75.0	Luque et al. (2012)
Coconut husk ash	Jatropha oil	–	1	250–500	12:1	0.75	45	7	99.8	Vadery et al. (2014)
Empty palm bunch ash	Waste cooking oil	KOH	–	–	5:1	2	60	17.3	–	Riadi et al. (2014)
Cocoa pod husk ash/MgO	Soybean oil	MgO	4	650	6:1	0.5	40	7	98.7	Ofori-Boateng and Lee (2013)
Fly ash/CaO-derived eggshell	Soybean oil	CaO	2	1000	7:1	5	70	1	96.9	Chakraborty et al. (2010)

CI: Chemical of impregnation. Ct: Holding time of calcination. CT: Temperature of calcination. M:O ratio: Methanol:oil molar ratio. t: Time of transesterification reaction. T: Temperature of transesterification reaction. CL: Catalyst loading. FAME: Fatty acid methyl esters.

12.6 CONCLUSION

Waste biomass is available in abundance, thus its untapped potential should be realized. The utilization of heterogeneous biomass-based catalysts, such as calcium rich and plant-based ones, derived from biomass appear to be a promising alternative to current biodiesel catalyst trends owing to its eco-friendliness and remarkable separation efficiency. A biomass-based catalyst can indirectly cut down time and energy consumption, which ultimately reduces the overall biodiesel production cost. Hence, continuous research is pivotal to address the current research gap in order to expand and commercialize the biodiesel industry to create a greener and more sustainable future. Apart from cost reduction, the exploitation of biomass for catalyst manufacture also provides a plausible answer to the biomass waste disposal issue that deteriorates with the ever-increasing rate of waste generation as a result of agricultural and other human activities.

REFERENCES

Abdullah, S. H. Y. S., H. M. H. Nur, A. Azman, et al. 2017. A review of biomass-derived heterogeneous catalyst for a sustainable biodiesel production. *Renewable and Sustainable Energy Reviews* 70:1040–1051.

Ajala, O. E., A. Folorunsho, E. O. Temitope, and M. A. Adejoke. 2015. Biodiesel: Sustainable energy replacement to petroleum-based diesel fuel: A review. *ChemBioEng Reviews* 2:145–156.

Asikin-Mijan, N., H. V. Lee, and Y. H. Taufiq-Yap. 2015. Synthesis and catalytic activity of hydration–dehydration treated clamshell derived CaO for biodiesel production. *Chemical Engineering Research and Design* 102: 368–377.

Boey, P., G. P. Maniam, S. A. Hamid, and D. M. H. Ali. 2011. Utilization of waste cockle shell (*Anadara granosa*) in biodiesel production from palm olein: Optimization using response surface methodology. *Fuel* 90: 2353–2358.

Boro, J., A. J. Thakur, and D. Deka. 2011. Solid oxide derived from waste shells of *Turbonilla striatula* as a renewable catalyst for biodiesel production. *Fuel Processing Technology* 92: 2061–2067.

Buasri, A., N. Chaiyut, Vorrada L., Phatsakon W., and S. Trongyong. 2013. Calcium oxide derived from waste shells of mussel, cockle, and scallop as the heterogeneous catalyst for biodiesel production. *Scientific World Journal* 2013: 460923.

Buasri, A., P. Worawanitchaphong, S. Trongyong, and V. Loryuenyong. 2014. Utilization of scallop waste shell for biodiesel production from palm oil: Optimization using Taguchi Method. *APCBEE Procedia* 8: 216–221.

Chakraborty, R., S. Bepari, and A. Banerjee. 2010. Transesterification of soybean oil catalyzed by fly ash and egg shell derived solid catalysts. *Chemical Engineering Journal* 165: 798–805.

Chakraborty, R., S. Bepari, and A. Banerjee. 2011. Application of calcined waste fish (*Labeo rohita*) scale as low-cost heterogeneous catalyst for biodiesel synthesis. *Bioresource Technology* 102: 3610–3618.

Chakraborty, R., Sukamal D., and S. K. Bhattacharjee. 2015. Optimization of biodiesel production from Indian mustard oil by biological tri-calcium phosphate catalyst derived from turkey bone ash. *Clean Technologies and Environmental Policy* 17: 455–463.

Chen, G. Y., R. Shan, J. F. Shi, C. Y. Liu, and B. B. Yan. 2015. Biodiesel production from palm oil using active and stable K doped hydroxyapatite catalysts. *Energy Conversion and Management* 98: 463–469.

Cho, Y. B. and G. Seo. 2010. High activity of acid-treated quail eggshell catalysts in the transesterification of palm oil with methanol. *Bioresource Technology* 101: 8515–8519.

Chouhan, A. P. S., and A. K. Sarma. 2013. Biodiesel production from *Jatropha curcas* l. oil using *Lemna perpusilla* torrey ash as heterogeneous catalyst. *Biomass and Bioenergy* 55: 386–389.

Chua, S. Y., L. Periasamy, C. M. H. Goh, et al. 2020. Biodiesel synthesis using natural solid catalyst derived from biomass waste: A review. *Journal of Industrial and Engineering Chemistry* 81: 41–60.

Correia, L. M., R. M. A. Saboya, N. S. Campelo, et al. 2014. Characterization of calcium oxide catalysts from natural sources and their application in the transesterification of sunflower oil. *Bioresource Technology* 151: 207–213.

Deka, D. C. and S. Basumatary. 2011. High quality biodiesel from yellow oleander (Thevetia peruviana) seed oil. *Biomass and Bioenergy* 35: 1797–1803.

Ghanei, R., R. K. Dermani, Y. Salehi, and M. Mohammadi. 2016. Waste animal bone as support for cao impregnation in catalytic biodiesel production from vegetable oil. *Waste and Biomass Valorization* 7: 527–532.

Gimbun, J., S. Ali, C. C. S. C. Kanwal, et al. 2013. Biodiesel production from rubber seed oil using activated cement. *Procedia Engineering* 53: 13–19.

Gryglewicz, S. 1999. Rapeseed oil methyl esters preparation using heterogeneous catalysts. *Bioresource Technology* 70: 249–253.

Hamid, S. A., P. Boey, and G. P. Maniam. 2009. Biodiesel production via transesterification of palm olein using waste mud crab (*Scylla serrata*) shell as a heterogeneous catalyst. *Bioresource Technology* 100: 6362–6368.

Harmas, M., T. Thomberg, H. Kurig, et al. 2016. Microporous–mesoporous carbons for energy storage synthesized by activation of carbonaceous material by zinc chloride, potassium hydroxide or mixture of them. *Journal of Power Sources* 326: 624–634.

Hu, S. Y., Y. Wang, and H. Y. Han. 2011. Utilization of waste freshwater mussel shell as an economic catalyst for biodiesel production. *Biomass and Bioenergy* 35: 3627–3635.

Jain, A., R. Balasubramanian, and M.P. Srinivasan. 2015. Production of high surface area mesoporous activated carbons from waste biomass using hydrogen peroxide-mediated hydrothermal treatment for adsorption applications. *Chemical Engineering Journal* 273: 622–629.

Jairam, S., P. Kolar, R. Sharma-Shivappa, J. A. Osborne, and J. P. Davis. 2012. KI-impregnated oyster shell as a solid catalyst for soybean oil transesterification. *Bioresource Technology* 104: 329–335.

Janaun, J. and N. Ellis. 2011. Role of silica template in the preparation of sulfonated mesoporous carbon catalysts. *Applied Catalysis A: General* 394: 25–31.

Konwar, L. J., J. Boro, and D. Deka. 2014. Review on latest developments in biodiesel production using carbon-based catalysts. *Renewable and Sustainable Energy Reviews* 29: 546–564.

Konwar, L. J., P. Maki-Arvela, E. Salminen, et al. 2015. Towards carbon efficient biorefining: Multifunctional mesoporous solid acids obtained from biodiesel production wastes for biomass conversion. *Applied Catalysis B: Environmental* 176–177: 20–35.

Kouzu, M., T. Kasuno, M. Tajika, et al. 2008. Calcium oxide as a solid base catalyst for transesterification of soybean oil and its application to biodiesel production. *Fuel* 87: 2798–2806.

Lee, H. W., H. Lee, Y. Kim, R. Park, and Y. Park. 2019. Recent application of biochar on the catalytic biorefinery and environmental processes. *Chinese Chemical Letters* 30: 2147–2150.

Li, H., S. L. Niu, C. M. Lu, M. Q. Liu, and M. Huo. 2014. Use of lime mud from paper mill as a heterogeneous catalyst for transesterification. *Science China Technological Sciences* 57: 438–444.

Ling, J. S. J., Y. H. Tan, N. M. Mubarak, et al. 2019. A review of heterogeneous calcium oxide based catalyst from waste for biodiesel synthesis. *SN Applied Sciences* 1: 810.

Liu, Z., A. Quek, S. K. Hoekman, and R. Balasubramanian. 2013. Production of solid biochar fuel from waste biomass by hydrothermal carbonization. *Fuel* 103: 943–949.

Liu, K., R. Wang, and Yu, M. 2018. An efficient, recoverable solid base catalyst of magnetic bamboo charcoal: Preparation, characterization, and performance in biodiesel production. *Renewable Energy*, 127: 531–538.

Luque, R., A. Pineda, J. C. Colmenares, et al. 2012. Carbonaceous residues from biomass gasification as catalysts for biodiesel production. *Journal of Natural Gas Chemistry* 21: 246–250.

Ma, F. R. and M. A. Hanna. 1999. Biodiesel production: A review. *Bioresource Technology* 70:1–15.

McCormick, R. L., M. S. Graboski, T. L. Alleman, A. M. Herring, and K. S. Tyson. 2001. Impact of biodiesel source material and chemical structure on emissions of criteria pollutants from a heavy-duty Engine. *Environmental Science & Technology* 35: 1742–1747.

Mittelbach, M. 2012. Advances in biodiesel catalysts and processing technologies. In *Advances in Biodiesel Production: Processes and Technologies*, ed. R. Luque and J. A. Melero, 133–153. London: Woodhead Publishing.

Nakatani, N., H. Takamori, K. Takeda, and H. Sakugawa. 2009. Transesterification of soybean oil using combusted oyster shell waste as a catalyst. *Bioresource Technology* 100:1510–1513.

Narowska, B., M. Kułażynski, M. Łukaszewicz, and E. Burchacka. 2019. Use of activated carbons as catalyst supports for biodiesel production. *Renewable Energy* 135: 176–185.

Ngamcharussrivichai, C., W. Wiwatnimit, and S. Wangnoi. 2007. Modified dolomites as catalysts for palm kernel oil transesterification. *Journal of Molecular Catalysis A: Chemical* 276: 24–33.

Ofori-Boateng, C. and K. T. Lee. 2013. The potential of using cocoa pod husks as green solid base catalysts for the transesterification of soybean oil into biodiesel: Effects of biodiesel on engine performance. *Chemical Engineering Journal* 220: 395–401.

Reza, M. T., M. H. Uddin, J. G. Lynam, S. K. Hoekman, and C. J. Coronella. 2014. Hydrothermal carbonization of loblolly pine: Reaction chemistry and water balance. *Biomass Conversion and Biorefinery* 4: 311–321.

Rezaei, R., M. Mohadesi, and G. R. Moradi. 2013. Optimization of biodiesel production using waste mussel shell catalyst. *Fuel* 109: 534–541.

Riadi, L., E. Purwanto, H. Kurniawan, and R. Oktaviana. 2014. Effect of bio-based catalyst in biodiesel synthesis. *Procedia Chemistry* 9:172–181.

Shan, R., L. Lu, Y. Shi, H. Yuan, and J. Shi. 2018. Catalysts from renewable resources for biodiesel production. *Energy Conversion and Management* 178: 277–289.

Smith, S. M., C. Oopathum, V. Weeramongkhonlert, et al. 2013. Transesterification of soybean oil using bovine bone waste as new catalyst. *Bioresource Technology* 143: 686–690.

Suryaputra, W., I. Winata, N. Indraswati, and S. Ismadji. 2013. Waste capiz (Amusium cristatum) shell as a new heterogeneous catalyst for biodiesel production. *Renewable Energy* 50: 795–799.

Tan, Y. H., M. O. Abdullah, and C. Nolasco-Hipolito. 2015. The potential of waste cooking oil-based biodiesel using heterogeneous catalyst derived from various calcined eggshells coupled with an emulsification technique: A review on the emission reduction and engine performance. *Renewable and Sustainable Energy Reviews* 47: 589–603.

Tan, Y. H., M. O. Abdullah, C. Nolasco-Hipolito, and N. S. A. Zauzi. 2017. Application of RSM and Taguchi methods for optimizing the transesterification of waste cooking oil catalyzed by solid ostrich and chicken-eggshell derived CaO. *Renewable Energy* 114: 437–447.

Tan, Y. H., M. O. Abdullah, J. Kansedo, et al. 2019. Biodiesel production from used cooking oil using green solid catalyst derived from calcined fusion waste chicken and fish bones. *Renewable Energy* 139: 696–706.

Tang, Z. E., S. Lim, Y. L. Pang, H. C. Ong, and K. T. Lee. 2018. Synthesis of biomass as heterogeneous catalyst for application in biodiesel production: State of the art and fundamental review. *Renewable and Sustainable Energy Reviews* 92: 235–253.

Vadery, V., B. N. Narayanan, R. M. Ramakrishnan, et al. 2014 Room temperature production of jatropha biodiesel over coconut husk ash. *Energy* 70: 588–594.

Vakros, J. 2018. Biochars and their use as transesterification catalysts for biodiesel production: A short review. *Catalysts* 8: 562.

Verziu, M., S. M. Coman, R. Richards, and V. I. Parvulescu. 2011. Transesterification of vegetable oils over CaO catalysts. *Catalysis Today* 167: 64–70.

Viriya-Empikul, N., P. Krasae, W. Nualpaeng, B. Yoosuk, and K. Faungnawakij. 2012. Biodiesel production over Ca-based solid catalysts derived from industrial wastes. *Fuel* 92: 239–244.

Wang, J. and S. Kaskel. 2012. KOH activation of carbon-based materials for energy storage. *Journal of Materials Chemistry* 22: 23710–23725.

Wang, J. X., K. T. Chen, B. Z. Wen, B. Y. H. Liao, and C. C. Chiing. 2012. Transesterification of soybean oil to biodiesel using cement as a solid base catalyst. *Journal of the Taiwan Institute of Chemical Engineers* 43: 215–219.

Wang, A., Li, H., Pan, H. Zhang, et al. 2018. Efficient and green production of biodiesel catalyzed by recyclable biomass-derived magnetic acids. *Fuel Processing Technology* 181: 259–267.

Xie, A., J. Dai, X. Chen, et al. 2016. Ultrahigh adsorption of typical antibiotics onto novel hierarchical porous carbons derived from renewable lignin via halloysite nanotubes-template and in-situ activation. *Chemical Engineering Journal* 304: 609–620.

Zabeti, M., W. M. A. W. Daud, and M. K. Aroua. 2009. Activity of solid catalysts for biodiesel production: A review. *Fuel Processing Technology* 90: 770–777.

13 Enzymatic Biodiesel Production

Challenges and Future Perspectives

Erika C. G. Aguieiras
Eliane P. Cipolatti
Martina C. C. Pinto
Jaqueline G. Duarte
Evelin Andrade Manoel
Denise M. G. Freire

CONTENTS

13.1 INTRODUCTION

The development of more sustainable and environment-friendly processes, based on principles of a circular economy, is a current need for the most different productive sectors, including the biofuels sector. Over recent years, the production of biodiesel through biotechnological routes has aroused enormous interest from academia and industry (Figure 13.1). Based on the available data, it has been observed that work regarding biodiesel and enzymes has exhibited a similar rising tendency. In this scenario, the use of lipases as biocatalysts to carry out such biotransformation is an environmentally attractive and, often, economically viable alternative.

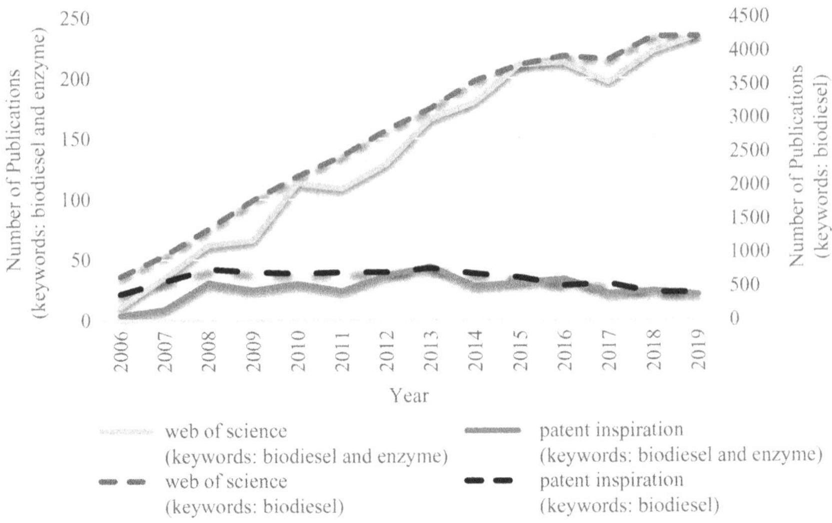

FIGURE 13.1 Evolution of biodiesel studies over recent years, based on two different databases: the Web of Science and patent inspiration.

Lipases ('glycerol ester hydrolases', E.C. 3.1.1.3) are hydrolytic enzymes that catalyze the hydrolysis reactions of 'triglycerides' (TAG), producing 'diglycerides' (DAG), 'monoglycerides' (MAG), 'free fatty acids' (FFA), and 'glycerol'. In aquo-restricted environments, these enzymes are able to catalyze synthesis reactions, including esterification and transesterification (Koskinen and Klibanov, 1996). Because of the high catalytic performance of these proteins, lipases are the third most commercialized biocatalysts, being largely used in detergent, food, paper and cellulose, animal feed, and the pharmaceutical industries. The market for microbial lipases was valued at USD425 million in 2018 and is projected to grow at 6.8% per year, reaching USD590 million by 2023 (Almeida et al., 2020).

The versatility of such enzymes, as well as the catalytic reaction under mild process conditions, the possibility of their reuse, the facility of separation steps, and the environmentally friendly process of using enzymes, has aroused the interest of researchers regarding its application in biodiesel production (Hama et al., 2018). Moreover, the robustness of lipases in the presence of FFA in feedstocks, since such enzymes catalyze both esterification and transesterification reactions, makes these proteins ideal to produce esters from partially hydrolyzed raw materials (acidity > 0.5%).

The literature reports the use of lipases from different sources including animals, vegetables, and microbial-based sources for biodiesel synthesis (Aguieiras et al., 2014; Rial et al., 2020; Sharma et al., 2001). However, microbial lipases are the most used due to their easy large-scale production, high versatility regarding distinct selectivity, and the possibility of genetic manipulation which aims at their production by heterologous expression with high productivity.

In this context, through 'protein engineering' tools, including directed evolution and rational design, it is possible to acquire lipase variants that are highly tolerant under industrially severe conditions, such as the presence of solvents, and high temperature and pressure, making it possible to design an ideal biocatalyst for a given process (Dror et al., 2015; Hama et al., 2018).

Metagenomics is another tool that enables the extraction of all microbial genomic DNA from certain environmental habitats, the construction of a metagenomic library, and the screening of novel thermo and solvent-tolerant lipase from the uncultivable components of microbial environments (Almeida et al., 2020; Hama et al., 2018; Sahoo et al., 2017).

The first paper reporting the use of lipase for biodiesel synthesis was described in 1990 by Mittelbach (1990). Since then, the number of publications and patents involving the application of lipases in biodiesel synthesis has grown. Most work has focused on some topics that may lead to high conversions to oil in 'fatty acid methyl esters' (FAME) and 'fatty acid ethyl esters' (FAEE), such as: the development of new biocatalysts aiming to diminish their cost and increase their activity and stability; the search for low-cost raw materials; the study of different reactor configurations; and the evaluation of the reaction parameters of temperature, enzyme concentration, the molar ratio of reagents, and the presence of a solvent. However, not only the scale-up of the biodiesel process but also work that focuses on eco-economic analyses related to enzymatic production of biodiesel is also scarce.

This chapter reviews the main aspects and challenges related to enzymatic biodiesel production and points out some future perspectives towards turning this process into an industrial reality. The development of new biocatalysts and research involving alternative feedstocks are also illustrated. A brief discussion regarding possible routes for obtaining enzymatic biodiesel, dependent on the use of raw materials, is presented. Attempts to scale up the enzymatic process, as well as work on economic analysis, are also presented and discussed.

13.2 TECHNOLOGICAL CHALLENGES FOR ENZYMATIC BIODIESEL PRODUCTION

Commercial lipases used in biodiesel production are mainly produced by heterologous expression and commercialized by companies such as Novozymes, Amano, Gist Brocades, etc. (Sharma et al., 2001; Yan et al., 2014a). Lipase B from the yeast '*Candida antarctica*' (CALB) is the most widely used lipase (Aarthy et al., 2014). Lipases from filamentous fungi, such as '*Rhizomucor miehei* lipase' (RML) (Aguieiras et al., 2017b; Rodrigues and Fernandez-Lafuente, 2010), *Rhizopus oryzae* lipase (ROL) (Canet et al., 2014), and '*Thermomyces lanuginosus* lipase' (TLL) (Fernandez-Lafuente, 2010) have also been largely used. The most used bacterial lipase is '*Burkholderia cepacia* lipase', formerly known as *Pseudomonas cepacia* (Amano Lipase PS) (Kumar et al., 2020; Nelson et al., 1996). Many of these enzymes are also commercially available in their immobilized form by Novozymes as Novozym 435 (CALB immobilized on a macroporous acrylic resin), 'Lipozyme RM IM' (RML immobilized on an anionic resin), and 'Lipozyme TL IM' (TLL immobilized on a gel of granulated silica) (Nielsen et al., 2008; Robles-Medina et al., 2009).

Although enzymatic biodiesel production is widely reported in the literature, some drawbacks, including the high cost of the biocatalysts, enzyme deactivation, and lower reaction rates, when compared to conventional processes, still limit the application of lipases to produce commodity products, such as biodiesel

(Freire et al., 2011). Studies to find solutions for such problems are focused on three ways: (1) biocatalysts, (2) reaction conditions, and (3) reactor configurations. Some of these topics are discussed more deeply in the next sections.

13.3 BIOCATALYSTS

The academic and industrial sectors have made an effort to develop higher active and stable biocatalysts to be subjected to different reaction conditions, whether application is at the small or large scale (Hartmann and Jung, 2010). Although the search for an ideal biocatalyst is one of the principal challenges related to the enzymatic production of biodiesel, the number of industrial uses of enzymes has significantly increased in recent years, mainly due to advances in protein engineering technology and environmental and economic demands.

The use of biocatalysts for biodiesel production can be divided into four main categories: free enzymes, 'immobilized lipases', 'whole-cells' and 'dry fermented solids' (DFS) obtained by 'solid-state fermentation' (SSF), as illustrated in Figure 13.2. Soluble enzymes have the advantages of a lower purchase cost, compared to immobilized enzymes, and practically no diffusional limitations, which improves the mass transfer of substrates and products (Andrade et al., 2019). However, free lipases, in soluble or powder formulations (obtained by freeze-drying), have the disadvantage of poor operational stability and easy deactivation. In addition, the formation of enzyme aggregates in organic media and the safety concerns regarding the use of enzyme powders, mainly on large-scale processes, are other issues related to the use of free lipases (Freire et al., 2011).

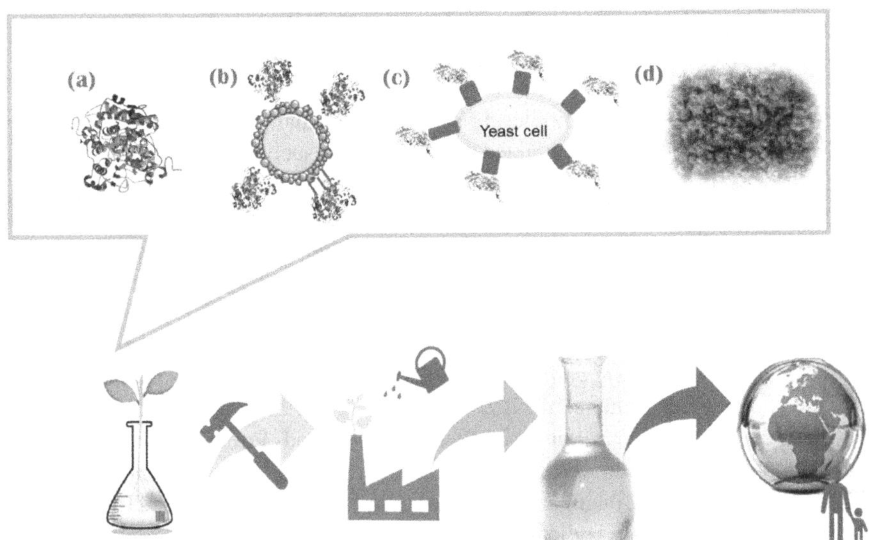

FIGURE 13.2 Distinct biocatalysts used in biodiesel production: (a) free lipase; (b) immobilized lipase; (c) whole-cell biocatalyst; (d) a dry fermented solid (here presented as the DFS obtained through growth of *Rhizomucor miehei* by SSF in cotton seed cake).

In this way, many technologies, comprising the production of more robust and low-cost biocatalysts, emerged which aimed at the development of more sustainable and economically viable processes.

13.3.1 Recent Advances in Lipase Immobilization

Enzyme immobilization technology is used as an important tool to improve enzyme properties (activity; stability, considering both operational and storage stabilities; specificity; and selectivity) and to permit the reuse of the biocatalysts in distinct reaction cycles, since the biocatalyst is insoluble on reaction media (Boudrant et al., 2020; Hartmann and Jung, 2010). Because of the recovery of the enzymes, it is possible to reduce the cost of the process, one of the main issues that limit the industrial production of biodiesel through enzymatic routes (DiCosimo et al., 2013). It is known that enzyme immobilization can occur in several ways, such as adsorption, covalent bonding, and encapsulation, as well as the association of these techniques.

Immobilized lipases have received special attention in the biotechnology field for biodiesel production (Zhong et al., 2020). Such biocatalysts are usually more stable under more severe reaction conditions (conditions in which free enzymes could not be active). Moreover, the use of immobilized biocatalysts reduces the number of reaction steps and avoids the continuous washings needed when biodiesel is synthesized through a chemical route, avoiding the generation of tons of effluent (Zhao et al., 2015).

In this context, the design of new supports and immobilized enzyme preparations is a relevant area of modern sciences and technologies (Cipolatti et al., 2014; Manoel et al., 2016; Pinto et al., 2014; Zaitsev et al., 2019). An emerging area, generically called 'support engineering', comprises the development of new polymeric materials in order to interact with specific enzymes, resulting in maximum performance biocatalysts. It is important to highlight that some features are required to use a material as support for enzymatic immobilization, such as insolubility in the substrate and product phases, mechanical stability, porosity, and chemical resistance. The biocatalysts' performance also depends on numerous factors, including support material, enzyme immobilization method, and the type of enzyme (Cipolatti et al., 2018, 2019; Manoel et al., 2016; Pinto et al., 2014).

Some studies have been conducted using natural polymers as supports for lipase immobilization (such as chitin, chitosan, gelatin, and cellulose). The employment of nanometric materials, including nanofibers, mesoporous nanocarriers, and magnetic nanoparticles, have been explored for the synthesis of distinct 'nanobiocatalysts'. Another polymerization technique, simultaneous suspension and emulsion polymerization reactions, have also been used for the synthesis of new core-shell supports, once it allows the production of porous micrometric particles and the functionalization of such particles, without the necessity of additional reaction steps (Cipolatti et al. 2018, 2019; Manoel et al., 2016; Pinto et al., 2014). For instance, Cipolatti et al. (2018) produced distinct core-shell supports, exhibiting different compositions, for the immobilization of CAL-B. The biocatalysts were used on esterification reactions aiming at the synthesis of biofuels.

TABLE 13.1
Biodiesel Synthesis Using Immobilized Lipases

Microorganism/ Source	Support	Substrate	Conversion Rate (%)	Conditions	References
Candida antarctica (fraction B)	Polyporous magnetic cellulose	Yellow horn seed oil	92.3	60°C, 2 h, pH 7.5	Zhang et al. (2020)
Porcine pancreas	*p*-nitrobenzyl cellulose xanthate	Soybean oil	96.5	2 h, 40°C, pH 7	Rial et al. (2020)
Pseudomonas cepacia	Polyvinyl alcohol/ sodium alginate	Castor oil	78.0	50°C, 24 h	Kumar et al. (2020)
Rhizopus oryzae	Alginate-polyvinyl alcohol beads	Sludge palm oil	91.3	40°C	Muanruksa and Kaewkannetra (2020)
Rhizopus oryzae	Fe_3O_4 superpara-magnetic nanoparticles	Oil from *Chlorella vulgaris*	69.8	45°C, 24 h	Nematian et al. (2020)
Candida antarctica B and *Rhizomucor miehei* (coimmobilized)	Silica	Palm oil	78.3	35.6°C, 33.5 h	Shahedi et al. (2019)

Obviously, many factors affect the use of lipases for biodiesel production. Table 13.1 summarizes some important factors related to the most recent studies in the area of immobilized lipases. It can be seen that the conversion rate depends on the substrate, the reaction conditions, the type of enzyme, and the support composition. It is important to highlight that the factors that directly interfere with biodiesel synthesis are not restricted to the ones exhibited in Table 13.1.

13.3.2 Whole-Cell Biocatalysts

Whole-cell technology is characterized by the application of the entire yeast cell surface and its use for anchoring enzymes of interest (Cherf and Cochran, 2015). The strategy is advantageous as it permits the production and immobilization of the protein in one step, avoiding problems related to substrate internalization and facilitating the recovery of the enzyme (Cherf and Cochran, 2015; Wachtmeister and Rother, 2016).

The term 'whole-cell' is used to refer to enzymes inside cells that are in a protected environment and are often more stable than when isolated (de Carvalho, 2011). In the present text, this term is used in a more specific way, making reference to the molecular biology strategy that uses recombinant proteins on the yeast cell surface, such as *Pichia pastoris* and *Saccharomyces cerevisiae*. This technology is based on the insertion of protein genes with an anchoring function, called 'yeast surface display' (YSD) (Figure 13.2c), which anchors proteins of interest to the cell surface. This strategy has proven to be valuable for a wide spectrum of biotechnology sectors (Jiang et al., 2007; Yang and Zhang, 2018).

In this direction, whole-cell biocatalysts have been gaining strength, as it presents itself as an alternative to reduce expenses due to purification and immobilization of the catalyst (Han et al., 2009; Liu et al., 2016). Cell-surface display techniques have been successfully applied as biocatalysts catalyzing waste oils into biodiesel (Liu et al., 2016). In work developed by Yan et al. (2014b), *P. pastoris* 'whole-cell catalyst' (WCC) with a functional intracellular expression of TLL was developed. The authors applied the biocatalyst in the conversion of waste cooking oils to biodiesel with an 82% yield within 84 h at a 6% dosage of whole-cells, and remaining at 78% activity after three batch cycles. They also stated that the WCC showed comparable reusability to Lipozyme TLIM (Yan et al., 2014b).

In another study, a *P. pastoris* display system was constructed to express CALB and RML as separately displayed on the yeast whole-cells to produce biodiesel in co-solvent media. These biocatalysts were applied in the methanolysis of refined and used vegetable oils, resulting in a conversion rate of 90% in 12 h in an agitated reactor. In addition, the developed biocatalyst could be reused 20 times with conversions greater than 85% (Jin et al., 2013).

The methanolysis of soybean oil using RML-displaying *P. pastoris* WCC reached 83.14% after 72 h and was considered by the authors as having good operational stability after ten batches of reaction (Huang et al., 2012). Studies with the use of recombinant *P. pastoris* WCC with intracellular overexpression of TLL were developed for biodiesel production from waste cooking oils and led to 82% biodiesel, which remained 78% active after three batch cycles (Yan et al., 2014b).

13.3.3 Dry Fermented Solids

The majority of commercial processes for the industrial production of enzymes are based on 'submerged fermentation' (SmF). SmF is an easier process compared to SSF, regarding the controlling and operation steps; however, it is more expensive, due to the required equipment and the need of an immobilization step after the product downstream, as free lipases have poor stability and cannot be reused. On the other hand, SSF has been highlighted in recent years, due to its advantages: simpler equipment, less energy and space requirement, easier downstream processing, and frequently higher productivity, resulting in lower costs for biomolecule production (Gutarra et al., 2009; Mitchell et al., 2019; Singhania et al., 2009).

SSF is very promising for lipase production, as the substrates used for growing microorganisms can be agro-industrial waste from vegetable oil production (consisting in the remaining deoiled solids from the pressing of oilseeds, such as cakes, brans, and fibers). These solid wastes are oil-rich by-products which induce the production of distinct lipases (Jain and Naik, 2018; Rodriguez et al., 2006). It is also important to highlight that the use of low-cost agro-industrial waste as a feedstock reinserts the waste into the production chain, reinforcing the circular economy bases and leading to an economically feasible enzymatic biodiesel synthesis.

It is noteworthy that countries presenting strong agro-industrial sectors, such as Brazil, India, and the United States, produce large amounts of such cakes from different vegetable oils, such as macauba (*Acrocomia aculeata*), cotton (*Gossypium* sp.), castor beans (*Ricinus communis*), babassu (*Attalea speciosa*), palm (*Elaeis guineensis*), and physic nut (*Jatropha curcas*) that can be applied for lipase production in SSF (Aguieiras et al., 2019; Aguieiras et al., 2014; Avila et al., 2020; Duarte et al., 2015; Godoy et al., 2009; Joshi and Khare, 2013;). Furthermore, SSF could be performed with inert supports, as perlite impregnated with a known media – usually used for the growth of the microorganisms in SmF (Martinez-Ruiz et al., 2018; Ooijkaas et al., 2000).

SSF produces a naturally immobilized lipase which avoids downstream steps, such as extraction and enzyme immobilization, reducing the costs of the biocatalyst (Sun and Xu, 2008). However, the fermented solids needed to be dried to form DFS in order to be efficiently applied in esterification and transesterification reactions aiming at the production of fatty-acid alkyl esters (FAAE) as FAME or FAEE. In the context of biodiesel production, some studies present the potential use of DFS produced by bacteria (Joshi and Khare, 2013; Liu et al., 2013, 2014) and filamentous fungi (Aguieiras et al., 2014, 2019; Avila et al., 2020; Fernandes et al., 2007).

Aguieiras et al. (2017a) reported the use of DFS for transesterification/esterification reactions employing acid oil from macauba pulp. The authors obtained an 85% conversion at 96 h of reaction and, following a second reaction cycle, it was possible to obtain higher conversions, achieving 91%. Fernandes et al. (2007) applied DFS from *B. cepacia* for transesterification reactions from corn oil; the time required to obtain 95% FAEE was 120 h. Other published reports notice that, to obtain high yields, enzymatic transesterification, compared to enzymatic esterifications, are longer (Hama et al., 2007, 2011; Salum et al., 2010; Watanabe et al., 2000).

Recently, de Oliveira et al. (2020) reported evaluating a comparison between the dried extract obtained from SmF and DFS, both with *Fusarium* sp. One step of the work was to perform a reaction between oleic acid (100 mmol/L in hexane) and methanol or ethanol (1: 5 molar ratio) by applying 30 U/reaction of both biocatalysts. The reaction with ethanol results in 75.1% FAEE applying DFS compared to 47.0% (SmF), and in the reaction with methanol they achieved 97.3% FAME using DFS, as against 41.8% (SmF).

Table 13.2 presents the most appropriate conditions for biodiesel production found in several studies. Based on the data, it is noted that DFS can be obtained by different substrates and several reaction conditions. Also, the reactions are often compared between several molar ratios of the substrates, keeping the concentration of oil/FFA constant, and the best conditions are often in excess of alcohol.

In addition to the data provided in Table 13.2, DFS could be reused for: up to six viable recycles in methyl ester synthesis at 40°C and maintain over 50% of activity (de Oliveira et al., 2020); four viable cycles at 40°C (Aguieiras et al., 2019), where FAEE could be obtained with conversions of over 84%; five cycles at 50°C; six cycles at 45°C (Soares et al., 2013); and nine cycles at 50 °C (Salum et al., 2010) – showing once more the superior advantages for the use of this biocatalyst.

Therefore, it is evident that distinct technologies have been developed in order to produce high performance biocatalysts. Many of them have shown promising results towards the enzymatic synthesis of biodiesel.

TABLE 13.2
Resume of Esterification/transesterification Studies Using DFS as a Biocatalyst

Biocatalyst (DFS)	Oil or FFA/alcohol/solvent	Reaction conditions	Conversion/time	Reference
Fusarium sp. in crambe cake	Oleic acid/methanol/n-hexane	Batch reactor; molar ratio 3:1 methanol/100 mM oleic acid in n-hexane at 40°C, 35 U of hydrolytic activity/reaction	99.6%/20 h	de Oliveira et al. (2020)
Rhizomucor miehei in cottonseed meal	Oleic acid/ethanol/solvent free reaction	Batch reactor; molar ratio 2:1 ethanol/oleic acid in n-hexane at 40°C, 369 U of hydrolitic activity/reaction	85%/4 h	Aguieiras et al. (2019)
Rhizomucor miehei in babassu cake	Acid macauba oil/ethanol (95%)/solvent free reaction	Batch reactor; molar ratio 6:1 of ethanol/oil (half ethanol addition at the beginning and at 24 h of reaction), at 40°C, 13 U of hydrolytic activity/g of oil	85%/96 h	Aguieiras et al. (2017a)
Rhizopus microsporus in perlite impregnated with an SmF media used to make it grow	Oleic acid/ethanol/n-hexane	CSTR, first fed at molar ratio 5:1 of ethanol/oleic acid and 2:1 after 2 h; 50 mM of oleic acid; flow rate 2 mL/min; substrates are pre-incubated 45°C	90%/from 2 to 14 h	Martinez-Ruiz et al. (2018)
B. cenocepacia in sugarcane bagasse/sunflower seed cake	Soybean oil/ethanol/*t*-butanol	Batch reactor; molar ratio 4:1 of ethanol:oil at 45°C with 14.1 U of hydrolytic activity/reaction	86%/96 h	Liu et al. (2013)
B. cepacia in a 1:1 (dry w/w) sugarcane bagasse/ sunflower seed meal	FFA from hydrolysis of soybean soapstock acid oil/ ethanol/ solvent-free system	PBR with media recirculation; molar ratio 3:1 of ethanol:FFA at 50°C with 1008 U of hydrolytic activity/reaction	93%/31 h	Soares et al. (2013)
Rhizopus microspores in 1:1 sugarcane bagasse/sunflower seed meal	Corn oil/ethanol/*n*-heptane	Batch reactor; molar ratio 3:1 of ethanol:oil at 44°C with 120 U of hydrolytic activity/reaction	91%/48 h	Zago et al. (2014)
R. miehei in babassu cake	FFA from hydrolysis of macauba oil (99.9% acidity)/ ethanol/solvent-free system	Batch reactor; molar ratio 2:1 of ethanol:FFA at 40°C with 151 U of hydrolytic activity/reaction	91%/8 h	Aguieiras et al. (2014)
B. cepacia in 1:1 sugarcane bagasse/sunflower seed meal	Soybean oil/ethanol/ solvent-free system	PBR with media recirculation; molar ratio 3:1 of ethanol/oil at 50°C with 72 U of hydrolytic activity/reaction	95%/46 h	Salum et al. (2010)
B. cepacia in 1:1 sugarcane bagasse/sunflower seed meal	Oleic acid/ethanol/*n*- heptane	Batch reactor; molar ratio 5:1 of ethanol:oleic acid at 37°C with 60 U of hydrolytic activity/ reaction	94%/18 h	Fernandes et al. (2007)

13.4 ENZYMATIC BIODIESEL PRODUCTION ROUTES

13.4.1 Transesterification/Esterification

Transesterification is the most used process for biodiesel synthesis, either from refined or alternative oils (Aguieiras et al., 2017a; Halim and Kamaruddin, 2008; Li et al., 2009; Watanabe et al., 2002). However, the literature about biodiesel production generally reports that a longer reaction time is required for enzymatic routes when compared to the chemical alkaline route (4 h).

Moreover, glycerol, a by-product of the complete conversion of TAG into esters, may be a problem since it is hydrophilic and insoluble in oil. Glycerol can be adsorbed onto the surface of the immobilized enzyme, leading to a decrease in lipase activity and operational stability, as observed in some studies (Aguieiras et al., 2015; Dossat et al., 1999; Hernandez-Martin and Otero, 2008; Talukder et al., 2009). In addition, glycerol can also extract methanol from the organic phase, decreasing the substrate concentration for enzyme, and its accumulation increases the viscosity of the reaction medium, diminishing the stirring efficiency and hampering the mass transfer of substrates and product in batch operations (Al-Zuhair, 2007; Hama et al., 2011; Hernandez-Martin and Otero, 2008).

The addition of organic co-solvents can be used to prevent the negative effects of glycerol on the lipase activity (Koskinen and Klibanov, 1996) and to reduce the viscosity of the reaction mixture. Tert-butanol is the most commonly employed solvent (Azocar et al., 2011; Li et al., 2006; Royon et al., 2007; Talukder et al., 2009; Wang et al., 2006), since this solvent improves the solubility of hydrophilic alcohols and hydrophobic oils and, at the same time, it is not used as a substrate by the lipase because it is a tertiary alcohol (Royon et al., 2007).

Besides conventional organic solvents, the literature also reports the use of alternative solvents including diesel (Kojima et al., 2004; Lara and Park, 2004), glycol ethers (glymes) (Tang et al., 2013), supercritical fluids (Rathore and Madras, 2007), and 'ionic liquids' (ILs) (de Diego et al., 2011; Ha et al., 2007). 'Supercritical fluids' (SCFs) have the properties of low viscosity, high diffusivity, and low surface tension, which increase the diffusion rates of substrates, enhancing the reaction rates (Rathore and Madras, 2007). 'Supercritical CO_2' ($SCCO_2$) is the most commonly used SCF because of its low-cost, non-toxicity, non-flammable properties, ease of availability, and mild critical properties (31.1°C and 7.38 MPa). Moreover, product separation can be easily achieved by reducing the pressure (Jackson et al., 2006; Rathore and Madras, 2007). The biodiesel yield of enzyme-catalyzed reactions in $SCCO_2$ ranges from 40 to 99% (Jackson et al., 2006; Lee et al., 2011b; Nagesha et al., 2004; Oliveira and Oliveira, 2001; Rathore and Madras, 2007) when using a high lipase concentration (Lee et al., 2011b; Rathore and Madras, 2007).

ILs have been reported as alternative solvents in enzymatic-catalyzed processes due to their negligible vapor pressure, non-flammability, high thermal stability, and the possibility of altering their properties according to specific reaction systems (de Diego et al., 2011; Ha et al., 2007). Some studies have reported the use of ILs combined with other approaches to further improve the biodiesel reaction process. An IL/$SCCO_2$ biphasic system, for instance, was employed in a continuous enzymatic reactor for the transesterification of triolein. A biodiesel yield up to 82% was attained in 12 cycles of 4 h (Lozano et al., 2011).

In another study, microwave irradiation was combined with an IL to achieve 92% of FAME in 6 h. Novozyme 435 activity in an IL was 1.8 fold higher than that in tert-butanol (Yu et al., 2011). It should be noted that the use of IL nowadays is unfeasible as it is too expensive for the large scale processes required to produce commodities products, such as biodiesel. Zhao et al. (2013) described 'deep eutectic solvents' (DESs), which are formed by a mixture of a solid organic salt (such as choline chloride) and a complexing agent (such as urea or glycerol), as novel promising solvents, since they exhibit many attractive properties of ILs and are available at much lower prices.

It should be highlighted that, for industrial applications, the reduction of volumetric capacity, the costs of the solvent itself, and the units of solvent recovery after the reaction must be considered. In this sense, solvent-free systems seem to be a better alternative for a large-scale process of enzymatic biodiesel production (Fjerbaek et al., 2009; Wang et al., 2006).

Other alternatives to the use of co-solvents in order to solve the problem of glycerol accumulation include the addition of hydrophilic substances (silica gel) for the partial removal of the glycerol during the reaction (Lee et al., 2011a), and alternative different reactor configurations including membrane reactor systems (Ko et al., 2012) and fixed bed reactors in series with a glycerol separation step before each column (Shimada et al., 2002; Watanabe et al., 2000; Halim and Kamaruddin, 2008).

13.4.2 Hydroesterification

For several lipases, the reaction using FFA is significantly faster than that observed with glycerides, since the smaller nucleophilic part of the former fits better in the enzyme active site (Nordblad et al., 2016). 'Hydroesterification', in which an acid oil (mono-, di-, and triglycerides) is first hydrolyzed to fatty acids which are, then, rapidly esterified with a short chain alcohol to obtain esters (biodiesel) (de Sousa et al., 2010), has been proposed as an alternative to solve the relatively long reaction times required for transesterification reactions.

In this process, high purity glycerin is formed in the hydrolysis step and water is produced as a by-product of the second step, which means that issues associated with the accumulation of glycerin are avoided. Hydroesterification can be carried out in an enzymatic/chemical-catalyzed hybrid process (Cavalcanti-Oliveira et al., 2011; Talukder et al., 2010a; Ting et al., 2008). In addition, both steps can be carried out enzymatically, i.e. an enzyme/enzyme hydroesterification process (Adachi et al., 2013; Aguieiras et al., 2014; Meng et al., 2011; Talukder et al., 2010b; Watanabe et al., 2007). Aguieiras et al. (2014), for example, selected a biocatalyst with high hydrolytic activity (obtained from dormant castor seeds) for the hydrolysis of macauba acid oil into FFA. The ethyl esters (91%) were produced in 8 h by the esterification of the FFA with ethanol, catalyzed by lipase-rich DFS from *R. miehei*.

The properties of the final fuel met important Brazilian standards and the biocatalyst used in the esterification step could be reused in ten cycles without loss of activity. This work shows how the two-step process can reduce the time required to obtain high ester yields in comparison to the single-stage transesterification/esterification process. Without considering the time required for FFA extraction from the first step, the total reaction time (hydrolysis + esterification) was 14 h.

The same authors obtained an ester content close to 80% in 48 h of reaction when the same raw material (macauba acid oil) was subjected to enzymatic ethanolysis (one reaction step) (Aguieiras et al., 2017a). Several works report the use of CALB for the esterification step, considering its high esterification activity (Adachi et al., 2013; Talukder et al., 2010b; Watanabe et al., 2007). The absence of glycerol improved the enzyme reuse several times as observed by Watanabe et al. (2007), which repeated both steps (hydrolysis and esterification) 40 times.

13.5 CONCLUSIONS AND FUTURE PERSPECTIVES

Biotechnology is a strategic area for the most varied sectors of industry, including the biofuel sectors, and represents a topic of significant technological and scientific relevance. In this way, many technologies have been developed in order to produce high performance biocatalysts, and many of them have shown promising results towards the enzymatic synthesis of biodiesel. Alternative sources for enzyme and biodiesel production contribute to the advancement of this application. Among different raw materials, the use of low-cost agro-industrial waste as feedstock reinserts itself in the production chain, reinforcing the circular economy base, and approaching an economically feasible enzymatic biodiesel synthesis.

Considering the economical aspects, there are still some limitations in the industrial application of enzymes for biodiesel production, namely: biomolecular advances, resulting in the synthesis of more active and stable enzymes; the development of new low-cost materials for enzyme production and also immobilization; the wide possibility of feedstocks to be used; and the design of reactors based in high intensification process mechanisms shows that this field still has numerous possibilities for development. Moreover, greater concerns about the environmental aspects strengthen the importance of greener biotechnology routes, which have been making these limitations less relevant, indicating that enzymatic biodiesel production is becoming a more attractive and economically viable pathway.

ACKNOWLEDGMENT

The authors would like to acknowledge FAPERJ, CAPES (project number 23038.004870/2015-11), and ANP (Agência Nacional de Petróleo, Gás Natural e Biocombustíveis, associated with the investment of resources of P, D&I from Sinochem Petróleo Brasil Ltda.) for financial support.

REFERENCES

Aarthy, M., P. Saravanan, M. K. Gowthaman, C. Rose, and N. R. Kamini. 2014. Enzymatic transesterification for production of biodiesel using yeast lipases: An overview. *Chemical Engineering Research and Design* 92: 1591–1601.

Adachi, D., S. Hama, K. Nakashima, et al. 2013. Production of biodiesel from plant oil hydrolysates using an *Aspergillus oryzae* whole-cell biocatalyst highly expressing *Candida antarctica* lipase B. *Bioresource Technology* 135: 410–416.

Aguieiras, E. C. G., E. D. Cavalcanti-Oliveira, A. M. de Castro, M. A. P. Langone, and D. M. G. Freire. 2014. Biodiesel production from *Acrocomia aculeata* acid oil by (enzyme/enzyme) hydroesterification process: Use of vegetable lipase and fermented solid as low-cost biocatalysts. *Fuel* 135: 315–321.

Aguieiras, E. C. G., E. D. Cavalcanti-Oliveira, and D. M. G. Freire. 2015. Current status and new developments of biodiesel production using fungal lipases. *Fuel* 159: 52–67.

Aguieiras, E. C. G., E. D. Cavalcanti-Oliveira, A. M. de Castro, M. A. P. Langone, and D. M. G. Freire. 2017a. Simultaneous enzymatic transesterification and esterification of an acid oil using fermented solid as biocatalyst. *Journal of the American Oil Chemists' Society* 94: 551–558.

Aguieiras, E. C. G., D. S. N. de Barros, H. Sousa, R. Fernandez-Lafuente, and D. M. G. Freire. 2017b. Influence of the raw material on the final properties of biodiesel produced using lipase from *Rhizomucor miehei* grown on babassu cake as biocatalyst of esterification reactions. *Renewable Energy* 113: 112–118.

Aguieiras, E. C. G., D. S. N. de Barros, R. Fernandez-Lafuente, and D. M. G. Freire. 2019. Production of lipases in cottonseed meal and application of the fermented solid as biocatalyst in esterification and transesterification reactions. *Renewable Energy* 130: 574–581.

Almeida, J. M., R. C. Alnoch, E. M. Souza, D. A. Mitchell, and N. Krieger. 2020. Metagenomics: Is it a powerful tool to obtain lipases for application in biocatalysis? *Biochimica et Biophysica Acta (BBA)-Proteins and Proteomics* 1868: 140320.

Al-Zuhair, S. 2007. Production of biodiesel: Possibilities and challenges. *Biofuels, Bioproducts & Biorefining* 1:57–66.

Andrade, T. A., M. Martin, M. Errico, and K. V. Christensen. 2019. Biodiesel production catalyzed by liquid and immobilized enzymes: Optimization and economic analysis. *Chemical Engineering Research and Design* 141: 1–14.

Avila, S. N. S., M. L. E. Gutarra, R. Fernandez-Lafuente, E. D. C. Cavalcanti, and D. M. G. Freire. 2020. Multipurpose fixed-bed bioreactor to simplify lipase production by solid-state fermentation and application in biocatalysis. *Biochemical Engineering Journal* 144: 1–7.

Azocar L., G. Ciudad, H. J. Heipieper, R. Munoz, and R. Navia. 2011. Lipase-catalyzed process in an anhydrous medium with enzyme reutilization to produce biodiesel with low acid value. *Journal of Bioscience and Bioengineering* 112: 583–589.

Boudrant, J., J. M. Woodley, and R. Fernandez-Lafuente. 2020. Parameters necessary to define an immobilized enzyme preparation. *Process Biochemistry* 90: 66–80.

Canet, A., M. D. Benaiges, and F. Valero. 2014. Biodiesel synthesis in a solvent-free system by recombinant rhizopus oryzae lipase: Study of the catalytic reaction progress. *Journal of the American Oil Chemists' Society* 91: 1499–1506.

Cavalcanti-Oliveira, E. D., P. R. da Silva, A. P. Ramos, D. A. G. Aranda, and D. M. G. Freire. 2011. Study of soybean oil hydrolysis catalyzed by *Thermomyces lanuginosus* lipase and its application to biodiesel production *via* hydroesterification. *Enzyme Research* 2011: 618692.

Cherf, G. M. and J. R. Cochran. 2015. Applications of yeast surface display for protein engineering. In *Yeast Surface Display*, ed. *B. Liu, Methods in Molecular Biology*, 1319: 155–175. New York, NY: Humana Press.

Cipolatti, E. P., M. J. A. Silva, M. Klein, et al. 2014. Current status and trends in enzymatic nanoimmobilization. *Journal of Molecular Catalysis B: Enzymatic* 99: 56–67.

Cipolatti, E. P., M. C. C. Pinto, J. de M. Robert, et al. 2018. Pilot-scale development of core-shell polymer supports for the immobilization of recombinant lipase B from *Candida antarctica* and their application in the production of ethyl esters from residual fatty acids. *Journal of Applied Polymer Science* 135: 46727.

Cipolatti, E. P., M. C. C. Pinto, R. O. Henriques, et al. 2019. Enzymes in green chemistry: State of the art in chemical transformations. In *Biomass, Biofuels, Biochemicals*

Advances in Enzyme Technology, ed. A. Pandey, R. S. Singh, R. R. Singhania, and C. Larroche, 137–151. Amsterdam: Elsevier.

De Carvalho, C. C. C. R. 2011. Enzymatic and whole cell catalysis: Finding new strategies for old processes. *Biotechnology Advances* 29: 75–83.

De Diego, T., A. Manjon, P. Lozano, and J. L. Iborra. 2011. A recyclable enzymatic biodiesel production process in ionic liquids. *Bioresource Technology* 102: 6336–6339.

De Oliveira, B. H., G. V. Coradi, P. de Oliva-Neto, and V. M. G. do Nascimento. 2020. Biocatalytic benefits of immobilized *Fusarium* Sp. (GFC) lipase from solid state fermentation on free lipase from submerged fermentation. *Industrial Crops and Products* 147: 112235.

De Sousa, J.S., E. D. Cavalcanti-Oliveira, D. A. G. Aranda, and D. M. G. Freire. 2010. Application of lipase from the physic nut (*Jatropha curcas* L.) to a new hybrid (enzyme/chemical) hydroesterification process for biodiesel production. *Journal of Molecular Catalysis B: Enzymatic* 65: 133–137.

DiCosimo, R., J. McAuliffe, A. J. Poulose, and G. Bohlmann. 2013. Industrial use of immobilized enzymes. *Chemical Society Reviews* 42: 6437–6474.

Dossat, V., D. Combes, and A. Marty. 1999. Continuous enzymatic transesterification of high oleic sunflower oil in a packed bed reactor: Influence of the glycerol production. *Enzyme and Microbiology Technology* 25: 194–200.

Dror, A., M. Kanteev, I. Kagan, et al. 2015. Structural insights into methanol-stable variants of lipase T6 from *Geobacillus stearothermophilus*. *Applied Microbiology and Biotechnology* 99: 9449–9461.

Duarte, J. G., L. L. S. Silva, D. M. G. Freire, M. C. Cammarota, and M. L. E. Gutarra. 2015. Enzymatic hydrolysis and anaerobic biological treatment of fish industry effluent: Evaluation of the mesophilic and thermophilic conditions. *Renewable Energy* 83: 455–462.

Fernandes, M. L. M., E. B. Saad, J. A. Meira, et al. 2007. Esterification and transesterification reactions catalyzed by addition of fermented solids to organic reaction media. *Journal of Molecular Catalalysis B: Enzymatic* 44: 8–13.

Fernandez-Lafuente, R. 2010. Lipase from *Thermomyces lanuginosus*: Uses and prospects as an industrial biocatalyst. *Journal of the Molecular Catalysis B: Enzymatic* 62: 197–212.

Fjerbaek, L., K. V. Christensen, and B. Norddahl. 2009. A review of the current state of biodiesel production using enzymatic transesterification. *Biotechnology and Bioengineering* 102: 1298–1315.

Freire, D. M. G., J. S. Sousa, and E. D. Cavalcanti-Oliveira. 2011. Biotechnological methods to produce biodiesel. In *Biofuels: Alternative Feedstocks and Conversion Processes*, ed. A. Pandey, 315–337. Amsterdam: Elsevier.

Godoy, M. G., M. L. E. Gutarra, F. M. Maciel, et al. 2009. Use of a low-cost methodology for biodetoxification of castor bean waste and lipase production. *Enzyme and Microbial Technology* 44: 317–322.

Gutarra, M. L. E., M. G. de Godoy, J. N. Silva, et al. 2009. Lipase production and *Penicillium simplicissimum* morphology in solid-state and submerged fermentations. *Biotechnology Journal* 4: 1450–1459.

Ha, S. H., M. N. Lan, S. H. Lee, S. M. Hwang, and Y.-M. Koo. 2007. *Lipase-catalyzed biodiesel* production from soybean oil in ionic liquids. *Enzyme and Microbial Technology* 41: 480–483.

Halim, S. F. A. and A. H. Kamaruddin. 2008. Catalytic studies of lipase on FAME production from waste cooking palm oil in a *tert*-butanol system. *Process Biochemistry* 43: 1436–1439.

Hama, S., H. Yamaji, T. Fukumizu, et al. 2007. Biodiesel-fuel production in a packed-bed reactor using lipase-producing *Rhizopus oryzae* cells immobilized within biomass support particles. *Biochemical Engineering Journal* 34: 273–278.

Hama, S., S. Tamalampudi, A. Yoshida, et al. 2011. Enzymatic packed-bed reactor integrated with glycerol-separating system for solvent-free production of biodiesel fuel. *Biochemical Engineering Journal* 55: 66–71.

Hama, S., H. Noda, and A. Kondo. 2018. How lipase technology contributes to evolution of biodiesel production using multiple feedstocks. *Current Opinion in Biotechnology* 50: 57–64.

Han, Z.-L., L. S.-Y. Han, S.-P. Zheng, and Y. Lin. 2009. Enhancing thermostability of a *Rhizomucor miehei* lipase by engineering a disulfide bond and displaying on the yeast cell surface. *Applied Microbiology and Biotechnology* 85: 117–126.

Hartmann, M. and D. Jung. 2010. Biocatalysis with enzymes immobilized on mesoporous hosts: The status quo and future trends. *Journal of Materials Chemistry* 20: 844–857.

Hernandez-Martin, E. and C. Otero. 2008. Different enzyme requirements for the synthesis of biodiesel: Novozym-435 and Lipozyme-TL IM. *Bioresource Technology* 99: 277–286.

Huang, D., S. Han, Z. Han, and Y. Lin. 2012. Biodiesel production catalyzed by *Rhizomucor miehei* lipase-displaying *Pichia pastoris* whole cells in an isooctane system. *Biochemical Engineering Journal* 63: 10–14.

Jackson, M. A., I. K. Mbaraka, and B. H. Shanks. 2006. Esterification of oleic acid in supercritical carbon dioxide catalyzed by functionalized mesoporous silica and immobilized lipase. *Applied Catalysis A: General* 310: 48–53.

Jain, R. and S. N. Naik. 2018. Biocatalysis and agricultural biotechnology adding value to the oil cake as a waste from oil processing industry: Production of lipase in solid state fermentation. *Biocatalysis and Agricultural Biotechnology* 15: 181–184.

Jiang, Z.-B., H.-T. Song, N. Gupta, L.-X. Ma, and Z.-B. Wu. 2007. Cell surface display of functionally active lipases from *Yarrowia lipolytica* in *Pichia pastoris*. *Protein Expression and Purification* 56: 35–39.

Jin, Z., S.-Y. Han, L. Zheng, et al. 2013. Combined utilization of lipase-displaying *pichia pastoris* whole-cell biocatalysts to improve biodiesel production in co-solvent media. *Bioresource Technology* 130: 102–109.

Joshi, C. and S. K. Khare. 2013. Purification and characterization of *Pseudomonas aeruginosa* lipase produced by SSF of deoiled Jatropha seed cake. *Biocatalysis and Agricultural Biotechnology* 2: 32–37.

Ko, M. J., H. J. Park, S. Y. Hong, and Y. J. Yoo. 2012. Continuous biodiesel production using in situ glycerol separation by membrane bioreactor system. *Bioprocess and Biosystems Engineering* 35: 69–75.

Kojima, S., D. Du, M. Sato, and E. Y. Park. 2004. Efficient production of fatty acid methyl ester from waste activated bleaching earth using diesel oil as organic solvent. *Journal of Bioscience and Bioengineering* 98: 420–424.

Koskinen, A. and A. Klibanov. 1996. *Enzymatic Reactions in Organic Media.* Dordrecht: Springer.

Kumar, D., T. Das, B. S. Giri, and B. Verma. 2020. Preparation and characterization of novel hybrid bio-support material immobilized from *Pseudomonas cepacia* lipase and its application to enhance biodiesel production. *Renewable Energy* 147: 11–24.

Lara, P. V. and E. Y. Park. 2004. Potential application of waste activated bleaching earth on the production of fatty acid alkyl esters using *Candida cylindracea* lipase in organic solvent system. *Enzyme Microbial and Technology* 34: 270–277.

Lee, M., J. Lee, D. Lee, et al. 2011a. Improvement of enzymatic biodiesel production by controlled substrate feeding using silica gel in solvent free system. *Enzyme and Microbial Technology* 49: 402–406.

Lee, J. H., S. B. Kim, S.W. Kang, et al. 2011b. Biodiesel production by a mixture of *Candida rugosa* and *Rhizopus oryzae* lipases using a supercritical carbon dioxide process. *Bioresource Technology* 102: 2105–2108.

Li, L., W. Du, D. Liu, L. Wang, and Z. Li. 2006. Lipase-catalyzed transesterification of rapeseed oils for biodiesel production with a novel organic solvent as the reaction medium. *Journal of Molecular Catalysis B: Enzymatic* 43: 58–62.

Li, N.-W., M.-H. Zong, and H. Wu. 2009. Highly efficient transformation of waste oil to biodiesel by immobilized lipase from *Penicillium expansum*. *Process Biochemistry* 44: 685–688.

Liu, Y., C. Li, X. Meng, and Y. Yan. 2013. Biodiesel synthesis directly catalyzed by the fermented solid of *Burkholderia cenocepacia via* solid state fermentation. *Fuel Processing Technology* 106: 303–309.

Liu, Y., C. Li, S. Wang, and W. Chen. 2014. Solid-supported microorganism of *Burkholderia cenocepacia* cultured via solid state fermentation for biodiesel production: Optimization and kinetics. *Applied Energy* 113: 713–721.

Liu, Z., S. H. Ho, T. Hasunuma, et al. 2016. Recent advances in yeast cell-surface display technologies for waste biorefineries. *Bioresource Technology* 215: 324–333.

Lozano, P., J. M. Bernal, and M. Vaultier. 2011. Towards continuous sustainable processes for enzymatic synthesis of biodiesel in hydrophobic ionic liquids/supercritical carbon dioxide biphasic systems. *Fuel* 90: 3461–3467.

Manoel, E. A., J. M. Robert, M. C. C. Pinto, et al. 2016. Evaluation of the performance of differently immobilized recombinant lipase B from *Candida antarctica* preparations for the synthesis of pharmacological derivatives in organic media. *RSC Advances* 6: 4043–4052.

Martinez-Ruiz, A., L. Tovar-Castro, H. S. Garcia, G. Saucedo-Castaneda, and E. Favela-Torres. 2018. Continuous ethyl oleate synthesis by lipases produced by solid-state fermentation by *Rhizopus Microsporus*. *Bioresource Technology* 265: 52–58.

Meng, Y., G. Wang, N. Yang, et al. 2011. Two-step synthesis of fatty acid ethyl ester from soybean oil catalyzed by *Yarrowia lipolytica* lipase. *Biotechnology for Biofuels* 4: 6.

Mitchell, D. A., M. H. Sugai-Guerios, and N. Krieger. 2019. Solid-state fermentation. In *Chemistry, Molecular Sciences and Chemical Engineering*. Amsterdam: Elsevier.

Mittelbach, M. 1990. Lipase catalyzed alcoholysis of sunflower oil. *Journal of the American Oil Chemists' Society* 67: 168–170.

Muanruksa, P. and P. Kaewkannetra. 2020. Combination of fatty acids extraction and enzymatic esterification for biodiesel production using sludge palm oil as a low-cost substrate. *Renewable Energy* 146: 901–906.

Nagesha, G. K., B. Manohar, and K. U. Sankar. 2004. Enzymatic esterification of free fatty acids of hydrolyzed soy deodorizer distillate in supercritical carbon dioxide. *Journal of Supercritical Fluids* 32: 137–145.

Nelson, L. A., T. A. Foglia, and W. N. Marmer. 1996. Lipase-catalyzed production of biodiesel. *Journal of the American Oil Chemists' Society* 73: 1191–1195.

Nematian, T., Z. Salehi, and A. Shakeri. 2020. Conversion of bio-oil extracted from *Chlorella vulga*ris microalgae to biodiesel *via* modified superparamagnetic nano-biocatalyst. *Renewable Energy* 146: 1796–1804.

Nielsen, P. M., J. Brask, and L. Fjerbaek. 2008. Enzymatic biodiesel production: technical and economical considerations. *European Journal of Lipid Science and Technology* 110: 692–700.

Nordblad, M., A. K. Pedersen, A. Rancke-Madsen, and J. M. Woodley. 2016. Enzymatic pretreatment of low-grade oils for biodiesel production. *Biotechnology and Bioengineering* 113: 754–760.

Oliveira, D. and J. V. Oliveira. 2001. Enzymatic alcoholysis of palm kernel oil in *n*-hexane and $SCCO_2$. *Journal of Supercritical Fluids* 19: 141–148.

Ooijkaas, L. P., F. J. Weber, R. M. Buitelaar, J. Tramper, and A. Rinzema. 2000. Defined Media and inert supports: Their potential as solid-state fermentation production systems. *Trends in Biotechnology* 18: 356–360.

Pinto, M. C. C., D. M. G. Freire, and J. C. Pinto. 2014. Influence of the morphology of core-shell supports on the immobilization of lipase B from *Candida antarctica*. *Molecules* 19: 12509–12530.

Rathore, V. and G. Madras. 2007. Synthesis of biodiesel from edible and non-edible oils in supercritical alcohols and enzymatic synthesis in supercritical carbon dioxide. *Fuel* 86: 2650–2659.

Rial, R. C., O. N. de Freitas, C. E. D. Nazario, and L. H. Viana. 2020. Biodiesel from soybean oil using *Porcine pancreas* lipase immobilized on a new support: p-nitrobenzyl cellulose xanthate. *Renewable Energy* 149: 970–979.

Robles-Medina, A., P. A. Gonzalez-Moreno, L. Esteban-Cerdan, and E. Molina-Grima. 2009. Biocatalysis: Towards ever greener biodiesel production. *Biotechnology Advances* 27: 398–408.

Rodrigues, R. C. and R. Fernandez-Lafuente. 2010. Lipase from *Rhizomucor miehei* as a biocatalyst in fats and oils modification. *Journal of Molecular Catalysis B: Enzymatic* 66: 15–32.

Rodriguez, J. A., J. C. Mateos, J. Nungaray, et al. 2006. Improving lipase production by nutrient source modification using *Rhizopus homothallicus* cultured in solid state fermentation. *Process Biochemistry* 41: 2264–2269.

Royon, D., M. Daz, G. Ellenrieder, and S. Locatelli. 2007. Enzymatic production of biodiesel from cotton seed oil using *t*-butanol as a solvent. *Bioresource Technology* 98: 648–653.

Sahoo, R. K., M. Kumar, L. B. Sukla, and E. Subudhi. 2017. Bioprospecting hot spring metagenome: Lipase for the production of biodiesel. *Environmental Science and Pollution Research* 24: 3802–3809.

Salum, T. F. C., P. Villeneuve, B. Barea, et al. 2010. Synthesis of biodiesel in column fixed-bed bioreactor using the fermented solid produced by *Burkholderia cepacia* LTEB11. *Process Biochemistry* 45: 1348–1354.

Shahedi, M., M. Yousefi, Z. Habibi, M. Mohammadi, and M. A. As'habi. 2019. Co-immobilization of *Rhizomucor miehei* lipase and *Candida antarctica* lipase B and optimization of biocatalytic biodiesel production from palm oil using response surface methodology. *Renewable Energy* 141: 847–857.

Sharma, R., Y. Chisti, and U. C. Banerjee. 2001. Production, purification, characterization, and applications of lipases. *Biotechnology Advances* 19: 627–662.

Shimada, Y., Y. Watanabe, A. Sugihara, and Y. Tominaga. 2002. Enzymatic alcoholysis for biodiesel fuel production and application of the reaction to oil processing. *Journal of Molecular Catalysis B: Enzymatic* 17: 133–142.

Singhania, R. R., A. K. Patel, C. R. Soccol, and A. Pandey. 2009. Recent advances in solid-state fermentation. *Biochemical Engineering Journal* 44: 13–18.

Soares, D., A. F. Pinto, A. G. Gonçalves, D. A. Mitchell, and N. Krieger. 2013. biodiesel production from soybean soapstock acid oil by hydrolysis in subcritical water followed by lipase-catalyzed esterification using a fermented solid in a packed-bed reactor. *Biochemical Engineering Journal* 81: 15–23.

Sun, S. Y. and Y. Xu. 2008. Solid-state fermentation for 'whole-cell synthetic lipase' production from *Rhizopus Chinensis* and identification of the functional enzyme. *Process Biochemistry* 43: 219–224.

Talukder, M. M. R., J. C. Wu, T. B. V. Nguyen, N. M. Fen, and Y. L. S. Melissa. 2009. Novozym 435 for production of biodiesel from unrefined palm oil: Comparison of methanolysis methods. *Journal of Molecular Catalysis B: Enzymatic* 60: 106–112.

Talukder, M. M. R., J. C. Wu, and L. P.-L. Chua. 2010a. Conversion of waste cooking oil to biodiesel via enzymatic hydrolysis followed by chemical esterification. *Energy & Fuels* 24: 2016–2019.

Talukder, M. M. R., J. C. Wu, N. M. Fen, and Y. L. S. Melissa. 2010b. Two-step lipase catalysis for production of biodiesel. *Biochemical Engineering Journal* 49: 207–212.

Tang, S., C. L. Jones, and H. Zhao. 2013. Glymes as new solvents for lipase activation and biodiesel preparation. *Bioresource Technology* 129: 667–671.

Ting, W.-J., C.-M. Huang, N. Giridhar, and W.-T. Wu. 2008. An enzymatic/acid catalyzed hybrid process for biodiesel production from soybean oil. *Journal of the Chinese Institute of Chemical Engineers* 39: 203–210.

Wachtmeister, J. and D. Rother. 2016. Recent advances in whole cell biocatalysis techniques bridging from investigative to industrial scale. *Current Opinion in Biotechnology* 42: 169–177.

Wang, L., W. Du, D. Liu, L. Li, and N. Dai. 2006. Lipase-catalyzed biodiesel production from soybean oil deodorizer distillate with absorbent present in *tert*-butanol system. *Journal of Molecular Catalysis B: Enzymatic* 43: 29–32.

Watanabe, Y., Y. Shimada, A. Sugihara, et al. 2000. Continuous production of biodiesel fuel from vegetable oil using immobilized *Candida Antarctica* lipase. *Journal of the American Oil Chemists' Society* 77: 355–360.

Watanabe, Y., Y. Shimada, T. Baba, et al. 2002. Methyl esterification of waste fatty acids with immobilized *Candida antarcti*ca Lipase. *Journal of Oleo Science* 51: 655–661.

Watanabe, Y., T. Nagao, Y. Nishida, Y. Takagi, and Y. Shimada. 2007. Enzymatic production of fatty acid methyl esters by hydrolysis of acid oil followed by esterification. *Journal of the American Oil Chemists' Society* 84: 1015–1021.

Yan, Y., X. Li, G. Wang, et al. 2014a. Biotechnological preparation of biodiesel and its high-valued derivatives: A review. *Applied Energy* 113: 1614–1631.

Yan, J., X. Zheng, and S. Li. 2014b. A novel and robust recombinant *Pichia pastoris* yeast whole cell biocatalyst with intracellular overexpression of a *Thermomyces lanuginosus* lipase: Preparation, characterization and application in biodiesel production. *Bioresource Technology* 151: 43–48.

Yang, Z. and Z. Zhang. 2018. Engineering strategies for enhanced production of protein and bio-products in *Pichia Pastoris*: A review. *Biotechnology Advances* 36: 182–195.

Yu, D., C. Wang, Y. Yin, et al. 2011. A synergistic effect of microwave irradiation and ionic liquids on enzyme-catalyzed biodiesel production. *Green Chemistry* 13: 1869–1875.

Zago, E., V. Botton, D. Alberton, et al. 2014. Synthesis of ethylic esters for biodiesel purposes using lipases naturally immobilized in a fermented solid produced using *Rhizopus microsporus*. *Energy & Fuels* 28: 5197–5203.

Zaitsev, S. Y., A. A. Savina, and I. S. Zaitsev. 2019. Biochemical aspects of lipase immobilization at polysaccharides for biotechnology. *Advances in Colloid and Interface Science* 272: 102016.

Zhang, H., T. Liu, Y. Zhu, et al. 2020. Lipases immobilized on the modified polyporous magnetic cellulose support as an efficient and recyclable catalyst for biodiesel production from yellow horn seed oil. *Renewable Energy* 145: 1246–1254.

Zhao, H., C. Zhang, and T. D. Crittle. 2013. Choline-based deep eutectic solvents for enzymatic preparation of biodiesel from soybean oil. *Journal of Molecular Catalysis B: Enzymatic* 85–86: 243–247.

Zhao, X., F. Qi, C. Yuan, W. Du, and D. Liu. 2015. Lipase-catalyzed process for biodiesel production: Enzyme immobilization, process simulation and optimization. *Renewable and Sustainable Energy Reviews* 44: 182–197.

Zhong, L., Y. Feng, G. Wang, et al. 2020. Production and use of immobilized lipases in/on nanomaterials: A review from the waste to biodiesel production. *International Journal of Biological Macromolecules* 19: 39322–39325.

14 Biodiesel Additives
Status and Perspectives

Mukul Tomar
Hansham Dewal
Lakhan Kumar
Naveen Kumar
Navneeta Bharadvaja

CONTENTS

14.1 ADDITIVES, TYPES, AND THEIR SELECTION

Additives are the chemicals or the foreign substances added to diesel, gasoline, and biodiesel to improve the properties and performance of these fuels. They play a significant role in order to meet strict emission norms and to enhance the physiochemical properties of the fuel. Additives can reduce the dependency on conventional fuel by resolving the challenges associated with the wide-reaching utilization of biodiesel fuel in the diesel engine (Tomar and Kumar, 2019). Additives have been utilized in the oil industries since the early 1990s. Therefore, some of the common obstacles which can be circumvented by their use are: (i) to provide protection against the corrosion of the fuel line, petroleum tank, and engine parts; (ii) to improve the cold flow properties and promote biodiesel blending with diesel; (iii) to reduce harmful emissions from engine combustion; (iv) to enhance the combustion process and performance characteristics; (v) to promote storage stability for long operating conditions.

The dosage of additives to the base fuel is dependent on the mixture's chemistry and the chemical composition of the base fuel. However, the optimum dosage to be blended with the fuel is identified by undergoing different experimental analyses. As already discussed, the additives can be classified into a wide-variety based on their origination, size, chemical compound, and state (solid or liquid) as shown in Figure 14.1. Therefore the selection of the appropriate additive to be mixed with biodiesel or its blend is a critical step. Factors include economic feasibility, fuel blending property, toxicity, additive solubility, flash point of the blend, viscosity of the blend, water solubility in the resultant blend, and water partitioning of the additive (Abe et al., 2015; Ali et al., 2013; Danilov, 2015; Madiwale et al., 2017).

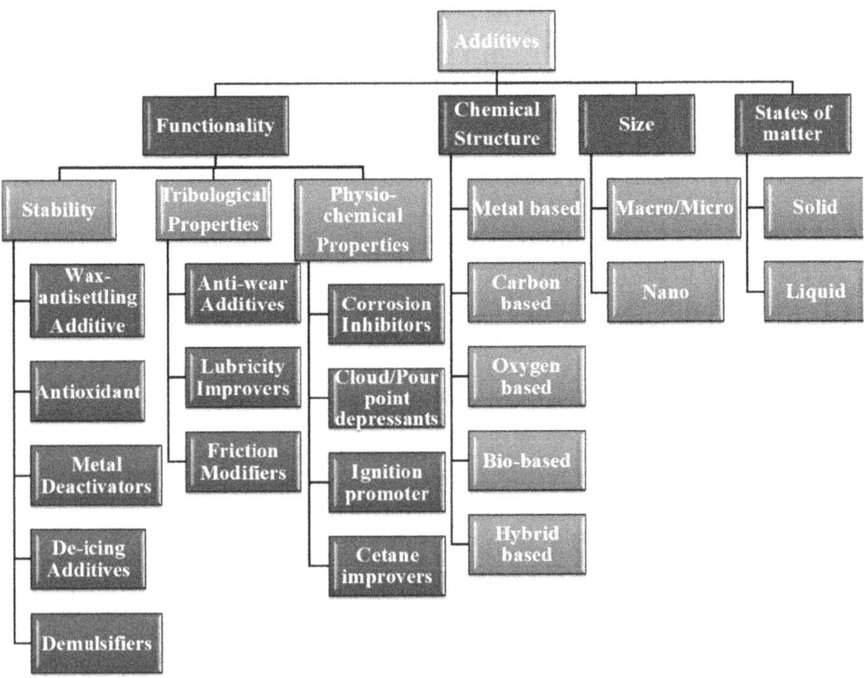

FIGURE 14.1 Classification of additives.

14.1.1 Categorization of Additives Based on Functionality

Every chemical or foreign substance serving the role of fuel additive has some unique features or functionality for which it is being mixed with the base fluid. In fuel industries, the additives are mainly differentiated on the basis of their role/function or remedies they are offering to the engines. Broadly, the additives are divided into three classes, i.e. on the basis of stability and the tribological and physiochemical properties of the fuel.

Though the commercial prospects of biodiesel have grown over the years, the longer term stability in diesel engines still remains a serious concern. The presence of unsaturated fatty acid, especially in nonedible oil feedstock, is the major hindrance which affects the oxidation stability of biodiesel (Sia et al., 2020). Therefore, the dosing of a small amount of antioxidants, such as 'butylated-hydroxytoluene' (BHT), 'alkyl/aryl-aminophenol, 2,6-di-tert-butyl-phenol' (2,6-dtbp), 'alkyl/aryl-phenylenediamine', neem extract (collected from the crushing of neem leaves), '2,4-dimethyl-6-tert-butyl-phenol', and 'ethylene-diamine', in biodiesel blends can improve their oxidation stability considerably (Borsato et al., 2014; Sindhi et al., 2013). Similarly there are certain additives which can resolve wax-settling issues that arise when operating engines with biodiesel in colder conditions. The 'fatty acid methyl esters' (FAMEs) present in biodiesel have a tendency to convert and accumulate together to form wax crystals at lower temperatures which limits the usage of biodiesel in colder regions.

Thus, various polymeric compounds containing N can be used as wax-settling additives to improve the cold flow properties of biodiesel and its blends. The corrosion of engine parts and the inner lines of the combustion cylinder after a certain period of time is the most common issue arising in diesel engines when operating on biobased fuels (Sorate and Bhale, 2015; Srivyas and Charoo, 2019). The metal deactivator 'N, N- disalicylidine 1,2- propane-diamine' which is available on the market can serve as a remedy to deactivate the oxidation of the upper surface of metal parts and to improve the oxidation and overall life of the diesel engine. In the same way, to get rid of poor ignition delay issues, the addition of chemicals such as alkyl-nitrates and nitrites, peroxides, and nitro-and nitroso compounds can improve overall engine performance and the emission characteristics of biodiesel and its blends to a greater extent. Table 14.1 lists these functional based additives, their details, their functions, and their treatment level in the base fuel.

14.1.2 Categorization of Additives Based on Chemical Compounds

Biodiesel is termed a 'mono-alkyl ester', having long-chain fatty acids, and is obtained from vegetable oils or animal fats. It is made up of different fatty esters having unique characteristics, all of which, in unison, determine the property of the biodiesel prepared. Properties such as ignition quality, cold flow, oxidation stability, heat of combustion, viscosity, and lubricity of biodiesel is highly dependent on the structure of the biodiesel. Therefore, taking into consideration the chemistry of

TABLE 14.1
List of Additives, Their Functions, and Examples

Additive	Role	Examples
Wax-antisettling	Restricts settling of paraffins such as n-alkane wax crystals	N-vinyltetrazoles, C-vinyltetrazoles
Antioxidants	Deactivates metals present in the fuel. Terminates chain reactions by disrupting radicalintermediates. Prevents decomposition of lubricant base stocks. Facilitates fuel blending.	'N-N' -Disalicylidene-1,2-propanediamine', 'butylated-hydroxytoluene' (BHT), '2,4-dimethyl-6-tert-butyl-phenol'
Metal deactivators	Deactivates the surface of metals that act as catalysts during oxidation	'N,N- disalicylidine 1,2- propane-diamine'
De-icing	Prevents the formation of ice in the fuel lines	'2-(2-methoxy ethoxy)' ethanol, APITOL 120, methyl carbitol
Demulsifiers	Restricts and terminates mist formation caused by water or insoluble compounds	'p-Dodecyl-phenol', '2,4-Di-t-butylphenol'
Anti-wear additives	Inhibits the wearing of engine parts	'zinc dithiophosphate' (ZDP), 'zinc dialkyl dithio phosphate' (ZDDP)
Lubricity improver	Enhances the lubricity and decreases the boiling point of the mixture in the fuel pump	Carboxyl-acids with 8–14 carbon numbers and their esters, vegetable oils
Friction modifiers	Reduces the friction, thereby imparting lower fuel consumption	Vegetable oils, nanoparticles, organo-molybdenum
Corrosion inhibitors	Protects fuel line from corrosion	Alkyl-succinates, amines salts, dimer acids, and salts of carboxylic acids
Cloud/pour point depressents	Restricts the free fatty acids from converting into wax crystals at lower temperatures. Decreases the initial temperature of paraffin crystallization.	Synthetic polyesters, monograde engine oil, turbine oil
Ignition promoter	Enhances starting ability. Reduces noise and vibrations. Reduces exhaust emissions.	'Amyl nitrate, hexyl nitrate', octyl nitrate, 'di-tertiary butyl peroxide'
Cetane improver	Improves cetane number and engine life. Easier cold starts. Lowers emissions and fuel consumption.	'2-ethylhexyl nitrate', octyl nitrate, isopropyl nitrate, amyl nitrate, and 'di-tert-butyl peroxide'

Source: Srivastava and Hancsok (2014).

biodiesel or its blends, the additives are distinguished on the grounds of the chemical compounds present in them. The higher the homogeneity of additives in the base fuel, the better will be the stability and engine performance. Metals such as aluminum, copper, manganese, iron, and cerium and their oxides have been promising alternatives to improving the performance and emissions of diesel engines (Hosseinzadeh-Bandbafha et al., 2018).

Furthermore, the advancement in the field of carbon based compounds such as carbon nanotubes and graphene is of particular interest regarding fuel additives. The unique physical properties, high mechanical strength, and thermal conductivity of carbon particles can improve the flash point, fire point, and cold flow properties of biodiesel fuel. Graphene-containing particles of sizes in the range 200–500 nm, when admixed with biodiesel and diesel-biodiesel blends, showed an improvement in the rate of combustion by 1.7 times, which is related to the higher thermal conductivity and emissivity of graphene additives (Soudagar et al., 2019).

Members of the alcohol group, such as ethanol, n-butanol, diethyl ether, and methanol, obtained from biological sources containing higher oxygen content, can be added in a smaller quantity to blends of biodiesel. The higher volatility, latent heat of vaporization, and higher flammability can improve the physical and thermal properties of biodiesel fuels. Also, the higher percentage of oxygen content in biobased additives helps in improving the cold flow properties of biodiesel in colder conditions by enhancing the combustion rate in the cylinder (Lawan et al., 2019). Hence, all these uncountable benefits make biobased additives a strong competitor in the group of additives.

14.1.3 Categorization of Additives Based on Size

Based on size, the additives are broadly classified as macro-, micro-, and nanoadditives. If the particle size of additives is above 100 micrometers, they are termed 'macroadditives'; the size of microadditives lies in the range 100 micrometers to 100 nanometers. However, challenges such as sedimentation and particle conglomeration arise with the usage of macro- and microadditives, which prohibit their mixing with biodiesel fuel, especially with the methyl esters that have poor flow properties. Hence, the addition of larger size additives is limited to a critical concentration in the base fluid. With the new benchmarks in the field of nanoscience, particles having a diameter below 100 nm can be easily mixed with biodiesel and its blends. Nanoparticles, having a higher surface: volume ratio, act as an excellent catalyst which significantly improves the fuel properties and engine characteristics of the diesel engine (Basu and Miglani, 2016; Shaafi et al., 2015). However, the mixing of nanoadditives in biodiesel is complex and is achieved with the help of an ultrasonication process. This sonication process involves the breakdown of nanoparticles with the aid of a high ultrasonic frequency, which disperses the particles inside the fuel. Many nanosized additives have been discovered by different researchers in the past with the aim of improving biodiesel stability, its properties, and its operation in compression ignition engines (Jayanthi and Rao, 2016; Karthikeyan and Silaimani, 2017; Kumar and Tomar, 2019; Prabu, 2017). The effect of these nanosized additives on fuel properties and engine characteristics are discussed in detail in later sections.

14.1.4 Categorization of Additives Based on State of Matter

On the basis of the state of matter, the additives are differentiated in terms of their physical state, i.e. solid/powder or liquid. The addition of additives to a wide variety of biodiesel feedstock sometimes give rise to unwanted issues such as agglomeration with respect to time. The formation of clusters inside the base fuel results in poor stability, lubrication properties of additives in the fuel, and also interrupts engine working characteristics to a greater extent. The nanoadditives, especially carbon based particles, more often tend to indicate agglomeration when mixed with biodiesel blends. However, there are different techniques, such as the addition of a small quantity of surfactants, which can resolve these troublesome concerns, though a smart technique utilizes the benefits of liquid chemicals, such as biobased additives. Nowadays, a step forward towards hybrid-based additives is also a popular approach in fuel economy. The hybrid additives make use of two different additive groups or states admixed together to form a new compound (Hazrat et al., 2019). The hybrid nanoparticles fill the gap associated with individual additives and can provide a persistent solution to the immeasurable issues associated with biodiesel and its blends.

14.2 EFFECT OF VARIOUS ADDITIVES ON BIODIESEL PROPERTIES

14.2.1 Effect of Additives on Fuel Properties

The fuel properties of biodiesel and conventional fuel are comparable to each other and can be used together as an alternative fuel for the diesel engine. The effect on fuel properties of chicken fat methyl ester (CFME) biodiesel mixed with a magnesium based additive were examined by Guru et al. (2010). The study showed that on decreasing the dosage of additive from 16 to 0 mmoL/L in the fuel blend, a significant increment in the value of the freezing point, flash point, and viscosity were observed. Also, a change in fuel properties confirms that the catalytic cracking effect of the additive results in the break-down of longer chains of hydrocarbons into smaller ones.

Cao et al., (2014) conducted an experimental study on 'waste cooking oil methyl ester' (WCOME) blends (B20) by adding an 'ethylene vinyl acetate copolymer' (EVAC) additive. EVAC has excellent cold flow improvers and decreases the formation of wax crystals in the fuel. The results showed considerable improvement in kinematic viscosity, oxidation stability, acid value, and the flash point of diesel-biodiesel blends. Also, as the concentration of additive increases from 0 to 0.1 wt% in the fuel blend, the values are upgraded of the above-mentioned fuel properties, i.e. 37.9% in viscosity, 8.19% in oxidation stability, 20% in acid value, and 4.72% in the flash point were reported. Similar research on mahua oil feedstock by Bhale et al. (2009) was carried out. Ethanol and kerosene were used as an additive to the biodiesel. The authors observed that increasing the concentration of ethanol in the biodiesel and its blends showed a significant reduction in fuel viscosity as compared to neat biodiesel. This is because ethanol acts as a wax depressant and restricts the formation of wax crystals in the fuel. However, on the other hand, a significant reduction in the flash/fire point was observed on using ethanol and kerosene as fuel additives.

Scientists have also explored the potentials of biobased fuel additives in biodiesel fuel. Joshi et al. (2011) examined the variation in fuel properties, i.e. kinematic viscosity, acid value, and flash point by adding 'ethyl levulinate (ethyl 4-oxopentanoate)' to cottonseed oil and poultry fat biodiesel. The results showed that on increasing the concentration of additive from 0 to 20 vol% in the biodiesel, the acid values of 'cotton seed methyl esters' (CSME) and 'poultry fat methyl esters' (PFME) were not significantly affected, though increasing the dosage has shown a concomitant decrease in the value of viscosity and flash point. The addition of ethanol as a fuel additive served a dual role for neem oil biodiesel and its blends, according to a study conducted by Sivalakshmi and Balusamy (2012). The results showed a substantial reduction in the calorific value and cetane index of fuel blends. However, viscosity reduced sharply at 40°C.

14.2.2 Effect of Additives on Cold Flow Properties

Cold flow properties determine how effectively fuel running in the engine can withstand cold weather conditions. Due to clogging of the fuel filters and pipelines at low ambient temperatures, these parameters play a significant role, especially in the colder regions. The cold flow properties for biodiesel can be illustrated in terms of its three properties, i.e. the 'cold filter plugging point' (CFPP), pour point, and cloud point. Biodiesel is highly prone to low-temperature operating problems due to the presence of long-chain, saturated fatty acids (Van Gerpen and He, 2014). The flow properties of biodiesel can be improved by several means, i.e. the utilizing of a light-oil blend, reducing the pressure distillation, adding additives, and winterization (Abe et al., 2015).

14.2.2.1 Cold Filter Plugging Point (CFPP)

This is defined as the minimum temperature at which a certain volume of pure biodiesel flows out through the standard filter within a time period of 60 seconds. In other words, it is the critical property that is used to predict the minimum temperature at which fuel will flow freely throughout the filters in a diesel engine. This is important as in cold countries a high cold filter plugging point will clog up vehicular engines more easily.

14.2.2.2 Pour Point

This is the lowest temperature at which liquid fuel becomes semisolid and loses its flow characteristics. The reason behind the loss of flow characteristics is that crude oil has a large paraffin content which crystallizes on decreasing the ambient temperature and forms a matrix of wax crystals. This matrix holds a large quantity of the liquid fraction of the crude oil within it, thus preventing the flow of liquid. The upper and lower pour points may sometimes be specified in order to indicate a temperature window within which the fuel will start to flow.

14.2.2.3 Cloud Point

This is defined as the lowest temperature of the fuel at which wax shows a cloudy appearance. The presence of solidified wax in the fuel chokes the fuel line and blocks the fuel filter system and fuel injectors. This term also plays a significant role in the storage stability of biodiesel.

Much research has been done in order to improve the cold flow properties of biodiesel, including modification of the chemical or physical properties of either the oil feedstock or biodiesel product, the usage of different types of additives in the biodiesel, and blending with petroleum products, i.e. diesel and gasoline. Traditionally, petroleum diesel additives can be defined as pour point (PP) depressants or wax crystal modifiers. These additives were created in order to improve the pumping ability of the crude oil that inhibits crystalline growth by eliminating the accumulation of large-sized crystals.

Several research studies on cold flow properties have been carried out in the past. The effect of a magnesium-based additive on the pour point of biodiesel of chicken fat methyl ester was examined (Guru et al., 2010). The authors successfully identified that on increasing the concentration of an additive in the fuel blend from 0 to 16 mmol/l, the pour point was reduced to 7°C. A similar study was conducted to examine the effect of an EVAC additive on WCOME used in an unmodified diesel engine (Cao et al., 2014). The authors found that on adding 0.04 wt% of EVAC in the blend of biodiesel and diesel B20 (20 vol% biodiesel + 80 vol% diesel), there were notable reductions in the values of cold flow properties, i.e. 8°C in the cloud point, 11°C in the CFPP, 10°C in the pour point, while compared with the neat diesel the values were 8°C in the cloud point, 10°C in the CFPP, and 10°C in the pour point, respectively. Therefore, the results confirmed that EVAC is an effective 'cold flow improver' (CFI) for both waste-cooking-oil-derived biodiesel and its blends.

To investigate the effect of additives on the pumping and injecting of biodiesel in 'compression ignition' (CI) engines in cold weather, a comparative study on mahua biodiesel was carried out (Bhale et al., 2009). Different concentrations of ethanol and kerosene (at 5%, 10%, 15%, and 20%) and lubrizol 7671 (at 0.5%, 1%, 1.5%, 2%, 2.5%, 3%, 3.5%, and 4%) were used. The results showed that at 20% ethanol concentration the cloud point of 'mahua methyl ester' (MME) was reduced from 18°C to 8°C with a further reduction of up to 5°C for a 20% dosing level of kerosene. In addition to this, a reduction of 11°C in the value of the pour point for 20% ethanol and a reduction of 15°C in the pour point value for 20% kerosene blended with biodiesel were observed. Reductions in the 'specific heat capacity' (CP) value by 4°C–5°C, the 'pour point' (PP) values by 3°C–4°C, and the 'cold filer plugging point' (CFPP) values by 3°C, respectively, were observed when 'ethyl levulinate' (EL), a biobased fuel additive, was mixed with cotton seed methyl ester and poultry fat methyl ester biodiesel (Joshi et al., 2011).

14.3 EFFECT OF VARIOUS ADDITIVES ON ENGINE COMBUSTION, PERFORMANCE, AND EMISSION CHARACTERISTICS

14.3.1 Effect of Additives on Combustion Characteristics

The combustion characteristics of any fuel are commonly governed by its chemical composition. One of the important combustion properties is how quickly the air–fuel mixture will ignite inside the engine within the optimum time. As the fuel ignition is a radical-driven reaction so the optimum dosage of fuel additives has a considerable impact on ignition properties, i.e. by enhancing or retarding the ability for ignition. In this section, we demonstrate the effect of using different kinds of additives on the combustion parameters of the biodiesel–diesel fuel blends inside the diesel engine.

The effect on combustion characteristics by adding a metal-based additive (Mg) in the fuel blend containing 90 vol% of diesel and 10 vol% of biodiesel (D90B10) was analyzed by Guru et al. (2010). A single cylinder four-stroke direct injection compression ignition engine was used for the analysis. The results illustrated an expeditious rise in cylinder pressure which is related to the combustion of fuel starting earlier with the use of the additive and the maximum engine cylinder gas pressure was also being slightly higher than that of neat diesel. The study also reported a significant increase in the cetane index of fuel blends with the addition of the additive. Hence, a short ignition delay was observed for B10 which gives a clear indication of a lower heat release rate in the premixed combustion as compared to diesel fuel.

The biobased additive ethanol had a notable impact on combustion characteristics of neem oil biodiesel. Sivalakshmi and Balusamy (2012) performed an experiment on a naturally aspirated, one-cylinder, four-stroke, direct injection diesel engine fueled with biodiesel derived from neem oil containing ethanol as an additive and diesel fuel. The experimental study reported that as the concentration of ethanol increases in the fuel blends, the peak pressure of the engine cylinder becomes higher. However, no significant change in ignition delay was observed for BE5 (5% ethanol in biodiesel) and BE10 blends when compared with neat biodiesel. The authors also confirmed that a negative heat release rate was observed due to the vaporization of the fuel that occurred at the beginning during the ignition delay. Increasing the quantity of ethanol in the fuel blend leads to the accumulation of the fuel during the premixed combustion stage, thereby resulting in higher rates of heat release for BE5 and BE10 blends. Further increasing the ethanol content in the fuel blend leads to a higher delay in ignition than that for neat biodiesel.

Apart from the analysis done on single cylinder diesel engines, Kivevele et al. (2011) successfully tested a four-cylinder 'turbocharged direct injection diesel engine' in order to find out the effect of using antioxidant additives, namely '2-tert butyl-4-methoxy phenol' ('butylated hydroxyanisole', BHA), '3, 4, 5-tri hydroxybenzoic acid' ('propyl gallate', PG), and '1, 2, 3 tri-hydroxy benzene' (pyrogallol, PY) in biodiesel derived from croton megalocarpus oil. The results revealed that in ideal conditions, no major change in the value of the peak pressure was observed with the addition of additives in the fuel blend as compared to neat diesel. However, at higher loads, the peak pressure recorded a higher value for fuel blends containing additives.

Moreover, the addition of additives in the B20 fuel blend showed a maximum rise in the peak of the heat release rate as compared to the other test fuels. Musthafa et al. (2018) carried out an experiment on a single-cylinder, four-stroke, water-cooled compression ignition engine fueled with a blend of palm-oil-derived biodiesel, 'di-tert-butyl peroxide' (DTBP), a cetane index improver additive at 1 vol%, and a mineral diesel in order to find out the effect of additives on the combustion phenomenon occurring inside the engine cylinder. The results showed that the maximum engine cylinder pressure was reduced when the additive is mixed in the B20 blend as compared to mineral diesel fuel. Similarly, the heat release rate was also at a maximum when the additive was added in the fuel blend as compared to the D100 (neat diesel).

14.3.2 Effect of Additives on Engine Performance Characteristics

Many research studies have been carried out in the past on a wide variety of biodiesel feedstocks by incorporating different kinds of additives in order to improve the performance and emissions characteristics of C.I. engines. However, the sample of suitable additives should be chosen wisely. In this section, a comprehensive overview on the effects of amalgamating different kinds of additives on engine performance is presented.

The investigation carried out on utilizing the assets of Mg additives in fuel blends containing B10 (10% biodiesel derived from chicken fat and 90% diesel) revealed a negligible change in the value of the torque of a single-cylinder 'direct injection' (DI) diesel engine (Guru et al., 2010). The dosage of the additive used for doped in the biodiesel has a concentration of about 12μmol. However, a significant reduction in the brake specific fuel consumption and engine exhaust gas temperature was observed. Similarly, engine performance was analyzed by Bhale et al. (2009) on a four-stroke water-cooled single-cylinder diesel engine. The blends of MME, ethanol, and diesel in different proportions were used for the investigation. The results showed substantial improvement in engine performance at full load for an MME–ethanol–diesel blend as compared to neat diesel.

Sathiyamoorthi and Sankaranarayanan (2016) performed an experiment on a 'common rail direct injection' (CRDI) compression ignition (DICI) engine using biodiesel derived from lemongrass oil. BHA and BHT were used as antioxidant additives for the analysis. The authors found that on increasing the amount of BHA in the fuel blend LGO25 (25% lemongrass-oil-derived biodiesel + 75% diesel), the brake specific fuel consumption (BSFC) value was reduced significantly. An almost similar trend of BSFC was observed in the case of adding a BHT additive to the fuel blend. Also, a rise in the curve of the brake thermal efficiency trend was observed when increasing the content of antioxidant additives in LGO25.

The effect of a fuel stabilizing additive, namely acetone, on the performance of a HATZ two-cylinder 'direct injection' (DI) diesel engine, fueled with diesel and biodiesel derived from castor oil by adding recycled 'expanded polystyrene' (EPS), was successfully examined by Calder et al. (2018). The results showed that, at all engine conditions, a higher concentration of biodiesel with EPS and acetone has a better 'brake specific energy consumption' (BSEC) than petroleum-derived diesel fuel. Also, for all blends of biodiesel, EPS and acetone showed higher BTEs than diesel. Therefore, the highest value of BTE was recorded for B50 (50% castor oil biodiesel + 50% diesel) with EPS and acetone additives.

14.3.3 Effect of Additives on Exhaust Emission Characteristics

The emissions associated with the combustion processes are increasing enormously due to the global rise in transportation and energy systems (Abdel-Rahman, 1998).

Controlling emissions levels has been the prime objective of the enormous quantity of research studies and developments carried out all over the world. In this section, a survey of the previous literature on the impact of various additives on engine emissions is presented in detail.

Guru et al. (2010) investigated the effects of an organic-based synthetic Mg additive with biodiesel derived from chicken fat and a diesel blend B10 in a one-cylinder direct injection diesel engine. The results showed that doping of about 12 μmol of an Mg additive in the fuel blend results in an increase of NOx emissions by 5%. However, other harmful emissions such as CO and HC were reduced by about 13%. The utilization of a 10 vol% of ethanol additives in mahua oil biodiesel results in a significant reduction in CO, HC, and NO_x emissions as compared to neat biodiesel (Bhale et al., 2009). However, with a further rise in the ethanol concentration, the emissions were found to be higher.

A similar investigation was conducted on biobased derived additive ethanol by Sivalakshmi and Balusamy (2012). The blends of neem oil biodiesel and diesel were used for analyzing the exhaust emission characteristics. The NO_x emissions were found to be initially higher, with the addition of ethanol in the fuel blend. However, above a certain limit, the NOx emissions showed a decreasing trend with respect to higher loads. Moreover, at a higher ethanol concentration, CO and HC emissions were found to be higher for the diesel–biodiesel blends as compared to neat diesel.

Sathiyamoorthi and Sankaranarayanan (2016) examined the effect of antioxidant additives on the emission characteristics of an LGO25 fuel blend. The results showed that CO emissions were increased by 5.8% at 500 ppm, 10.1% at 1,000 ppm, and 14.8% at 2,000 ppm with the addition of a BHA antioxidant in an LGO25 fuel blend. Similarly, for a BHT antioxidant additive, the CO emissions were increased by 8.5% at 500 ppm, 13.2% at 1,000 ppm, and 16.60% at 2,000 ppm at full load conditions. This can be justified by the decrease in ignition delay of fuel blends with the addition of antioxidant additives and thereby encouraging the formation of CO emissions. A similar trend is observed for unburned hydrocarbon emissions, which were also found to be slightly higher with the adding of the antioxidants in the LGO25 fuel blend. However, the percentage of NOx emissions were slightly lower due to the usage of antioxidant additives.

The effect of a hybrid additive on engine emissions was successfully tested by Calder et al. (2018). The authors observed that, on adding EPS in a canola oil–biodiesel fuel blend with or without using acetone additive showed about 20%–30% less NOx emissions as compared to that of the diesel fuel. The logic behind the decreasing NOx emissions was the lower heating value of the fuel blend in the study. The acetone was used as a stabilizer in the fuel blends. Its presence decreases the calorific value and increases the viscosity which results in the poor atomization of the fuel blends at low engine speed. The results also illustrated that EPS dissolved in B50 blends with or without using acetone generates more smoke emissions as compared to diesel. Moreover, acetone added to B20 and B50 blends showed a reduction of about 58% in CO emissions as compared to neat diesel. Table 14.2 summarizes the effect of using different kinds of additives on engine combustion, performance, and emissions of biodiesel and its blends under various engine operating conditions.

TABLE 14.2
Variation in Engine Characteristics of Different Biodiesel Fuel by the Addition of Additives of Different Concentration

Biodiesel	Additive	Additive Concentration	Engine Specification	RPM	HRR	Peak Pressure	HC	CO	NOx	BP	BTE	References
CFME (B10)	Mg	12 μmol	Air-cooled DI engine, 18:1, 1-cylinder	2200 (at full load condition)	↑	↑	—	↓	↑	↓	↓	Guru et al. (2010)
MME	Ethanol	20 vol%	1-cylinder, water-cooled, 17.5:1	1500 (at different loading condition)	—	—	↑	↓	↓	↑	↑	Bhale et al. (2009)
NOME	Ethanol	5–10 vol%	Single-cylinder DI engine, 16.5:1	1500 (at various loading conditions)	↑	↑	↑	↓	↓ (At higher dose)	—	↑	Sivalakshmi and Balusamy (2012)
LGO	BHA	500–2000 ppm	Single cylinder, DI, air cooled,17.5:1	1500	—	—	↑	↑	↓	↑	↑	Sathiyamoorthi and Sankaranarayanan (2016)
	BHT	500–2000 ppm			—	—	↑	↑	↓	↑	↑	
COME[a] + EPS	Acetone	50 ml/L	Twin cylinder air-cooled, 20.5:1	1000–3000 (at various loading conditions)	—	—	—	↓	↓	—	↑	Calder et al. (2018)
COME[b]	PY	1000 ppm	TDI, 19.5:1	3000 (at various loading conditions)	—	—	—	↑	↑	—	↑	Kivevele et al. (2011)
POME	DTBP	1 vol%	Single cylinder, water-cooled, 14:1–18:1	1500	↑	↓	↑	↓	↓	—	↑	Musthafa et al. (2018)
HOME	Alumina nanoparticle	20–60 ppm		1500	↑	↑	↓	↓	↑	—	↑	Abdel-Rahman (1998)
JOME	MWCNT	10–50 ppm	1-cylinder, DI, air-cooled	1500–2500	↑	↑	↓	↓	↓	—	↑	El-Seesy et al. (2017)
MOME	TiO_2	100–200 ppm	1-cylinder, DI, air-cooled	1100 (at different loading conditions)	↑	↑	↓	↓	↓	—	—s	Yuvarajan et al. (2017)

CFME = chicken fat methyl ester, MME = mahua methyl ester, NOME = neem oil methyl ester, LGO = lemongrass oil, COME[a] = canola oil methyl ester, EPS = expanded polystyrene, COME[b] = croton oil methyl ester, POME = palm oil methyl ester, HOME = honge oil methyl ester, Mg = magnesium, BHA = butylated hydroxyanisole, BHT = butylated hydroxytoluene, PY = pyrogallol, DTBP = di-tert-butyl peroxide, JOME = jojoba oil methyl ester, MOME = mustard oil methyl ester, MWCNT = multi-walled carbon nanotubes, TiO_2 = titanium dioxide, DI = direct injection.

14.4 RECENT TRENDS AND ADVANCEMENTS IN THE FIELD OF ADDITIVES

As the world is moving towards sustainable development, transition towards eco-friendly additives has supported the need to explore the potentials of biobased fuel additives.

A wide range of research has been carried out over the past few years in the fuel industry to examine how environmentally benign additives can contribute to reducing the environmental impact of petroleum-based additives. Conventional biodiesel additives especially lubricant additives which are made up of synthetic esters are harmful when released into the environment and also very expensive to produce (Eisentraeger et al., 2002). Vegetable oils, such as soybean, sunflower, jojoba, and castor oil, have noteworthy properties which can improve the biodegradability and renewability of lubricity-based additives on the world market (Singh et al., 2017).

However, due to poor oxidative and hydrolytic stability, poor cold flow properties, and an affinity towards gumming, there remains considerable room for improvement in the widespread use of these lubricants. Similarly, the addition of bioglycerol-based fuel additives in biodiesel can assist in improving the engine performance and emission characteristics of diesel engines. Due to the burgeoning biodiesel production all around the globe, the valorization of waste glycerol is of vital importance for the sustainability of biodiesel industries. Triacetin, a by-product obtained from the acetylation of triglycerides and methyl acetate, has proved to be a promising oxygenated fuel additive for CI engines. The increase in oxygen content can reduce the knocking phenomenon in diesel engines. Also, a significant increase in brake thermal efficiency and a reduction in CO, CO_2, HC, and PM were observed with the addition of triacetin in biodiesel (Cornejo et al., 2017).

The introduction of different biologically based techniques to produce numerous green additives has been shown to improve the performance and extend the lifetime of biodiesel fuels (Arjanggi and Kansedo, 2019; Kumaravel et al., 2019). However, the widespread use of these additives is still in its developing stages due to the adverse reaction conditions and critical process parameters, which results in poor product yield. Thus, scientists are searching for new and economical pathways for the large scale production of these renewable fuel additives which can improve fuel economy, power, ignitability, and combustion.

14.5 CONCLUSION

Over the last ten years, biodiesel has emerged as an attractive alternative fuel for diesel engines because of its environmental sustainability and origination in renewable resources. In search of cleaner fuels and making biodiesel usage economically viable, additives are becoming an indispensable tool in the global trade. Additives cover a wide range of subjects and are categorized into various types according to their size, chemical compounds, state of matter, and functionality.

As already discussed in this chapter, the amalgamation of additives in biodiesel and its blends has a significant effect on fuel properties, such as viscosity, fire point, flash point, and calorific value, which furthermore influences the combustion, performance, and emission characteristics of biodiesel fuel. Various oxygenated fuel additives improve the combustion process and lower the in-cylinder pressure due to the higher latent heat of vaporization.

Similarly, ignition promoter additives improve the ignition delay characteristics and lower the chance of knocking in diesel engines. Some multi-functional additives also serve a dual role, such as decreasing the delay in ignition, improving the premixed combustion duration, and providing combustion stability. Recent advancements in the field of biobased additives have also set a new benchmark in the era of fuel additives.

Thus, as energy sources are upgraded towards cleaner and more renewable technology, the additives shared on the world market will likely increase over the next ten years. Despite the fact that a variety of additives addressing the shortcomings of biodiesel have been developed, the matter cannot be considered as finally settled. Also, very limited studies are available on the aspects of wear mechanisms, endurance, and durability on biodiesel engines due to the complexity and time-intensive nature of the tasks, thus opening a window of opportunity for further research in the domain of additives.

REFERENCES

Abdel-Rahman, A. A. 1998. On the emissions from internal-combustion engines: A review. *International Journal of Energy Research* 22: 483–513.

Abe, M., S. Hirata, H. Komatsu, K. Yamagiwa, and H. Tajima. 2015. Thermodynamic selection of effective additives to improve the cloud point of biodiesel fuels. *Fuel* 171: 94–100.

Ali, O. M., R. Mamat, and C. K. M. Faizal. 2013. Review of the effects of additives on biodiesel properties, performance, and emission features. *Journal of Renewable and Sustainable Energy* 5: 4792846.

Arjanggi, R. D. and J. Kansedo. 2019. Recent advancement and prospective of waste plastics as biodiesel additives: A review. *Journal of the Energy Institute* 2019: 08.005.

Basu, S. and A. Miglani. 2016. Combustion and heat transfer characteristics of nanofluid fuel droplets: A short review. *International Journal of Heat and Mass Transfer* 96: 482–503.

Bhale, P. V., N. V. Deshpande, and S. B. Thombre. 2009. Improving the low temperature properties of biodiesel fuel. *Renewable Energy* 34: 794–800.

Borsato, D., J. R. de Moraes Cini, H. C. da Silva, et al. 2014. Oxidation kinetics of biodiesel from soybean mixed with synthetic antioxidants BHA, BHT and TBHQ: Determination of activation energy. *Fuel Processing Technology* 127: 111–116.

Calder, J., M. M. Roy, and W. Wang. 2018. Performance and emissions of a diesel engine fueled by biodiesel-diesel blends with recycled expanded polystyrene and fuel stabilizing additive. *Energy* 149: 204–212.

Cao, L., J. Wang, C. Liu, et al. 2014. Ethylene vinyl acetate copolymer: A bio-based cold flow improver for waste cooking oil derived biodiesel blends. *Applied Energy* 132: 163–167.

Cornejo, A., I. Barrio, M. Campoy, J. Lazaro, and B. Navarrete. 2017. Oxygenated fuel additives from glycerol valorization. Main production pathways and effects on fuel properties and engine performance: A critical review. *Renewable and Sustainable Energy Reviews* 79: 1400–1413.

Danilov, A. M. 2015. Progress in research on fuel additives (review). *Petroleum Chemistry* 55: 169–179.

Eisentraeger, A., M. Schmidt, H. Murrenhoff, W. Dott, and S. Hahn. 2002. Biodegradability testing of synthetic ester lubricants: Effects of additives and usage. *Chemosphere* 48: 89–96.

El-Seesy, A. I., A. K. Abdel-Rahman, M. Bady, and S. Ookawara. 2017. Performance, combustion, and emission characteristics of a diesel engine fueled by biodiesel-diesel mixtures with multi-walled carbon nanotubes additives. *Energy Conversion and Management* 135: 373–393.

Guru, M., A. Koca, O. Can, C. Cinar, and F. Sahin. 2010. Biodiesel production from waste chicken fat based sources and evaluation with Mg based additive in a diesel engine. *Renewable Energy* 35: 637–643.

Hazrat, M. A., M. G. Rasul, M. M.K. Khan, N. Ashwath, and T. E. Rufford. 2019. Emission characteristics of polymer additive mixed diesel-sunflower biodiesel fuel. *Energy Procedia* 156: 59–64.

Hosseinzadeh-Bandbafha, H., M. Tabatabaei, M. Aghbashlo, M. Khanali, and A. Demirbas. 2018. A comprehensive review on the environmental impacts of diesel/biodiesel additives. *Energy Conversion and Management* 174: 579–614.

Jayanthi, P. and S. Rao. 2016. Effects of nanoparticles additives on performance and emissions characteristics of a DI diesel engine fuelled with biodiesel. *International Journal of Advances in Engineering & Technology* 9: 689–695.

Joshi, H., B. R. Moser, J. Toler, W. F. Smith, and T. Walker. 2011. Ethyl levulinate: A potential bio-based diluent for biodiesel which improves cold flow properties. *Biomass and Bioenergy* 35: 3262–3266.

Karthikeyan, S. and S. M. Silaimani. 2017. An emission analysis on the role of TiO_2 additives in kusum oil methyl esters on working characteristics of CI Engine. *Indian Journal of Chemical Technology* 24: 393–399.

Kivevele, T. T., L. Kristof, A. Bereczky, and M. M. Mbarawa. 2011. Engine performance, exhaust emissions and combustion characteristics of a CI engine fuelled with croton *Megalocarpus* methyl ester with antioxidant. *Fuel* 90: 2782–2789.

Kumar, N., and M. Tomar. 2019. Influence of nanoadditives on ignition characteristics of kusum (*Schleichera Oleosa*) biodiesel. *International Journal of Energy Research* 43: 3223–3236.

Kumaravel, S. T., A. Murugesan, C. Vijayakumar, and M. Thenmozhi. 2019. Enhancing the fuel properties of tyre oil diesel blends by doping nano additives for green environments. *Journal of Cleaner Production* 240: 118–128.

Lawan, I., W. Zhou, Z. Nasiru Garba, et al. 2019. Critical Insights into the effects of bio-based additives on biodiesels properties. *Renewable and Sustainable Energy Reviews* 102: 83–95.

Madiwale, S., A. Karthikeyan, and V. Bhojwani. 2017. A comprehensive review of effect of biodiesel additives on properties, performance, and emission. *IOP Conference Series: Materials Science and Engineering* 197: 012015.

Musthafa, M. M., T. A. Kumar, T. Mohanraj, and R. Chandramouli. 2018. A comparative study on performance, combustion and emission characteristics of diesel engine fuelled by biodiesel blends with and without an additive. *Fuel* 225: 343–348.

Prabu, A. 2017. Nanoparticles as additive in biodiesel on the working characteristics of a DI diesel engine. *Ain Shams Engineering Journal.* 9: 2343–2349.

Sathiyamoorthi, R. and G. Sankaranarayanan. 2016. Effect of antioxidant additives on the performance and emission characteristics of a DICI engine using neat lemongrass oil-diesel blend. *Fuel* 174: 89–96.

Shaafi, T., K. Sairam, A. Gopinath, G. Kumaresan, and R. Velraj. 2015. Effect of dispersion of various nanoadditives on the performance and emission characteristics of a CI engine fuelled with diesel, biodiesel and blends: A review. *Renewable and Sustainable Energy Reviews* 49: 563–573.

Sia, C. B., J. Kansedo, Y. H. Tan, and K. T. Lee. 2020. Evaluation on biodiesel cold flow properties, oxidative stability and enhancement strategies: A review. *Biocatalysis and Agricultural Biotechnology* 24: 101514.

Sindhi, V., V. Gupta, and K. Sharma, et al. 2013. Potential applications of antioxidants: A review. *Journal of Pharmacy Research* 7: 828–835.

Singh, Y., A. Farooq, A. Raza, M. A. Mahmood, and S. Jain. 2017. Sustainability of a non-edible vegetable oil based bio-lubricant for automotive applications: A review. *Process Safety and Environmental Protection* 111: 701–713.

Sivalakshmi, S. and T. Balusamy. 2012. Influence of ethanol addition on a diesel engine fuelled with neem oil methyl ester. *International Journal of Green Energy* 9: 218–228.

Sorate, K. A. and P. V. Bhale. 2015. Biodiesel properties and automotive system compatibility issues. *Renewable and Sustainable Energy Reviews* 41: 777–798.

Soudagar, M. E. M., N. N. Nik-Ghazali, M. A. Kalam, et al. 2019. The effects of graphene oxide nanoparticle additive stably dispersed in dairy scum oil biodiesel-diesel fuel blend on CI engine: Performance, emission and combustion characteristics. *Fuel* 257: 116015.

Srivastava, S. P. and J. Hancsok. 2014. *Fuels and Fuel-Additives*. Hoboken, NJ: John Wiley & Sons.

Srivyas, P. D. and M. S. Charoo. 2019. Effect of lubricants additive: Use and benefit. *Materials Today: Proceedings* 18: 4773–4781.

Tomar, M. and N. Kumar. 2019. Influence of nanoadditives on the performance and emission characteristics of a CI engine fuelled with diesel, biodiesel, and blends: A review. *Energy Sources, Part A: Recovery, Utilization, and Environmental Effects* 2019: 1623347.

Van Gerpen, J. H. and B. B. He. 2014. Biodiesel and renewable diesel production methods. Advances in biorefineries: Biomass and waste supply chain exploitation. In *Advances in Biorefineries*, ed. K. Waldron, 441–475. Cambridge: Woodhead Publishing.

Yuvarajan, D, M. D. Babu, N. Beemkumar, and P. A. Kishore. 2017. Experimental investigation on the influence of titanium dioxide nanofluid on emission pattern of biodiesel in a diesel engine. *Atmospheric Pollution Research* 9: 47–52.

15 Qualitative Characterization of Biodiesel Fuels

Basics and Beyond

Niraj Kumar
Md. Abul Kalam

CONTENTS

15.1 INTRODUCTION

Energy is essential for the existence of mankind. The past few decades have witnessed a multifold upsurge in energy requirements globally. This increase is essentially driven by exponential growth in technology and intensive economic growth as well as an unrestricted rise in the population. Energy demand is expected to rise further with each passing year. It has been estimated by the US 'Energy Information Administration' (EIA) hat the global energy requirement will increase by 28% from 2015 to 2040. Most of this growth is expected to come from the two Asian giants, China and India (EIA, 2017).

Conventional energy resources are under severe pressure to meet demand. Fossil fuels are a major source of energy. The most widely used fossil fuels by the industrialized and developing countries are crude oil, coal, and natural gas. Among these, crude oil is the most utilized for the conversion of energy, while the consumption of coal is second, followed by natural gas. The prime consumer of petroleum fuels is world surface transport. The limited known petroleum reserves and the concentration of such reserves in certain regions of the world have created an undesirable situation. Among petroleum fuels, diesel is used in transportation and power generation. The overbearing dependence on fossil fuels has created a disquieting situation particularly related to trade and the environment. The unwanted pollutants from 'compression engine' (CI) engines are the oxides of carbon (CO_x), the oxides of nitrogen (NO_x), soot, and 'unburned hydrocarbons' (UHC). The impact of these undesirable emissions can have a wider impact, including global warming, degraded air quality, acid rain, ozone layer exhaustion, and deforestation. These events are aggravated by the emissions generated by diesel engines, mainly smoke and NO_x.

Biodiesel has evolved as a potential alternative to diesel due to its ability to mitigate environmental degradation. Biodiesel can be produced from different feedstocks depending on crops that are compliant to a local climate. Intense research in recent years has shown the viability of biodiesel as a technical and commercial alternative to petrodiesel due to its inherent characteristics, such as renewability, biodegradability, enhanced lubricating ability, high cetane number, and availability in domestic regions, as well as having an environmentally friendly emission profile. The acceptability of biodiesel may look straightforward from the perspective of its properties, but that perception is somewhat deceptive since its suitability is hindered by some shortcomings, including oxidative instability, inferior cold flow properties, unsaturation, higher cost, lower volatility, water absorbency, and lower energy content (Mahmudul et al., 2017; Ong et al., 2013). The performance of an engine in many ways depends upon the properties of the fuel. The properties of biodiesel are a function of many factors, including the feedstock quality, the fatty acid composition of the feedstock, the process of conversion, as well as the catalyst and alcohol used, the post-production parameters, and the storage condition and duration. This implies that a probable variation in the properties of biodiesel is always possible and which may be anticipated by a number of variables. This chapter is an attempt to analyze and summarize these issues with the help of relevant data and technical information.

15.2 CHARACTERISTICS AND PROPERTIES OF BIODIESEL

15.2.1 Combustion Properties

15.2.1.1 Cetane Number

The 'cetane number' (CN) is a non-linear dimensionless parameter, which is a measure of the ignition quality of diesel engine fuel and which bears an inverse relation to ignition delay (ID). A high CN of a fuel is a most desirable quality which is a factor in the smoother and longer life of an engine and is directly or indirectly related to engine starting ability, emission profile, combustion, etc. A poor CN of a fuel may lead to diesel knock, inferior operation of engines – such as misfiring, engine deposits, piston tarnishing, and hard starting especially in cold climatic conditions – and poor tail pipe emissions.

Any new fuel, in order to qualify as diesel engine fuel, must possess a minimum CN set by different authorities. The minimum value of the CN for neat biodiesel is prescribed as 47 and 51 in the 'American Society for Testing and Materials' (ASTM) and European standard, respectively. The CN of biodiesel varies with the variation in the feedstock source and is much higher than that of diesel (Mahmudul et al., 2017; Ong et al., 2013). However, some values can be higher or even lower, as shown in Table 15.1. This large variation can be due to one effect or the cumulative effect of several factors, such as chemical composition, oil extraction and processing methodology, and climate condition of the feedstock cultivation area. The CN of saturated compounds such as myristic acid (C14:0), palmitic acid (C16:0), and stearic acid (C18:0) is much higher than those of unsaturated compounds and increases with chain length (Kumar et al., 2013).

15.2.1.2 Flash Point

The 'flash point' (FP) of a fuel is a measure of the minimum temperature at which the air–fuel mixture can exhale vapor to form an ignitable mixture in the vicinity of the surface of the fluid which will burn when it comes in contact with a spark or flame. In this process, a visible flame, either unsustainable or sustainable, can be detected. This is the property which shall be considered in assessing the overall flammability and safety of a fuel.

The FP of biodiesel is measured with ASTM D93 or EN 14214 standards which limit the flash point at min 100 and 120, respectively. The higher FP set for biodiesel is to ensure the removal of excess methanol during the production process as very small amounts of residue can reduce the FP significantly and may affect the durability of some of the constituents. The FP of biodiesel is significantly higher (159°C) than for petrodiesel fuel (58°C), which largely reflects the boiling points of the individual constituents present. In the case of petrodiesel, the branched and lower molecular weight components, which possess lower boiling points, lead to a reduction of the FP. Biodiesel, due to its higher boiling point (BP), can be considered a combustible fuel rather than a flammable fuel and hence a safer fuel to transport. However, the FP is a function of the storage as it can deteriorate as a consequence of the poor stability of esters (Kumar, 2017).

TABLE 15.1
Properties of Different Biodiesel

Oil Name	Density at 15°C (kg/m³)	Viscosity at 40°C (mm²/s)	Cetane Number	Iodine Number	Calorific Value (MJ/kg)	Acid Value (mg KOH/g)	Pour Point (°C)	Flash Point (°C)	Cloud Point (°C)	Cold Filter Plugging Point (°C)	Copper Strip Corrosion (3 h at 50°C)	Sulphur (%) (m/m)	Sulphated ash (%) (m/m)	Oxidation Stability (h, 110°C)
ASTM D6751	880	1.9–6.0	Min 47	–	–	Max 0.50	–15 to –16	Min 100–170	–3 to –12	19	Max 3	Max 0.05	Max 0.02	3
EN 14214	860–900	3.5–5.0	Min 51	Max 120	35	Max 0.5	–	>120	–	Max +5	Min 1	10	Max 0.02	6
Palm ME	864.42	4.5	54.6	54	–	0.24	15	135	16	12	1a	0.003	0.002	10.3
Coconut ME	807.3	2.726	–	–	–	0.106	–	114.8	0	–4	1b	3.2	0.006	3.55
Sunflower ME	880	4.439	49	–	–	0.027	–	160	3.4	–3	1a	0.2	0.005	0.9
Soybean ME	913.8	4.038	37.9	128–143	39.76	0266	2	76	9	11	1b	0.8	0.005	2.1
Peanut ME	848.5	4.42	53.59	67.45	40.1	0.28	–8	166	0	–	–	0	–	2
Rapeseed ME	882	4.439	54.4	–	37	–	–12	170	–3.3	–13	–	–	–	7.6
Safflower ME	888.5	5.8	56	–	38.122	–	–	148	–5	–	–	–	–	–
Mustard ME	931	6.13	55	–	43.42	0.37	–	–	3.2	–5	1a	–	–	–
Olive ME	–	4.5	57	–	–	0.19	–	178	0	–6	1a	1.9	0.005	3.3
Tobacco seed ME	888.5	4.23	51.6	136	44.6	0.3	–	165.4	–	–5	1a	–	0.0004	0.8
Neem ME	868	5.213	–	–	39.81	0.649	2	76	9	11	1b	473.8	>0.005	7.1
Calophyllum ME	888.6	7.724	51.9	85	–	0.76	–	151	38	–	1b	16	–	–
Rubber ME	–	5.81	–	–	36.5	–	–8	130	4	–	–	–	–	–
Mahua ME	874	5	65	–	37	0.41	6	208	–	–	–	164.8	–	–
Beef tallow ME	877	4.824	58.8	–	8	0.147	9	150	12	14	1a	7	0.005	1.6
Jatropha ME	879.5	4.8	51.6	104	39.23	0.4	2	135	2.7	0	1a	1.2	0.009	3.2
Pongamia ME	931	6.13	55	–	43.42	0.42	3	95	7	–	–	–	–	–
Cotton seed ME	876.7	4.11	55	–	40.430	0.19	6	153	7	1	1a	1.9	0.005	1.85
Jatropha ME	868.8	3.91	58.2	105	0.00	0.24	2.0	161.5	3.0	0.00	1a	8.01	0.003	9.40
Sunflower ME	856.9	4.15	56.5	103	40.179	0.14	–3	162.0	1.2	–2.5	1a	14.33	0.003	3.44
Canola ME	856.9	4.15	57.2	107	40.490	0.38	1.7	163.5	2.5	1.0	1a	13.97	0.009	4.22

Source: Mahmudul et al. (2017); Ong et al. (2013). ME: Methyl ester.

15.2.2 Physical Properties

15.2.2.1 Specific Gravity

'Specific gravity' (SG) can be defined by ASTM D4439 as the ratio of the density of a substance to the density of a standard substance in a prespecified condition. It is notable that the properties of fuel which are injection characteristics have to be precisely optimized in order to achieve proper combustion. Further, they can be efficiently applied as a precursor for estimating some of the important properties, whose direct measurement is difficult as well as having poor repeatability and reproducibility.

The SG of biodiesel is marginally higher than that of petrodiesel due to factors such as molecular weight, the free fatty acid (FFA) content, and the presence of unsaturation and water content. Among biodiesel fuels, the one having a lower molecular weight and a high degree of unsaturation exhibits a higher SG. Further, biodiesels with longer chain constituents have smaller values of density (Kumar et al., 2013).

15.2.2.2 Heat of Combustion

The 'heat of combustion' or 'calorific value' of a fuel is an indication of the energy chemically bound in it. Further, it signifies the amount of heat generated by combustion of fuel inside an engine that provides power to perform useful work. It is largely acknowledged that consumption of fuel is directly related to its volumetric calorific value.

Biodiesel has a lower calorific value due to the presence of 10–12% oxygen by weight. The degree of unsaturation in biodiesel obtained from different sources causes a difference in the carbon:hydrogen ratio. Hence, greater levels of unsaturation result in lower calorific values despite having a similar chain length. The energy content of biodiesel is directly related to chain length, since longer chain esters have more carbon but a similar number of oxygen atoms. Further, biodiesel fuels with larger ester head groups (such as ethyl, propyl, or butyl) generally possess a higher heat of combustion as a consequence of their larger carbon to oxygen ratios (Moser, 2009).

15.2.2.3 Distillation Curve

The 'distillation curve' for different fuels is unique. This curve is obtained using the ASTM D86 and helps to estimate fractional composition based on boiling points. A specific volume of fuel (0, 10, 50, 90, and 100%) boils off at a certain temperature. These temperatures help to establish the distillation curve. The volatility of a fuel has a great impact on the performance of diesel engines. A lower volatility indicates high distillation end points, longer combustion duration, as well as poor combustion, while 'vapor lock' is likely to occur with a highly volatile fuel.

As mixtures of a few comparatively similar compounds, fat and oil esters have a narrower as well as a higher boiling range relative to those of diesel. The boiling range of diesel is between 159°C and 336°C, while biodiesel exhibits a range of 293°C to 356°C. Thus, the distillation range for diesel and biodiesel is approximately 177°C and 63°C, respectively (Yang, 2008).

15.2.3 Flow Properties

15.2.3.1 Low Temperature Flow Properties

The 'cloud point' (CP), 'pour point' (PP), and 'cold filter plugging point' (CFPP) are the key parameters that largely determine the low temperature characteristic of esters. The CP refers to the temperature at which the separation of wax starts and the diameter of crystal forms is larger than 0.5 μm, while the PP is the minimum temperature below which the fluid fails to flow. The CFPP is the lowest temperature at which a certain volume of fuel entirely flows under certain conditions through a standardized filtration device within a specified interval. Biodiesel fuels show inferior 'cold flow properties'. The low-temperature performance of biodiesel is largely determined by the molecular structure as well as the nature of the feedstock and strongly depends upon the degree of saturation.

The unsaturated fatty esters, as a consequence of their much lower melting points, behave as solvents. The saturated esters which dissolve in a solvent, crystallize upon cooling at a comparatively higher temperature than that of an unsaturated compound. Unfortunately, the CP of a biodiesel derived from something of a highly saturated nature can rise up to an undesirable level and severely limit its blending with diesel (Lopes et al., 2008).

15.2.3.2 Viscosity and Surface Tension

The 'viscosity' of a fluid is a measure of its resistance to flow. Viscosities of biodiesels can be determined with the help of standards such as ASTM D445 or ISO 3104. The viscosity and surface tension are key properties, and their values vary in the range of 3.9–5.8 mm^2/s and 25–30 mN/m, respectively. These properties are higher in comparison to petrodiesel fuels and have pronounced effects on injection characteristics as well as altering the combustion pattern. The spray penetration is deeper, faster, and has a narrow spray plume angle along with a larger 'Sauter mean diameter' for biodiesel injection compared to those of petrodiesel. The variation in these properties is largely a function of the ester chain length, its nature (*cis* or *trans*), the degree of unsaturation, and alcohol moiety. Further, FFAs or compounds with hydroxy groups exhibit considerably higher viscosity (Knothe and Steidley, 2005b).

15.2.4 Storage and Stability

15.2.4.1 Oxidative Stability

A standard specification ASTM D6751 for B100 biodiesel and ASTM D7467 for biodiesel blends have been developed for the commercial distribution of biodiesel. Additionally, the standard for diesel D975 has been revised to incorporate 5% biodiesel mixing to meet the D6751 specification. However, EN 14112 is the currently followed standard for the stability of biodiesel. An inferior oxidative susceptibility of biodiesel and its blends restricts the wide applicability of biodiesel, which eventuates on aerobic contact during storage and handling.

Due to the environmental concerns, it is necessary to degrade fuels. However, degradation process also degrades the vital properties of biodiesel and seriously diminishes its acceptability.

'Biodiesel oxidation' involves a multi-step reaction. The primary products formed during the primary reactions are conjugated diene and hydroperoxides, which further chemically react to produce several secondary oxidation compounds, including shorter chain fatty acids, aldehydes, aliphatic alcohols, formic acid, formate esters, and species with higher molecular weights. Free radicals are formed through hydrogen abstraction and these reactions are accelerated when exposed to propitious conditions such as presence of light, heat, peroxides, and transition metal. These radicals in the latter stage produce peroxides after reacting with oxygen (Moser, 2009).

A wide variation in stability is reported among different biodiesels, depending upon the inbuilt natural antioxidant and its 'fatty acid methyl esters' (FAME) composition, which include a number of double bonds, the orientation of double bonds, the length of the carbon chains, and the types of the ester head groups. In addition, purification of biodiesel by distillation, which removes some of the natural antioxidants such as tocopherol, is also one of the reasons for lowering its stability. The oxidation of biodiesel can adversely alter fuel quality by affecting its properties and may lead to some operative issues.

15.2.4.2 Iodine Number

The 'iodine number' (IN) is a measure of the amount of unsaturation of an oil, fat, or wax, measured in grams of iodine absorbed by 100 g of a sample when formally adding iodine to the sample. The IN for biodiesel is higher than that of petrodiesel. It indicates the susceptibility of the oil or fat to undergo polymerization and the ability to form engine deposits. A limiting value of IN is specified as 120, 130, and 140 in EN 14214, EN 14213, and the South African standard, respectively, while it is not addressed in American and Australian standards. It decreases when higher alcohol is used in the production of biodiesel since the IN is the molecular weight dependent. The IN in some cases can be misleading as the same IN can be obtained from an infinite number of fatty acid profiles as well as different fatty acid structures, despite the propensity for oxidation to have extreme values (Knothe, 2002; Knothe et al., 2004).

15.2.5 Chemical Properties

Biodiesel can be derived from a number of feedstocks depending upon its domestic availability. Animal fats and plant oils are comprised mostly of fatty esters of glycerol (triacylglycerides or triglycerides). The triacylglycerides can be mono-, di-, or triesters of glycerol. Fatty acids may be classified as saturated (no carbon-carbon double bonds), monounsaturated (one C=C double bond), and polyunsaturated (two or more double bonds). FFA profiles of different biodiesel fuels are shown in Figure 15.1. The amount of each fatty acid in triglyceride and biodiesel largely determines their characteristics. In general, fatty acids or carboxylic acids are straight-chain compounds with carbon atoms ranging from three to eighteen. However, the chain length can be more or less particular for tropical oils, which is supplemented with lauric acid. The chain length and amount of unsaturation largely determine the chemical composition of fat and oil esters.

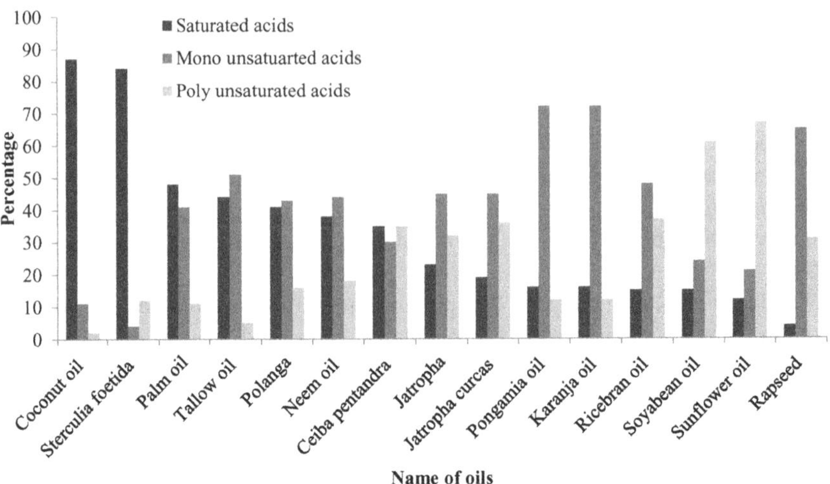

FIGURE 15.1 FFA profiles of different biodiesels.

Source: Mahmudul et al. (2017); Ong et al. (2013).

Biodiesel contains nearly 10–12% of oxygen by weight, which lowers its heat of combustion. The carbon content of biodiesel is nearly 15% lower and nearly the same as hydrogen in comparison to petrodiesel fuel on a weight basis. Biodiesel is essentially sulfur free. The amount of sulfur prescribed in ASTM D5453 is as low as 0.00011% by mass (1 ppm), whereas petrodiesel contains around 0.02% (200 ppm). This is advantageous for after-treatment devices and helps in reducing SO_x emission. Unlike petrodiesel, biodiesel is essentially non-aromatic, which helps to reduce emissions. Most of the biodiesel fuels possess a substantial amount of saturated long chain fatty acids which are quite comparable to long chain paraffins. Further, the amount of unsaturation/saturation affects some vital properties.

15.3 FACTORS AFFECTING BIODIESEL PROPERTIES

15.3.1 Alcohol

The application of 'straight vegetable oils' (SVOs) in 'compression ignition' (CI) engines is not new and is as old as the diesel engine itself. Since then, many researchers have experimented with SVOs in diesel engines with limited success. Some of the problems encountered with SVO utilization are the plugging of the fuel injectors and lines, carbon deposition on engine parts such as piston rings, the piston crown, and the cylinder head, dilution of lubricating oil, and fouling on the piston heads. These problems can be attributed to inferior properties of SVOs such as poor cold flow properties, high viscosity, high density, and low volatility. These properties can be amended through chemical modification such as transesterification, which converts vegetable oils (used or fresh) and animal fats into 'fatty acid alkyl esters' (FAAEs) or biodiesel. Transesterification involves the reaction of glycerides with an alcohol in the presence of some suitable catalyst.

The nature of the alcoholic head group of biodiesel has clear and pronounced effects on the yield as well as the properties of biodiesel. Branched and linear alcohols are employed in transesterification. However, methanol is widely used due to its advantages over other alcohols, such as lower price, its polar nature, and its shorter chain length. Shorter alcohols are more efficient in the transesterification reaction and shorter alkyl esters due to its polar nature imparting a sufficiently amphiphilic nature which leads to the head-to-head orientation of molecules, while a head-to-tail arrangement with much larger molecular spacing is reported for larger alkyl esters and non-polar head groups are detected due to the shielding of forces between more polar portions of the head group. Further, a larger head group creates rotational disorder in the hydrocarbon tail group.

Cold flow properties such as CFPP, CP, and PP in biodiesel can be improved by increasing the length of the hydrocarbon chain of alcohol moiety as well as by the branching of alcohols in FAAEs. In fact, branched-chain alcohols act as crystal growth inhibitors and help reduce the PPs of biodiesel fuels. Introduction of branches into linear, long-chain esters attenuate intermolecular associations, hence reducing crystallization temperatures (Malins et al., 2014). The crystallization temperatures of isopropyl esters of lard and tallow are comparable to those for methyl esters of soybean oil in spite of their higher saturation level. Canola and soybean biodiesel with branched head groups have shown reduction in CP by 3°C and 9°C, respectively (Knothe et al., 2005a; Lee et al., 1995).

Other properties, such as the heat of combustion, are directly related to and increase the chain length of FAAEs as shown in Figures 15.2 and 15.3. Among tested fuels, the lowest value was recorded for FAME1 (39.91 MJ/kg) of normal soybean oil, which is similar to the base oil, and the highest was obtained from '4-Methylpentan-2-ol' (FAMIBE6), which shows that the 'hydrophobic organic compounds' (HOC) increases with the increase in the hydrocarbon chain length. Figure 15.3 clearly shows that esters with branched alcohols have lower densities than those of esters with straight chain alcohols.

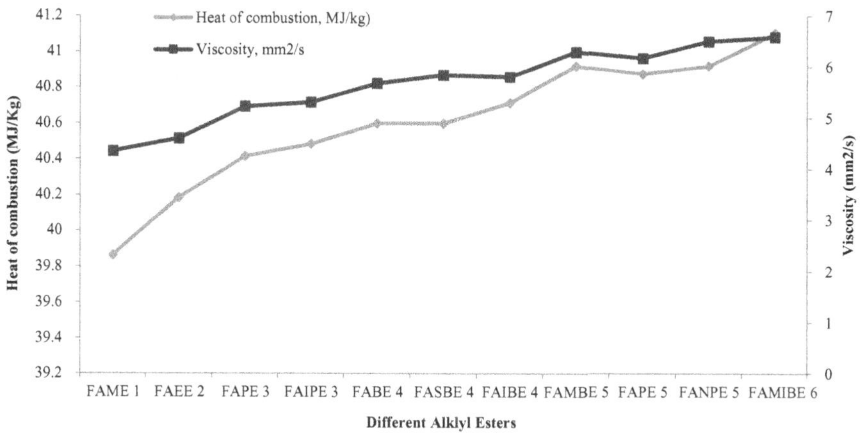

FIGURE 15.2 Heat of combustion and viscosity for different alkyl esters.

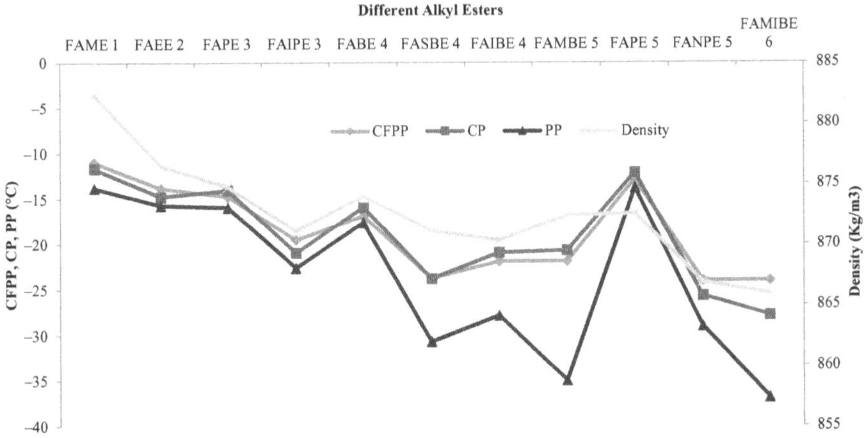

FIGURE 15.3 CFPP, CP, PP, and density for different alkyl esters.

Further, the recorded density was highest for FAME1 (883 kg/m^3) and lowest for FAMIBE6 (866 kg/m^3) while viscosity was lowest for FAME1 and increases with the increasing number of carbon atoms in the hydrocarbon chain of alcohol moiety. Esters with longer head groups have shown improved ignition quality. This signifies that increase in energy density and viscosity and decrease in density are observed with the increasing number of carbon atoms in the hydrocarbon chain length in alcohol moiety (Malins et al., 2014).

15.3.2 Fatty Acid Composition

Biodiesel is a mixture of different fatty esters usually derived from the parent oil, and each ester component determines the key properties of the individual fuel. The presence of a double bond, the location of unsaturation, the chain length, and the branching of the chain greatly influence the vital properties of biodiesel, such as the CN, heat of combustion, cold flow, oxidative stability, viscosity, and lubricity. A decrease in unsaturation (chemical structure) and an increase in chain length result in an increase in the CN. In fact, the presence of one long chain is enough to increase the CN despite the other moiety being branched.

The CN is higher for saturated compounds, such as myristic acid (C14:0), palmitic acid (C16:0), and stearic acid (C18:0), than for those of unsaturated acids. This phenomenon can be attributed to the formation of low-CN compounds during pre-combustion for less saturated esters. However, reduction in the CN can be achieved through the branching of aliphatic hydrocarbons, while the branching of the ester at the head group produces a CN that is similar to that for methyl esters (Canakci and van Gerpen, 2001; Knothe et al., 1998).

Viscosity is another key property which, in general, increases with the carbon chain length and decreases with its degree of unsaturation for esters. Further, double bond configuration is also an instrumental factor as a *cis* double-bond configuration exhibits a viscosity lower than that of a *trans* (Knothe and Steidley, 2005a). The cold flow properties of an ester are a strong function of the level of saturation. Saturated esters dissolve in unsaturated esters, where the latter act as solvents due to a comparatively lower melting point. Any decrease in temperature results in the crystallization of saturated fatty compounds in the mixture earlier than that of unsaturated esters. Hence, esters containing higher saturated fatty compounds show inferior cold flow properties. CFPPs are strongly correlated with the chain length and increase with increasing length (Lopes et al., 2008).

The presence of unsaturation greatly affects cold flow properties; however, the ratio of mono to di-unsaturated ester has negligible effect. A higher CN and higher cloud point can be observed for esters produced from saturated sources such as animal fats and palm and coconut oils. The formation of solid crystals in biodiesel has a greater influence on viscosity, volatility, flowability and, filterability (Smith et al., 2010).

The FFA also influences the oxidative stability of biodiesel and is inferior to petrodiesel. It can be linked to the presence of unsaturation and the number and position of double bonds. The oxidative stability increases with decreasing unsaturation as 'cotton oil methyl ester' (COME) (10.9% unsaturated fatty acid) exhibits superior stability compared to those of palm methyl ester (PME) (53.6% unsaturated fatty acid) and Jatropha methyl ester (JME) (78.9% unsaturated fatty acid). The reactivity of polyunsaturated esters is more pronounced than that of monounsaturated as a consequence of the higher number of reactive bis-allylic sites (-CH=CH-CH2-CH=CH-) present in polyunsaturated esters. The stability of such sites is poor for radical formation. Higher reactivity is due to the attraction of electrons adjacent to the double bond at either side of the methylene group; however, exposure to a radical initiator can easily remove it.

Further, linoleic acid with one bis-allylic position at C-11 and linolenic acids with two bis-allylic positions at C-11 and C-14 are more prone to oxidation in comparison to allylic positions. It is further elucidated that the oxidative reaction of esters decreases as the position of unsaturation moves closer to the ester head group and the hydroxyl groups are removed. Further, *cis*-isomers exhibit inferior stability compared to that of *trans*. However, the reactivity of conjugated *trans* unsaturations are higher than *cis* unsaturations. Though oxidative stability is related to unsaturation, the structural indices allylic position equivalent (APE) and bis-allylic position equivalent (BAPE) can provide a preferable indication for stability (Ramos et al., 2009; Shahabuddin et al., 2012. The degradation of biodiesel over time has a greater influence on some of the key properties.

The presence of oxygen moieties (10–12% by weight) in biodiesel leads to a deterioration in the calorific value. Further, the carbon:hydrogen ratio in biodiesel depends upon the degree of unsaturation and influences the heat content of the fuel. That is the reason why the longer carbon chain length of biodiesel results in an increase in the calorific value.

15.3.3 Contaminants

Despite the simplicity of the biodiesel conversion process, there is always a probability of detecting contaminants in the produced biodiesel, which may influence some of the properties. Different contaminants generally found in biodiesel are moisture content, free glycerin, bound glycerin, alcohol, free fatty acids, soaps, traces of metals, catalysts, unsaponifiable matter, and the products of oxidation. Trace metals assist in the formation of radicals as catalysts and promote the metal-mediated initiation reaction which leads to accelerating the free radical, causing deterioration in the stability of biodiesel. The exposure of metals to biodiesel deteriorates the stability due to the fact that metals act as a catalyst and stimulate the production of linoleate. Further, the reactivity of the transition metals, such as copper, nickel, manganese, cobalt, and iron, are higher, while aluminum, zinc, and tin have the least impact (Dodos et al., 2009; Fazal et al., 2014).

Fuel filter clogging can be encountered at temperatures significantly above the fuel's CP for biodiesel operated engines owing to the presence of unconverted partial acylglycerols, such as saturated 'monoacylglycerols' (MAGs) or 'free steryl glucosides' (FStGs), which promote the formation of a solid residue. However, cold flow properties are not sensitive to solid residue formation. In general, unsaponifiable matter, if present in the concentration below 2%, does not affect the CP and PP. However, the presence of high levels of bound glycerin such as saturated 'monoglycerides' (MGs) and 'diglycerides' (DGs) can catalyze crystallization and lead to an increase in viscosity. The CP of biodiesel increases with increasing concentration of saturated MGs or DGs. A saturated MG as low as 0.5% can significantly increase the CP. The saturated MGs have more pronounced effects as they have very low solubility in methyl esters and their insolubility increases with a decrease in temperature. Interestingly, the presence of DGs may hinder crystal formation by MGs.

However, the effect of SMGs is more pronounced compared to that of water and SG. Over time or upon heating, a metastable phase of 'saturated monoglycerides' (SMGs) is detected, which can transform into a more stable, less soluble polymorph. More saturated biodiesel exhibits a slower rate of phase transformation because of a slower rate of reaction. A very small concentration (above 0.24 wt%) of SMG may lead to failure of the cold soak filtration test. The presence of water can reduce the repercussions of SMGs. Further, saturated MAGs may form solid residues during extended storage because of the low solubility in ester.

The presence of MAGs (below 0.10 mass%) has been shown to have a higher sensitivity for CFPP compared to that of CP or FP. MGs are detected in biodiesel mainly because of the incomplete reaction of fats or oils during the transesterification reaction. At lower temperatures, MGs separate out of biodiesel as a slimy gum that quickly clogs paper filters. This phenomenon can be attributed to the partial solubility of MGs in biodiesel, while SG, which is naturally present in biodiesel as soluble fatty acid esters, also contribute to filter clogging as it crystallizes and agglomerates over time (Chupka et al., 2012; Dunn, 2009, 2012).

15.4 EFFECTS OF BIODIESEL PROPERTIES ON THE PERFORMANCE AND EMISSIONS OF DIESEL ENGINES

Biodiesel, unlike petrodiesel (which is a mixture of paraffinic, naphthenic, and aromatic hydrocarbons), is mainly a mixture of mono-alkyl ester of saturated and unsaturated long chain fatty acids, and thus distinguishes between basic properties and affects combustion, performance, and emissions of a diesel engine, which can be observed from the start of injection to combustion.

During injection, biodiesel show a shorter 'injection delay' compared to that of diesel. This can be attributed to a faster pressure rise due to a higher density and bulk modulus, resulting in the faster actuation of the fuel injection pump. The higher bulk modulus of compressibility causes a higher speed of sound, which leads to faster pressure-wave propagation. In addition, higher sound velocity and viscosity facilitate this phenomenon. However, this effect is more conspicuous in a conventional pump-line-nozzle fuel injection system and the probability of such an effect in a modern common rail fuel injection system is rather arguable as they are not sensitive to the bulk modulus of compressibility.

Further, a lower ignition delay is also reported for biodiesel, despite the larger temperature fall during evaporation in the preparation phase. This phenomenon may be attributed to inbuilt oxygen in biodiesel which improves ignitability. Further, high-molecular-weight biodiesel, when injected into a cylinder at the prevailing high temperature, may undergo a complex reaction and is believed to be split into smaller compounds of high volatility, which ignite earlier and reduce the delay period. During injection, biodiesel experiences higher friction with a nozzle surface due to higher viscosity and density, leading to a lower velocity of injection. In addition, biodiesel has a higher viscosity and surface tension as well as a lower Weber number in comparison to diesel.

All these factors contribute to a higher 'Sauter mean diameter' (SMD) for biodiesel. This effect is more pronounced with fatty acid, which has high carbon numbers (≥C16 saturates); those with mainly C18 carbon numbers have moderate SMDs. A higher SMD of biodiesel has greater implication for the performance of diesel engines as it leads to a reduction in the fuel spray angles and increases spray penetration, which may lead to inferior fuel–air mixing and fuel impingement. Inferior vaporization of fuel also contributes to longer penetration (Kumar et al., 2016; Lee et al., 2005).

Diesel combustion is generally divided into ignition delay, premixed combustion, controlled combustion, and after burning. The premixed combustion phase is largely governed by the characteristics of spray as well as ignition delay. Poor atomization and inferior fuel–air mixing of biodiesel lead to lower peak pressure. Further, peak pressure occurs earlier for biodiesel compared to that of petrodiesel, mainly due to a shorter ID and the availability of fuel-bound oxygen. The combustion for biodiesel starts earlier than diesel and exhibits a lower 'heat release rate' (HRR). The highest temperature for biodiesel (B80) is observed at round 6° CA 'before top dead center' (bTDC), while it is 1° CA earlier for petrodiesel. A lower mass of biodiesel accumulated due to a shorter ID and led to a lower premixed combustion heat release (Gumus, 2010).

In general, biodiesel utilization decreases the brake power (BP) of an engine. This can be attributed to the lower heat of combustion and the inferior combustion of biodiesel. However, some power recovery is also proclaimed for biodiesel as a consequence of in-built oxygen and a higher density, bulk modulus, and viscosity. More mass for the same volume of fuel is injected due to the higher density. At the same time, the higher viscosity of biodiesel helps in reducing the leakage, while in-built oxygen assists in better combustion, especially for the fuel-rich zone. However, power recovery does not compensate for loss of power. It is also observed that the BP for biodiesel produced with ethyl alcohol is slightly lower than that for the corresponding methyl alcohol. Further, power produced from more saturated biodiesels is anticipated to be higher than for those of unsaturated biodiesels, due to the higher calorific value, high CN, and higher viscosity.

Besides its negative contribution, higher viscosity does reduce leakage. Another important performance parameter, 'brake specific fuel consumption' (BSFC) for biodiesel fuels, as expected, is higher than that of petrodiesel due to the unifying effects of the inferior properties of biodiesel, such as higher density and viscosity and a lower calorific value (Chauhan et al., 2012; Kumar et al., 2013). However, some contradictory results are also reported. Opposed to BSFC, the 'brake thermal efficiency' (BTE) for biodiesel is less than that of petrodiesel as a consequence of inferior air–fuel mixing and spray characteristics as well as poor viscosity, volatility, and calorific value. In addition, the earlier commencement of combustion due to the higher CN of biodiesel results in the proliferation of compression work and heat loss, leading to a decrease in engine efficiency (Kumar et al., 2016; Rao et al., 2007).

Biodiesel is oxygenated fuel and is considered as an environmental friendly fuel. The application of biodiesel reduces CO emissions due to its lower C: H ratio of the biodiesel and oxygen content which facilitates better combustion and provides necessary oxygen to convert CO to CO_2. However, a large variation in this is reported in the literature which can be seen in Table 15.1. CO_2 emissions by biodiesel-fueled engines are generally higher than that of petrodiesel. The presence of fuel-bound oxygen provides enough oxidizer, especially to the oxygen starving pocket, thereby converting most of the carbon into carbon dioxide. However, the lower presence of carbon in biodiesel reduces CO_2 emissions.

Exhaust 'particulate matter' (PM) is the most complex diesel emission and is generally fractionated into three components, 'dry soot' (DS), sulfate, and 'soluble organic fraction' (SOF). PM is a cluster of solid carbon spheres formed in the local fuel-rich zone of the heterogeneous mixture of fuel and air due to the unavailability of oxygen. Soot can have up to 5,000 solid carbon spheres, whose diameters can vary from 9 nm to 90 nm. In diesel engines, soot forms in several steps: inception, growth, coagulation, and oxidation. Biodiesel as a fuel has strong effects on the amount, size, as well as the reactivity of the soot emitted by diesel engines.

The formation of diesel engine soot is greatly governed by spray and injection characteristics, the state of turbulence, air movement, prevailing pressure, and the temperature in the combustion chamber. Excess PM formed during the initial combustion stage oxidizes later in the combustion process at the boundary of the diffusive flame due to the prevailing high temperature and availability of oxygen. In the case of biodiesel combustion, the extra oxygen is provided by the fuel itself. Soot formation starts from the chemical initiation of combustion and fuel decomposition which produces small radicals.

Biodiesel considerably alters the nucleation step and initial formation of 'polycyclic aromatic hydrocarbons' (PAHs), thereby reducing the formation of soot. Further, extended combustion duration due to shorter ID allows more time for combustion and therefore reduced PM emission. The application of biodiesel also reduces the rate of surface growth due to a lower aromatic content, resulting in lower soot emission. Overall improvement in combustion is due to the presence of oxygenated fatty acids in biodiesel and a lower content of C-C bonds as well as aromatic reduction of PM emissions. Interestingly, biodiesel combustion reduces overall PM emissions but significantly increases the emission of finer PM.

The soot emitted by biodiesel contains a higher proportion of 'organic carbon' (OC) and a lower proportion of 'elemental carbon' (EC), hence the OC:EC ratio increases with an increasing proportion of biodiesel in fuel. Further, reactivity of the soot produced from biodiesel combustion is higher than that of diesel, mainly due to the highly disordered soot structure. It can also be attributed to the more amorphous internal primary structure of the primary soot particles as well as the higher surface to volume ratio due to the smaller size. Further, an increase in volatile fractions also increases the oxidative reactivity of soot (Czerwinski et al., 2013). Song et al. (2006) have proposed that the high reactivity of the biodiesel soot can be attributed to the fast and capsule type oxidation of biodiesel through internal burning. Hence, the reactivity is the result of structural transformation of the soot conglomerate surfaces.

No unanimous agreement has been found in the reported results related to biodiesel effects on NO_x emissions. Some results illustrate the effects of biodiesel properties in reducing NO_x emissions. The duration of premixed combustion is a function of CN. Biodiesel due to its higher CN reduces ignition delay and hence reduces the size of the premixed combustion which leads to a gentler rise in combustion pressure, allowing additional time for cooling through heat transfer and dilution. This decreases localized gas temperatures and hence lower NO_x formation. In addition, biodiesel which is free from aromatic and polyaromatic hydrocarbons helps in lowering the flame temperature and NO_x production. However, a large group of scientists have agreed that biodiesel application as fuel actually increases NO_x emissions. The reasons cited are the advanced start of combustion, rapid rate of burning, lowering of radiative heat transfer, lower adiabatic flame temperature, and system response issues.

It is well known that the presence of soot enhances radiation from the flame zone and assists in cooling diffusion flame temperatures, causing reduction in NO_x emissions. Biodiesel combustion reduces soot formation, thereby increasing flame temperatures by reducing soot radiation from the combustion zone. Biodiesel fuels mostly have higher unsaturation which leads to higher NO_x emissions. The prevailing high temperature in the combustion chamber provides a conducive environment for NO_x formation. It is also proposed that other possible reasons for the increase in NO_x emissions could be due to the increase in NO formation via the prompt NO mechanism rather, than the Zeldovich mechanism. Higher levels of certain hydrocarbon radicals are formed during the premixed phase, especially in the fuel rich zone, due to the presence of unsaturation in biodiesel. These radicals are believed to increase NO_x formation via a prompt NO pathway. The inbuilt oxygen of biodiesel also helps in increasing NO_x emissions by providing the necessary oxygen to monoatomic nitrogen present at high temperatures for conversion into NO_x.

Furthermore, the presence of unsaturation in biodiesel increases the adiabatic flame temperature and hence the increase of NO_x. It has also been shown that load is one of the factors that influence NO_x formation as biodiesel produces higher NO_x emissions at a higher load and the reverse for a lower load. However, the NO_x emissions from ethyl esters is higher than the corresponding methyl ester as a consequence of the higher bulk modulus of ethyl ester than that of methyl ester. The available data reflects the many theories that seek to understand the predominant experimental data showing that biodiesel contributes toward increased NO_x formation. In summary, these data can be grouped in terms of fuel compressibility effect, decreased radiative heat transfer, higher adiabatic flame temperature, and combustion phasing theories (Kumar, 2019; Saxena et al., 2018, 2019).

15.5 CONCLUSION AND SUMMARY

The properties of biodiesel depend largely on the choice of feedstock which often is dependent on a domestic source. The fatty acid compositions of the parent oil or fat, the nature of the alcoholic head group of the biodiesel, and the contaminants present determine the characteristics of the produced biodiesel. Intense research on biodiesel has outlined the significance of biodiesel characteristics as the expected performance and emission characteristics of engines are by and large governed by these properties. In general, unsaturated esters have lower CN, lower density, lower viscosity, inferior oxidative stability, and superior cold flow properties than those of saturated esters.

Further, superior cold flow properties can be achieved through increasing the length of the hydrocarbon chain of alcohol moiety as well as the branching of alcohols. The performance of biodiesel fuel engines is comparable to that of petrodiesel. However, a significant improvement in emissions except NO_x can be achieved by employing biodiesel as fuel. The presence of unsaturation and a shorter chain length tends to produce higher NO_x emissions. Some inferior properties and a large variation in these properties may hinder the wide acceptability of biodiesel. Hence, future research should focus on streamlining biodiesel properties and the genetic modification of feedstock.

REFERENCES

Canakci, M. and J. van Gerpen. 2001. Biodiesel production from oils and fats with high free fatty acids. *Transactions of the American Society of Agricultural Engineers* 44: 1429–1436.

Chauhan, B. S., N. Kumar, and H. M. Cho. 2012. A study on the performance and emission of a diesel engine fueled with Jatropha biodiesel oil and its blends. *Energy* 37: 616–622.

Chupka, G. M., L. Fouts, and R. L. McCormick. 2012. Effect of low-level impurities on low-temperature performance properties of biodiesel. *Energy & Environmental Science* 5: 8734–8742.

Czerwinski, J., P. D. Eggenschwiler, and N. Heeb, et al. 2013. Diesel emissions with DPF & SCR and toxic potentials with biodiesel (RME) blend fuels. *SAE Technical Paper* 2013-01-0523.

Dodos, G. S., F. Zannikos, and S. Stournas. 2009. Effect of metals in the oxidation stability and lubricity of biodiesel fuel. *SAE Technical Paper* 2009-01-1829.

Dunn, R. O. 2009. Effects of minor constituents on cold flow properties and performance of biodiesel. *Progress in Energy and Combustion Science* 35: 481–489.

Dunn, R. O. 2012. Effects of monoacylglycerols on the cold flow properties of biodiesel. *Journal of the American Oil Chemists' Society* 89: 1509–1520.

EIA. 2017. *EIA Projects 28% Increase in World Energy Use by 2040.* Washington, DC: Energy Information Administration.

Fazal, M.A., M. R. Jakeria, and A. S. M. A. Haseeb. 2014. Effect of copper and mild steel on the stability of palm biodiesel properties: A comparative study. *Industrial Crops and Products* 58: 8–14.

Gumus, M. 2010. A comprehensive experimental investigation of combustion and heat release characteristics of a biodiesel (hazelnut kernel oil methyl ester) fueled direct injection compression ignition engine. *Fuel* 89: 2802–2814.

Knothe, G. 2002. Structure indices in FA chemistry. How relevant is the iodine value? *Journal of the American Oil Chemists Society* 79: 847–854.

Knothe, G. and K. R. Steidley. 2005a. Kinematic viscosity of biodiesel fuel components and related compounds. Influence of compound structure and comparison to petrodiesel fuel components. *Fuel* 84: 1059–1065.

Knothe, G., and K. R. Steidley. 2005b. Lubricity of components of biodiesel and petrodiesel. The origin of biodiesel lubricity. *Energy Fuels* 19: 192–200.

Knothe, G., M. O. Bagby, and T. W. Ryan. 1998. Precombustion of fatty acids and esters of biodiesel. A possible explanation for differing cetane numbers. *Journal of the American Oil Chemists' Society* 75: 1007–1013.

Knothe, G., J. Krahl, and J. van Gerpen. 2004. *The Biodiesel Handbook.* Champaign, IL: AOCS Press.

Kumar, N. 2017. Oxidative stability of biodiesel: Causes, effects and prevention. *Fuel* 190: 328–350.

Kumar, N. 2019. Study of oxygenated eco-fuel applications in CI engine, gas turbine and jet engine. In *Advanced Biofuels: Applications, Technologies and Environmental Sustainability*, ed. A. K. Azad and M. Rasul, 405–441. Sawston, Cambridge: Woodhead Publishing.

Kumar, N., Varun, and S. R. Chauhan. 2013. Performance and emission characteristics of biodiesel from different origins: A review. *Renewable and Sustainable Energy Reviews* 21: 633–658.

Kumar, N., Varun, and S. R. Chauhan. 2016. Evaluation of the effects of engine parameters on performance and emissions of diesel engine operating with biodiesel blend. *International Journal of Ambient Energy* 37: 121–135.

Lee, I., L. A. Johnson, and E. G. Hammond. 1995. Use of branched-chain esters to reduce the crystallization temperature of biodiesel. *Journal of the American Oil Chemists' Society* 72: 1155–1160.

Lee, C. S., S. W. Park, and S. Kwon. 2005. An Experimental study on the atomization and combustion characteristics of biodiesel-blended fuels. *Energy Fuels* 19: 2201–2208.

Lopes, J. C. A., L. Boros, and M. A. Krahenbuhl, et al. 2008. Prediction of cloud points of biodiesel. *Energy and Fuels* 22: 747–752.

Mahmudul, H. M., F. Y. Hagos, and R. Mamat, et al. 2017. Production, characterization and performance of biodiesel as an alternative fuel in diesel engines: A review. *Renewable and Sustainable Energy Reviews* 72: 497–509.

Malins, K., V. Kampars, and R. Kampare, et al. 2014. Properties of rapeseed oil fatty acid alkyl esters derived from different alcohols. *Fuel* 137: 28–35.

Moser, B. R. 2009. Biodiesel production, properties, and feedstocks. *In Vitro Cellular & Developmental Biology - Plant* 45: 229–266.

Ong, H. C., A. S. Silitonga, and H. H. Masjuki, et al. 2013. Production and comparative fuel properties of biodiesel from non-edible oils: *Jatropha curcas, Sterculia foetida* and *Ceiba pentandra. Energy Conversion and Management* 73: 245–255.

Ramos, M. J., C. M. Fernandez, A. Casas, L. Rodriguez, and A. Perez. 2009. Influence of fatty acid composition of raw materials on biodiesel properties. *Bioresource Technology* 100: 261-268.

Rao, G. L. N., B. D. Prasad, S. Sampath, and K. Rajagopal. 2007. Combustion analysis of diesel engine fueled with jatropha oil methyl ester-diesel blends. *International Journal of Green Energy* 4: 645–658.

Saxena, V., N. Kumar, and V. K. Saxena. 2018. Biodiesel synthesis from *Acacia concinna* seed oil: A comprehensive study. *Energy Sources, Part A: Recovery, Utilization, and Environmental Effects* 40: 2009–2020.

Saxena, V., N. Kumar, and V. K. Saxena. 2019. Multi-objective optimization of modified nanofluid fuel blends at different TiO_2 nanoparticle concentration in diesel engine: Experimental assessment and modeling. *Applied Energy* 248: 330–353.

Shahabuddin, M., M. A. Kalam, H. H. Masjuki, M. M. K. Bhuiya, and M. Mofijur. 2012. An experimental investigation into biodiesel stability by means of oxidation and property determination. *Energy* 44: 616–622.

Smith, P. C., Y. Ngothai, Q. D. Nguyen, and B. K. O'Neill. 2010. Improving the low-temperature properties of biodiesel: Methods and consequences. *Renewable Energy* 35: 1145–1151.

Song, J., M. Alam, A. L. Boehman, and U. Kim. 2006. Examination of the oxidation behavior of biodiesel soot. *Combustion and Flame* 146: 589–604.

Yang, S. T. 2008. Bioenergy. In *Renewable Energy Focus Handbook*, 464-482. Cambridge, MA: Academic Press.

16 Use of Biodiesel Fuels in Diesel Engines

Anh Tuan Hoang
Van Viet Pham
Xuan Phuong Nguyen

CONTENTS

16.1 INTRODUCTION

Owing to their continuous depletion, fossil fuels are currently an unsustainable source of energy. They also trigger problems for the environment such as global warming and climate change. Therefore, biobased fuels with renewable, nontoxic, biodegradative characteristics should be an ecofriendly choice to replace fossil-based fuels. Biobased fuels could be considered bioproducts generated from biosources either for transportation or for combustion, which can be extracted from agricultural or forest products and which can even originate from the reusable and biodegradable portion of industrial and municipal wastes. Based on an analysis made at the end of 2017, around 50% of total renewable energy came from renewable bioenergy, which delivered four times the overall amount from the sun, solar cells, and wind power. Biofuels are forecast to lead the growth in the consumption of renewable energy in the period 2018–2023. Indeed, nearly 30% of the increase in bioenergy is bound to come from advanced biofuel generation, in liquid, solid, or gaseous form, which is believed to be suitable for use as fuel in internal combustion engines.

Biofuels are viable renewable energy supply sources because of their ecofriendly properties. Biodiesel is divided into four categories depending on the type of input feedstocks, according to the 'European Academies Science Advisory Council' (EASAC). Biodiesel is derived from first-generation edible oils, second-generation nonedible oils, and third-generation waste oils, which are naturally renewable and also locally found. Biodiesel of the fourth generation requires synthetic biology technology, though the research is still at the infancy level. As shown by the 'American Society for Testing and Materials' (ASTM), monoalkyl esters of long-chain lipids are labeled 'biodiesel' and classified as B100 and are generated using methanol and a catalyst through the process of transesterification of triglycerides, in which edible oils, nonedible oils, and waste oils or animal fats are considered as suitable feedstocks for producing 'fatty acid alkyl esters' (FAAE) . In a transesterification reaction, there are three categories of catalysts, namely strong alkaline, strong acid, and enzyme, which are usable, as well as methanol, a type of alkanol, which is used in most cases to manufacture biodiesel. Therefore, biodiesel is also referred to as 'fatty acid methyl ester' (FAME). In the biodiesel production process, glycerol (glycerin) is generated as a by-product.

Regarding the use of biodiesel for engines, there are a few drawbacks associated with higher 'nitrogen oxides' (NO_x) levels, low-temperature cold start issues, inferior calorific value, considerable copper strip degradation, as well as fuel piping challenges caused by high viscosity. For cold climates, the main concerns when using biodiesel regards cold starting and quality issues. Several other drawbacks relating to the use of biodiesel for diesel engines are shown in Figure 16.1. The aim of this chapter is,

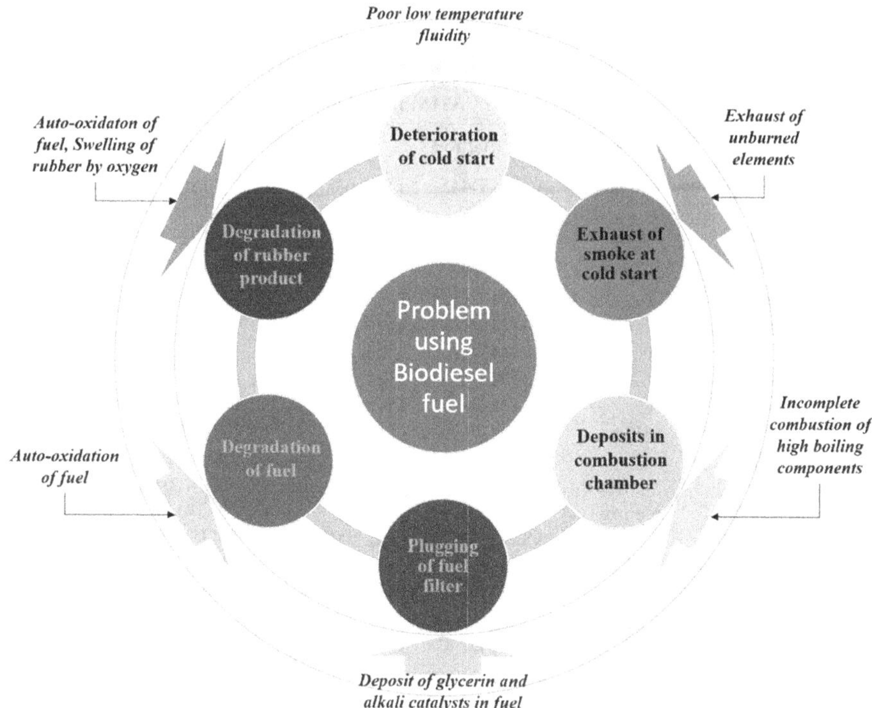

FIGURE 16.1 Drawbacks relating to the use of biodiesel for diesel engines.

therefore, to offer a review of using biodiesel as an alternative fuel regarding engine efficiency, engine performance, combustion and emission characteristics, corrosiveness, and the accumulation of deposits in engines.

16.2 BIODIESEL PROPERTIES

Advances in the quality of biodiesel are growing at a global level. Since biodiesel is made from various plants, its properties and qualities are different. Due to this reason, the standardization of biodiesel quality is essential to maintain the best possible efficiency of the engine. Some of the critical chemical and physical attributes of the fuel are kinematic viscosity (mm^2/s), oxidation stability, glycerin (% mass), cetane number (CN), sulfur content (% mass), pour point (°C), acid number (mg KOH/g oil), boiling point (°C), cloud point (°C), flash point (°C), density (kg/m^3), heating value (MJ/kg), ash content (% mass), water and sediment quality, copper corrosion, distillation rate, and carbon residues. The physicochemical properties of biodiesel fuels with standard limitations according to the test methods are described in Table 16.1.

TABLE 16.1
Physicochemical Properties of Biodiesel Fuels with Standard Limitations

Property specification	Units	ASTM D6751		EN14214	
		Test method	Limits	Test method	Limits
Kinematic viscosity at 40°C	mm^2/s	ASTM D445	1.9 to 6.0	EN ISO 3104	3.5 to 5.0
Density at 15°C	kg/m^3	ASTM D1298	880	EN ISO 3675	832 to 982
Flash point (FP)	°C	ASTM D93	Min: 30; Max: 90	EN ISO 3679	Max: 120
Boiling point (BP)	°C	ASTM D7398	100 to 615	—	—
Cloud point (CP)	°C	ASTM D2500	–3 to 12	—	—
Pour point (PP)	°C	ASTM D97	–15 to 16	—	—
Cetane number (CN)		ASTM D613	Min: 47	EN ISO 5165	Min: 51
Filter plugging point	°C	ASTM D6371	Max: +5	EN-14214	—
Calorific value (CV)	MJ/kg	ASTM D6751	—	EN-14213	Min: 35
Oxidation stability		ASTM D6751	Min: 3 h	EN-14112	Min: 6 h
Acid number (AN)	mg KOH/g	ASTM D664	Max: 0.5	EN-14104	Max: 0.5
Iodine number (IN)	mg I_2/g	ASTM D6751	—	EN-14111	120
Carbon residue	% mass	ASTM D4530	Max: 0.05	EN ISO 10370	Max: 0.3
Sulfate ash content	% mass	ASTM D874	Max: 0.002	EN ISO 3987	Max: 0.02
Water and sediment content	% volume	ASTM D2709	Max: 0.05% vol	EN ISO 12937	500 mg/kg
Copper strip corrosion		ASTM D130	Class 3	EN ISO 2160	Class 1

16.2.1 Viscosity

Viscosity is the most crucial fuel parameter because it signifies the capacity of a liquid substance to circulate; hence it involves the function of the fuel injection devices as well as spray atomization, especially at low temperatures, because viscosity is inversely proportional to temperature. The viscosity of biodiesel is two to four times higher than that of petrodiesel. This is because of its massive molecular mass and complex chemical structure. The standards ASTM D445, EN ISO 3104, or ISO 3104/P25 are used for assessing the kinematic viscosity of biodiesel fuel, which varies from 1.9 mm^2/s up to 6 mm^2/s as shown by ASTM requirements, and scales from 3.5 mm^2/s to 5 mm^2/s (Yusuf et al., 2011) under the 'European standards' (EN) requirements.

16.2.2 Density

Density is a physical property that is used to measure the exact quantity of fuel required to ensure sufficient combustion. Biodiesel is typically denser and less compressible than petrodiesel, which thus poses one of the most significant barriers toward widespread biodiesel use. Biodiesel density can be determined according to the specifications ASTM D1298 and EN ISO 3675. Based on such standards, biodiesel density is tested in a temperature range of 15°C to 20°C (Torres-Jimenez et al., 2011). The variation of densities for biodiesel generated from various feedstocks is typically between 832 kg/m^3 and 982 kg/m^3.

16.2.3 Flash Point

The 'flash point' (FP) is the temperature at which, when exposed to flames, the fuel will fire; the volatility of the fuel is inversely proportional to the FP. This is the method defined in ASTM D93, EN ISO 3679, and P21 for assessing the FP. Biodiesel has an FP higher than 150°C while petrodiesel fuel has it at 55–66°C. Biodiesel's FP is greater than the permissible petrodiesel limit for safe transport, processing, and storing purposes. However, the FPs of biodiesel are considerably smaller than those of vegetable oils. The FP reaches a maximum in ASTM D93 at 90°C, and is 120°C in EN ISO 3679 (Antolin et al., 2002).

16.2.4 Boiling Point

The boiling point (BP) is the temperature at which the vapor pressure of the element equals the surrounding pressure. If any element has a greater BP than this it is a sign of that element's lower volatility. The BP is dependent on the type of bond between the element's molecules. Gas chromatography is applied with the assistance of the ASTM D7398 standard to figure out the range of the BP, which is 100–615°C.

16.2.5 Cloud Point, Pour Point, and Cold Filter Plugging Point

An important quality criterion is the attributes of biodiesel at low temperatures. Partial or entire solidification of the fuel can trigger fuel lines and filters to be blocked, resulting in fuel starvation, difficulties starting, and damage to the engine because of inadequate lubrication.

Biodiesel's 'cloud point' (CP) is the degree at which wax crystals become observable when the biodiesel is chilled. Different feedstocks are used in the manufacture of biodiesel, and these have various fatty acid compositions, resulting in CP differences in the generated biodiesel. In ASTM D2500, the standard framework for assessing the CP (for biodiesel) is defined within the range of –3°C to 12°C (Moser and Vaughn, 2010a).

The 'cold filter plugging point' (CFPP) is used for testing the fuel's usability in cold flow. The CFPP is the minimum temperature at which a normal, precise filter discharges the sample fuel. The CFPP and CP are used for defining the filterability limit; the sample fuel CFPP value is less than the CP. The standards of ASTM D6371 and EN-14214 are applied to the biodiesel CFPP.

The 'pour point' (PP) is the temperature at which the solution's total wax is adequate to gel the liquid, hence it is thus considered as the minimum temperature at which the fuel can flow at. ASTM D2500, EN ISO 23015, and D97 processes are used for calculating the PP. Compared with petrodiesel, biodiesel has a greater PP.

16.2.6 Cetane Number

The 'cetane number' (CN) is one of the essential parameters identified during the selection process of biodiesel as an alternative fuel. The CN is the element that directly affects the delay in ignition. Normally, the superior quality of the ignition is always linked to an increased CN value. The higher CN indicates the fuel's capacity for self-igniting more quickly upon delivery into the combustion chamber. Biodiesel has a greater CN than petrodiesel, suggesting greater efficiency in combustion. The CN of fuels defined by ASTM D613 and EN ISO 5165 is 47 min and 51 min, respectively (Lapuerta et al., 2008).

16.2.7 Calorific Value

The 'heating value' or 'calorific value' (CV) of the fuel represents the amount of energy that is released when the fuel quantity is burnt. Fuel with a higher CV is desirable for an internal combustion engine. Biodiesel CV has been reported to be lower than that of petrodiesel fuel because of the 10–11% oxygen content in it (Moser, 2009). The CV is not designated in the biodiesel specifications ASTM D6751 and EN-14214, but it is recommended in EN-14213 (for heating purposes) with a minimum value of 35 MJ/kg (Skagerlind et al., 1995).

16.2.8 Stability of Oxidation

The stability of oxidation is a vital parameter for measuring the extent of the biodiesel response to air and the oxidation level. The quantity of bis-allylic sites found in unsaturated biodiesel substances affects the fuel's 'oxidation stability', which is determined by the biodiesel period, the composition of FAME, and the storage conditions. The molecular structure of biodiesel fuels makes it more degradable than petrodiesel fuel to oxidative depletion. Adding additives into biodiesel enhances and improves its quality. The 'Rancimat method' (EN ISO 14112) is the specification for oxidation stability in standards ASTM D6751 and EN-14214. For ASTM D6751, a minimum stability of oxidation at 110°C is 3 h, while in EN-14214 a more stringent limit of 6 h or more is set (Moser and Vaughn, 2010b).

16.2.9 Acid Number

The 'acid number' (AN) is characterized by the amount of 'free fatty acid' (FFA) contained in a fuel sample. The FFA would be the saturated or unsaturated monocarboxylic acid normally found in fats, oils, or greases. The greater acid value corresponds to a larger quantity of FFA. A higher AN causes the problem of corrosion in the engine's fuel injection system. The acid value is measured by mg 'potassium hydroxide' (KOH) essential to neutralize 1 g of FAME. Acid values are measured using ASTM D664 and EN-14104. Both criteria accept a total acid content of 0.50 mg of KOH/g for biodiesel (Agarwal, 2007).

16.2.10 Iodine Number

The 'iodine number' (IN) is an indicator of the number of double bonds present in biodiesel, and is used to determine the extent of 'biodiesel unsaturation'. This IN index greatly affects the oxidation stability and the polymerization of glycerides as well as deposit development in diesel engine injectors. The IN has been found to be strongly associated with the viscosity of biodiesel, the CN, and the CFPP. ASTM D6751 does not have an IN criterion, although EN-14111 specifies a total IN of 120 mg I_2/g (Atabani et al., 2013).

16.2.11 Carbon Residue

The 'carbon residue' (CR) of the fuel reflects the amount of carbon depositing fuel after burning. The CR is associated with FFA, more unsaturated fatty acids, glycerides, polymers, and other inorganic impurities. The limited range of the CR standard on the basis of ASTM D4530 is a maximum of 0.05 % (m/m) and of EN ISO10370 is a maximum of 0.30% (m/m) (Murugesan et al., 2009).

16.2.12 Sulfate Ash Content

The 'sulfate ash content' comprises synthetic pollutants such as abrasive solids, catalyst residues, and the concentration of soluble metal soaps present in a fuel. The amount of sulfate ash in biodiesel is lower than that of petrodiesel (0.02% m/m). Biodiesel would thus not pose any challenges when used as a fuel in diesel engines. Additionally, the level of sulfur oxide emitted during combustion is dependent on the level of sulfur in the fuel available. Biodiesel fuels derived from vegetable oils have less sulfur in comparison to petrodiesel. The maximum sulfate ash content standard based on ASTM D874is 0.002% mass and EN ISO 3987 is 0.02% mass.

16.2.13 Water and Sediment Content

The purity of the biodiesel is reflected by the content of water and the sediments available in it. Biodiesel is generally regarded as insoluble in water absorbing far more water than petrodiesel fuel. Biodiesel can include up to 1,500 ppm of soluble water while petrodiesel fuel typically absorbs only around 50 ppm (Atabani et al., 2012). With the occurrence of water in biodiesel, its CV is reduced; the existence of water is considered as one of causes of corroding engine components. The content of water and sediments is evaluated on the basis of the standards ASTM D2709 and EN ISO 12937. The water content and the sediment for biodiesel requirements are a total of 0.05% volume under both mentioned standards (van Gerpen, 2005).

16.2.14 Copper Strip Corrosion

The assessment of the corrosion characteristics of copper strips is a qualitative approach to assessing 'biodiesel corrosiveness', which might be caused by some sulfur compounds; this parameter is thus correlated with AN. In a fuel bath, a copper strip is warmed up to 50°C for 3 h to evaluate the rate of corrosion as compared to standard strips (Atabani et al., 2012). ASTM D130 and EN ISO 2160 standards are used to determine the corrosiveness of biodiesel to copper strips, the levels being class 3 and class 1, respectively.

16.3 PERFORMANCE AND EMISSION CHARACTERISTICS

16.3.1 Engine Performance

In addition to biodiesel properties, engine performance is found to depend greatly on the air turbulence degree in the combustion process, the quality of air–fuel mixing, injection pressure, the start of combustion, as well as other factors. It may also vary with the conditions of engine operation, such as engine speed and load. Normally, the blends of biodiesel–petrodiesel or pure biodiesel fuel (B100) can be directly used for diesel engines without any modifications. The influences of blends of biodiesel–petrodiesel on engine efficiency can be determined through engine power/torque, 'brake thermal efficiency' (BTE), 'brake specific fuel consumption' (BSFC), and emission characteristics.

16.3.1.1 Brake Power

The total produced power in a diesel engine depends on the amount of fuel pumped which can be efficiently combusted in the combustion chamber. Because of the lower CV of biodiesel in comparison to petrodiesel fuel, the maximum produced power in an unmodified diesel engine fueled with biodiesel or its blends is generally lower than that of petrodiesel fuel. Further, the output brake power/torque of diesel engines fueled with biodiesel or its blends derived from various feedstocks is mostly lower than those of petrodiesel fuel, barring a few exceptions. The extent of output power reduction changes according to various type of biodiesel. At all engine speeds, biodiesel's torque and power outputs are lower compared to petrodiesel fuel because the CV of biodiesel is 8–10% lower than that of petrodiesel fuel. Even so, fixed fuel volume injection at maximum torque can also lead to lower torque and power outputs from the biodiesel-fueled engine (Kawano et al., 2006). However, in some cases – rice-bran biodiesel with B5, B10, and B20 is an example – the diesel engine's maximum torque output is either equivalent to or slightly better than petrodiesel fuel (Sinha and Agarwal, 2005).

It is a fact that the torque output is marginally declined in a 'direct injection' (DI) diesel engine for higher biodiesel blends in comparison to petrodiesel fuel at lower engine speeds, although the torque is recorded at nearly the same at higher engine speeds (Figure 16.2). This is mostly because the CV of biodiesel is fairly low compared to petrodiesel fuel, although biodiesel density is higher than that of diesel fuel. Regarding mixed brake power/torque output from different tests, the differences of overall engine performance are insignificant in the case of using a lower biodiesel ratio, though such a difference, nonetheless, tends to increase with an increasing proportion of biodiesel. This stresses the necessity to optimize the biodiesel mixing proportion in terms of achieving either similar or slightly increased engine performance compared to petrodiesel fuel, which might be obtained by tuning/

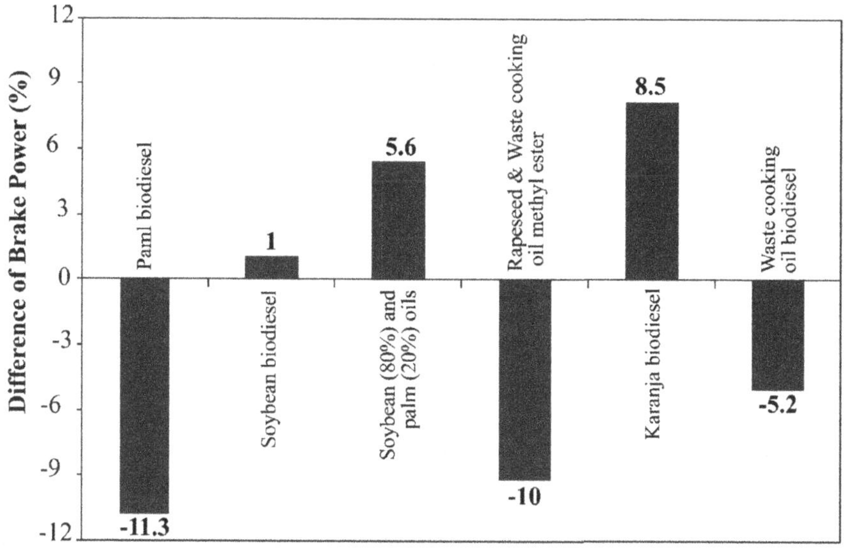

FIGURE 16.2 Comparison of percentage variation of brake power when using biodiesel and its blends.

recalibrating the engine's fuel injection system. As a result, it is not a severe problem to be addressed when introducing biodiesel as a potential fuel on a large-scale application in transportation.

16.3.1.2 Brake Specific Fuel Consumption

The 'brake specific fuel consumption' (BSFC) is the ratio of the consumed fuel mass to the engine-generated braking power. Normally, an increased BTE contributes to a reduced BSFC, while a reduced fuel CV is thought to increase the BSFC. Typically, the BSFC is increased in the context of biodiesel-fueled engines in comparison to petrodiesel-fueled engines, although there are some exceptions. A higher biodiesel mass is thus supposed to release an equal amount of heat to diesel fuel in the cylinder due to the higher density but lower biodiesel CV, as opposed to petrodiesel fuel, though an insignificant difference is detected between the BSFC of engines operated with B20 and petrodiesel fuel (Canakci and van Gerpen, 2003). These BSFC results are mainly due to a lower CV, higher viscosity and density of biodiesel, 'fuel injection pressure' (FIP), and the fuel injection technology employed.

In general, the BSFC of diesel engines running on biodiesels or its blends is slightly higher compared to that of petrodiesel fuel, in most experiments, and the BSFC depends greatly on the injection strategy and fuel properties. However, lower biodiesel blends increase the BSFC due to molecular oxygen in biodiesel components, which enhances the combustion process in the cylinder, resulting in the improvement of the BTE. Additionally, biodiesel that is produced by using a peroxidation process is found to have higher oxygen content and a higher number of saturated carbon-to-carbon bonds. Nonetheless, the application of a peroxidation process is believed to reduce the CV and CN of biodiesel. Therefore, the BSFC of engines running on both biodiesels produced with and without a peroxidation process is higher than that of diesel fuel in the reverse order of the CV.

Improving the 'compression ratio' (CR) for diesel engines is found to reduce the BSFC for both petrodiesel fuel and biodiesel, but the CR revealed considerable advantages for biodiesel in compared to petrodiesel fuel (Hoang et al., 2019a). This obtained result on higher engine performance for biodiesel compared to petrodiesel fuel is due to the comparatively low volatility and higher kinematic viscosity of biodiesel, its efficiency being surprisingly better at a higher CR. Moreover, some other factors such as operating conditions and biodiesel residues are also thought to have significant effects on the BSFC of diesel engines. The comparison of the percentage variation of BSFC using biodiesel and petrodiesel is illustrated in Figure 16.3. A rural mode of operation, for instance, shows a minimum BSFC when compared to urban and motorway modes. Generally, the increase in engine speed and load causes a decrease of the BSFC. Moreover, the BSFC of diesel engines also depends dramatically on the biodiesel type and engine type.

For example, *Gracilaria verrucosa*-based biodiesel blends display lower fuel consumption than the blends of rice-bran-based biodiesel at all injection timings: the reduction of fuel consumption is due to the difference in heating value. The BSFC for *Chlorella variabilis*-based biodiesel is about 6% lower than that of Jatropha-based biodiesel. In other experiments, the fuel consumption of diesel engine driving power generators running on blends of waste frying-oil-based biodiesel and petrodiesel fuel is also found to increase as the biodiesel ratio is decreased. In short, the increase in

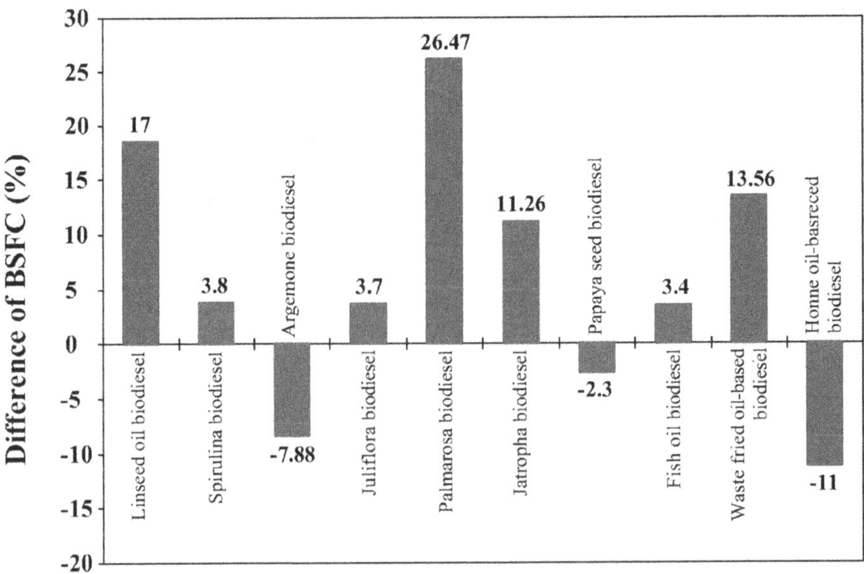

FIGURE 16.3 Comparison of percentage variation of BSFC using biodiesel and its blends.

biodiesel proportion in biodiesel–petrodiesel fuel blends reduces the fuel consumption. The average result of fuel consumption for B5 (5% volume of biodiesel) is 3.10 kg/h, while it is 2.64 kg/h of fuel consumption for B20 (20% volume of biodiesel) (Lue et al., 2001).

16.3.1.3 Brake Thermal Efficiency

The 'brake thermal efficiency' (BTE) is known as the ratio of brake power obtained to the fuel power supplied to the engine, and generally provides a basis to compare the effectiveness of fuel possessing different CVs. Normally, the BTE of an engine fueled with biodiesel, particularly at high load, is higher in comparison with petrodiesel fuel because the oxygen content of biodiesel is considered as improving the combustion efficiency, even for fuel-rich zones. This trend is due to the earlier 'start of combustion' (SOC) of biodiesel at lower loads which prolongs the combustion duration. In general, the combustion duration is shorter at higher engine loads. Nonetheless, at lower engine speeds and loads, the BTE for biodiesel is lower than diesel fuel because the vaporization characteristics of biodiesel is relatively inferior at lower temperatures in the cylinder (Jindal et al., 2010).

Moreover, the BTE can be further improved by adding a small proportion of alcohol. Biodiesel–ethanol and biodiesel–methanol blends are found to have a superior BTE at all operating conditions compared to petrodiesel fuel. This is because a small amount of alcohol in the blends is favorable for reducing the viscosity and density of the fuel, which enhances the atomization, resulting in combustion improvement. However, in the case of blends containing a higher fraction of alcohol, the cooling effect resulting from the higher latent heat of the vaporization of alcohols is a dominant factor, causing a lower BTE. Additionally, the BTE of biodiesel and its blends can also be improved by utilizing a higher FIP, which has positive effects on the biodiesel spray atomization and vaporization.

In general, engine speeds, engine loads, and percentages of biodiesel–petrodiesel blends have both positive and negative impacts on the BTE, although 20% (v/v) biodiesel blended with petrodiesel fuel can obtain an optimal BTE (Ramanathan et al., 2009). In summary, biodiesel from different feedstocks and their blends with petrodiesel fuel and/or alcohols can deliver a slightly higher BTE than petrodiesel fuel, except very few cases which have a lower BTE for biodiesel. Figure 16.4 shows a comparison of the percentage variation of the BTE using biodiesel and its blends compared to petrodiesel fuel.

16.3.1.4 Exhaust Gas Temperature

The 'exhaust gas temperature' (EGT) is a critical parameter that indicates the temperature of combustion in the cylinder. The findings of previous studies reported that the use of biodiesel resulted in an increase and decrease in the EGT. Biodiesel with a longer 'ignition delay' and higher viscosity is thought to contribute to higher combustion temperatures. This is also related to a high CN of biodiesel, which has impacts on reducing the premixing time and increasing the combustion efficiency (Hirkude and Padalkar, 2014). The increase in the biodiesel ratio in the blends decelerates the heating generation process during combustion and thus lowers the gas temperature.

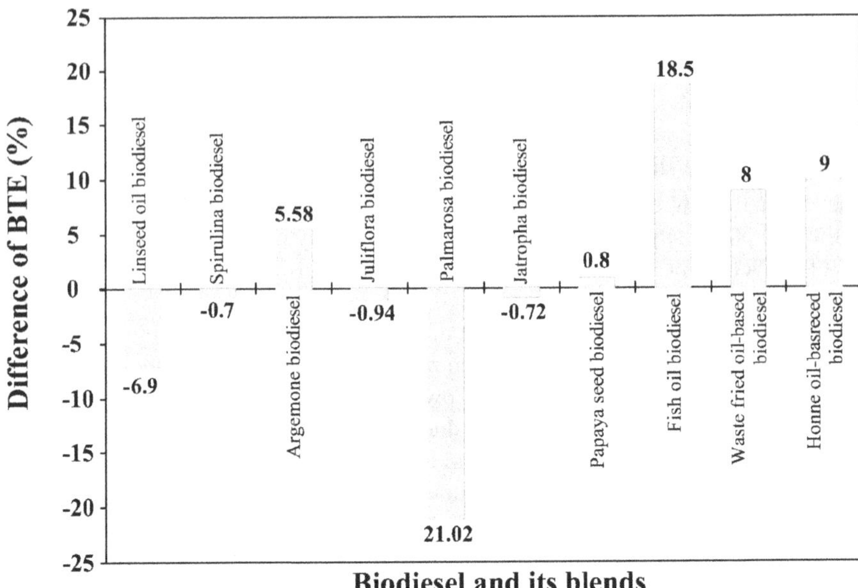

FIGURE 16.4 Comparison of percentage variation of BTE using biodiesel and its blends.

16.3.2 Combustion and Emission Characteristics

16.3.2.1 Combustion Characteristics

The engine performance and emission characteristics depend strongly on the combustion characteristics. Some basic indicators such as 'in-cylinder pressure', 'ignition delay' (ID) , combustion duration, and 'heat release rate' (HRR) are considered as core factors used to explain engine performance. Combustion analysis is helpful in understanding the influence of fuel properties on the characteristics of engine performance and emissions. Typically, the combustion characteristics of an engine are studied by analyzing the history correlation of in-cylinder pressure and crank angle.

The first issue worth mentioning is that the impact of biodiesel on ID is specified as the period between the beginning of injection and the SOC, and is determined by the 'crank angle degree unit' (°CA). Analysis of this parameter is significant for the use of various fuels. Atomization and vaporization of fuel happen to the injected fuel during this time, resulting in easy mixing with the compressed air, and the 'in-cylinder temperature' increases to the 'auto-ignition point'. The chemical reactions take place gradually, then they are intensified once there is inflammation. In general, biodiesel has a shorter ID than petrodiesel fuel. In the load event, a shortened ID with increased load is noted. The higher temperature of the combustion chamber walls and decreased dilution of the exhaust gas at higher loads could be explanations for this. Additionally, the ID is increased with increasing engine speeds. This is due to the oxygen content in biodiesel, which not only increases the ignitibility and separates the heavier fatty acid substances found in biodiesel into smaller compounds, produces more volatile matter, but also promotes earlier ignition.

For certain, the fuel's CN, engine speed, and engine load provide an impact on the ID of a diesel engine, as the CN of the fuel is normally inversely proportional to ID. This fact shows that the engine's ID is reduced by about 2°CA with increased biodiesel content in the test fuels and consequently reduces combustion duration (Hoang and Pham, 2019a), as the earlier SOC is due to a higher CN for biodiesel. Biodiesel is found to have a higher CN of 65 in comparison to a petrodiesel fuel of 51 (Eknath and Ramchandra, 2014). A further possible reason might be because of the complicated and rapid chemical reaction during the premixing combustion. Furthermore, a high 'in-cylinder gas temperature' induces the biodiesel to encounter thermal cracking, and light compounds could form, resulting in an initial auto-ignition. Similarly, low density and the CV of biodiesel are explanations for the shorter premixing combustion delay.

For biodiesel and its blends, the 'peak in-cylinder pressure' is higher than that of diesel fuel at low engine loads, yet they are the same as petrodiesel fuel at high engine loads. An explanation for this may be the rise in the ID with reducing engine loads. Combustion comes later with petrodiesel fuel than with biodiesel and its blends, due to the longer ID, which allows more time to mix air and fuel, resulting in low in-cylinder pressure. The biodiesel ratio in the biodiesel–petrodiesel blends does indeed have a considerable influence on in-cylinder pressure. The employment of biodiesel–petrodiesel blends in various ratios and pure biodiesel identifies a noticeable difference in the 'peak in-cylinder pressure' with increased biodiesel content (Hasan and Rahman, 2017). By contrast, there is also a slight decrease in in-cylinder pressure for biodiesel fuel with an increase in biodiesel content in its blends. These trends may result from the decreased CV of biodiesel fuel in comparison to petrodiesel fuel.

Typically, the biodiesel–petrodiesel blends provide a lower peak HRR in comparison to petrodiesel fuel. The reason for this is that the longer ID of petrodiesel fuel compared to biodiesel–petrodiesel blends or biodiesel has prompted more air and fuel to blend well. The maximum HRR is increased as engine loads increase from low to medium, but the trend of HRR for biodiesel is different at high engine loads. The rise in the maximum HRR at high loads is attributable to the fraction of fuel burned during the premixed combustion phasing with an increase in the injected fuel's oxygen fraction. However, the HRR is decreased when the proportion of biodiesel in the fuel is increased. Additionally, the physical characteristics of biodiesel are different from petrodiesel fuel, including kinematic viscosity, density, heating value, and CN, which affect significantly the HRR during the premix combustion phasing. As a result, higher kinematic viscosity, density, CN, and a lower heating value are considered as the main causes of a lower HRR (Shahabuddin et al., 2013). Premixed combustion HRR trends with different blends are plotted in Figure 16.5.

16.3.2.2 Emission Characteristics

Globally, the emission regulations for engines are becoming increasingly stringent. The emissions from diesel engines include hydrocarbons (HCs), CO, NO_x, and 'particulate matter' (PM) (Moser and Vaughn, 2010b). The well-known 2002 historical study by the US 'Environmental Protection Agency' (EPA) found that emissions of HC, CO, and PM decreased, but NO_x rose marginally, when

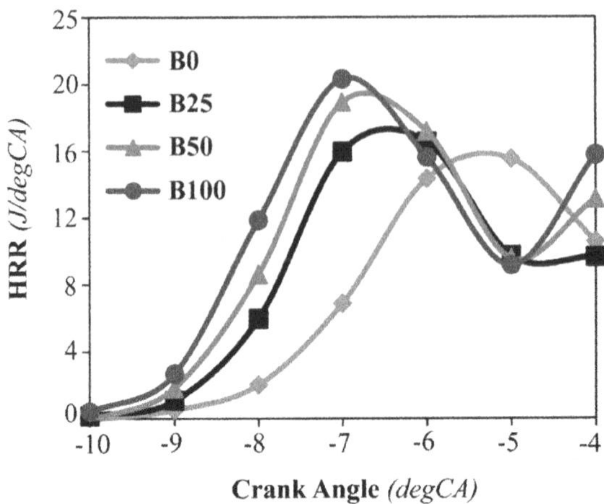

FIGURE 16.5 HRR trends using biodiesel and its blends.

concentrations of biodiesel blend were increased (Agarwal et al., 2017). NO_x and PM are the two sources of emissions that are being used with limitation as they negatively affect human health and living environment. For heavy-duty vehicles, the NO_x limits from 'Euro-III' to 'Euro-IV emission regulations' were reduced from 5.0 g/kWh to 3.5 g/kWh. Likewise, the NO_x limit has been lowered from 0.5 g/km to 0.25 g/km from Euro-III to Euro-IV on the basis of emission legislation for passenger cars and light commercial vehicles as well. These limits for NO_x and PM are also lowered in 'EURO-V regulations' to 0.18 g/km and 0.005 g/km, respectively (Hoang and Pham, 2019b). With the introduction of the EURO-V and 'EURO-VI emission regulations', PM concentration limits (6.10^{11} particles/km) will take effect in addition to PM emissions. Nevertheless, the amount of emissions ranges from engine to engine and depends on operating conditions, fuel quality, and engine design.

Nitrogen oxides are by far the most dangerous contaminants from engines and depend on the 'in-cylinder combustion temperature' and duration of this high temperature (above 1700 K) environment, chemical fuel composition, and oxygen accessibility in high-temperature areas. Along with the addition of biodiesel to petrodiesel, physical parameters, including the combustion chamber structure, oxygen concentration in the high-temperature combustion zones, together with the chemical structure of the burning mixture, also change. There are usually two different outcomes on NO_x emissions from biodiesel and its blends, and it is assumed that these emissions depend greatly on operational conditions and fuel properties.

One experimental group shows comparatively high NO_x emissions from biodiesel, and another experimental group generally implies lower NO_x emissions. These emissions from biodiesel are equivalent to petrodiesel fuel at part loads. They rise slightly due to increased temperature in the combustion chamber and the availability of an oxygen component. Moreover, higher NO_x emissions are attributed to a comparatively earlier onset of fuel injection in biodiesel blends, as opposed to petrodiesel fuel, mainly given the differences in the compressibility bulk modulus. There is still little explanation for increased NO_x emissions; yet some influencing factors, like fuel type and quality, fuel spray characteristics, operating conditions, and engine technology, are among the viable explanations.

NO_x emissions are heavily reliant on the equivalence ratio, concentration of oxygen, and temperature of the burning gas; and they are increased once biodiesel and blends are utilized. Mostly, the rise of these emissions results from the greater oxygen content of biodiesel or its blends, leading to the flame zone being similar to the stoichiometric of the charging air. In addition, the CN and the injection characteristics also influence the NO_x emissions of diesel engines running on biodiesel or its blends. The use of biodiesel leads to a more advanced and faster overall combustion event in modern high-speed diesel engines, contributing to elevated in-cylinder temperatures and enhanced formation of NO_x.

Normally, unsaturated biodiesel (with a higher iodine value) creates comparatively higher NO_x emissions in an engine fitted with an electronic injector. The impact of biodiesel unsaturation is generally lower in engines equipped with a 'common rail direct injection' (CRDI) system compared to engines installed with a unit injector. This indicates that a higher bulk modulus of biodiesel compressibility is not the only explanation for growing NO_x emissions. However, the effects of additives presented in biodiesel on NO_x emissions have not been thoroughly investigated. Furthermore, with rising engine loads leading to higher combustion chamber temperature, NO_x emissions are observed to rise. In some cases, the decrease in NO_x emission is nevertheless due to a lower heating value of biodiesel compared to petrodiesel fuel and a lower ID, which forces hot gases to stay in the combustion chamber at high temperature to generate less NO_x.

It is further discovered that adding longer chain alcohols lowers NO_x emissions in the range of 27.44, 19.27, and 15.05% for pentanol, butanol, and propanol compared to a 50% biodiesel blend. An explanation for this phenomenon is the lower CN of alcohol. In addition, NO_x emissions generally decrease with rising FIP, retarding the 'start of injection' (SOI), and employing split injection, although this trend is not regular and significant. Additionally, the lower heating values and higher CN of biodiesel reduce the ID and in-cylinder temperatures, leading to a decrease in the total NO_x concentrations in the combustion. Figure 16.6 shows a comparison of the percentage variation of NO_x emission using biodiesel and diesel.

Emissions of smoke and PM are an indication of inefficient combustion. PM emission includes three factors: sulfates, absorbed heavy-hydrocarbon, and condensed heavy-hydrocarbon. The emission of PM is mostly induced by the involvement of sulfur and ash in the fuel. The key parameters governing the degree of PM emission in a diesel engine are the 'diffusion flame combustion', oxygen ratio, and engine loads. The increased oxygen content of biodiesel boosts the process of combustion, though the sulfur content is insignificant. As oxygenated fuels accelerate the process of soot oxidation and trim down the local fuel-rich regions inside the spray cone, the accumulation of PM is minimized during the process of combustion.

Petrodiesel fuel with zero oxygen content shows a peak PM emission while the highly oxygenated (14.23%) biodiesel is reported to have the lowest PM emission. Consequently, the oxygen content of biodiesel minimizes soot precursors even under high load pressures, hence minimizing PM emissions. This decrease is due to the oxygen enrichment in biodiesel that compensates for the deficiency of oxygen in

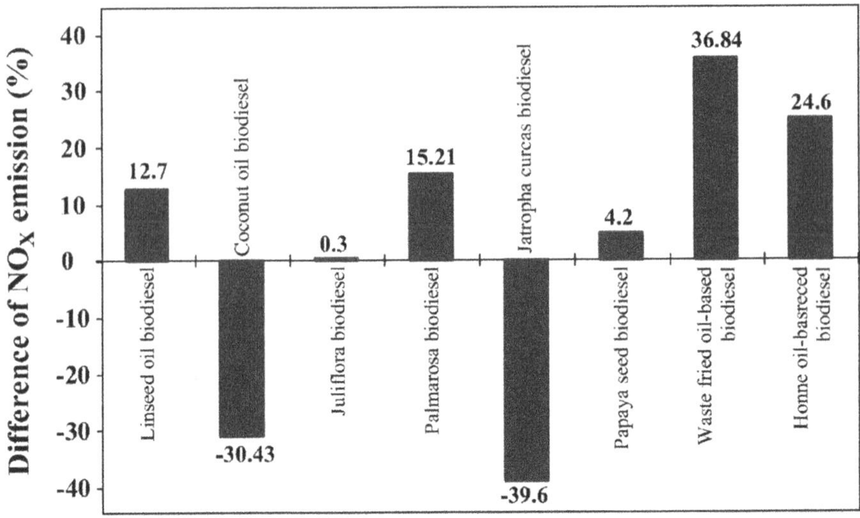

FIGURE 16.6 Comparison of percentage variation of NO_x emission using biodiesel and petrodiesel.

fuel-rich regions; the complete oxidation of biodiesel in fuel-rich regions in the combustion chamber results in a reduction in smoke emission. Additionally, increased 'injection pressure' (IP) for biodiesel is also found to decrease smoke density in emissions.

Results obtained are attributed to the shorter combustion period and the reduced fuel-rich zone, resulting in lower smoke levels at higher IP. Furthermore, the addition of oxygenated components, such as alcohol in biodiesel, also results in a significant decrease in PM at medium and high loads due to the blend's enriched oxygen content. On the other hand, the presence of oxygenated alcohol generates free radicals, such as OH-, H-, and O_2^-, which foster oxidation and suppress the development of soot precursors. In general, adding biodiesel into petrodiesel fuel results in a loss of PM mass output and smoke opacity, though the concentration of PM rises because of an increase in the number concentration of nuclei mode particles having a smaller size. The increase in these particles is believed to increase the formed soluble organic fraction because of relatively inferior evaporation characteristics of biodiesel in comparison with diesel fuel. Therefore, a lower fraction of biodiesel blends is found to be successful in reducing PM concentration.

The chemical composition of PM produced by biodiesel-fueled engines needs to be properly studied to understand their toxicity risk and formation mechanism. This will further help to propose changes in engine control strategies to reduce PM emissions along with other controlled pollutants when using biodiesel or its blends.

CO is the result of the intermediate combustion of fuels created by conditions such as a shortage of oxidants, inefficient mixing of air and fuel, and insufficient time for post-oxidation. It is generally assumed that CO emissions drastically drop when petrodiesel fuel is substituted with a biodiesel–petrodiesel blend (Figure 16.7); this

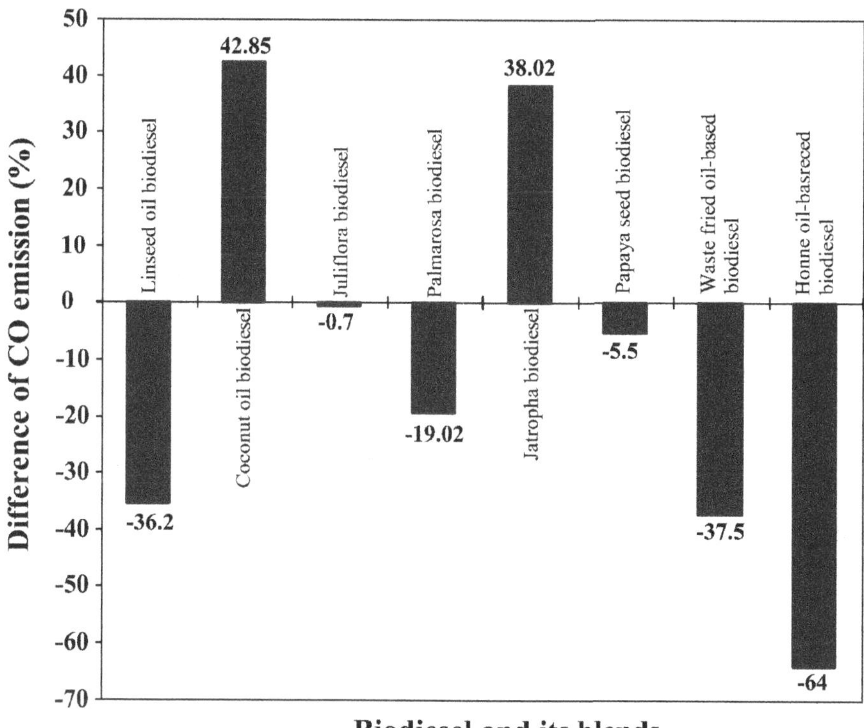

FIGURE 16.7 Comparison of percentage variation of CO emission using biodiesel and petrodiesel.

decrease in the amount of CO emissions becomes larger with an increase in the percentage of biodiesel in the blend. The primary reasons for lower emissions from biodiesel are the increased oxygen content in biodiesel that transforms CO to CO_2 and the higher CN. Evidence shows that, for biodiesel, an oxygen content of about 10.37–12.25% and a higher CN enhance combustion, which enhances the conversion of CO into CO_2, and thus reduces CO emissions.

In general, engine loads, engine speeds, the ratio of biodiesel in blends, and the engine type have significant effect on the CO emission of diesel engines. CO emissions increase with increasing engine speed. Similarly, a rising proportion of biodiesel in blends has an effect on lowering CO emissions. In an 'emission test cycle' (ETC) cycle, biodiesel showed the lowest CO emission in a countryside zone with the lowest engine load (Hoang and Pham, 2019b). Adding higher-order alcohols significantly increases CO emissions by an average of 39.95, 38.83, and 12.6% for propanol, butanol, and pentanol, respectively. However, for pentanol, it increases the higher concentration of local oxygen in the cylinder, which lowers the formation of CO compared to other alcohol blends (Uyumaz, 2018).

The emissions of unburned hydrocarbons (HCs) are the notable effects of incomplete combustion of fuel molecules within the engine cylinder. The main factors affecting HC emissions are engine operating conditions, combustion chamber structure, and fuel properties. Normally, HCs include all categories of hydrocarbon compounds produced by the engine, which cannot be assessed separately. Thus, they are clubbed together and identified as HC emissions equal to C_1, C_3, or C_6. The HC emission is due either to the mixing of air and fuel that are too lean to ignite automatically or to sustain propagating flames or to an air–fuel mixture that is too rich for auto-ignition. For biodiesel, higher oxygen content and higher CN are thought to encourage complete combustion, leading to lower HC emissions.

In addition, the methyl ester element in biodiesel leads to faster evaporation and more stable combustion; thus total emissions of HC are considerably lower. An explanation for this is based on the various hydrocarbon compositions found in heavy fuel oil as opposed to biodiesel. The EPA (2002) study addressing the effect of biodiesel on contaminants from heavy-duty diesel engines without 'exhaust gas recirculation' (EGR) and an 'after-treatment system' showed a general trend of a 65% and 21% reduction in HC emissions for B100 and B20, respectively (McCormick et al., 2005). In the case of changing loads, HC emissions for biodiesel blends are significantly reduced compared to petrodiesel fuel at part loads and full load. For B100, HC emissions are further reduced at lower loads in the case of retarded injection timings in comparison to the case of standard injection timing. Moreover, the use of B100 also reduces HC emissions at medium loads ($\lambda = 2–3$). Adding a small proportion of alcohol to blends of biodiesel further reduces HC emissions.

HC emissions are also restricted by engine operating conditions, such as increasing FIP and retarding injection timings. Comparison of the percentage variation of HC emission using biodiesel and petrodiesel is illustrated in Figure 16.8. The recent trend indicates that the use of biodiesel and blends contributes to a significant decrease in HC and CO emissions in lower FIP engines. However, this benefit of lower CO and HC levels in modern CRDI diesel engines running at comparatively high FIP is rather restricted because these engines are controlled by an 'electronic control unit' (ECU) and can be thus extensively optimized for fuel properties. For this reason, it is necessary to recalibrate the ECU to improve the emission advantage of biodiesel and its blends in comparison to petrodiesel.

16.4 ENGINE DURABILITY AND CORROSIVENESS CHARACTERISTICS

16.4.1 Engine Durability

Prior to the actual large-scale introduction of biodiesel as an alternative diesel fuel in the transport sector, questions raised about its compatibility with materials in current engines and fuel injection systems, due to the substantially different chemical compositions of biodiesel compared to petrodiesel fuel, have to be addressed. Durability testing of the biodiesel-using engine is therefore very important. In engine studies as well as in vehicular field trial studies, these durability tests must be conducted over a long period. Focusing on durability testing is actually much harder and more expensive because of a variety of factors that measure engine performance and emissions.

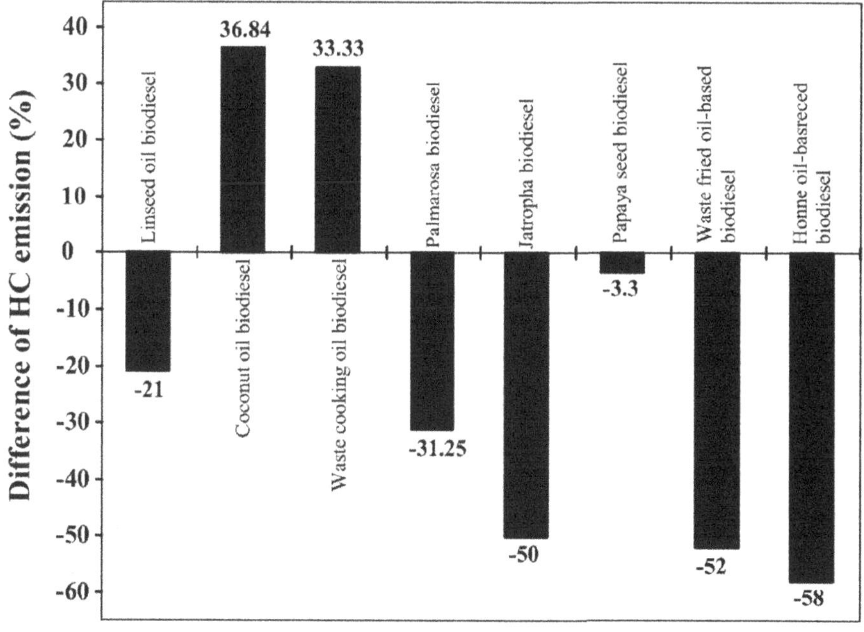

FIGURE 16.8 Comparison of percentage variation of HC emission using biodiesel and petrodiesel.

When testing the durability of diesel engines running on biodiesel or its blends, in comparison with diesel fuel, the assessment should be performed for injector coking, lubricating oil dilution, carbon deposits, ring sticking, fuel pump failure, and wear of other engine components (such as the cylinder liner and the big end bearing). After the durability test, surface roughness profiles at different locations as well as 'scanning electron microscopy' (SEM) are conducted. The statistics show no unusual wear in biodiesel–petrodiesel blends while using less than 15% of biodiesel. Moreover, retardation of injection timing is recommended for more than 15% of biodiesel to slow the rate of pressure rise, which effectively minimizes piston erosion.

16.4.2 Corrosiveness Characteristics and Inhibitors

The corrosion of metals exposed to biodiesel, and biodiesel degradation, are ongoing problems, as it takes a long time to test the corrosion, degradation, and durability of mechanical parts of engines fueled with biodiesel. Most engine pieces and fuel system parts are made of aluminum, copper, iron, and stainless-steel-based metals and alloys, and these materials are susceptible to corrosion. Aluminum/aluminum alloys can be utilized in small diesel engines for certain parts, such as piston components, cylinder heads, and engine blocks; copper and its alloys are used for fuel pump components and injector modules; nozzles, fuel filters, valve bodies, and pump rings are made of stainless steel. Generally, in the engine system, the corrosion phenomenon of metal mechanical parts can be divided into three groups, namely perforation, cosmetic, and edge corrosion. Chemical corrosion, though, is the main process; the corrosion products are metal oxides and the salts of fatty acids – the latter substance is the result of the reaction between metal oxides in the metal oxide layer and the fatty acids in biodiesel (Dharma et al., 2016).

Furthermore, it is worth mentioning galvanic corrosion, which occurs due to the very low conductivity of biodiesel, which thus has been deemed an electrolyte along with the polarized covalent bond of hydrogen-oxygen, hydrogen-carbon, and carbon-oxygen. Biodiesel is more polarized than petrodiesel fuel, which is attributable to the presence of the oxygen compound in biodiesel as compared to petrodiesel fuel with its oxygen electron negativity. The electrolytic dissociation of acid, water, or ester in biodiesel into oxygen-containing radicals or ion groups serves as a prime factor in the corrosiveness of biodiesel. Furthermore, the development of microorganisms, such as aerobic and anaerobic forms, is also known to be the key cause of increased acidity with a pH as much as 3.5 (Hoang et al., 2019b). Currently, ASTM D130 and ASTM D93 specifications have been used to calculate the corrosion degree of biodiesel. Nevertheless, such methods, along with a titration approach for evaluating the total acid number value, are not effective in assessing the degree of corrosion of independent organic acids.

In general, the degree of corrosion depends on unsaturated molecules, free fatty acid, the hygroscopic nature of biodiesel, and the types of materials that are in contact with the biodiesel. Consequently, the analysis of biodiesel color may be the easiest way to assess and check the corrosion of the above-mentioned metals/alloys. Other methods include the 'static immersion test' (SIT), where the immersion in various temperature conditions are the notable versions. SIT is usually more appropriate for test conditions, in which the specimens are measured pre and post-test. At the final stage of SIT, the corrosion behaviors of biodiesel-exposing metals are evaluated using a corrosion rate assessment.

Overall, the corrosion characteristics in copper biodiesel are the strongest, followed by aluminum, with carbon steel reported to have a lower resilience to corrosion than stainless steel but to be stronger than aluminum. Corrosion triggers serious consequences, so finding corrosion inhibitors is really vital to minimize and delay the corrosion phenomenon for metallic mechanical parts in engines. The common corrosion inhibitors are the amino-amines, oxyalkylated amines, diamines, primary amines, dodecyl benzene sulfonic acids, imidazolines, naphthenic acid, and phosphate esters. However, some metal corrosion inhibitors in biodiesel differ from those in petrodiesel fuel due to their many separate biodiesel physicochemical compounds (mainly fatty acid mono-alkyl esters) compared to petrodiesel fuel (mainly hydrocarbons). The corrosion inhibitors can only prolong corrosion start-up time without actually preventing it. In general, the inhibitors' working function is to build a stable layer of metallic oxide, which is tough to remove, on the metal surface when added to biodiesel.

16.5 DEPOSITS

16.5.1 Deposit Formation Mechanism

The formation process of deposits is divided into two classes – high molecular liquid substances and particles – and is recognized as a time function and in terms of physical conditions associated with temperature, fuel properties, flow rate, and concentration gradients. General deposits are described as a heterogeneous combination of carbon residue and a carbonaceous mixture blended with organic material in complicated oxidation processes to form the resin material. Typically, fuel properties (such as unsaturated bonds and ash content) and the release of lubrication oil into the combustion chamber by thermal cracking, along with incomplete combustion, are the main causes of deposit formation.

The presence of deposits throughout the engine has a huge effect on the fuel spray process, engine power, specific fuel consumption, efficiency of fuel use, heat transfer capability to the coolant, engine performance, and emission characteristics. In addition, there are some problems in relation to the engine system, such as the fuel supply system, the charging system, and the exhaust system. These effects could be indicated by: lower fuel rate, air-flow restriction, the negative effects on the cold-start and warm-up process, decrease of working volume for the combustion chamber, decrease of engine durability, and decrease of catalyst reactivity capacity, causing engine knock and a decrease of HC and CO emissions. Figure 16.9 illustrates the deposit formation mechanism in a diesel engine combustion chamber.

It can be seen from the above that the composition and characteristics of deposits have significant reliance on the composition of used fuels, the combustion temperatures of diesel engines, and the use of additives. Deposits with porous structures are seen as the main culprit that play a key role in HC emission generation. In addition, the porous structure of deposits has a noticeable impact on heat transfer and combustion over convection and radiation. Furthermore, the variability of the

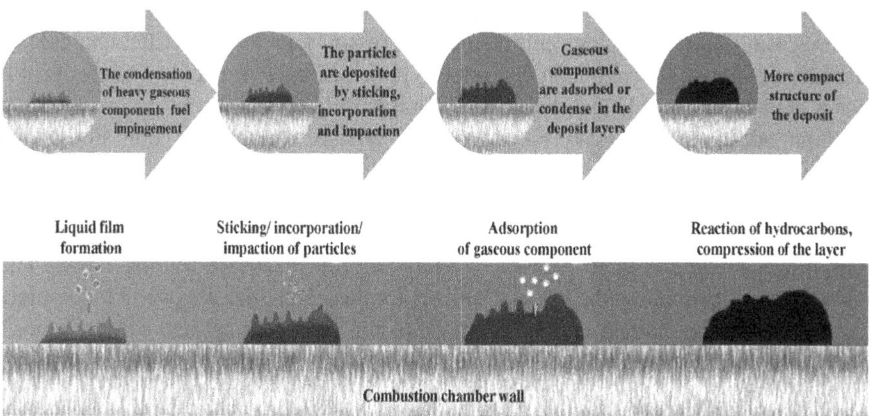

FIGURE 16.9 Deposit formation mechanism on combustion chamber wall.

chemical structure of deposits over the running time of diesel engines, which leads to deposit conductivity and the porous features of the internal microstructure, is directly related to intractability and difficult removal. In addition to the porous structure of deposits, the impervious attribute of oxidation as well as the burn-off of graphitic and condensed deposits also makes them difficult to be removed from diesel engines. For deposits in the injector, the adhesive is perhaps the main element in the inner injector, which sticks to plungers and internal valves, while unburnt fuel usually occurs at the tip of the injector, where deposits are found in the spray hole.

16.5.2 Deposits from the Use of Biodiesel

Deposits are formed first on the lowest temperature part of the combustion chamber, which is the nose of the injector, followed by the rings, cylinder wall, cylinder head, and piston crown. Deposits on the engine components result in the reducing of the surface area for heat transfer, which shortens engine lifetime and leads to deterioration of the lubricant oil. A combination of the lighter fraction evaporation in the fuel content and degradation has been proposed as a trigger for sticky deposition development. Moreover, this process of forming deposits is attributed to soot and PM, unburned fuel, volatilized lubrication oil, and pollutants. Deposits in the combustion chamber, piston, valves, and injectors already easily form, although diesel engine fuel injection equipment systems that are highly optimized do satisfy the various injection pressures required for many fuel types. In addition, biodiesel is classified as in the high and narrow distillation range, which prompts the condensation and accumulation of a thin liquid fuel film on the surface of the combustion chamber wall, resulting in ever-increasing development of deposits. In addition, poor atomization and low evaporation of biodiesel (due to high density, high kinematic viscosity, and surface tension), leading to larger fuel droplets and heterogeneity of the air–fuel mixture, are other causes of deposit formation (Hoang and Pham, 2019a).

The problem of pollution control that complies with tighter and more rigorous legislation indicates the importance in the approach of improving the combustion and injection events. The deposits on the spray tip are thus believed to be the main culprit in increasing toxic emissions and disturbing the spray pattern. A visual example of evaluating deposit formation on the pintle-type-based injector needle was carried out for a ‘Ricardo Comet V diesel engine’ fueled with biodiesel in two cases, with and without additives, to assess the effects of deposit build-up on emissions. The ‘emission test cycle (ECE) 15 test cycle may be utilized to calculate PM emissions after 1,000 km, which shows an increase in PM emissions proportional to the deposits. In other experiments, the levels of HC, smoke, and soot were recorded in proportional correlation to deposit development and a nozzle fouling index. The obtained results relate to the use of biodiesel for diesel engines are due to incomplete combustion (as explained in Section 16.2), piston ring sticking, and carbon deposit formation (because of the high FFA and AN) (Hoang and Pham, 2019b).

Furthermore, lower thermal stability and higher viscosity and density in comparison with diesel fuel are causes of an increase in deposit formation in diesel engines. However, the combustion process for different types of biodiesel is also different from their properties, resulting in a different degree in the deposit formation

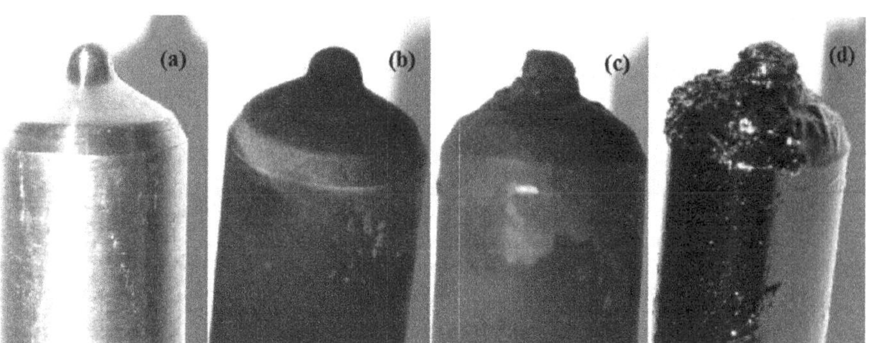

FIGURE 16.10 Optical investigation of deposit formation level on injectors: (a) before the test, (b) for diesel fuel, (c) B30, (d) B100.

in/on engine components. Experiments show a larger deposit formation level outside the injector nozzle of a diesel engine while using 100% biodiesel in comparison to diesel fuel (illustrated in Figure 16.10) (Hoang and Le, 2019; Hoang et al., 2019a). In general, the use of biodiesel is found to increase deposit formation in/on the components of engines, with the increasing degree of deposit being proportional to the increase in biodiesel percentage in its blends with petrodiesel fuel.

16.6 CONCLUSION

Biodiesel is made of oxygenated ester compounds produced from various feedstocks, with fatty acids as the main components which originate from vegetable oils, animal fats, and others. Biodiesel is widely utilized as an alternative fuel to petrodiesel fuel in diesel engines because of its comparable properties. The use of biodiesel or biodiesel–petrodiesel fuel blends for diesel engines may have significant effects on engine performance and emissions, deposit formation, problems relating to engine durability, and corrosiveness to engine components.

A mixture of petrodiesel fuel with a small portion of biodiesel (< 15% volume) has been proved to have better engine performance and emissions, except NO_x emissions, and it does not cause any noticeable issues for components, systems, and lubricating oil in diesel engines. Nonetheless, it is worth mentioning that the considerable problems when using biodiesel with a higher fraction must be addressed. Higher NO_x emissions could be solved through the application of advanced technologies, such as ‘exhaust gas recirculation’ (EGR) or exhaust gas-assisted fuel reforming. The use of additives and inhibitors could be highly effective in the reduction of deposit formation and dealing with lubrication oil degradation and corrosion of engine components.

In closing, biodiesel should be considered as a potential green fuel with renewable and friendly environmental characteristics. The application of advanced supporting technologies for modern engines aiming to use biodiesel will satisfy well the strict regulations on emissions and contribute to improving air quality and the living environment.

REFERENCES

Agarwal, A. K. 2007. Biofuels (alcohols and biodiesel) applications as fuels for internal combustion engines. *Progress in Energy and Combustion Science* 33: 233–271.

Agarwal, A. K., J. G. Gupta, and A. Dhar. 2017. Potential and challenges for large-scale application of biodiesel in automotive sector. *Progress in Energy and Combustion Science* 61: 113–149.

Antolin, G., F. V. Tinaut, and Y. Briceno, et al. 2002. Optimisation of biodiesel production by sunflower oil transesterification. *Bioresource Technology* 83: 111–114.

Atabani, A. E., A. S. Silitonga, and I. A. Badruddin, et al. 2012. A comprehensive review on biodiesel as an alternative energy resource and its characteristics. *Renewable and Sustainable Energy Reviews* 16: 2070–2093.

Atabani, A. E., A. S. Silitonga, and H. C. Hong, et al. 2013. Non-edible vegetable oils: A critical evaluation of oil extraction, fatty acid compositions, biodiesel production, characteristics, engine performance and emissions production. *Renewable and Sustainable Energy Reviews* 18: 211–245.

Canakci, M. and J. H. van Gerpen. 2003. Comparison of engine performance and emissions for petroleum diesel fuel, yellow grease biodiesel, and soybean oil biodiesel. *Transacions of the ASAE* 46: 937–944.

Dharma, S., H. C. Ong, H. H. Masjuki, A. H. Sebayang, and A. S. Silitonga. 2016. An overview of engine durability and compatibility using biodiesel-bioethanol-diesel blends in compression-ignition engines. *Energy Conversion and Management* 128: 66–81.

Eknath, R. D. and J. S. Ramchandra. 2014. Effect of compression ratio on energy and emission of VCR diesel engine fuelled with dual blends of biodiesel. *Journal of Engineering Science and Technology* 9: 620–640.

Hasan, M. M. and M. M. Rahman. 2017. Performance and emission characteristics of biodiesel-diesel blend and environmental and economic impacts of biodiesel production: A review. *Renewable and Sustainable Energy Reviews* 74: 938–948.

Hirkude, J. and A. S. Padalkar. 2014. Experimental investigation of the effect of compression ratio on performance and emissions of CI engine operated with waste fried oil methyl ester blend. *Fuel Processing Technology* 128: 367–375.

Hoang, A. T. and A. T. Le. 2019. A review on deposit formation in the injector of diesel engines running on biodiesel. *Energy Sources, Part A: Recovery, Utilization, and Environmental Effects* 41: 584–599.

Hoang, A. T. and V. V. Pham. 2019a. A study of emission characteristic, deposits, and lubrication oil degradation of a diesel engine running on preheated vegetable oil and diesel oil. *Energy Sources, Part A: Recovery, Utilization, and Environmental Effects* 4: 611–625.

Hoang, A. T. and V. V. Pham. 2019b. Impact of Jatropha oil on engine performance, emission characteristics, deposit formation, and lubricating oil degradation. *Combustion Science and Technology* 191: 504–519.

Hoang, A. T., A. T. Le, and V. V. Pham. 2019a. A core correlation of spray characteristics, deposit formation, and combustion of a high-speed diesel engine fueled with Jatropha oil and diesel fuel. *Fuel* 244: 159–175.

Hoang, A. T., M. Tabatabaei, and M. Aghbashlo. 2019b. A review of the effect of biodiesel on the corrosion behavior of metals/alloys in diesel engines. *Energy Sources, Part A: Recovery, Utilization, and Environmental Effects* 42: 2923–43.

Jindal, S., B. P. Nandwana, N. S. Rathore, and V. Vashistha. 2010. Experimental investigation of the effect of compression ratio and injection pressure in a direct injection diesel engine running on Jatropha methyl ester. *Applied Thermal Engineering* 30: 442–448.

Kawano, D., H. Ishii, Y. Goto, A. Noda, and Y. Aoyagi. 2006. Application of biodiesel fuel to modern diesel engine. *SAE Technical Paper* 2006-01-0233.

Lapuerta, M., O. Armas, and J. Rodriguez-Fernandez. 2008. Effect of biodiesel fuels on diesel engine emissions. *Progress in Energy and Combustion Science* 34: 198–223.

Lue, Y.-F., Y.-Y. Yeh, and C.-H. Wu. 2001. The emission characteristics of a small DI diesel engine using biodiesel blended fuels. *Journal of Environmental Science and Health, Part A: Toxic/Hazardous Substances and Environmental Engineering* 36: 845–859.

McCormick, R. L., C. J. Tennant, and R. R. Hayes, et al. 2005. Regulated emissions from biodiesel tested in heavy-duty engines meeting 2004 emission standards. *SAE Technical Paper* 2005-01-2200.

Moser, B. R. 2009. Biodiesel production, properties, and feedstocks. *In Vitro Cellular & Developmental Biology- Plant* 45: 229-266.

Moser, B. R. and S. F. Vaughn. 2010a. Coriander seed oil methyl esters as biodiesel fuel: Unique fatty acid composition and excellent oxidative stability. *Biomass and Bioenergy* 34: 550–558.

Moser, B. R. and S. F. Vaughn. 2010b. Evaluation of alkyl esters from *Camelina sativa* oil as biodiesel and as blend components in ultra low-sulfur diesel fuel. *Bioresource Technology* 101: 646–653.

Murugesan, A., C. Umarani, and T. R. Chinnusamy, et al. 2009. Production and analysis of bio-diesel from non-edible oils: A review. *Renewable and Sustainable Energy Reviews* 13: 825–834.

Ramanathan, A R, G. R. Kannan, K. R. Reddy, and S. Velmathi. 2009. The performance and emissions of a variable compression ratio diesel engine fuelled with bio-diesel from cotton seed oil *Journal of Engineering and Applied Sciences* 4: 72–87.

Shahabuddin, M., A. M. Liaquat, H. H. Masjuki, M. A. Kalam, and M. Mofijur. 2013. Ignition delay, combustion and emission characteristics of diesel engine fueled with biodiesel. *Renewable and Sustainable Energy Reviews* 21: 623–632.

Sinha S. and A. K. Agarwal. 2005. Performance evaluation of a biodiesel (rice bran oil methyl ester) fuelled transport diesel engine. *SAE Technical Paper* 2005-01-1730.

Skagerlind, P., M. Jansson, B. Bergenstahl, and K. Hult. 1995. Binding of *Rhizomucor miehei* lipase to emulsion interfaces and its interference with surfactants. *Colloids and Surfaces B: Biointerfaces* 4: 129–135.

Torres-Jimenez, E., M. S. Jerman, and A. Gregorc, et al. 2011. Physical and chemical properties of ethanol-diesel fuel blends. *Fuel*, 90: 795–802.

Uyumaz, A. 2018. Combustion, performance and emission characteristics of a DI diesel engine fueled with mustard oil biodiesel fuel blends at different engine loads. *Fuel* 212: 256–267.

Van Gerpen, J. 2005. Biodiesel processing and production. *Fuel Processing Technology* 86: 1097–1107.

Yusuf, N. N. A. N., S. K. Kamarudin, and Z. Yaakub. 2011. Overview on the current trends in biodiesel production. *Energy Conversion and Management* 52: 2741–2751.

17 Biodiesel Promotion Policies

A Global Perspective

Shashi Kumar Jain

Sunil Kumar

A. Chaube

CONTENTS

17.1 INTRODUCTION

As energy is essential for human development, today's world faces a dual challenge: to provide reliable and affordable energy to a growing population, while mitigating the effect of climate change. A significant portion of the world's population, which inhabits developing and under-developed countries, is energy deprived and living in dire circumstances. Hence, the provision of cheap and clean energy is a dual challenge that will have ramifications for every nation's economic, energy security, and environmental goals.

As shown in Table 17.1, primary energy consumption grew at a rate of 2.3% in 2018, almost double its ten-year average of 1.5% per year, and the fastest since 2010 (IEA, 2019a). In terms of fuel, energy consumption growth was driven by natural gas, which contributed more than 40% of the increase. All fuels grew faster than their ten-year averages, apart from renewables, although these still accounted for the second largest increment to energy growth. China, the USA, and India together accounted for more than two-thirds of the global increase in energy demand, with US consumption expanding at its fastest rate for 30 years. Carbon emissions grew by 2.0%, the fastest growth for seven years (BP, 2019).

17.2 CONSUMPTION OF OIL

Oil continues to play a leading role in the world's energy mix, with growing demand driven by commercial transportation and feedstocks for the chemicals industry. It is predicted that commerce and trade will drive transportation energy consumption up more than 25% from 2017 to 2040. Global transportation demand is driven by differing trends for commercial transportation and light-duty passenger vehicles. As economic activity expands, especially in developing regions, commercial transportation is expected to grow. The majority of the growth comes from heavy-duty trucking as a result of goods movement. Passenger vehicle ownership is expected to expand as a result of the dramatic growth in the middle class and increased urbanization, leading

TABLE 17.1
Total Primary Energy Demand in the World

	Energy Demand (Mtoe)	Growth Rate (%)	Shares (%)	
	2018	2017–2018	2000	2018
Total primary energy demand	14 301	2.30	100	100
Coal	3 778	0.70	23	26
Oil	4 488	1.20	37	31
Gas	3 253	4.60	21	23
Nuclear	710	3.30	7	5
Hydro	364	3.10	2	3
Biomass and waste	1 418	2.50	10	10
Other renewables	289	14.00	1	2

Source: IEA (2019a). Mtoe: Million tons of oil equivalent.

TABLE 17.2
Oil Energy Demand in the World

	Oil Primary Energy Demand	Growth Rate
	2018	2017–2018
United States	20	2.70
China	13	3.50
India	5	4.50
Europe	15	0.10
Rest of the World	46	0.20
WORLD	99	1.30

Source: IEA (2019a).

to increased passenger vehicle travel. The fuel mix continues to evolve with more alternatives, such as electric vehicles.

Aviation demand will see the highest annual growth rate at 2.2% from 2017 to 2040 due to both rising economic activity as well as rapid growth of the middle class, specifically in emerging economies (ExxonMobil, 2019). As seen in Table 17.1, oil energy is one of the major contributors to meeting the primary energy demand of the world. In order to mitigate the harmful effects of oil consumption, as a policy matter suitable techno-economically feasible energy alternatives must be harnessed and used globally. Table 17.2 clearly shows that the major onus for identifying and using viable alternatives to oil energy must address the fact that all nations are increasing their energy as well as their oil demand.

17.3 CARBON EMISSIONS

Climate experts expect global carbon emissions from fossil fuels and cement production to rise in 2020, from an estimated 36.8 billion tonnes of CO_2 last year (*Guardian*, 2020). Table 17.3 shows very alarming data as economic development in the USA, China, India, and other parts of the world has led to positive growth in CO_2 emissions. Hence, the global energy system needs imminent transformation. The present energy supply system mainly based on fossil fuels has to be based, instead, on renewable energy. This can eventually lead to fulfilment of the United Nations' SDG 7 which ensures access to affordable, reliable, sustainable, and modern energy for all – not for just some.

17.4 ENERGY INTENSITY

Global energy intensity, defined as the ratio of primary energy supply to GDP, is the indicator used to track progress on global energy efficiency. The original target was an annual reduction of 2.6% until 2030, although the world has fallen short of this goal since it was announced, such that the required rate of improvement has risen to 2.7% after an improvement of only 1.7% in 2017. The further slowdown in 2018, with an improvement in energy intensity of only 1.2% according to IEA analysis, means that from 2019 to 2030 global energy intensity must improve by 2.9%

TABLE 17.3
CO_2 Emissions in the World

	Total CO_2 Emissions (Million Tonnes)	Growth Rate (%)
	2018	2017–2018
United States	4 888	3.10
China	9 481	2.50
India	2 299	4.80
Europe	3 956	–1.30
Rest of the World	11 249	1.10
WORLD	33 143	1.70

Source: IEA (2019a).

TABLE 17.4
Energy Intensity Improvements in the World

Energy Intensity Improvements (toe per $1000, PPP)		Growth Rate (%)
	7.099 pt	2017–2018
United States	0.112	0.80
China	0.125	–2.90
India	0.092	–3.10
Europe	0.079	–1.60
Rest of the World	0.109	–1.10
WORLD	0.108	–1.30

Source: IEA (2019a). Toe: Tons of oil equivalent.

annually to satisfy SDG 7.3. Meeting this objective will require an important step up in the implementation and expansion of energy efficiency policies (IEA, 2019b).

In terms of energy security, importing countries reduced their exposure to oil market instability through technical efficiency improvements. Technical efficiency gains continue to deliver cuts in energy-related emissions. Between 2015 and 2018, technical efficiency improvements reduced energy-related carbon emissions by 3.5 gigatonnes of carbon dioxide ($GtCO_2$), roughly the equivalent of the energy-related emissions of Japan over the same period. This is helping to bring the world closer to an emissions trajectory consistent with achieving global climate change goals. Table 17.4 shows that huge scope exists for improvement in energy intensity across various countries in the world. Localized renewable energy generation and utilization offers a great opportunity for improvement in energy intensity.

17.5 POLICY PARALYSIS IN THE BIOFUEL SECTOR

As per the available statistics, diesel (739 MMT) and gasoline (683 MMT) are both consumed substantially throughout the world (IEA, 2019c). Diesel consumption is almost on a par with gasoline consumption. Figure 17.1 shows actual as well as projected

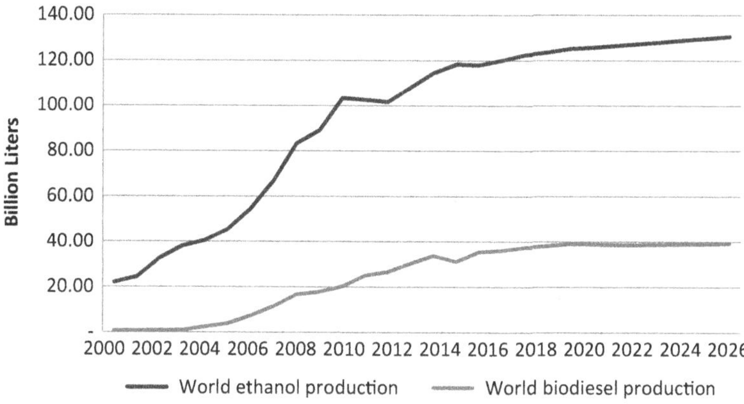

FIGURE 17.1 World ethanol and biodiesel production. (*Source:* OECD/FAO (2019)).

biodiesel production throughout the world. As can be seen, one-third of biodiesel production as compared to ethanol production shows complete policy paralysis. A thorough analysis of prevailing biofuel policies, production, and consumption statistics in biodiesel producing countries has been conducted to obtain a precise look at this state of affairs.

17.6 OBSERVATIONS FROM GLOBAL BIODIESEL PRODUCTION PROJECTS

The extent to which biofuels have reached the end consumers varies significantly country by country as well as region by region. The reasons for these differences are complex and include a variety of policy and market issues. While biofuels offer significant environmental, technical, and socio-economic advantages, their prices are sometimes higher than their petroleum product equivalents. As a result, biofuels have been successfully implemented only in those countries that have recognized the value of those benefits and have made appropriate policy decisions to support them. Standardization is one of the key issues in the development and adoption of new products. For the producers and distributors of biodiesel blended fuel, standards for all the possible blend percentages are a vital necessity. The enforcing agencies and authorities require approved standards for the evaluation of performance, safety, and environmental risks. The development of engines is mainly based on the properties of the fuel.

17.6.1 Standards for Biodiesel Blends

Diesel engines cater to a wider variety of applications. Suitable standards for biodiesel blends must be worked upon on a priority basis for its wider acceptance on a national as well as an international level. Major oil consuming countries like China, India, and Japan have not come out with any technical specifications of biodiesel blended fuel. In this case it will be indeed difficult to implement usage of such fuel on a mass scale basis.

The development of a new fuel standard is a complex and long-lasting task, even on the national level. Closer coordination and cooperation are required at the international level for the further development of biodiesel fuel in international markets.

17.6.2 Economic Sustainability of Feedstock

Major oil consuming as well as diesel consuming nations do not have an economic sustainability model for the biodiesel feedstock supply chain. This is the main reason for the poor economic viability and the failure of biodiesel to grab a bigger chunk of the biofuel market globally. The European Union is one the largest consumers of diesel and is itself trying to establish a trade-off between in-house developed biodiesel and imported biodiesel from countries such as Indonesia, Argentina, and Malaysia. Countries like China, India, and Japan do not have a clear-cut policy regarding the identification of suitable feedstock for producing biodiesel.

Populous countries like India and under-developed countries in Africa must come up with a social security and rural development model for the development of suitable biodiesel feedstock at the local level. This will ensure energy independence and economic development in these regions. The proper involvement of actual stakeholders is mandatory for the real success of these social programs.

17.6.3 Wastelands

The large availability of wasteland has often been discussed in many studies. It must be carefully noted that it is impossible to cultivate any biodiesel feedstock without the proper input of the type of land. Hence, the emphasis must be on the identification of suitable land with specific inputs for cultivating a proper crop to act as a feedstock for biodiesel. With institutional support, this will in turn be a revenue generation model for poor rural people.

17.6.4 Policy Interventions

Deploying new technology allows society to do more with less. A policy, like tax incentives, can spur the development of new technology, though these technologies ultimately need to compete without subsidies to reach a large enough scale to impact global markets. Demand for energy begins with the numerous choices that consumers make in their daily lives, such as lower energy costs and lower emissions. Consumer preferences can also be altered over time by policies that incentivize choices, as in the case of standalone diesel generator sets in rural areas that must be operated with biodiesel blended fuel. The biodiesel must be made up of local feedstock. A local level cooperative body must be made accountable for these initiatives. This would help in establishing the proper supply chain of feedstocks.

17.6.5 Energy Demand in the Marine and Aviation Sector

Aviation demand will see the highest annual growth rate at 2.2% from 2017 to 2040 due to both rising economic activity as well as the rapid growth of the middle class, specifically in emerging economies (ExxonMobil, 2019).

In the 'sustainable development scenario' (SDS), low-carbon fuels will meet 7% of international shipping and 9% of aviation fuel demand in 2030. However, current biofuel consumption is minimal in both these subsectors. Some progress has been made in aviation. Flights using biofuel blends have surpassed 200,000; a continuous biofuel supply is available at six airports; and policy support was enhanced in the United States and Europe in 2018. Still, aviation biofuel production of about 15 million litres in 2018 accounted for less than 0.01% of aviation fuel demand. This means that very significant market development is needed to deliver the aviation biofuel production required to be on the SDS trajectory in 2030. To scale up biofuel consumption, market and policy frameworks must be devised that reflect the international nature of these sectors. This task falls within the remit of the 'International Civil Aviation Organization and the International Maritime Organization'.

Domestic aviation and navigation fall under the jurisdiction of national governments, however, and policy measures that close cost premiums with fossil fuels (e.g. consumption subsidies or carbon pricing) can be employed to increase the economic viability of biofuel use (IEA, 2019d).

17.6.6 Energy Demand for Commercial Vehicles

'Manufacturing commercial vehicles' (MCVs) will include 70% alternative fuels, and 'heavy commercial vehicles' (HCVs) approximately 20%, mostly biofuels, due to the need for high energy density fuels in long-haul trucks. This would require a rapid acceleration in the early 2020s of both alternate fuels into the heavy-duty fleet as well as infrastructure build-out to support the alternatives (ExxonMobil, 2019).

17.7 OBSERVATIONS AND RECOMMENDATIONS FROM A STATUS REPORT ON THE IMPLEMENTATION OF EXISTING BIODIESEL PROMOTION POLICIES

Table 17.5 presents a comprehensive review of biodiesel supporting policies prevailing in the major oil consuming countries of the world. It is clearly shown that biodiesel promotion policies are either non-existent or had failed in most countries of the world. The real motive of the review was to ascertain the seriousness of federal agencies in executing the sustainable development of biodiesel as part of an energy security initiative. Apart from the serious initiatives of the EU, the USA, Canada, Argentina, and Brazil towards sustainable development of biodiesel, most of the countries in the world are non-committed.

17.7.1 Greenhouse Gas Reduction Initiatives

Notably, only the EU has specified its emission reduction targets as part of reducing 'greenhouse gas (GHG) emissions' . It is worth mentioning that the largest economies, like the USA and China, are not able to set up and meet any GHG reductions. Major oil consuming countries like Argentina, India, Japan, and Canada must also emphasize the proper use of biodiesel blends for meeting GHG reduction targets.

TABLE 17.5
Biodiesel Policy Status Report

S. No.	Country	Production Capacity	Production	Blending %	Consumption	Exports	GHG Reduction Targets	Specific Policy Available	Type of Feedstock	Energy Security Concerns	Povert Alleviation & Rural Development	Reference
1	Argentina	5 billion liters	2.5 billion liters	9,6	1.3 billion liters	1.2 billion liters	Not specified	Biofuels Law (Law 26093) of 2006	Edible	Not specified	Not a major issue	USDA, 2019a
2	India	650 million liters	85 million liters	0,14	Not specified	Not specified	Not specified	National Biofuel Policy 2018	Non-edible	Critical to energy security	Major issue	USDA, 2019b
3	Canada	500 million liters	460 million liters	2	Not specified	340 million liters	Not specified	Renewable Fuels Regulations 2010	Edible	Not specified	Not a major issue	USDA, 2019c
4	China	Not specified	Not specified	0,2	Not specified	Not specified	Not specified	Not specified	Used cooking oil	Critical but not specified	Not a major issue	USDA; 2019d
5	Japan	Not specified	Not specified	0,3	Not specified	Not specified	Not specified	Not specified	Not specified	Critical but not specified	Not a major issue	USDA; 2019e
6	US	2520 million gallons	1725 million gallons	10 - 20%	1975 million gallons	2719 barrels	Not specified	Renewable Fuel Standard (RFS) 2005	Vegetable - soybean, corn and canola	Not specified	Not a major issue	IEA, 2018
7	Europe	21181 million liters	14442 million liters	7,7	16854 million liters	664 million liters	6%	Renewable Energy Directive II (RED II)	Vegetable - UCO, soybean, rapeseed, palm, sunflower	Economic tradeoff between imports and inhouse production is major concern	Not a major issue	USDA; 2019f

UCO: Used cooking oils.

17.7.2 Biodiesel Blending Percentage

Brazil has also announced plans to progressively scale up its biodiesel mandate from 11% to 15%. Major oil consuming nations are highly conservative about increasing their biodiesel blending percentage mandate. A very low penetration of biodiesel in Japan, China, and India shows the lackluster concern of policy makers in these countries towards its sustainable development. In the marine sector, the use of biodiesel should be promoted to ascertain the technical feasibility of keeping economics in consideration.

17.7.3 Usage of Higher Percentage Biodiesel Blends

Vehicles adapted to high biofuel blend levels or unblended biofuel usage must be developed and used, especially for energy intensive usage. Drop-in biofuels can be used unblended or at high blend shares without modifications to engines, maintenance regimes, or fuel supply infrastructure.

17.7.4 Production Capacity

Most countries are reporting under-utilization of production capacity. This itself represents poor planning and execution amongst policy makers, business houses, and federal agencies. Biodiesel production projects can be viable only if the larger interests of society are taken into consideration before implementation. The minimum in-house requirement of biodiesel fuel must be ensured before coming up with fully fledged production units.

Certain countries are setting up plantations in other countries, such as underdeveloped African nations, as part of their supply chain for feedstock and improving the livelihoods of poor people. This is also a contentious issue. In the longer run, this should not lead to land grabbing in poor countries or the setting up of colonial rule.

17.7.5 Sustainable Development of Biodiesel Production

Sustainable development is essential to ensure that scaling up of biodiesel consumption delivers tangible social, economic, and environmental benefits, including lifecycle GHG emission reductions.

Policy makers must establish frameworks to ensure only sustainable biofuels receive policy support. Adherence to sustainability criteria can be demonstrated by third-party certification of biofuel supply chains.

In Mexico and South Africa transport biofuel industries are at an infancy stage. Therefore, market development and technology leapfrogging are needed to get on track with the SDS.

Rapidly increasing transport fuel demand ably supported with policy initiatives for biodiesel production is a means to raise energy security while ensuring demand for strategically important agricultural commodities as a feedstock.

Biodiesel from non-food-crop feedstocks as well as 'used cooking oils' (UCO) need to command a more substantial share of biodiesel consumption in the 'safety data sheet' (SDS). This is because they mitigate land use change concerns and generally offer higher lifecycle GHG emission reductions than conventional biodiesel.

The European Union, the United States, and Brazil have established frameworks to ensure biofuel sustainability, but there is a need for other countries to ensure that rigorous sustainability governance is linked to biofuel policy support (IEA, 2019d).

17.7.6 Scaling up Advanced Biofuels Is Essential

Technologies to produce biodiesel and 'hydrotreated vegetable oil' (HVO) from waste oil and animal fat feedstocks are technically mature and provided 8% of all biofuel output in 2018. However, production of novel advanced biofuels from other technologies is still modest, with progress needed to improve technology readiness. These technologies are important, nevertheless, as they can utilize feedstocks with high availability and limited other uses (e.g. agricultural residues and municipal solid waste) (IEA, 2019d).

17.7.7 Policy Support to Commercialize Advanced Biofuels

Most of the countries around the world are not able to follow the policy guidelines for production and usage of biodiesel. This can be due to unrealistic estimates regarding cost, availability of cheap and abundant feedstock, tax benefits for parity with petro-products, improper support from established petro-product suppliers/producers, and projections of high yields. Policy guidelines must be able to facilitate the technology learning and production scale-up necessary to reduce costs.

Relevant policies should include advanced biofuel quotas and financial derisking measures, e.g. loan guarantees from development banks. These would be particularly effective in those countries which possess significant feedstock resources.

Countries and regions should consider policies that specify reductions in fuel lifecycle carbon intensity (such as California's Low Carbon Fuel Standard), which are effective in boosting demand for biodiesel and HVO from waste oil, fat, and grease feedstocks, as well as biomethane. They could also support the deployment of novel advanced biofuels once production costs fall (IEA, 2019d).

REFERENCES

BP. 2019. *BP Statistical Review of World Energy*. London: British Petroleum.

ExxonMobil. 2019. *Outlook for Energy: A Perspective to 2040* Irving, TX: ExxonMobil.

Guardian. 2020. *Carbon Emissions from Fossil Fuels Could Fall by 2.5bn Tonnes in 2020*. 12 April 2020.

IEA. 2018. *United States – 2018 update*. Paris: International Energy Agency.

IEA. 2019a. *Global energy & CO2 status Report*. Paris: International Energy Agency.

IEA. 2019b. *SDG7: Data and Projections. Access to Affordable, Reliable, Sustainable and Modern Energy for All*. Paris: International Energy Agency.

IEA. 2019c. *Oil Information 2019: A Comprehensive Reference on Current Developments in Oil Supply and emand*. Paris: International Energy Agency.

IEA. 2019d. *Tracking Transport: Transport Biofuels*. Paris: International Energy Agency.

OECD/FAO. 2019. *OECD-FAO Agricultural Outlook*. Paris: Organisation for Economic Co-operation and Development.

USDA. 2019a. *Argentina Biofuels Annual 2019*. Washington, DC: USDA Foreign Agricultural Service.

USDA. 2019b. *India Biofuels Annual 2019*. Washington, DC: USDA Foreign Agricultural Service.

USDA. 2019c. *Canada Biofuels Annual 2019*. Washington, DC: USDA Foreign Agricultural Service.

USDA, 2019d. *China Biofuels Annual 2019*. Washington, DC: USDA Foreign Agricultural Service.

USDA, 2019e. *Japan Biofuels Annual 2019*. Washington, DC: USDA Foreign Agricultural Service.

USDA: 2019f. *EU Biofuels Annual 2019*. Washington, DC: USDA Foreign Agricultural Service.

Part IV

Glycerol

18 Glycerol
A Scientometric Review of the Research

Ozcan Konur

CONTENTS

18.1 INTRODUCTION

Crude oils have been primary sources of energy and fuels, such as petrodieselfuels (Busca et al., 1998; Khalili et al., 1995; Rogge et al., 1993; Schauer et al., 1999). However, significant public concerns about the sustainability, price fluctuations, and adverse environmental impact of crude oils have emerged since the 1970s (Ahmadun et al., 2009; Atlas, 1981; Babich and Moulijn, 2003; Kilian, 2009; Perron, 1989). Thus, biooils (Bridgwater and Peacocke, 2000; Czernik and Bridgwater, 2004; Gallezot, 2012; Mohan et al., 2006) and biooil-based biodiesel fuels (Chisti, 2007; Hill et al., 2006; Lapuerta et al., 2008; Mata et al., 2010; Zhang et al., 2003) have emerged as alternatives to crude oils and crude oil-based petrodiesel fuels in recent decades. Nowadays, both biodiesel and petrodiesel fuels are being used extensively at the global scale (Konur, 2021a–ag).

Glycerol, a by-product of biodiesel fuels, has been used to produce further biofuels and biochemicals, thus reducing the production cost of biodiesel fuels (Behr et al., 2008; Carrettin et al., 2002, 2003; da Silva et al., 2009; Dasari et al., 2005; Johnson and Taconi, 2007; Pagliaro et al., 2007; Rep et al., 2000; Thompson and He, 2006; Yazdani and Gonzalez, 2007, 2008; Zhou et al., 2008).

However, for the efficient progression of the research in this field, it is necessary to develop efficient incentive structures for the primary stakeholders and to inform these stakeholders about the research (Konur, 2000, 2002a–c, 2006a–b, 2007a–b; North, 1991a–b).

Scientometric analysis offers ways to evaluate the research in a respective field (Garfield, 1955, 1972). This method has been used in a number of research fields (Konur, 2011, 2012a–n, 2015, 2016a–f, 2017a–f, 2018a–b, 2019a–b). However, there has been no scientometric study of this field.

This chapter presents a study of the scientometric evaluation of the research in this field using two datasets. The first dataset includes the 100-most-cited papers ($n = 100$ sample papers) whilst the second dataset includes population papers ($n =$ over 6,900 papers) published between 1980 and 2019.

The data on the indices, document types, authors, institutions, funding bodies, source titles, 'Web of Science' subject categories, key words, research fronts, and citation impact are presented and discussed.

18.2 MATERIALS AND METHODOLOGY

The search for the literature was carried out in the 'Web of Science' (WOS) database in January 2020. It contains the 'Science Citation Index-Expanded' (SCI-E), the Social Sciences Citation Index' (SSCI), the 'Book Citation Index-Science' (BCI-S), the 'Conference Proceedings Citation Index-Science' (CPCI-S), the 'Emerging Sources Citation Index' (ESCI), the 'Book Citation Index-Social Sciences and Humanities' (BCI-SSH), the 'Conference Proceedings Citation Index-Social Sciences and Humanities' (CPCI-SSH), and the 'Arts and Humanities Citation Index' (A&HCI).

The keywords for the search are collated from the screening of abstract pages for the first 1,000 highly cited papers. This keyword set is provided in the Appendix.

Two datasets are used for this study. The highly cited 100 papers comprise the first dataset ($n = 100$ sample papers) whilst all the papers form the second dataset (population dataset, $n =$ over 6,900 papers).

The data on the indices, document types, publication years, institutions, funding bodies, source titles, countries, 'Web of Science' subject categories, citation impact, keywords, and research fronts are collated from these datasets. The key findings are provided in the relevant tables and figure, supplemented with explanatory notes in the text. The findings are discussed, a number of conclusions are drawn, and a number of recommendations for further study are made.

18.3 RESULTS

18.3.1 Indices and Documents

There are over 7,900 papers in this field in the 'Web of Science' as of January 2020. This original population dataset is refined for the document type (article, review, book chapter, book, editorial material, note, and letter) and language (English), resulting in over 6,900 papers comprising over 87.9% of the original population dataset.

The primary index is the SCI-E for both the sample and population papers. About 98.3% of the population papers are indexed by the SCI-E database. Additionally 4.1, 1.5, and 0.3% of these papers are indexed by the CPCI-S, ESCI, and BCI-S databases, respectively. The papers on the social and humanitarian aspects of this field are relatively negligible with 0.1 and 0.0% of the population papers indexed by the SSCI and A&HCI, respectively.

Brief information on the document types for both datasets is provided in Table 18.1. The key finding is that article types of documents are the primary documents for both the sample and population papers, whilst reviews form 17% of the sample papers.

18.3.2 Authors

Brief information about the 13-most-prolific authors with at least three sample papers each is provided in Table 18.2. Around 350 and 17,200 authors contribute to the sample and population papers, respectively.

The most-prolific author is 'Keiichi Tomishige' with nine sample papers, working primarily on the 'hydrogenolysis' of glycerol for 'propanediol' production. The other prolific authors are 'Shuichi Koso', 'Kimio Kunimori', and 'Tomohisha Miyazawa' with six, five, and five sample papers, respectively.

On the other hand, a number of authors have a significant presence in the population papers: 'An-Ping Zeng', 'Zhaoyin Hou', 'Ahmad Zuhairi Abdullah', 'Franck Dumeignil', 'Zhi-Long Xiu', 'Sunghoon Park', 'Laura Prati', 'Waldemar

TABLE 18.1
Document Types

	Document Type	Sample Dataset (%)	Population Dataset (%)	Difference (%)
1	Article	82	95.9	–13.9
2	Review	17	2.8	14.2
3	Book chapter	0	0.3	–0.3
4	Proceeding paper	4	4.1	–0.1
5	Editorial material	0	0.1	–0.1
6	Letter	0	0.2	–0.2
7	Book	0	0.0	0.0
8	Note	1	0.9	0.1

TABLE 18.2
Authors

	Authors	No. of sample papers (%)	No. of population papers (%)	Surplus (%)	Institution	Country	Research front	Product
1	Tomishige, Keiichi	9	0.3	8.7	Univ. Tsukuba	Japan	Hydrogenolysis	Propanediol
2	Koso, Shuichi	6	0.1	5.9	Univ. Tsukuba	Japan	Hydrogenolysis	Propanediol
3	Kunimori, Kimio	5	0.1	4.9	Univ. Tsukuba	Japan	Hydrogenolysis	Propanediol
4	Miyazawa, Tomohisha	5	0.1	4.9	Univ. Tsukuba	Japan	Hydrogenolysis	Propanediol
5	Aggelis, George	4	0.2	3.8	Agr. Univ. Athens	Greece	Microbial conversion	Propanediol
6	Gonzalez, Ramon	4	0.2	3.8	Rice Univ.	USA	Microbial fermentation	Fuels
7	Nakagawa, Yoshinao	4	0.2	3.8	Univ. Tsukuba	Japan	Hydrogenolysis	Propanediol
8	Davis, Robert J.	3	0.2	2.8	Univ. Virginia	USA	Hyrogenolysis, oxidation	Ethylene glycol
9	Deckwer, Wolf-Dieter	3	0.2	2.8	Soc. Biotechnol. Res.	Germany	Microbial fermentation	Propanediol
10	Kimura, Hiroshi	3	0.2	2.8	AOP Catalyst Lab.	Japan	Oxidation	Chemicals
11	Papanikolaou, Seraphim	3	0.4	2.6	Agr. Univ. Athens	Greece	Microbial conversion	Propanediol
12	Shinmi, Yasunori	3	0.1	2.9	Univ. Tsukuba	Japan	Hydrogenolysis	Propanediol
13	Yazdani, Syed Shams	3	0.1	2.9	Rice Univ.	USA	Microbial fermentation	Fuels

Rymowicz', 'N. Lingaiah', 'Alberto Villa', 'Chao Wang', 'Chengxi Zhang', 'Pedro Luis Arias', 'Binlin Dou', 'Ping Chen', 'Anita Rywinska', 'K. V. R. Chary', 'Graham J. Hutchings', 'Karen Trchounian', 'Yannick Pouilloux', and 'Lina Wang' have at least 0.3% of the population papers each.

The most-prolific institution for these top authors is the 'University of Tsukuba' of Japan. The other prolific institutions with two authors are the 'Agricultural University of Athens' of Greece and 'Rice University' of the USA. Thus, in total, six institutions house these top authors.

It is notable that none of these top researchers are listed in the 'Highly Cited Researchers' (HCR) in 2019 (Clarivate Analytics, 2019; Docampo and Cram, 2019).

The most-prolific country for these top authors is Japan with seven. The other prolific countries with three and two authors, respectively are the USA and Greece. Thus, in total, four countries contribute to these top papers.

There are four key topical research fronts for these top researchers: 'hydrogenolysis of glycerol', 'microbial conversion of glycerol', 'oxidation of glycerol', and 'fermentation of glycerol' with six, five, three, and two sample papers, respectively. On the other hand, these top authors focus on the production of 'propanediol', 'fuels', 'chemicals', and 'ethylene glycol' with nine, two, one, and one authors, respectively.

It is further notable that there is a significant gender deficit among these top authors as all of them are male (Lariviere et al., 2013; Xie and Shauman, 1998).

The author with the most impact is 'Keiichi Tomishige' with an 8.7% publication surplus. The other authors with the most impact are 'Shuichi Koso', 'Kimio Kunimori', and 'Tomohisha Miyazawa' with 5.9, 4.9, and 4.9% publication surpluses, respectively.

On the other hand, the authors with the least impact are 'Seraphim Papanikolaou', 'Wolf-Dieter Decker', 'Hiroshi Kimura', and 'Robert J. Davis' with at least a 2.8% publication surplus each.

18.3.3 Publication Years

Information about publication years for both datasets is provided in Figure 18.1. This figure shows that 2, 11, 74, and 23% of the sample papers and 4.3, 7.2, 13.4, and 74.8% of the population papers were published in the 1980s, 1990s, 2000s, and 2010s, respectively.

Similarly, the most-prolific publication years for the sample dataset are 2007, 2008, 2010, and 2005 with 19, 14, 11, and 8 papers, respectively. On the other hand, the most-prolific publication years for the population dataset are 2016, 2017, 2018, 2015, and 2019 with 8.7, 8.6, 8.6, 8.3, and 8.3% of the population papers, respectively. It is notable that there is a sharply rising trend for the population papers in the 2000s and 2010s.

18.3.4 Institutions

Brief information on the top 11 institutions with at least 3% of the sample papers each is provided in Table 18.3. In total, around 140 and 4,100 institutions contribute to the sample and population papers, respectively.

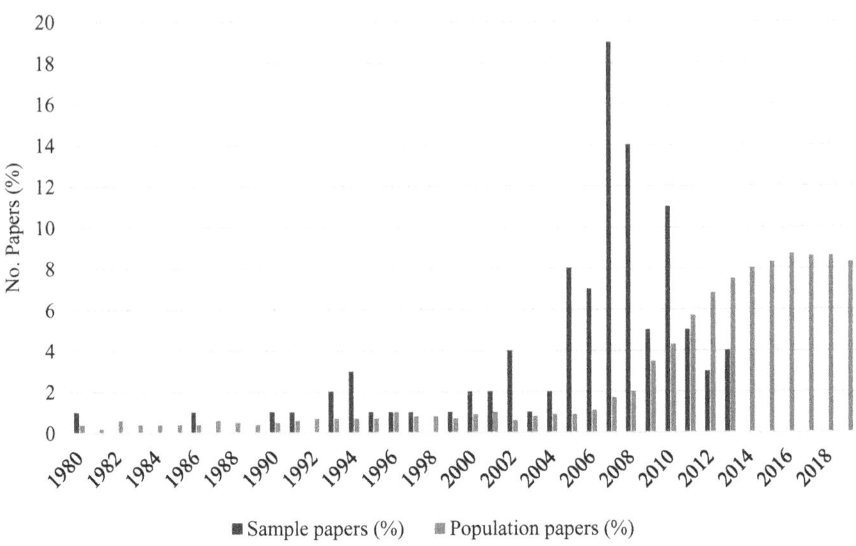

FIGURE 18.1 Research output between 1980 and 2019.

TABLE 18.3
Institutions

	Institution	Country	No. of sample papers (%)	No. of population papers (%)	Difference (%)
1	Natl. Sci. Res. Ctr.	France	8	3.3	4.7
2	Univ. Tsukuba	Japan	8	0.2	7.8
3	Rice Univ.	USA	5	0.2	4.8
4	Superior Counc. Sci. Invs.	Spain	4	1.9	2.1
5	Natl. Res. Counc.	Italy	4	1.1	2.9
6	Agric. Univ. Athens	Greece	3	0.4	2.6
7	Chinese Acad. Sci.	China	3	3.2	–0.2
8	Soc. Biotechnol. Res	Germany	3	0.2	2.8
9	Tohoku Univ.	Japan	3	0.4	2.6
10	Univ. Milan	Italy	3	0.7	2.3
11	Univ. Virginia	USA	3	0.2	2.8

These top institutions publish 47.0 and 11.8% of the sample and population papers, respectively. The top institutions are the 'National Scientific Research Center' of France and the 'University of Tsukuba' of Japan with eight and eight sample papers and 4.7 and 7.8% publication surpluses, respectively. These top institutions are followed by 'Rice University', the 'Superior Council of Scientific Investigations' of Spain, and the 'National Research Council' of Italy with five, four, and four sample papers, respectively.

The most-prolific countries for these top institutions are the USA, Italy, and Japan with two sample papers each. The other countries are China, France, Germany, Greece, and Spain. In total, eight countries house these institutions.

The institutions with the most impact are the 'University of Tsukuba', 'Rice University', and the 'National Scientific Research Center' of France with 7.8, 4.8, and 4.7% of the sample papers, respectively. On the other hand, the institutions with the least impact are the 'Chinese Academy of Sciences', the 'Superior Council of Scientific Investigations' of Spain, and the 'University of Milan' with –0.2, 2.1, and 2.3% publication surpluses, respectively.

It is notable that some institutions have a heavy presence in the population papers: the 'Council of Scientific Industrial Research' and the 'Indian Institute of Technology' of India, 'Dalian University of Technology' of China, the 'University of Sao Paulo' of Brazil, 'Jiangnan University' of China, the 'National Council for Scientific and Technical Investigations' of Argentina, the 'Russian Academy of Sciences', and 'Tsinghua University' of China with at least a 0.9% presence in the population papers each.

18.3.5 Funding Bodies

Brief information about the top seven funding bodies with at least 2% of the sample papers each is provided in Table 18.4. It is significant that only 27 and 67% of the sample and population papers declare any funding, respectively. Around 50 and 4,100 funding bodies fund the research for the sample and population papers, respectively.

TABLE 18.4
Funding Bodies

	Institution	Country	No. of Sample papers (%)	No. of Population papers (%)	Difference (%)
1	National Natural Science Foundation	China	3	11.1	–8.1
2	New Energy and Industrial Technology Development Organization	Japan	3	0.4	2.6
3	National Institutes of Health	USA	3	1.2	1.8
4	National Scientific Research Center	France	2	0.7	1.3
5	Engineering Physical Sciences Research Council	UK	2	0.6	1.4
6	European Union	EU	2	2.9	–0.9
7	Ministry of Education Culture Sports Science and Technology	Japan	2	1.3	0.7

The top funding bodies are the 'National Natural Science Foundation' of China, the 'New Energy and Industrial Technology Development Organization' of Japan, and the 'National Institutes of Health' of the USA with three sample papers each.

The most-prolific country for these funding bodies is Japan with two. In total, these bodies are from five countries only. It is notable that some top funding agencies have a heavy presence in the population studies. Some of them are the 'National Council for Scientific and Technological Development' and 'CAPES' of Brazil, the 'Council of Scientific Industrial Research' of India, the 'Fundamental Research Funds for the Central Universities' of China, the 'Ministry of Science and Innovation' of Spain, the 'Sao Paulo Research Foundation' of Brazil, the 'National Basic Research Program' of China, the 'National Science Foundation' of the USA, the 'Natural Sciences and Engineering Research Council' of Canada, the 'Department of Health Human Services' of the USA, the 'Ministry of Education and Science' of Spain, and the 'Department of Science Technology' of India with at least 1.0% of the population papers each. These funding bodies are from Brazil, India, China, Canada, the USA, and Spain.

The funding bodies with the most impact are the 'New Energy and Industrial Technology Development Organization' of Japan and the 'National Institutes of Health' of the USA with 2.6 and 1.8% publication surpluses, respectively. On the other hand, the funding bodies with the least impact are the 'National Natural Science Foundation' of China and the 'European Union' with –8.9 and –0.9% publication deficits, respectively.

18.3.6 Source Titles

Brief information about the top 16 source titles with at least two sample papers each is provided in Table 18.5. In total, 51 and over 1,000 source titles publish the sample and population papers, respectively. On the other hand, these top 16 journals publish 65.0 and 21.8% of the sample and population papers, respectively.

The top journal is the 'Journal of Catalysis', publishing 13 sample papers with an 11.4% publication surplus. This top journal is followed by 'Green Chemistry' and 'Applied Catalysis A General' with 11 and 8 sample papers, respectively.

Although these journals are indexed by 13 subject categories, the top category is 'Chemistry Physics' with seven journals. The other most-prolific subject categories are 'Biotechnology Applied Microbiology' and 'Engineering Chemical' with five journals each. Additionally, 'Energy Fuels', 'Chemistry Multidisciplinary', and 'Green Sustainable Science Technology' index four, two, and two journals, respectively.

The journals with the most impact are the 'Journal of Catalysis', 'Green Chemistry', and 'Applied Catalysis A General' with 11.4, 9.1, and 5.7% publication surpluses, respectively. On the other hand, the journals with the least impact are 'Bioresource Technology', 'Biotechnology Advances', and the 'International Journal of Hydrogen Energy' with –1.7, –0.1, and 0.0% publication deficits, respectively.

It is notable that some journals have a heavy presence in the population papers. Some of them are 'Industrial Engineering Chemistry Research', 'RSC Advances', 'Fuel', 'Chemical Engineering Journal', the 'Journal of the American Oil Chemists

TABLE 18.5
Source Titles

	Source Title	WOS Subject Category	No. of Sample Papers (%)	No. of Population Papers (%)	Difference (%)
1	Journal of Catalysis	Chem. Phys., Eng. Chem.	13	1.6	11.4
2	Green Chemistry	Chem. Mult., Green Sust. Sci. technol.	11	1.9	9.1
3	Applied Catalysis A General	Chem. Phys., Env. Sci.	8	2.3	5.7
4	Applied Catalysis B Environmental	Chem. Phys., Eng. Env., Eng. Chem.	4	1.8	2.2
5	Renewable Sustainable Energy Reviews	Green Sust. Sci. Technol., Ener. Fuels	4	0.4	3.6
6	International Journal of Hydrogen Energy	Chem. Phys., Electrochem., Ener. Fuels	3	3.0	0
7	Catalysis Today	Chem. Appl., Chem. Phys., Eng. Chem.	3	1.7	1.3
8	Biotechnology and Bioengineering	Biot. Appl. Microb.	3	0.6	2.4
9	Bioresource Technology	Agr. Eng., Biot. Appl. Microb., Ener. Fuels	2	3.7	–1.7
10	Applied Microbiology and Biotechnology	Biot. Appl. Microb.	2	1.4	0.6
11	Catalysis Communications	Chem. Phys.	2	1.1	0.9
12	Energy Fuels	Ener. Fuels, Eng. Chem.	2	0.8	1.2
13	Journal of Molecular Catalysis A Chemical	Chem. Phys.	2	0.7	1.3
14	Process Biochemistry	Biochem. Mol. Biol., Biot. Appl. Microb., Eng. Chem.	2	0.6	1.4
15	Angewandte Chemie International Edition	Chem. Mult.	2	0.1	1.9
16	Biotechnology Advances	Biot. Appl. Microb.	2	0.1	–0.1

Society', 'Catalysis Science Technology', 'Fuel Processing Technology', 'Chemsuschem', 'Catalysis Letters', 'Renewable Energy', 'Energy Conversion and Management', and the 'Journal of Chemical and Engineering Data' with at least a 0.7% presence in the population papers each.

18.3.7 Countries

Brief information about the top 12 countries with at least three sample papers each is provided in Table 18.6. In total, around 25 and 110 countries contribute to the sample and population papers, respectively.

The top country is the USA, publishing 21.0 and 11.8% of the sample and population papers, respectively. Japan, China, France, and Germany follow the USA with 15, 14, 13, and 12% of the sample papers, respectively. The other prolific countries are Spain and the UK, publishing seven sample papers each.

On the other hand, the European and Asian countries represented in Table 18.6 publish altogether 52 and 32% of the sample papers, whilst they publish 28.0 and 27.1% of the population papers, respectively.

It is notable that the publication surplus for the USA and these European and Asian countries is 9.2, 24.0 and 4.9%, respectively. On the other hand, the countries with the most impact are the USA, Japan, France, and Germany with 9.2, 8.7, 7.2, and 6.7% publication surpluses, respectively. Furthermore, the countries with the least impact are Brazil, China, Malaysia, and Spain with –5.0, –3.9, 0.1, and 0.3% publication deficits/surpluses, respectively.

It is also notable that some countries have a heavy presence in the population papers. The major producers of the population papers are India, South Korea, Canada, Poland, Thailand, Portugal, Iran, the Netherlands, Australia, Argentina, Russia, Sweden, Romania, Taiwan, and Turkey with at least 1.0% of the population papers each.

TABLE 18.6
Countries

	Country	No. of Sample Papers (%)	No. of Population Papers (%)	Difference (%)
1	USA	21	11.8	9.2
2	Japan	15	6.3	8.7
3	China	14	17.9	–3.9
4	France	13	5.8	7.2
5	Germany	12	5.3	6.7
6	Spain	7	6.7	0.3
7	UK	7	4.1	2.9
8	Greece	5	1.5	3.5
9	Italy	5	3.6	1.4
10	Belgium	3	1.0	2.0
11	Brazil	3	8.0	–5.0
12	Malaysia	3	2.9	0.1
	Europe-7	52	28.0	24.0
	Asia-3	32	27.1	4.9

18.3.8 'Web of Science' Subject Categories

Brief information about the top 14 'Web of Science' subject categories with at least four sample papers each is provided in Table 18.7. The sample and population papers are indexed by 24 and 101 subject categories, respectively.

For the sample papers, the top subject is 'Chemistry Physical' with 45.0 and 26.3% of the sample and population papers, respectively. This top subject category is followed by 'Engineering Chemical' and 'Biotechnology Applied Microbiology' with 28 and 21% of the sample papers, respectively. The other prolific subject categories are 'Chemistry Multidisciplinary', 'Energy Fuels', and 'Green Sustainable Science Technology' with 17, 17, and 15 sample papers, respectively.

It is notable that the publication surplus is most significant for 'Chemistry Physical', 'Green Sustainable Science Technology', 'Engineering Chemical', and 'Environmental Sciences' with 18.7, 9.5, 5.5, and 5.4% publication surpluses, respectively. On the other hand, the subjects with least impact are 'Chemistry Applied', 'Biochemistry Molecular Biology', 'Energy Fuels', and 'Electrochemistry' with –2.7, –2.5, –1.8, and –1.6% publication deficits, respectively. This latter group of subject categories are under-represented in the sample papers.

Additionally, some subject categories also have a heavy presence in the population papers: 'Food Science Technology', 'Thermodynamics', 'Polymer Science', 'Chemistry Analytical', 'Chemistry Organic', 'Physics Atomic Molecular Chemical', 'Biochemical Research Methods', 'Physics Applied', 'Nanoscience Nanotechnology', 'Biophysics', 'Multidisciplinary Sciences', 'Mechanics', and 'Physics Condensed Matter' have at least a 1.0% presence in the population papers each.

TABLE 18.7
Web of Science Subject Categories

	Subject	No. of Sample Papers (%)	No. of Population Papers (%)	Difference (%)
1	Chemistry Physical	45	26.3	18.7
2	Engineering Chemical	28	22.5	5.5
3	Biotechnology Applied Microbiology	21	18.5	2.5
4	Chemistry Multidisciplinary	17	15.0	2.0
5	Energy Fuels	15	16.8	–1.8
6	Green Sustainable Science Technology	15	5.5	9.5
7	Environmental Sciences	10	4.6	5.4
8	Engineering Environmental	6	5.2	0.8
9	Chemistry Applied	5	7.7	–2.7
10	Materials Science Multidisciplinary	5	4.2	0.8
11	Agricultural Engineering	4	4.6	–0.6
12	Biochemistry Molecular Biology	4	6.5	–2.5
13	Electrochemistry	4	5.6	–1.6
14	Microbiology	4	4.1	–0.1

18.3.9 Citation Impact

These sample and population papers received about 29,000 and 182,000 citations, respectively as of January 2020. Thus, the average number of citations per paper for these papers are about 290 and 26 respectively. On the other hand the h-index is 100 and 153 for the sample and population datasets, respectively.

18.3.10 Keywords

Although a number of keywords are listed in the Appendix for the datasets related to this field, some of them are more significant for the sample papers.

The most-prolific keyword for the keyword set related to glycerol is '*glycerol*' with 121 occurrences. The other prolific keywords are 'glycerin' and 'by-product*' with three papers each.

On the other hand, the most-prolific keyword related to other fields is '*cataly*' with 50 occurrences. The other most-prolific keywords are 'hydrogen' (17), 'hydrogenolysis' (15), 'oxidation' (13), 'fermentation' (10), 'chemical*' (10), 'dehyrat*' (10), 'acrolein' (9), 'fuel*' (9), 'propanediol' (8), 'reforming' (8), 'conversion' (7), 'valu*' (6), 'carbonate' (6), 'carbon source' (5), and 'ethanol' (5).

18.3.11 Research Fronts

Brief information about the key research fronts is provided in Table 18.8. There are eight major topical research fronts for these sample papers: 'catalytic conversion' of

TABLE 18.8
Research Fronts

	Research Front	No. of Sample Papers (%)
I	**Glycerol conversion**	
	Catalytic conversion	50
	Microbial conversion	20
	Hydrogenolysis	13
	Oxidation	13
	Fermentation	10
	Dehydration	10
	Reforming	8
	Other methods	21
II	**Products**	
	Hydrogen	20
	Chemicals	10
	Acrolein	9
	Fuels	9
	Propanediol	8
	Carbonate	6
	Ethanol	5
	Other products	20
III	**Characterization**	**8**

glycerol (50) (Chaminand et al., 2004; Corma et al., 2008; Miyazawa et al., 2006; Zhou et al., 2008) , 'microbial conversion' of glycerol (20) (Papanikolaou and Aggelis, 2002; Rep et al., 2000; Wang et al., 2001), 'hydrogenolysis' of glycerol (13) (Chaminand et al., 2004; Dasari et al., 2005; Maris and Davis, 2007), 'oxidation' of glycerol (13) (Carrettin et al., 2002, 2003), 'fermentation' of glycerol (10) (Wang et al., 2001; Yazdani and Gonzalez, 2007), 'dehydration' of glycerol (10) (Chai et al.,2007; Katryniok et al., 2010), 'reforming' of glycerol (8) (Wen et al., 2008; Zhang et al., 2007), and 'other methods' for the conversion of glycerol (21) such as 'trans-esterification' (Ochoa-Gomez et al., 2009) or 'pyrolysis' (Buhler et al., 2002) of glycerol.

On the other hand, there are eight primary research fronts for the products of glycerol conversion: 'hydrogen' (20) (Wen et al., 2008; Zhang et al., 2007), 'chemicals' in general (10) (Corma et al., 2008; Zhou et al., 2008), 'acrolein' (9) (Chai et al., 2007; Corma et al., 2008; Katryniok et al., 2010), 'fuels' in general (9) (Yazdani and Gonzalez, 2007), 'propanediol' (8) (Nakagawa et al., 2010; Papanikolau et al., 2008), 'carbonate' (6) (Ochoa-Gomez et al., 2009), 'ethanol' (Yazdani and Gonzalez, 2008), and 'other products' (20) such as 'hydroxybutyrate' (Cavalheiro et al., 2009) or 'docosahexaenoic acids' (Chi et al., 2007).

There are also eight papers related primarily to the characterization of glycerol (Birge, 1986; Cheng, 2008; Thompson and He, 2006).

18.4 DISCUSSION

The size of the research in this field has increased to over 6,900 papers as of January 2020. It is expected that the number of the population papers in this field will exceed 15,000 by the end of the 2020s.

The research has developed more in the technological aspects of this field than the social and humanitarian pathways, as evidenced by the negligible number of population papers in the indices of the 'Web of Science', SSCI, and A&HCI.

The article types of documents are the primary documents for both datasets and reviews are over-represented by 14.2% in the sample papers, whilst articles are under-represented by 13.9% (Table 18.1). Thus, the contribution of reviews by 17% of the sample papers in this field is highly exceptional (cf. Konur, 2011, 2012a–n, 2015, 2016a–f, 2017a–f, 2018a–b, 2019a–b).

Thirteen authors from six institutions have at least three sample papers each (Table 18.2). Seven, three, and two of these authors are from Japan, the USA, and Greece, respectively.

There are four key topical research fronts for these top researchers: 'hydrogenolysis of glycerol', 'microbial conversion of glycerol', 'oxidation of glycerol', and 'fermentation of glycerol'. On the other hand, these top authors focus on the production of 'propanediol', 'fuels' in general, 'chemicals' in general, and 'ethylene glycol'.

It is significant that there is ample 'gender deficit' among these top authors as all of them are male (Xie and Shauman, 1998; Lariviere et al., 2013).

The population papers have built upon the sample papers, primarily published in the 2000s and to a lesser extent in the 2010s (Figure 18.1). Following this rising trend, particularly in the 2000s and 2010s, it is expected that the number of papers will reach to 15,000 papers by the end of the 2020s, more than doubling the current size.

The engagement of the institutions in this field at the global scale is significant as around 140 and 4,100 institutions contribute to the sample and population papers, respectively.

Eleven top institutions publish 47.0 and 11.8% of the sample and population papers, respectively (Table 18.3). The top institutions are the 'National Scientific Research Center' of France and the 'University of Tsukuba' of Japan.

The most-prolific countries for these top institutions are the USA, Italy, and Japan. It is notable that some institutions with a heavy presence in the population papers are under-represented in the sample papers.

It is significant that only 27 and about 67% of the sample and population papers declare any funding, respectively. It is notable that the most-prolific country for these funding bodies is Japan (Table 18.4). It is further notable that some top funding agencies for the population studies do not enter this top funding body list.

However, the presence of Chinese funding bodies in this top funding body table is notable. This finding is in line with studies showing the heavy research funding in China; the NSFC is the primary funding agency (Wang et al., 2012).

The sample and population papers are published by 51 and over 1,000 journals, respectively. It is significant that 16 top journals publish 65.0 and 21.8% of the sample and population papers, respectively (Table 18.5).

The top journal is the 'Journal of Catalysis'. This is followed by 'Fuel', 'Green Chemistry', and 'Applied Catalysis A General'.

The top categories for these journals are 'Chemistry Physics', 'Biotechnology Applied Microbiology' and 'Engineering Chemical'. It is notable that some journals with a heavy presence in the population papers are relatively under-represented in the sample papers.

In total, around 25 and 110 countries contribute to the sample and population papers, respectively. The top country is the USA (Table 18.6). This finding is in line with studies arguing that the USA is not losing ground in science and technology (Leydesdorff and Wagner, 2009).

The other prolific countries are Japan, China, France, Germany, Spain, and the UK. These findings are in line with studies showing heavy research activity in these countries in recent decades (Bordons et al., 2015; Bornmann et al., 2017; Hu et al., 2018; van Raan et al., 2011; Zhou and Leydesdorff, 2006).

On the other hand, the European and Asian countries represented in Table 18.6 publish altogether 52 and 32% of the sample papers, whilst they publish 28.0 and 27.1% of the population papers, respectively. These findings are in line with studies showing that both European and Asian countries have superior publication performance in science and technology (Bordons et al., 2015; Okubo et al., 1998; Youtie et al., 2008).

It is notable that the publication surplus for the USA and these European and Asian countries is 9.2, 24.0, and 4.9%, respectively. On the other hand, the countries with the most impact are the USA, Japan, France, and Germany. Furthermore, the countries with the least impact are Brazil, China, Malaysia, and Spain.

China's presence in this top table is notable. This finding is in line with China's efforts to be a leading nation in science and technology (Zhou and Leydesdorff, 2006; Guan and Ma, 2007; Youtie et al., 2008).

It is also notable that some countries have a heavy presence in the population papers. The major producers of the population papers are India, South Korea, Canada, Poland, Thailand, Portugal, Iran, the Netherlands, Australia, Argentina, Russia, Sweden, Romania, Taiwan, and Turkey (Hassan et al., 2012; Huang et al., 2006; Leydesdorff and Zhou, 2005; Prathap, 2017).

The sample and population papers are indexed by 24 and 101 subject categories, respectively. For the sample papers, the top subject is 'Chemistry Physical' with 45.0 and 26.3% of the sample and population papers, respectively (Table 18.7). This top subject category is followed by 'Engineering Chemical', 'Biotechnology Applied Microbiology', 'Chemistry Multidisciplinary', 'Energy Fuels', and 'Green Sustainable Science Technology'.

It is notable that the publication surplus is most significant for 'Chemistry Physical', 'Green Sustainable Science Technology', 'Engineering Chemical', and 'Environmental Sciences'. On the other hand, the subjects with least impact are 'Chemistry Applied', 'Biochemistry Molecular Biology', 'Energy Fuels', and 'Electrochemistry'. This latter group of subject categories are under-represented in the sample papers.

These sample and population papers receive about 29,000 and 182,000 citations, respectively as of January 2020. Thus, the average numbers of citations per paper for these papers are about 290 and 26 respectively. On the other hand, the h-index is 100 and 153 for the sample and population datasets, respectively. Hence, the citation impact of these top 100 papers in this field has been significant.

Although a number of keywords are listed in the Appendix for the datasets related to this field, some of them are more significant for the sample papers.

The most-prolific keyword for the keyword set related to glycerol is '*glycerol*'. On the other hand, the most-prolific keyword related to other fields is '*cataly*' with 50 occurrences. The other most-prolific keywords are 'hydrogen', 'hydrogenolysis', 'oxidation', 'fermentation', 'chemical*', 'dehyrat*', 'acrolein', 'fuel*', 'propanediol', 'reforming', 'conversion', 'valu*', 'carbonate', 'carbon source', and 'ethanol'. These keywords are related to the conversion of glycerol to biofuels and biochemicals through a number of methods such as hydrogenolysis. As expected, these keywords provide valuable information about the pathways of the research in this field.

There are eight major topical research fronts for these sample papers: 'Catalytic conversion' of glycerol, 'microbial conversion' of glycerol', 'hydrogenolysis' of glycerol, 'oxidation' of glycerol, 'fermentation' of glycerol, 'dehydration' of glycerol, 'reforming' of glycerol, and 'other methods' for conversion of glycerol such as 'transesterification' of glycerol.

On the other hand, there are eight primary research fronts for the products of glycerol conversion: 'hydrogen', 'chemicals' in general, 'acrolein', 'fuels' in general, 'propanediol', 'carbonate', 'ethanol', and 'other products' such as 'hydroxybutyrate' or 'docosahexaenoic acids'. There are also eight papers related primarily to the characterization of glycerol.

The key emphasis in these research fronts is the exploration of the structure–processing–property relationships of glycerol biomass and bioproducts obtained from glycerol (Konur and Matthews, 1989; Scherf and List, 2002; Rogers and Hopfinger, 1994; Cheng and Ma, 2011).

18.5 CONCLUSION

This chapter has mapped the research on the production of biofuels and biochemicals from glycerol, a by-product of biodiesel fuels, using a scientometric method.

The size of over 6,900 population papers shows the public importance of this interdisciplinary research field. However, it is significant that the research has developed more in the technological aspects in this field, rather than the social and humanitarian pathways.

Articles dominate both the sample and population papers. The population papers, primarily published in the 2010s, build on these sample papers, primarily published in the 2000s.

The data presented in the tables and figure show that a small number of authors, institutions, funding bodies, journals, keywords, research fronts, subject categories, and countries have shaped the research in this field.

It is notable that the authors, institutions, and funding bodies from the USA, Japan, China, France, Germany, Spain, and the UK dominate the research in this field. Furthermore, it is also notable that some countries have a heavy presence in the population papers. The major producers of the population papers are India, South Korea, Canada, Poland, Thailand, Portugal, Iran, the Netherlands, Australia, Argentina, Russia, Sweden, Romania, Taiwan, and Turkey. Additionally, China and Brazil are under-represented significantly in the sample papers.

These findings show the importance of the progression of efficient incentive structures for the development of the research in this field as in others. It seems that some countries (such as the USA, Japan, China, France, Germany, Spain, and the UK) have efficient incentive structures for such development, contrary to India, South Korea, Canada, Poland, Thailand, Portugal, Iran, the Netherlands, Australia, Argentina, Russia, Sweden, Romania, Taiwan, and Turkey.

It further seems that although the research funding is a significant element of these incentive structures, it might not be a sole solution for increasing the incentives in this field, as in the case of India, South Korea, Canada, Poland, Thailand, Portugal, Iran, the Netherlands, Australia, Argentina, Russia, Sweden, Romania, Taiwan, and Turkey.

On the other hand, it seems there is more to do to reduce the significant gender deficit in this field as in other fields of science and technology (Xie and Shauman, 1998; Lariviere et al., 2013).

The data on the research fronts, keywords, source titles, and subject categories provide valuable evidence for the interdisciplinary (Lariviere and Gingras, 2010; Morillo et al., 2001) nature of the research in this field.

There is ample justification for the broad search strategy employed in this study due to the interdisciplinary nature of this research field, as evidenced by the top subject categories. The search strategy employed in this study is in line with those employed for related and other research fields (Konur, 2011, 2012a–n, 2015, 2016a–f, 2017a–f, 2018a–b, 2019a–b). It is particularly noted that only 15.0 and 16.7% of the sample and population papers are indexed by the 'energy fuels' subject category, respectively.

There are eight major topical research fronts for these sample papers: 'catalytic conversion' of glycerol, 'microbial conversion' of glycerol, 'hydrogenolysis' of glycerol, 'oxidation' of glycerol, 'fermentation' of glycerol, 'dehydration' of glycerol, 'reforming' of glycerol, and 'other methods' for conversion of glycerol such as 'transesterification' of glycerol.

On the other hand, there are eight primary research fronts for the products of glycerol conversion: 'hydrogen', 'chemicals' in general, 'acrolein', 'fuels' in general, 'propanediol', 'carbonate', 'ethanol', and 'other products' such as 'hydroxybutyrate' or 'docosahexaenoic acids'. There are also eight papers related primarily to the characterization of glycerol.

It is recommended that further scientometric studies are carried out for each of these research fronts, building on the pioneering studies in these fields.

ACKNOWLEDGMENTS

The contribution of the highly cited researchers in the fields of production of biofuels and biochemicals from glycerol is greatly acknowledged.

18.A APPENDIX

The keyword set for glycerol

Syntax: 1 NOT 2

18.A.1 Keywords

(TI=(glycerol or bioglycerol or glycerin* or bioglycerin* or "bio-glycerol" or (*diesel and ("by-product*" or "co-product*")))) AND (TI=(*hydrogen* or ethanol or *hydroxybutyrate* or docosahexaenoic or chemical* or valu* or derivative* or "carbon source*" or *fuel* or *oxidation* or *fermentation or characterization or viscosity or spectroscopy or solvent* or acrolein* or eutectic* or glycerate or lipid* or pyrolysis or *dehydrat* or valor* or *energy or *propanediol* or *carbonate or carbonylation or biocomponent* or carboxylation or dihydroxyacetone or "glycerol*production" or "citric acid*" or reforming or hydrodeoxygenation or "succinic acid*" or esterification or neutron or yarrowia or transesterification or 'trans-esterification' or "nitrite reduction*" or "tert-butylation" or *cataly* or saccharomyces or clostrid* or *acetylation or "h-2 production" or yeast* or products or "propylene-glycol" or "glyceric acid*" or conversion or escherichia or syngas or *dehydroxylation or gasification or substrate* or "ionic liquid*" or acetal* or etherification or "synthesis gas" or "water mixture*" or *butanediol* or klebsiella or "lactic acid*" or supercool* or chromatography or clostrid* or "by-product*" or electrode* or carbonatation or glycols or delignification or biogas or erythritol or *glyceride* or "aqueous mixture*" or succinate* or triacetylglycerol or glycerochem* or solketal or "anaerobic digestion" or acetol or triacylglycerol* or *additive* or thermograv* or "propylene glycol" or biorefinery or biopropanol* or olefin* or "*propionic acid*" or bombardment* or mannitol or deoxygenation or oligomerization or purification or carbonylation or butanol or combustion or "poly*hydroxyalkanoate*" or "alkyl-aromatic*" or rhodotorula or electrolysis or "acrylic acid*" or pseudomonas or lactobacillus or biopolyol* or dehydrogenation or Rhodosporidium or zeolite* or citrobacter* or Gluconobacter* or polyurethane* or Propionibacterium or Zobellella) OR WC=(energy*)).

18.A.2 Excluding terms

NOT TI=(*phosphat* or *protein* or dna or fibrosis or *mucosa* or geo* or pelag* or vitam* or arach* or *porin* or neur* or channel* or tetraether* or nerve* or tissue* or muscle* or human* or kinase* or sebac* or renal or cancer or sperm* or mice or adipose or diacyl* or *brain or rumen or GlpF or enol* or rat or mouse or turgor or food* or agr* or *starch or gluten* or disease* or chitosan or wine* or man) OR WC=(clin* or reprod* or critical* or gastro* or neur* or "engineering biomed*" or dermat* or ophth* or periph* or urol* or dent* or otorhin* or obstet* or pharm* or infect* or immun* or cardiac* or pathol* or medic* or hemat* or geochem* or geol* or veter* or pediat* or allerg* or optic* or respir* or toxic* or endoc* or parasit* or sport* or "materials science bio*" or physiol* or oncol* or surge* or androl* or "chemistry med*" or zool* or entomol* or fish* or hort*).

REFERENCES

Ahmadun, F. R., A. Pendashteh, and L. C. Abdullah, et al. 2009. Review of technologies for oil and gas produced water treatment. *Journal of Hazardous Materials* 170: 530–551.

Atlas, R. M. 1981. Microbial degradation of petroleum hydrocarbons: An environmental perspective. *Microbiological Reviews* 45: 180–209.

Babich, I. V. and J. A. Moulijn. 2003. Science and technology of novel processes for deep desulfurization of oil refinery streams: A review. *Fuel* 82: 607–631.

Behr, A., J. Eilting, K. Irawadi, J. Leschinski, and F. Lindner. 2008. Improved utilisation of renewable resources: New important derivatives of glycerol. *Green Chemistry* 10: 13–30.

Birge, N. O. 1986. Specific-heat spectroscopy of glycerol and propylene-glycol near the glass transition. *Physical Review B* 34: 1631–1642.

Bordons, M., B. Gonzalez-Albo, J. Aparicio, and L. Moreno. 2015. The influence of R & D intensity of countries on the impact of international collaborative research: Evidence from Spain. *Scientometrics* 102: 1385–1400.

Bornmann, L., J. Bauer, J., and E. M. Schlagberger. 2017. Characteristics of highly cited researchers 2015 in Germany. *Scientometrics* 111: 543–545.

Bridgwater, A. V. and G. V. C. Peacocke. 2000. Fast pyrolysis processes for biomass. *Renewable & Sustainable Energy Reviews* 4: 1–73.

Buhler, W., E. Dinjus, H. J. Ederer, A. Kruse, and C. Mas. 2002. Ionic reactions and pyrolysis of glycerol as competing reaction pathways in near- and supercritical water. *Journal of Supercritical Fluids* 22: 2237–22: 2253.

Busca, G., L. Lietti, G. Ramis, and F. Berti. 1998. Chemical and mechanistic aspects of the selective catalytic reduction of NO_x by ammonia over oxide catalysts: A review. *Applied Catalysis B-Environmental* 18: 1–36.

Carrettin, S., P. McMorn, P. Johnston, K. Griffin, and G. J. Hutchings. 2002. Selective oxidation of glycerol to glyceric acid using a gold catalyst in aqueous sodium hydroxide. *Chemical Communications* (7): 696–697.

Carrettin, S., P. McMorn, and P. Johnston, et al. 2003. Oxidation of glycerol using supported Pt, Pd and Au catalysts. *Physical Chemistry Chemical Physics* 5: 1329–1336.

Cavalheiro, J. M. B. T., M.C. M. D. de Almeida, C. Grandfils, and M. M. R. da Fonseca. 2009. Poly(3-hydroxybutyrate) production by *Cupriavidus necator* using waste glycerol. *Process Biochemistry* 44: 509–515.

Chai, S. H., H. P. Wang, Y. Liang, and B. Q. Xu. 2007. Sustainable production of acrolein: Investigation of solid acid-base catalysts for gas-phase dehydration of glycerol. *Green Chemistry* 9: 1130–1136.

Chaminand, J., L. Djakovitch, and P. Gallezot, et al. 2004. Glycerol hydrogenolysis on heterogeneous catalysts. *Green Chemistry* 6: 359–361.

Cheng, N. S. 2008. Formula for the viscosity of a glycerol-water mixture. *Industrial & Engineering Chemistry Research* 47: 3285–3288.

Cheng, Y. Q. and E. Ma. 2011. Atomic-level structure and structure–property relationship in metallic glasses. *Progress in Materials Science* 56: 379–473.

Chi, Z. Y., D. Pyle, Z. Y. Wen, C. Frear, and S. L. Chen. 2007. A laboratory study of producing docosahexaenoic acid from biodiesel-waste glycerol by microalgal fermentation. *Process Biochemistry* 42: 1537–1545.

Chisti, Y. 2007. Biodiesel from microalgae. *Biotechnology Advances* 25: 294–306.

Clarivate Analytics. 2019. *Highly cited researchers: 2019 Recipients*. Philadelphia, PA: Clarivate Analytics. https://recognition.webofsciencegroup.com/awards/highly-cited/2019/ (accessed January, 3, 2020).

Corma, A., G. W. Huber, L. Sauvanauda, and P. O'Connor. 2008. Biomass to chemicals: Catalytic conversion of glycerol/water mixtures into acrolein, reaction network. *Journal of Catalysis* 257: 163–171.

Czernik, S. and A. V. Bridgwater. 2004. Overview of applications of biomass fast pyrolysis oil. *Energy & Fuels* 18: 590–598.

Da Silva, G. P., M. Mack, and J. Contiero. 2009. Glycerol: A promising and abundant carbon source for industrial microbiology. *Biotechnology Advances* 27: 30–39.

Dasari, M. A., P. P. Kiatsimkul, W. R. Sutterlin, and G. J. Suppes. 2005. Low-pressure hydrogenolysis of glycerol to propylene glycol. *Applied Catalysis A-General* 281: 225–231.

Docampo, D. and L. Cram. 2019. Highly cited researchers: A moving target. *Scientometrics* 118: 1011–1025.

Gallezot, P. 2012. Conversion of biomass to selected chemical products. *Chemical Society Reviews* 41: 1538–1558.

Garfield, E. 1955. Citation indexes for science. *Science* 122: 108–111.

Garfield, E. 1972. Citation analysis as a tool in journal evaluation. *Science* 178: 471–479.

Guan, J. C. and N. Ma. 2007. China's emerging presence in nanoscience and nanotechnology: A comparative bibliometric study of several nanoscience 'giants'. *Research Policy* 36: 880–886.

Hassan, S. U., P. Haddawy, P. Kuinkel, A. Degelsegger, and C. Blasy. 2012. A bibliometric study of research activity in ASEAN related to the EU in FP7 priority areas. *Scientometrics* 91: 1035–1051.

Hill, J., E. Nelson, D. Tilman, S. Polasky, and D. Tiffany. 2006. Environmental, economic, and energetic costs and benefits of biodiesel and ethanol biofuels. *Proceedings of the National Academy of Sciences of the United States of America* 103: 11206–11210.

Hu, Z., G. Lin, T. Sun, T., and X. Wang. 2018. An EU without the UK: mapping the UK's changing roles in the EU scientific research. *Scientometrics* 115: 1185–1198.

Huang, M. H., H. W. Chang, and D. Z. Chen. 2006. Research evaluation of research-oriented universities in Taiwan from 1993 to 2003. *Scientometrics* 67: 419–435.

Johnson, D. T. and K. A. Taconi. 2007. The glycerin glut: Options for the value-added conversion of crude glycerol resulting from biodiesel production. *Environmental Progress* 26: 338–348.

Katryniok, B., S. Paul, V. Belliere-Baca, P. Rey, and F. Dumeignil, 2010. Glycerol dehydration to acrolein in the context of new uses of glycerol. *Green Chemistry* 12: 2079–2098.

Khalili, N. R., P. A. Scheff, and T. M. Holsen. 1995. PAH source fingerprints for coke ovens, diesel and gasoline-engines, highway tunnels, and wood combustion emissions. *Atmospheric Environment* 29: 533–542.

Kilian, L. 2009. Not all oil price shocks are alike: Disentangling demand and supply shocks in the crude oil market. *American Economic Review* 99: 1053–1069.

Konur, O. 2000. Creating enforceable civil rights for disabled students in higher education: An institutional theory perspective. *Disability & Society* 15: 1041–1063.

Konur, O. 2002a. Access to nursing education by disabled students: Rights and duties of nursing programs. *Nurse Education Today* 22: 364–374.

Konur, O. 2002b. Assessment of disabled students in higher education: Current public policy issues. *Assessment and Evaluation in Higher Education* 27: 131–152.

Konur, O. 2002c. Access to employment by disabled people in the UK: Is the Disability Discrimination Act working? *International Journal of Discrimination and the Law* 5: 247–279.

Konur, O. 2006a. Participation of children with dyslexia in compulsory education: Current public policy issues. *Dyslexia* 12: 51–67.

Konur, O. 2006b. Teaching disabled students in Higher Education. *Teaching in Higher Education* 11: 351–363.

Konur, O. 2007a. A judicial outcome analysis of the Disability Discrimination Act: A windfall for the employers? *Disability & Society* 22: 187–204.

Konur, O. 2007b. Computer-assisted teaching and assessment of disabled students in higher education: The interface between academic standards and disability rights. *Journal of Computer Assisted Learning* 23: 207–219.

Konur, O. 2011. The scientometric evaluation of the research on the algae and bio-energy. *Applied Energy* 88: 3532–3540.

Konur, O. 2012a. Evaluation of the research on the social sciences in Turkey: A scientometric approach. *Energy Education Science and Technology Part B: Social and Educational Studies* 4: 1893–1908.

Konur, O. 2012b. Prof. Dr. Ayhan Demirbas' scientometric biography. *Energy Education Science and Technology Part A: Energy Science and Research* 28: 727–738.

Konur, O. 2012c. The evaluation of the biogas research: A scientometric approach. *Energy Education Science and Technology Part A: Energy Science and Research* 29: 1277–1292.

Konur, O. 2012d. The evaluation of the educational research: A scientometric approach. *Energy Education Science and Technology Part B: Social and Educational Studies* 4: 1935–1948.

Konur, O. 2012e. The evaluation of the global energy and fuels research: A scientometric approach. *Energy Education Science and Technology Part A: Energy Science and Research* 30: 613–628.

Konur, O. 2012f. The evaluation of the research on the Arts and Humanities in Turkey: A scientometric approach. *Energy Education Science and Technology Part B: Social and Educational Studies* 4: 1603–1618.

Konur, O. 2012g. The evaluation of the research on the biodiesel: A scientometric approach. *Energy Education Science and Technology Part A: Energy Science and Research* 28: 1003–1014.

Konur, O. 2012h. The evaluation of the research on the bioethanol: A scientometric approach. *Energy Education Science and Technology Part A: Energy Science and Research* 28: 1051–1064.

Konur, O. 2012i. The evaluation of the research on the biofuels: A scientometric approach. *Energy Education Science and Technology Part A: Energy Science and Research* 28: 903–916.

Konur, O. 2012j. The evaluation of the research on the biohydrogen: A scientometric approach. *Energy Education Science and Technology Part A: Energy Science and Research* 29: 323–338.

Konur, O. 2012k. The evaluation of the research on the microbial fuel cells: A scientometric approach. *Energy Education Science and Technology Part A: Energy Science and Research* 29: 309–322.

Konur, O. 2012l. The scientometric evaluation of the research on the production of bioenergy from biomass. *Biomass and Bioenergy* 47: 504–515.

Konur, O. 2012m. The scientometric evaluation of the research on the deaf students in higher education. *Energy Education Science and Technology Part B: Social and Educational Studies* 4: 1573–1588.

Konur, O. 2012n. The scientometric evaluation of the research on the students with ADHD in higher education. *Energy Education Science and Technology Part B: Social and Educational Studies* 4: 15471562.

Konur, O. 2015. Current state of research on algal biodiesel. In *Marine Bioenergy: Trends and Developments*, S. K. Kim, and C. G. Lee, ed., 487–512. Boca Raton, FL: CRC Press.

Konur, O. 2016a. Scientometric overview in nanobiodrugs. In *Nanoarchitectonics for Smart Delivery and Drug Targeting*, A. M. Holban and A.M. Grumezescu, ed., 405–428. Amsterdam: Elsevier.

Konur, O. 2016b. Scientometric overview regarding nanoemulsions used in the food industry. In *Emulsions: Nanotechnology in the Agri-Food Industry*, A. M. Grumezescu, ed., 689–711. Amsterdam: Elsevier.

Konur, O. 2016c. Scientometric overview regarding the nanobiomaterials in antimicrobial therapy. In *Nanobiomaterials in Antimicrobial Therapy,* A. M. Grumezescu, ed., 511–535. Amsterdam: Elsevier.

Konur, O. 2016d. Scientometric overview regarding the nanobiomaterials in dentistry. In *Nanobiomaterials in Dentistry,* A. M. Grumezescu, ed., 425–453. Amsterdam: Elsevier.

Konur, O. 2016e. Scientometric overview regarding the surface chemistry of nanobiomaterials. In *Surface Chemistry of Nanobiomaterials,* A. M. Grumezescu, ed., 463–486. Amsterdam: Elsevier.

Konur, O. 2016f. The scientometric overview in cancer targeting. In *Nanoarchitectonics for Smart Delivery and Drug Targeting,* A. M. Holban and A. Grumezescu, ed., 871–895. Amsterdam; Elsevier.

Konur, O. 2017a. Recent citation classics in antimicrobial nanobiomaterials. In *Nanostructures for Antimicrobial Therapy,* A. Ficai and A. M. Grumezescu, ed., 669–685. Amsterdam: Elsevier.

Konur, O. 2017b. Scientometric overview in nanopesticides. In *New Pesticides and Soil Sensors,* A. M. Grumezescu, ed. 719–744. Amsterdam: Elsevier.

Konur, O. 2017c. Scientometric overview regarding oral cancer nanomedicine. In *Nanostructures for Oral Medicine*, E. Andronescu, A. M. Grumezescu, ed., 939–962. Amsterdam: Elsevier.

Konur, O. 2017d. Scientometric overview regarding water nanopurification. In *Water Purification,* A. M. Grumezescu, ed., 693–716. Amsterdam: Elsevier.

Konur, O. 2017e. Scientometric overview in food nanopreservation. In *Food Preservation,* A. M. Grumezescu, ed., 703–729. Amsterdam: Elsevier.

Konur, O. 2017f. The top citation classics in alginates for biomedicine. In *Seaweed Polysaccharides: Isolation, Biological and Biomedical Applications*, J. Venkatesan, S. Anil, S. K. Kim, ed., 223–249. Amsterdam: Elsevier.

Konur, O. 2018a. Scientometric evaluation of the global research in spine: An update on the pioneering study by Wei et al. *European Spine Journal* 27: 525–529.

Konur, O. 2018b. Bioenergy and biofuels science and technology: Scientometric overview and citation classics. In *Bioenergy and Biofuels*, O. Konur, ed., 3–63. Boca Raton: CRC Press.

Konur, O. 2019a. Cyanobacterial bioenergy and biofuels science and technology: A scientometric overview. In *Cyanobacteria: From Basic Science to Applications*, ed. A. K. Mishra, D. N. Tiwari and A. N. Rai, 419–442. Amsterdam: Elsevier.

Konur, O. 2019b. Nanotechnology applications in food: A scientometric overview. In *Nanoscience for Sustainable Agriculture*, R. N., Pudake, N. Chauhan, and C. Kole, ed., 683–711. Cham: Springer.

Konur, O., ed. 2021a. *Handbook of Biodiesel and Petrodiesel Fuels: Science, Technology, Health, and Environment.* Boca Raton, FL: CRC Press.

Konur, O., ed. 2021b. *Handbook of Biodiesel and Petrodiesel Fuels: Science, Technology, Health, and Environment. Volume 1. Biodiesel Fuels: Science, Technology, Health, and Environment.* Boca Raton, FL: CRC Press.

Konur, O., ed. 2021c. *Handbook of Biodiesel and Petrodiesel Fuels: Science, Technology, Health, and Environment. Volume 2. Biodiesel Fuels based on the Edible and Nonedible Feedstocks, Wastes, and Algae: Science, Technology, Health, and Environment.* Boca Raton, FL: CRC Press.

Konur, O., ed. 2021d. *Handbook of Biodiesel and Petrodiesel Fuels: Science, Technology, Health, and Environment. Volume 3. Petrodiesel Fuels: Science, Technology, Health, and Environment.* Boca Raton, FL: CRC Press.

Konur, O. 2021e. Biodiesel and petrodiesel fuels: Science, technology, health, and environment. In *Handbook of Biodiesel and Petrodiesel Fuels: Science, Technology, Health, and Environment. Volume 1. Biodiesel Fuels: Science, Technology, Health, and Environment*, ed. O. Konur. Boca Raton, FL: CRC Press.

Konur, O. 2021f. Biodiesel and petrodiesel fuels: A scientometric review of the research. In *Handbook of Biodiesel and Petrodiesel Fuels: Science, Technology, Health, and Environment. Volume 1. Biodiesel Fuels: Science, Technology, Health, and Environment*, ed. O. Konur. Boca Raton, FL: CRC Press.

Konur, O. 2021g. Biodiesel and petrodiesel fuels: A review of the research. In *Handbook of Biodiesel and Petrodiesel Fuels: Science, Technology, Health, and Environment. Volume 1. Biodiesel Fuels: Science, Technology, Health, and Environment*, ed. O. Konur. Boca Raton, FL: CRC Press.

Konur, O. 2021h Nanotechnology applications in the diesel fuels and the related research fields: A review of the research. In *Handbook of Biodiesel and Petrodiesel Fuels: Science, Technology, Health, and Environment. Volume 1. Biodiesel Fuels: Science, Technology, Health, and Environment*, ed. O. Konur. Boca Raton, FL: CRC Press.

Konur, O. 2021i. Biooils: A scientometric review of the research. In *Handbook of Biodiesel and Petrodiesel Fuels: Science, Technology, Health, and Environment. Volume 1. Biodiesel Fuels: Science, Technology, Health, and Environment*, ed. O. Konur. Boca Raton, FL: CRC Press.

Konur, O. 2021j. Characterization and properties of biooils: A review of the research. In *Handbook of Biodiesel and Petrodiesel Fuels: Science, Technology, Health, and Environment. Volume 1. Biodiesel Fuels: Science, Technology, Health, and Environment*, ed. O. Konur. Boca Raton, FL: CRC Press.

Konur, O. 2021k. Biomass pyrolysis and pyrolysis oils: A review of the research. In *Handbook of Biodiesel and Petrodiesel Fuels: Science, Technology, Health, and Environment. Volume 1. Biodiesel Fuels: Science, Technology, Health, and Environment*, ed. O. Konur. Boca Raton, FL: CRC Press.

Konur, O. 2021l. Biodiesel fuels: A scientometric review of the research. In *Handbook of Biodiesel and Petrodiesel Fuels: Science, Technology, Health, and Environment. Volume 1. Biodiesel Fuels: Science, Technology, Health, and Environment*, ed. O. Konur. Boca Raton, FL: CRC Press.

Konur, O. 2021m. Glycerol: A scientometric review of the research. In *Handbook of Biodiesel and Petrodiesel Fuels: Science, Technology, Health, and Environment. Volume 1. Biodiesel Fuels: Science, Technology, Health, and Environment*, ed. O. Konur. Boca Raton, FL: CRC Press. 23.1.2020

Konur, O. 2021n. Propanediol production from glycerol: A review of the research. In *Handbook of Biodiesel and Petrodiesel Fuels: Science, Technology, Health, and Environment. Volume 1. Biodiesel Fuels: Science, Technology, Health, and Environment*, ed. O. Konur. Boca Raton, FL: CRC Press. 17.2.2020

Konur, O. 2021o. Edible oil-based biodiesel fuels: A scientometric review of the research. *In Handbook of Biodiesel and Petrodiesel Fuels: Science, Technology, Health, and Environment. Volume 2. Biodiesel Fuels based on the Edible and Nonedible Feedstocks, Wastes, and Algae: Science, Technology, Health, and Environment,* ed. O. Konur. Boca Raton, FL: CRC Press.

Konur, O. 2021p. Palm oil-based biodiesel fuels: A review of the research. In *Handbook of Biodiesel and Petrodiesel Fuels: Science, Technology, Health, and Environment. Volume 2. Biodiesel Fuels based on the Edible and Nonedible Feedstocks, Wastes, and Algae*, ed. O. Konur. Boca Raton, FL: CRC Press.

Konur, O. 2021q. Rapeseed oil-based biodiesel fuels: A review of the research. In *Handbook of Biodiesel and Petrodiesel Fuels: Science, Technology, Health, and Environment. Volume 2. Biodiesel Fuels based on the Edible and Nonedible Feedstocks, Wastes, and Algae,* ed. O. Konur. Boca Raton, FL: CRC Press.

Konur, O. 2021r. Nonedible oil-based biodiesel fuels: A scientometric review of the research. In *Handbook of Biodiesel and Petrodiesel Fuels: Science, Technology, Health, and Environment. Volume 2. Biodiesel Fuels based on the Edible and Nonedible Feedstocks, Wastes, and Algae: Science, Technology, Health, and Environment*, ed. O. Konur. Boca Raton, FL: CRC Press. 19.1.2020

Konur, O. 2021s. Waste oil-based biodiesel fuels: A scientometric review of the research. In *Handbook of Biodiesel and Petrodiesel Fuels: Science, Technology, Health, and Environment. Volume 2. Biodiesel Fuels based on the Edible and Nonedible Feedstocks, Wastes, and Algae: Science, Technology, Health, and Environment*, ed. O. Konur. Boca Raton, FL: CRC Press.

Konur, O. 2021t. Algal biodiesel fuels: A scientometric review of the research. In *Handbook of Biodiesel and Petrodiesel Fuels: Science, Technology, Health, and Environment. Volume 2. Biodiesel Fuels based on the Edible and Nonedible Feedstocks, Wastes, and Algae: Science, Technology, Health, and Environment*, ed. O. Konur. Boca Raton, FL: CRC Press.

Konur, O. 2021u. Algal biomass production for biodiesel production: A review of the research. In *Handbook of Biodiesel and Petrodiesel Fuels: Science, Technology, Health, and Environment. Volume 2. Biodiesel Fuels based on the Edible and Nonedible Feedstocks, Wastes, and Algae*, Ed. O. Konur. Boca Raton, FL: CRC Press. 23.2.2020

Konur, O. 2021v. Algal biomass production in wastewaters for biodiesel production: A review of the research. In *Handbook of Biodiesel and Petrodiesel Fuels: Science, Technology, Health, and Environment. Volume 2. Biodiesel Fuels based on the Edible and Nonedible Feedstocks, Wastes, and Algae*, ed. O. Konur. Boca Raton, FL: CRC Press. 23.2.2020

Konur, O. 2021x. Algal lipid production for biodiesel production: A review of the research. In *Handbook of Biodiesel and Petrodiesel Fuels: Science, Technology, Health, and Environment. Volume 2. Biodiesel Fuels based on the Edible and Nonedible Feedstocks, Wastes, and Algae*, Ed. O. Konur. Boca Raton, FL: CRC Press. 15.5.2020

Konur, O. 2021y. Crude oils: A scientometric review of the research. In *Handbook of Biodiesel and Petrodiesel Fuels: Science, Technology, Health, and Environment. Volume 3. Petrodiesel Fuels: Science, Technology, Health, and Environment*, ed. O. Konur. Boca Raton, FL: CRC Press.

Konur, O. 2021z. Petrodiesel fuels: A scientometric review of the research. In *Handbook of Biodiesel and Petrodiesel Fuels: Science, Technology, Health, and Environment. Volume 3. Petrodiesel Fuels: Science, Technology, Health, and Environment*, ed. O. Konur. Boca Raton, FL: CRC Press.

Konur, O. 2021aa. Bioremediation of petroleum hydrocarbons in the contaminated soils: A review of the research. In *Handbook of Biodiesel and Petrodiesel Fuels: Science, Technology, Health, and Environment. Volume 3. Petrodiesel Fuels: Science, Technology, Health, and Environment*, ed. O. Konur. Boca Raton, FL: CRC Press.

Konur, O. 2021ab. Desulfurization of diesel fuels: A review of the research. In *Handbook of Biodiesel and Petrodiesel Fuels: Science, Technology, Health, and Environment. Volume 3. Petrodiesel Fuels: Science, Technology, Health, and Environment*, ed. O. Konur. Boca Raton, FL: CRC Press.

Konur, O. 2021ac. Diesel fuel exhaust emissions: A scientometric review of the research. In *Handbook of Biodiesel and Petrodiesel Fuels: Science, Technology, Health, and Environment. Volume 3. Petrodiesel Fuels: Science, Technology, Health, and Environment*, ed. O. Konur. Boca Raton, FL: CRC Press.

Konur, O. 2021ad. The adverse health and safety impact of diesel fuels: A scientometric review of the research. In *Handbook of Biodiesel and Petrodiesel Fuels: Science, Technology, Health, and Environment. Volume 3. Petrodiesel Fuels: Science, Technology, Health, and Environment*, ed. O. Konur. Boca Raton, FL: CRC Press.

Konur, O. 2021ae. Respiratory illnesses caused by the diesel fuel exhaust emissions: A review of the research. In *Handbook of Biodiesel and Petrodiesel Fuels: Science, Technology, Health, and Environment. Volume 3. Petrodiesel Fuels: Science, Technology, Health, and Environment*, ed. O. Konur. Boca Raton, FL: CRC Press.

Konur, O. 2021af. Cancer caused by the diesel fuel exhaust emissions: A review of the research. In *Handbook of Biodiesel and Petrodiesel Fuels: Science, Technology, Health, and Environment. Volume 3. Petrodiesel Fuels: Science, Technology, Health, and Environment*, ed. O. Konur. Boca Raton, FL: CRC Press.

Konur, O. 2021ag. Cardiovascular and other illnesses caused by the diesel fuel exhaust emissions: A review of the research. In *Handbook of Biodiesel and Petrodiesel Fuels: Science, Technology, Health, and Environment. Volume 3. Petrodiesel Fuels: Science, Technology, Health, and Environment*, ed. O. Konur. Boca Raton, FL: CRC Press.

Konur, O. and F. L. Matthews. 1989. Effect of the properties of the constituents on the fatigue performance of composites: A review. *Composites* 20: 317–328.

Lapuerta, M., O. Armas, and J. Rodriguez-Fernandez. 2008. Effect of biodiesel fuels on diesel engine emissions. *Progress in Energy and Combustion Science* 34: 198–223.

Lariviere, V. and Y. Gingras. 2010. On the relationship between interdisciplinarity and scientific impact. *Journal of the American Society for Information Science and Technology* 61: 126–131.

Lariviere, V., C. Ni, Y. Gingras, B. Cronin, and C. R. Sugimoto. 2013. Bibliometrics: Global gender disparities in science. *Nature News* 504: 211–213.

Leydesdorff, L. and C. Wagner. 2009. Is the United States losing ground in science? A global perspective on the world science system. *Scientometrics* 78: 23–36.

Leydesdorff, L. and P. Zhou. 2005. Are the contributions of China and Korea upsetting the world system of science? *Scientometrics* 63: 617–630.

Maris, E. P. and R. J. Davis. 2007. Hydrogenolysis of glycerol over carbon-supported Ru and Pt catalysts. *Journal of Catalysis* 249: 328–337.

Mata, T. M., A. A. Martins, and N. S. Caetano. 2010. Microalgae for biodiesel production and other applications: A review. *Renewable & Sustainable Energy Reviews* 14: 217–232.

Miyazawa, T., Y. Kusunoki, Y., K. Kunimori, K., and K. Tomishige. 2006. Glycerol conversion in the aqueous solution under hydrogen over Ru/C plus an ion-exchange resin and its reaction mechanism. *Journal of Catalysis* 240: 213–221.

Mohan, D., C. U. Pittman, and P. H. Steele. 2006. Pyrolysis of wood/biomass for bio-oil: A critical review. *Energy & Fuels* 20: 848–889.

Morillo, F., M. Bordons, and I. Gomez. 2001. An approach to interdisciplinarity through bibliometric indicators. *Scientometrics* 51: 203–222.

Nakagawa, Y., Y. Shinmi, S. Koso, and K. Tomishige. 2010. Direct hydrogenolysis of glycerol into 1, 3-propanediol over rhenium-modified iridium catalyst. *Journal of Catalysis* 272: 191–194.

North, D. C. 1991a. *Institutions, Institutional Change and Economic Performance*. Cambridge, Mass.: Cambridge University Press.

North, D.C. 1991b. Institutions. *Journal of Economic Perspectives* 5: 97–112.

Ochoa-Gomez, J. R., O. Gomez-Jimenez-Aberasturi, and B. Maestro-Madurga, et al. 2009. Synthesis of glycerol carbonate from glycerol and dimethyl carbonate by transesterification: Catalyst screening and reaction optimization. *Applied Catalysis A-General* 366: 315–324.

Okubo, Y., J. C. Dore, T. Ojasoo, and J. F. Miquel. 1998. A multivariate analysis of publication trends in the 1980s with special reference to South-East Asia. *Scientometrics* 41: 273.

Pagliaro, M., R. Ciriminna, H. Kimura, M. Rossi, and C. della Pina. 2007. From glycerol to value-added products. *Angewandte Chemie-International Edition* 46: 4434–4440.

Papanikolaou, S. and G. Aggelis. 2002. Lipid production by *Yarrowia lipolytica* growing on industrial glycerol in a single-stage continuous culture. *Bioresource Technology* 82: 43–49.

Papanikolaou, S., S. Fakas, and M. Fick, et al., 2008. Biotechnological valorisation of raw glycerol discharged after bio-diesel (fatty acid methyl esters) manufacturing process: Production of 1, 3-propanediol, citric acid and single cell oil. *Biomass & Bioenergy* 32: 60–71.

Perron, P. 1989. The great crash, the oil price shock, and the unit root hypothesis. *Econometrica: Journal of the Econometric Society* 57: 1361–1401.

Prathap, G. 2017. A three-dimensional bibliometric evaluation of recent research in India. *Scientometrics* 110: 1085–1097.

Rep, M., M. Krantz, J. M. Thevelein, and S. Hohmann. 2000. The transcriptional response of *Saccharomyces cerevisiae* to osmotic shock: Hot 1p and Msn 2p/Msn4p are required for the induction of subsets of high osmolarity glycerol pathway-dependent genes. *Journal of Biological Chemistry* 275: 8290–8300.

Rogers, D. and A. J. Hopfinger. 1994. Application of genetic function approximation to quantitative structure-activity relationships and quantitative structure-property relationships. *Journal of Chemical Information and Computer Sciences* 34: 854–866.

Rogge, W. F., L. M. Hildemann, M. A. Mazurek, G. R. Cass, and B. R. T. Simoneit. 1993. Sources of fine organic aerosol. 2. Noncatalyst and catalyst-equipped automobiles and heavy-duty diesel trucks. *Environmental Science & Technology* 27: 636–651.

Schauer, J. J., M. J. Kleeman, G. R. Cass, and B. R. T. Simoneit. 1999. Measurement of emissions from air pollution sources. 2. C_1 through C_{30} organic compounds from medium duty diesel trucks. *Environmental Science & Technology* 33: 1578–1587.

Scherf, U. and E. J. List. 2002. Semiconducting polyfluorenes-towards reliable structure–property relationships. *Advanced Materials* 14: 477–487.

Thompson, J. C. and B. B. He. 2006. Characterization of crude glycerol from biodiesel production from multiple feedstocks. *Applied Engineering in Agriculture* 22: 261–265.

Van Raan, A. F., T. N., Van Leeuwen, and M. S. Visser. 2011. Severe language effect in university rankings: particularly Germany and France are wronged in citation-based rankings. *Scientometrics* 88: 495–498.

Wang, Z. X., J. Zhuge, H. Y. Fang, and B. A. Prior. 2001. Glycerol production by microbial fermentation: A review. *Biotechnology Advances* 19: 201–223.

Wang, X., D. Liu, K. Ding, K., and X. Wang. 2012. Science funding and research output: A study on 10 countries. *Scientometrics* 91: 591–599.

Wen, G. D., Y. P. Xu, H. J. Ma, Z. S. Xu, and Z. J. Tian. 2008. Production of hydrogen by aqueous-phase reforming of glycerol. *International Journal of Hydrogen Energy* 33: 6657–6666.

Xie, Y. and K. A. Shauman. 1998. Sex differences in research productivity: New evidence about an old puzzle. *American Sociological Review* 63: 847–870.

Yazdani, S. S. and R. Gonzalez. 2007. Anaerobic fermentation of glycerol: A path to economic viability for the biofuels industry. *Current Opinion In Biotechnology* 18: 213–219.

Yazdani, S. S. and R. Gonzalez. 2008. Engineering *Escherichia coli* for the efficient conversion of glycerol to ethanol and co-products. *Metabolic Engineering* 10: 340–351.

Youtie, J, P. Shapira, and A. L. Porter. 2008. Nanotechnology publications and citations by leading countries and blocs. *Journal of Nanoparticle Research* 10: 981–986.

Zhang, Y., M. A. Dube, D. D. McLean, and M. Kates. 2003. Biodiesel production from waste cooking oil: 1. Process design and technological assessment. *Bioresource Technology* 89: 1–16.

Zhang, B. C., X. L. Tang, Y. Li, Y. D. Xu, and W. J. Shen. 2007. Hydrogen production from steam reforming of ethanol and glycerol over ceria-supported metal catalysts. *International Journal of Hydrogen Energy* 32: 2367–2373.

Zhou, C. H. C., J. N. Beltramini, Y. X. Fan, and G. Q. M. Lu. 2008. Chemoselective catalytic conversion of glycerol as a biorenewable source to valuable commodity chemicals. *Chemical Society Reviews* 37: 527–549.

Zhou, P. and L. Leydesdorff. 2006. The emergence of China as a leading nation in science. *Research Policy* 35: 83–104.

19 Hydrogen-Rich Syngas Production from Biodiesel-derived Glycerol

An Overview of the Modeling and Optimization Strategies

Bamidele Victor Ayodele
Siti Indati Mustapa
May Ali Alsaffa

CONTENTS

19.1 INTRODUCTION

The utilization of energy derived from fossil fuel is often associated with challenges of greenhouse gas emissions which have significantly result in global warming (Mardani et al., 2019). This phenomenon has increased the quest for alternative, cleaner, and renewable sources of energy in the past decades (Muhammad et al., 2018). Amongst the several renewable energy sources, biodiesel has been projected as a possible candidate to favorably compete with conventional petrodiesel (Mekhilef et al., 2011). Hence, the global production of biodiesel has increased drastically over

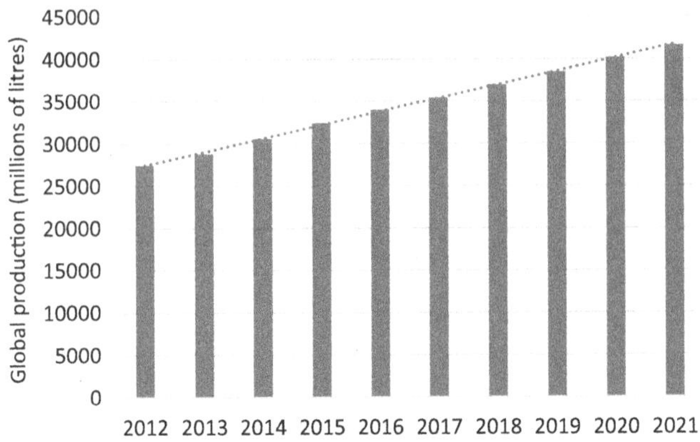

FIGURE 19.1 Biodiesel outlook from 2012 to 2021. (*Source:* OECD/FAO (2012)).

the years, as shown in Figure 19.1. However, biodiesel is produced concurrently with glycerol which has been a major challenge in the production line (Quispe et al., 2013). According to Yang et al. (2012), 10% (w/w) 1.05 pounds of glycerol is produced for an equivalent gallon of biodiesel. Hence, research focus has been on the sustainable ways of utilizing biodiesel-derived glycerol for the production of value-added chemicals (Ayodele et al., 2020).

Glycerol has been proven as a viable feedstock to produce chemical intermediates such as hydrogen-rich syngas by various thermo-catalytic and bioconversion processes (Seadira et al., 2017; Vaidya and Rodrigues, 2009). These chemical intermediates can be employed to produce other value-added chemicals (Tan et al., 2013). The conversion of glycerol to hydrogen-rich syngas can be achieved through the various thermo-catalytic process such as 'steam reforming', 'dry reforming', 'autothermal reforming', 'supercritical water reforming', 'pyrolysis', 'photo-catalysis', as well as 'microbial bioconversion' using a microorganism (Ayodele et al., 2019, 2020).

Glycerol conversion to hydrogen-rich gas has been reported to be influenced by several parameters, such as the reaction temperature, pressure, water-to-glycerol molar feed ratio, and irradiation time (Estahbanati et al., 2017; Mangayil et al., 2015). The individual and interaction effects of these parameters on glycerol conversion and hydrogen-rich syngas production have been investigated using various modeling techniques (Adeniyi and Ighalo, 2019; Galera and Ortiz, 2015; Ghasemzadeh et al., 2018). In addition, the optimum conditions of these parameters that could maximize glycerol conversion and the formation of the desired product have been investigated using various optimization strategies (Estahbanati et al., 2017; Zamzuri et al., 2017).

The main objective of this chapter is to present an overview of the various modeling and optimization strategies that have been employed for hydrogen-rich syngas production from glycerol conversion.

19.2 TECHNOLOGICAL ROUTES FOR HYDROGEN-RICH SYNGAS PRODUCTION FROM GLYCEROL

Glycerol can be converted to hydrogen-rich syngas using various technological routes, such as reforming, pyrolysis, and fermentative processes, as depicted in Figure 19.2 (Monteiro et al., 2018). The reforming and pyrolysis of glycerol are thermo-catalytic processes that are performed under high temperature (> 500°C) using various types of catalysts (Dieuzeide et al., 2015; Shahirah et al., 2017). The details of each of these processes are discussed subsequently.

The reforming of glycerol using steam is one of the most established processes for producing hydrogen from hydrocarbon. The steam reforming of glycerol is a thermo-catalytic process which involves the reaction of glycerol and steam in the presence of a catalyst, as represented in Equation (19.1) (Dieuzeideh et al., 2015). Primarily, the steam reforming of the glycerol reaction in Equation (19.1) can be described as a summation of the glycerol decomposition reaction (Equation 19.2) and the water-gas-shift reaction (Equation 19.3) (Cheng et al., 2011).

$$C_3H_8O_3(g) + 3H_2O(g) \leftrightarrow 3CO_2(g) + 7H_2(g) \tag{19.1}$$

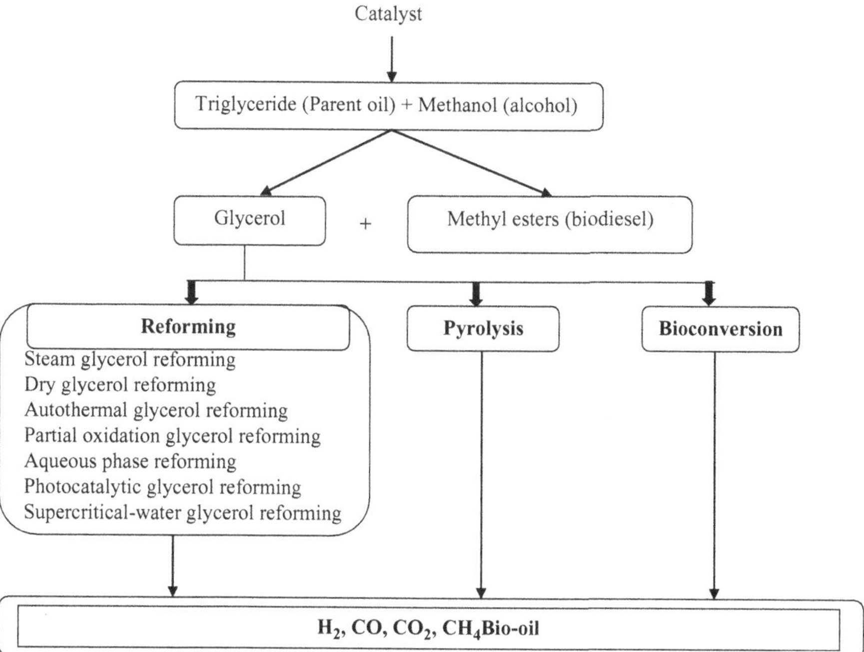

FIGURE 19.2 The various technological pathways for conversion of glycerol to hydrogen-rich syngas

$$C_3H_8O_3(g) \leftrightarrow 3CO(g) + 4H_2(g) \quad (19.2)$$

$$CO(g) + H_2O(g) \leftrightarrow CO_2(g) + H_2(g) \quad (19.3)$$

The steam reforming of the glycerol reaction is often influenced by several parameters, such as the feed composition (water/glycerol), the feed rate, the nature of the catalyst, the reaction temperature, and the gas hourly space velocity. Bobadilla et al. (2015) investigated the effect of the reaction temperature, water/glycerol molar ratio, and space velocity on the glycerol steam reforming of hydrogen. The study revealed that the hydrogen yield was strongly influenced by the reaction temperature. An increase in the hydrogen yield was observed with a corresponding increase in the reaction temperature up until 650°C. This is an expected trend for a temperature-dependent gas-phase reaction as stipulated by the Arrhenius theory. The authors also reported that the water to glycerol molar concentration in the feed significantly influenced hydrogen production. The hydrogen yield was reported to decrease with an increase in the glycerol concentration in the feed. The decrease in the hydrogen yield was attributed to the reduction in the activity of the catalysts as a result of coke deposition from the unconverted glycerol. Similarly, an increase in the space velocity reportedly results in a corresponding decrease in the hydrogen yield.

The nature of the catalysts used for steam reforming of glycerol has been reported to influence hydrogen production. As reported by Ayodele et al. (2020), the make-up of the catalysts significantly influences the yield of the hydrogen produced by the steam reforming reaction, as shown in Figure 19.3. However, the performance of the catalyst in terms of the product yield is not solely dependent on the chemical composition but also on parameters such as the preparation method, the activation temperature, and the reaction conditions (Ayodele et al., 2020).

The metal loading of the catalyst has also been reported to have a significant effect on catalytic performance. As reported by Senseni et al. (2016a), the variation of Ni

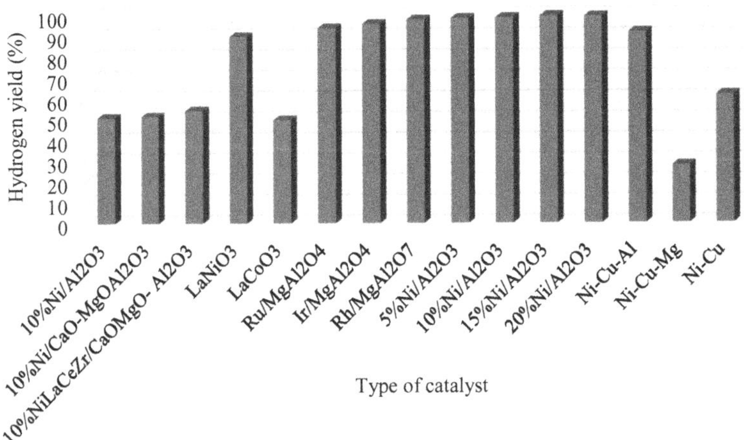

FIGURE 19.3 Hydrogen yield obtained for various catalysts used for steam reforming of glycerol.

loading from 5–20% has an impact on the performance of Ni/Al_2O_3 during steam reforming of glycerol at 750°C. A decrease in the hydrogen yield was observed with an increase in Ni loading. An increase in the metal loading could result in sintering which is one of the major causes of Ni-based catalysts.

The reforming of glycerol using carbon dioxide (CO_2) is often referred to as the 'dry reforming process' (Arif et al., 2019). This process entails the utilization of CO_2 and glycerol for the production of hydrogen-rich syngas. The process offers the advantage of employing both CO_2, a principal component of greenhouse gases, and glycerol, a by-product of biodiesel production, as feedstock for producing renewable hydrogen. The main reaction of the dry reforming of glycerol is represented in Equation 19.4. Besides the main reaction, other competing side reactions such as glycerol decomposition, water gas shift reaction, methanation, and carbon gasification play significant roles in the product formation (Roslan et al., 2020).

$$C_3H_8O_3(g) + CO_2(g) \leftrightarrow 4CO(g) + 3H_2(g) + H_2O(g) \tag{19.4}$$

Factors such as the nature of catalysts, the catalyst synthesis method, the type of reactor, and the reaction conditions have also been reported to significantly influence hydrogen production by the dry reforming of glycerol (Ayodele et al., 2020). Most of the studies conducted on the dry reforming of glycerol reported a temperature range of 700–850°C for hydrogen-rich gas production. One major advantage of the dry reforming of glycerol is that the hydrogen:carbon monoxide (H_2:CO) ratio obtained from the reaction is often close to unity, making it suitable as a potential feedstock for the production of liquid hydrocarbon by 'Fischer–Tropsch synthesis' (Pakhare and Spivey, 2014). However, one of the major drawbacks of the process like other reforming reactions is catalyst deactivation by carbon deposition and sintering. Efforts are being made by researchers to develop various strategies that can mitigate the adverse effects of catalyst deactivation on the performance of the various dry reforming methods of glycerol catalysts.

'Aqueous phase reforming' is an emerging technological route to produce hydrogen-rich syngas from glycerol (Manfro et al., 2013). This process involves the conversion of glycerol in water using a suitable catalyst as shown in Equation 19.5. Unlike the other reforming processes, aqueous phase reforming does not require high thermal energy. Studies have shown that aqueous phase reforming of glycerol often occurs at a pressure and temperature range of 15–50 bar and 200–300°C, respectively (Ayodele et al., 2020). These reaction conditions do not allow the occurrence of catalyst deactivation. Besides the main reaction in Equation 19.5, a water-gas shift reaction also occurs as a side reaction. However, the reaction conditions of aqueous phase reforming often favor the production of hydrogen with low CO content from the water-gas shift reaction.

$$C_3H_8O_3(g) + H_2O(g) \leftrightarrow 3CO(g) + 7H_2(g) \tag{19.5}$$

Aqueous phase reforming of glycerol in a batch reactor over a Pt/Al_2O_3 catalyst has been reported by Seretis and Tsiakaras (2016). The effect of the reaction temperature (200–240°C), the reaction time (30–240 min), the glycerol concentration (1 and 10%), and the Pt loading (0.5–5.0%) on the product distribution was investigated. The analysis of the gaseous products after each reaction run revealed the presence of H_2, CO, CO_2, and CH_4 in the outlet stream. The findings showed that maximum yields of the gaseous products were obtained using 1 g of 5% Pt/Al_2O_3, 1 wt% of glycerol solution, 240°C, and a 4 h reaction time. Optimization strategies such as the response surface methodology can be employed to determine the interaction effects of all the process parameters and the optimum conditions at which the gaseous products and the glycerol conversion can be maximized.

'Partial oxidation reforming' is a process that involves the use of an appropriate amount of oxygen for reforming glycerol into hydrogen-rich syngas (Rennard et al., 2009). Unlike the steam and dry reforming of glycerol, which are endothermic reactions, the partial oxidation reforming of the glycerol reaction occurs exothermically. However, a hot zone is usually created within the reactor as a result of the excessive heat generated from the exothermic reaction. Hence, due to the sensitivity of the partial oxidation reaction, an appropriate control measure is often put in place during the experimental phase. Moreover, materials that could withstand high temperatures are often considered for building a reactor for the partial oxidation reforming process. One major limitation of the partial oxidation reforming of glycerol is the low hydrogen yield that is usually obtained from this process.

Partial oxidation reforming of glycerol can also be conducted under autothermal conditions using oxygen and steam. Liu and Lin (2014) demonstrated the 'autothermal partial oxidation'-based glycerol reforming using $LaMnO_3$- and $LaNiO_3$-coated monolith perovskite catalysts. The study revealed that the hydrogen-rich syngas was produced by the autothermal partial oxidation reaction over the $LaMnO_3$ and $LaNiO_3$ catalysts. However, $LaMnO_3$ showed superior activity in the autothermal reforming compared with the $LaNiO_3$ catalyst.

Studies have shown that parameters such as the reaction temperature, steam:carbon ratio, oxygen:carbon ratio, and the gas hourly space velocity significantly influence the yield of the hydrogen-rich syngas during the partial oxidation autothermal reforming of glycerol. Liu and Lin (2014) reported an operational range of 300–700°C, 0.4–1.5, and 0.1–0.3 for inlet temperature, steam:carbon ratio, and oxygen:carbon ratio, respectively. The study revealed that the gaseous products which include H_2, CO, CO_2, and CH_4 increased with increased inlet temperature. This trend was attributed to the high rate of conversion of the glycerol as the temperature increases. On the other hand, the steam:carbon ratio was found to affect the yield negatively. The hydrogen yield obtained from the partial autothermal reforming of glycerol was observed to decrease with an increase in the steam:carbon ratio. The yield of hydrogen was however found to increase with an increase in the oxygen:carbon ratio. Applying appropriate optimization strategies could help to determine the optimum conditions to obtain the maximum yields of the hydrogen-rich syngas from autothermal reforming.

Just like the aqueous phase reforming of glycerol, another emerging reforming process is the 'supercritical water reforming' of glycerol which employs the conditions of supercritical water (pressure > 221 bar and temperature > 374°C) for the conversion of glycerol to hydrogen-rich syngas (Markocic et al., 2013). Parameters such as inlet concentration of the feed, the nature of the catalysts, the pressure, temperature, reaction time, and type of reactor are vital in obtaining the desired product distribution in the supercritical water reforming of glycerol. However, these parameters need to be optimized using various strategies that can help to determine the optimum conditions that could maximize hydrogen-rich gas production.

Other than using thermo-catalytic means for converting glycerol to hydrogen-rich syngas, which require high thermal energy to initiate the reaction, researchers are tapping into the abundance of solar energy and readily available photocatalysts to effect this conversion (Vaiano et al., 2018). The 'photocatalytic reforming' of glycerol employs a photo-induced hole, which facilitates the conversion of the glycerol, and induced electrons that reduce the H^+ to H_2. In comparison with the thermo-catalytic routes, there is no issue of deactivation with the photocatalysts since the reaction readily occurs at low temperatures. According to Li et al. (2009), factors such as irradiation time, initial concentration of glycerol in the feed, and pH have been reported to significantly influence the rate of hydrogen production from the photocatalytic reforming of glycerol. The rate of hydrogen production was found to increase with an increase in the irradiation time and glycerol concentration in the feed. Similarly, the rate was found to increase as the pH increases from 4 to 8 and subsequently to decrease from a pH greater than 8. Optimum conditions for the photocatalytic reforming of glycerol can be obtained for hydrogen production using the appropriate optimization strategies.

Besides reforming, 'pyrolysis' is another thermal process that can be employed to convert glycerol to hydrogen-rich syngas (Shahirah et al., 2017). Typically, pyrolysis of glycerol often produced char, biooil, and gaseous products which are mainly hydrogen-rich syngas. The formation of the various categories of products depends on the reaction conditions. The formation of biooil is favored at a temperature range of 400–600°C while a temperature above 750°C favors the formation of gaseous products. The gaseous product distribution is dependent on whether the pyrolysis process is catalyzed or not. Non-catalyzed glycerol pyrolysis often depends on parameters such as temperature, the flow rate of the carrier gas, and the particle diameter of the materials used for packing. Studies have shown that the yield and selectivity of gaseous products obtained in the non-catalytic glycerol pyrolysis are often lower compared to the catalytic pyrolysis of glycerol. The low yield of the gaseous products obtained for non-catalytic glycerol pyrolysis can be attributed to the higher rate of converting glycerol to biooil. Just like other thermo-catalytic processes, the catalytic pyrolysis of glycerol is also constrained by catalyst deactivation and reactor blockage. Applying the appropriate optimization of the catalyst synthesis and the pyrolysis reaction conditions for the glycerol conversion could mitigate these challenges.

The challenges of high thermal energy requirement and catalyst deactivations could be mitigated using the method of converting glycerol through biological means. The bioconversion process entails the use of microorganisms to metabolize glycerol to hydrogen-rich syngas (Garlapati et al., 2016). The bioconversion process is usually performed in a batch reactor using microorganisms such as *Citrobacter*, *Clostridium*, *Enterobacter*, *Klebsiella*, *Thermotoga*, and *Bacillus* spp. Although the yield of hydrogen obtained from the bioconversion of glycerol is far lower compared with that obtained from the thermo-catalytic process, bioconversion is more environmentally friendly and does not require high thermal energy. The gaseous product distributions in the bioconversion of glycerol are strongly dependent on parameters such as the N source, glycerol concentration, and pH of the medium. The yield of hydrogen from the bioconversion of glycerol can be improved using appropriate optimization strategies.

The various technological routes for converting glycerol to hydrogen-rich syngas can be improved by applying appropriate modeling and optimization strategies. Thosemodeling that have been employed are discussed in the subsequent section.

19.3 MODELING AND OPTIMIZATION STRATEGIES TO IMPROVE GLYCEROL CONVERSION

19.3.1 Modeling Strategies for Glycerol Conversion Processes

Modeling a process provides a platform to screen and explore alternatives that could facilitate making strategic decisions in the eventuality of scaling-up. Process modeling could reduce the rigors and the cost associated with building an experimental plant, which is time-consuming. Moreover, model parameters can be optimized by integrating the experimental data into the process models, which would enable the mitigation of unanticipated risks and help in the elimination of wastages, thereby focusing on the judicious use of available resources. Through process modeling, the process's objective functions can be tailored to maximized profit, optimize material utilization, and minimize energy use. Moreover, operational flexibility could be incorporated in the process models to cater for changes in the feed stream and to help reduce undesired products. The results obtained by process modeling could help in the proper planning of experiments to reduce cost and experimentation time. Various strategies have been applied for modeling glycerol conversion to hydrogen-rich syngas (Table 19.1). Some of these strategies include mathematical modeling (Caglar et al., 2015), kinetic modeling (Adhikari et al., 2009), process modeling using software (Unlu and Hilmioglu, 2020), and artificial neural network modeling (Estahbanati et al., 2017).

Modeling glycerol conversion mathematically entails the representation of the entire process using mathematical terminologies which can help to analyze how the different components of the process influence its performance. Mathematical modeling can be expressed from first principles using physical laws, stochastically based on data distributions and averages, and empirically based on the historical pattern of the data. The detailed stages involved in formulating a mathematical model is depicted in Figure 19.4. The modeling cycle includes specifying the problem, setting up a metaphor, formulating the mathematical model, solving the mathematical problem, interpreting the solution, comparing the solution with reality, and finally employing the solution to analyze real-world problems such as predicting, making decisions, and improving an existing process design. The mathematical modeling approach has been employed by Caglar et al. (2015) for the steam reforming of glycerol in an adiabatic microchannel reformer. The modeling was formulated on the finite element technique to investigate the effect of wall thickness, reactor wall

TABLE 19.1
Summary of Literature on the Application of Various Modeling Strategies for Glycerol Conversion to Hydrogen-rich Syngas

Production Method	Modeling Technique	Input Parameters	Target	Relative Importance	References
Photo-catalytic valorization	ANN	Glycerol concentration, catalyst loading, Pt content, initial pH of solution	H_2 yield	Glycerol = 9%, catalyst loading = 36%, Pt% = 26%, and pH = 29%	Estahbanati et al. (2017)
Steam reforming	ANN	Reactor pressure, sweep factor	Glycerol conversion, total H_2, H_2 recovery, H_2, selectivity, CO_2 selectivity, CO selectivity	Sweep factor 41%, Reactor pressure 59%	Ghasemzadeh et al. (2018)
Steam reforming	Mathematical	Reaction temperature, water to glycerol feed ratios,	H_2 yield	N/A	Silva et al. (2015)
Supercritical water reforming	Kinetics	Reaction temperature, molar concentration of glycerol	Rate of hydrogen formation and glycerol conversion	N/A	Byrd et al. (2008)
Steam reforming	Kinetics	Reaction temperature and pressure, molar concentration of glycerol and steam	Rate of hydrogen formation and glycerol conversion	N/A	Cheng et al. (2011)
Steam reforming	Kinetics	Reaction temperature and pressure, molar concentration of glycerol and steam	Rate of hydrogen formation and glycerol conversion	N/A	Adhikari et al. (2009)
Steam reforming	Mathematical	Reactor length, reactor wall thickness	Glycerol conversion	N/A	Caglar et al. (2015)
Steam reforming	Aspen Plus	Temperature, weight hourly space velocity	H_2 mole fraction	N/A	Hajjaji et al. (2014b)
Steam reforming	Aspen Plus	Temperature, pressure, water to glycerol molar feed ratio	H_2 mole fraction, glycerol conversion	N/A	Unlu et al. (2020)
Steam reforming	Aspen Plus	Reaction temperature, reactor pressure, steam to glycerol molar ratio	H_2 molar composition, CO molar composition, CO_2 molar composition and CH_4 molar composition	N/A	Adeniyi and Ighalo (2019)
Supercritical water gasification	Aspen Plus	Reaction temperature, feed concentration, pressure	H_2 yield, CO yield, CH_4 yield	N/A	Reddy et al. (2016)

ANN: Artificial neural network

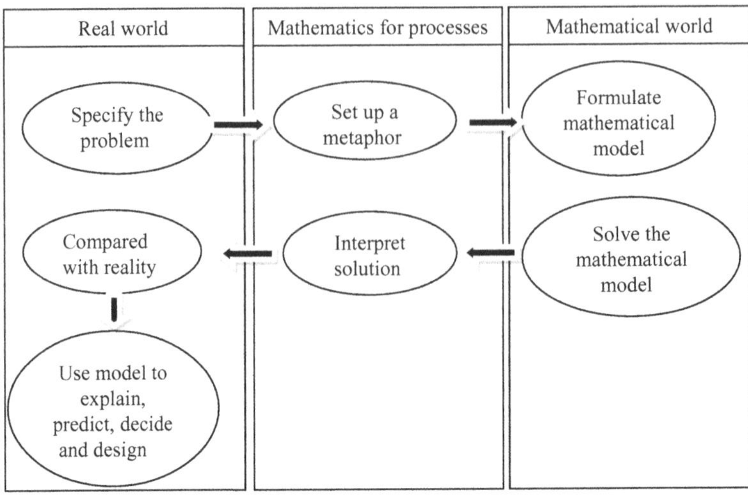

FIGURE 19.4 Stages involved in mathematical modeling. (*Source:* Adapted from Aarts (2010)).

materials, and inlet temperature on glycerol conversion by steam reforming to hydrogen. The findings revealed that the microchannel architecture could help to facilitate the fast and uniform transfer of the tangible heat of the feed stream to the catalyst layer. In a similar study, Jin et al. (2016) used the mathematical approach for modeling glycerol gasification in supercritical water in a tubular reactor. The mathematical formulation helps to represent the detailed flow, temperature, and distribution of the chemical species in the reactor. The mathematical models were incorporated into computational fluid dynamics to simulate the effects of the various parameters on the glycerol conversion process.

Glycerol conversion to hydrogen-rich syngas can also be modeled using kinetic-modeling techniques. The concept of kinetic modeling used for glycerol conversion reactions entails the procedure of analyzing the individual elementary reactions during the various reaction stages. The various side reactions as well as the complex intermediates during the glycerol conversion reaction are often represented in detailed kinetic modeling. A schematic representation of the stages involvedmodeling is depicted in Figure 19.5. The first step modelingis designing the appropriate measurement techniques to acquire the necessary data neededmodelingmodeling. This can be achieved using the design of experiment approach. The data obtained are subsequently used for the kinetic modeling which entails the estimation of the various kinetic parameters. During the modeling, the appropriate kinetic approach needs to be identified for each of the reaction steps. The various kinetic parameters estimated-modeling can subsequently be used for the reactor modeling and design.

Several authors have used various kinetic approaches, such as power-law, Eley-Rideal, and Langmuir–Hinshelwood, for modeling glycerol conversion to hydrogen-rich syngas. Kinetic modeling of hydrogen-rich syngas production by glycerol pyrolysis over an Sm-Ni/Al_2O_3 catalyst using the power-law and Langmuir–Hinshelwood

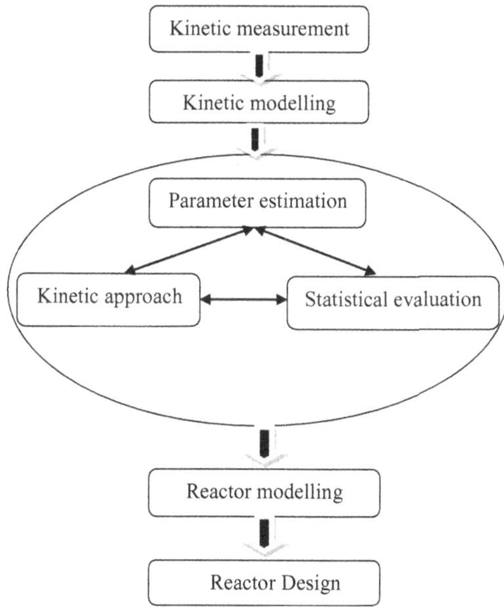

FIGURE 19.5 Stages involved in the kinetic modeling of a process.

approaches has been reported by Shahirah et al. (2016). The parameter estimation resulted in activation energies of 23.4 kJ/mol and 45.24 kJ/mol for the power-law and Langmuir–Hinshelwood models, respectively. Based on the parameter estimation, the authors proposed a single site associated with molecular adsorption for the glycerol pyrolysis. Similarly, the autothermal reforming of glycerol to hydrogen-rich syngas has been modeled using a Langmuir–Hinshelwood kinetic approach. Based on the kinetic analysis, the study revealed that the mechanism steps of autothermal glycerol reforming were a function of the non-dissociative adsorption of glycerol and the dissociative adsorption of steam.

Process modeling is also one of the strategies that have been employed in modeling glycerol conversion to hydrogen-rich syngas. This technique employs computer software to build an interconnected component of the process which is subsequently solved in order to allow for the prediction of the process in a steady or dynamic state. In addition, the software often has inbuilt algorithms that enable the economic analysis of the entire process. The use of Aspen Plus for the modeling of hydrogen production from glycerol steam reforming has been reported by Unlu and Hilmioglu (2020). The effects of process parameters such as reaction temperature, water to glycerol molar feed ratio, and the reactor pressure were investigated. Based on the modeling results, the optimum conditions of glycerol steam reforming were attained at a reaction temperature of 500°C, a water to glycerol molar feed ratio of 9:1, and a reactor pressure of 1 bar. The optimum conditions were employed for the reactor design.

Galera and Ortiz (2015) employed Aspen Plus modeling for the techno-economic analysis of hydrogen and power production from the supercritical water reforming of glycerol. The process flowsheet depicted in Figure 19.6 was employed for calculating the energy and mass balance of the glycerol conversion process. In addition to the material and energy balance calculation of the process, the exergy analysis of the process can be calculated using Aspen Plus. Hajjaji et al. (2014a) performed the energy and exergy analysis of hydrogen production by the autothermal reforming of glycerol. Based on the flowsheet analysis, the optimized conditions of glycerol autothermal reforming facilitates the minimization of the methane and carbon dioxide, thereby mitigating the formation and deposition of coke on the catalyst. The energy analysis of the reforming process revealed that there was a huge recovery of the useful product where most of the energy is fed into the process, whereas an exergetic efficiency of 57% was estimated for the process based on the exergy analysis. The process modeling enables the utilization of minimum energy consumption for the generation of a stipulated amount of hydrogen from autothermal glycerol reforming.

The advent of machine learning has increased the use of artificial intelligence techniques for modeling chemical processes. One of the commonly used techniques for modeling hydrogen-rich syngas production from glycerol conversion is an artificial neural network (ANN). ANN modeling works on imitation of the biological neuron system whereby a huge number of interconnected units arranged in layers are processed. As shown in Figure 19.7, a typical ANN configuration consists of the input layer ($x_1 - x_3$), a hidden layer, and the outer layers. The interconnected layers are associated with weights (w) and bias (b). For predictive modeling using an ANN, the input signals and each of the associated weights are multiplied to obtain a combined hidden layer which is subsequently activated by a function to produce a corresponding output. Before the ANN modeling, necessary data are acquired using an appropriate experimental design.

As shown in Figure 19.8, the first step in ANN modeling is the specification of the input parameters. This is followed by optimizing the hidden neurons by determining which one could minimize the prediction errors. Using the optimized neuron, the datasets are trained, tested, and validated to predict the output. MATLAB is commonly used for the ANN modeling. Several authors have employed an ANN approach for the predictive modeling of hydrogen-rich syngas from glycerol reforming. Ghasemzadeh et al. (2018) employed an ANN for modeling hydrogen production by the steam reforming of glycerol in a Pd-Ag membrane reactor. The input layer consists of the reaction pressure and sweep factor while the output layer consists of glycerol conversions, total hydrogen yields, hydrogen recovery, hydrogen selectivity, CO_2 selectivity, and CO selectivity. The authors employed a feed-forward back-propagation algorithm with an optimized topology of 2-10-6 and a Sigmoid activation function for the ANN modeling. The results revealed that the predicted outputs were in good agreement with the observed values with a high value of coefficient of determination (an R^2 of 0.9998). The sensitivity analysis revealed that both the reaction pressure and the sweep factor significantly influence the various outputs.

Similarly, Estahbanati et al. (2017) employed an ANN for modeling the photocatalytic valorization of glycerol to hydrogen. The authors employed a three-layer feed-forward back-propagated algorithm with an optimized topology of 4-4-1 for the ANN modeling. The input layer consists of the pH, glycerol concentration, catalyst

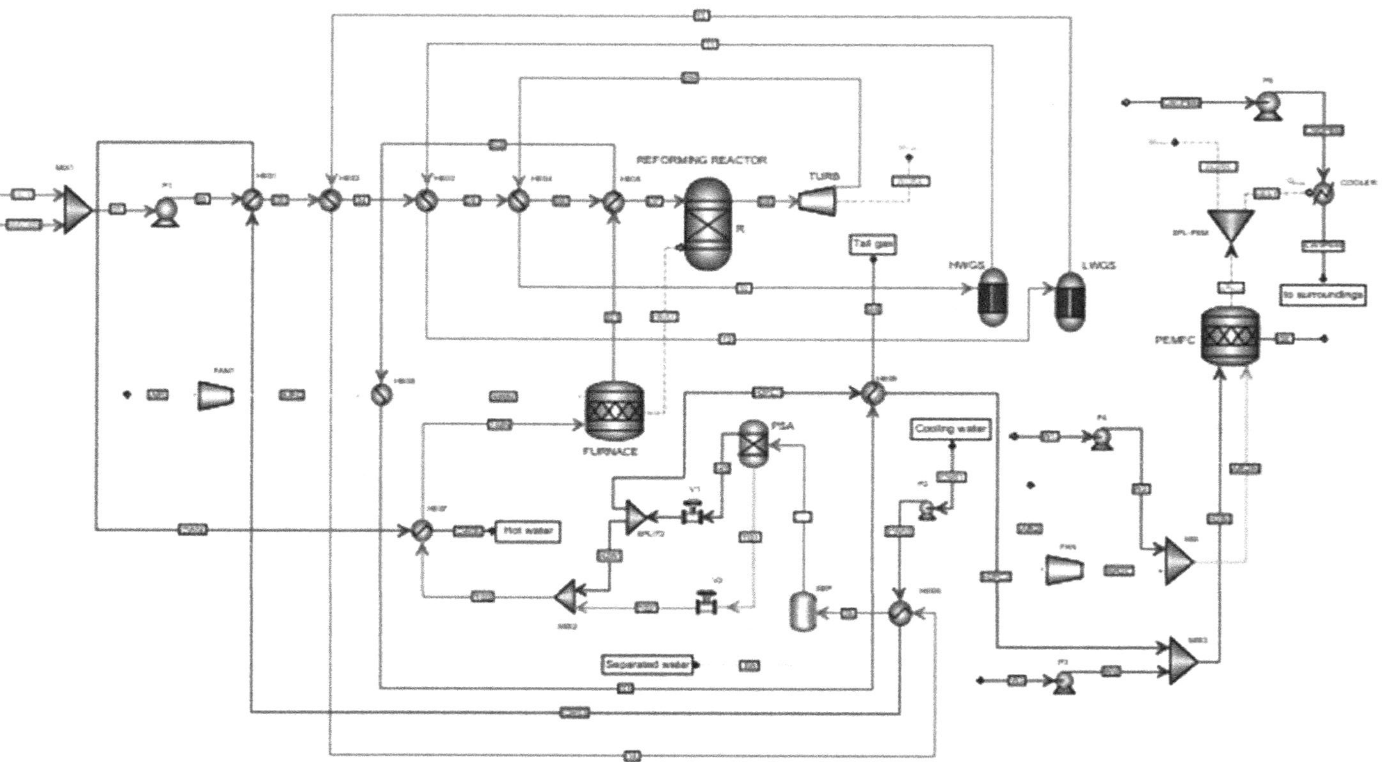

FIGURE 19.6 Process flowsheet for producing hydrogen and power from the supercritical water reforming of glycerol. (*Source:* Ortiz et al. (2013), reprinted with permission from Elsevier).

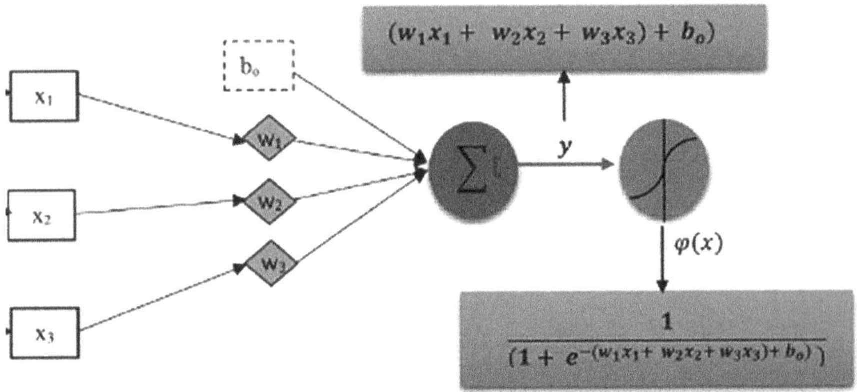

FIGURE 19.7 Schematic representation of an ANN configuration. (*Source:* Alsaffar et al. (2020), reprinted with permission from Elsevier).

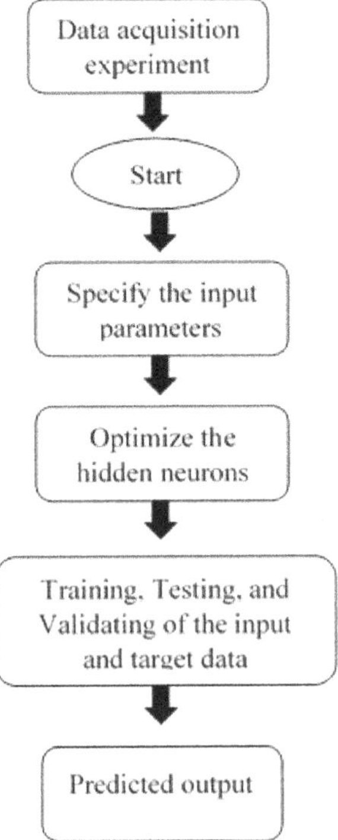

FIGURE 19.8 Steps involved in ANN modeling.

loading, and Pt content, while the hydrogen production rate was the output variable. The findings revealed that the ANN model accurately predicted the rate of hydrogen production with a high R^2 and a low mean square error. Moreover, mathematical modeling of the glycerol conversion to hydrogen-rich syngas has been reported by Silva et al. (2015) and Caglar et al. (2015), while the use of the kinetic modeling approach has been reported by Byrd et al. (2008), Adhikari et al. (2009), and Cheng et al. (2011) using various strategies such as power-law and Langmuir–Hinshelwood. Modeling using computer software such as Aspen Plus has been widely reported by Hajjaji et al. (2014a), Reddy et al. (2016), Adeniyi and Ighalo (2019), and Unlu et al. (2020) for the steam reforming of glycerol. The effect of the reaction temperature, reactor pressure, and water-to-glycerol molar feed ratio on the product distributions were simulated using the Aspen Plus software.

19.3.2 Optimization Strategies for Glycerol Conversion Processes

In order to address the challenges of catalyst deactivation, low product yield, and glycerol conversions and reactor blockage during various thermo-catalytic glycerol conversion techniques, the process parameters can be optimized. One major strategy that has been widely investigated for the optimization of the various glycerol conversion processes is the response surface methodology' (RSM). The RSM is a combination of mathematical and statistical approaches for optimizing the responses of a process based on its interaction with the input parameters. As shown in Figure 19.9, a series of steps are involved in employing the RSM for process optimization. The first step is to set the goal of the optimization that is the objective function. Based on the optimization goals, the necessary input parameters and the response(s) suitable to meet the objectives of the optimization are identified. Thereafter, the significance of the input parameters on the responses is investigated using a screening experiment which can be designed using full factorial design. Based on the analysis of the screening experiment, the significant factors are subsequently used for the optimization study.

For the optimization study, the design of experiment techniques such as the central composite design, central composite rotational design, and the Box–Behnken design could be employed. The experiment design usually consists of treatment combinations of the various factors (input parameters) which are made up of a series of experimental runs to determine each of the responses. The results obtained from the experimental runs are subsequently analyzed to determine the validity of the response surface model. The various input parameters are optimized using the desirability functions. This is followed by validating the RSM model using the set of optimum conditions of the input parameters in an experimental run to obtain new responses which can be compared with the predicted values.

As shown in Table 19.2, several authors have employed RSM techniques to optimize various glycerol conversion processes. Chookaew et al. (2014) reported the use of 'central composite design' (CCD) and RSM strategies to optimize the hydrogen yield from the bioconversion of glycerol using *Klebsiella sp. TR1*. The maximum H_2

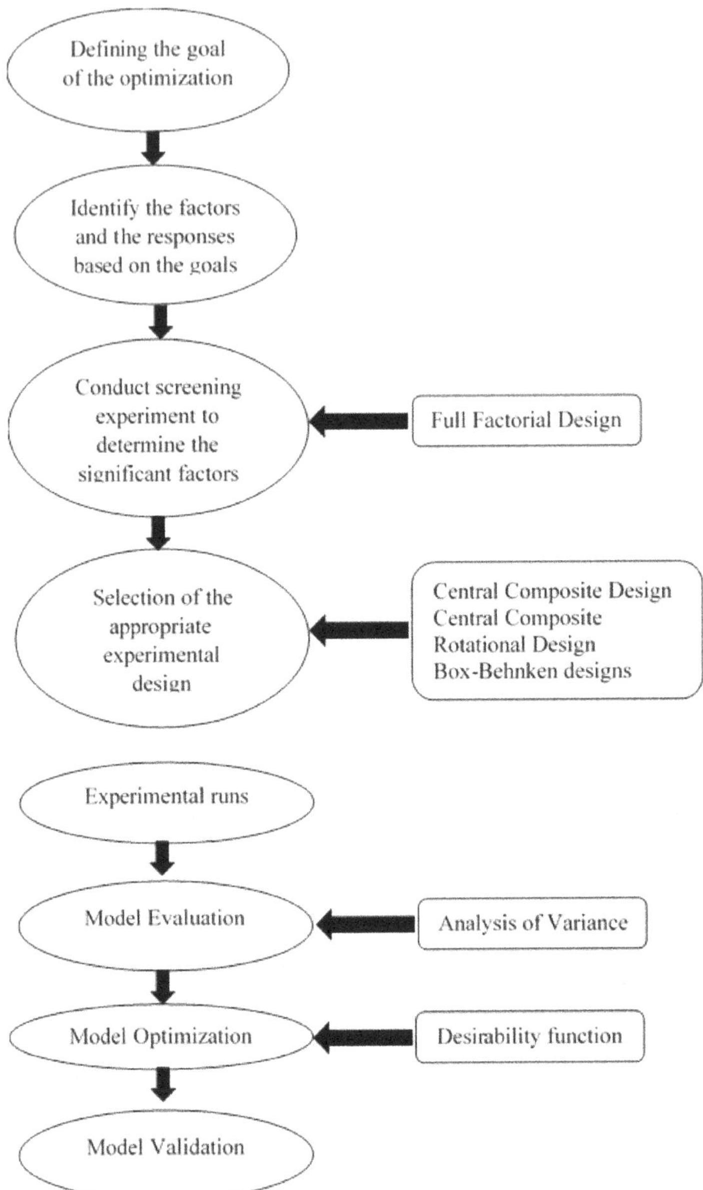

FIGURE 19.9 Steps involved in response surface optimization.

yield of 0.26 mol/glycerol was obtained at the optimum conditions of 11.14 g/L glycerol, 2.47 g/L KH_2PO_4, and 6.03 g/L NH_4Cl. Besides the optimization of the bioconversion of glycerol, the optimization of steam reforming and supercritical water reforming of glycerol using CCD, 'Box-Behnken design' (BBD), and RSM have been reported by Yang et al. (2012), Senseni et al. (2016b), and Zamzuri et al. (2017). The interaction effects of the reaction temperature, feed flow rate, water to

TABLE 19.2
Summary of the Literature on the Optimization of Glycerol Conversion Processes using RSM Strategies

Production Method	Experimental Design	Factors	Responses	Optimum Condition	Reference
Bioconversion	CCD	Crude glycerol, KH_2PO_4, NH_4Cl	H_2 yield	11.14 g/L glycerol, 2.47 g/L KH_2PO_4, and 6.03 g/L NH_4Cl, 0.26 mol H_2/mol glycerol	Chookaew et al. (2014)
Photo-catalytic valorization	BBD	Glycerol concentration, catalyst loading Pt content, initial pH of solution	H_2 yield	Glycero concentration = 50 % v/v, catalyst loading = 3.9 g/L, Pt loading 3.1% and initial pH of solution = 4.5, hydrogen yield = 321 umol	Estahbanati et al. (2017)
Bioconversion	CCD	NH_4Cl (g/L), K_2HPO_4 (g/L), KH_2PO_4 (g/L), $MgCl_2.6H_2O$, KCl	H_2 yield	(NH_4Cl, 4.40 g/L; K_2HPO_4, 1.6 g/L; KH_2PO_4, 2.27 g/L; MgCl2.6H2O,1.0 g/L; KCl,1.0 g/L; Na-acetate.$3H_2O$, 1.0 g/L and tryptone, 2.0 g/L)	Mangayil et al. (2015)
Bioconversion	BBD	Light intensity, glycerol concentration, glutamate concentration	H_2 yield, nitrogenate activity	H_2/glycerol = 6.69 mol at the optimal point (30 mM glycerol, 4.5 mM glutamate, 175 W/m^2)	Ghosh et al. (2012)
Steam reforming	CCD	Reaction temperature, feed flow rate, water to glycerol molar ratio	H_2 selectivity and glycerol conversion	H2 selectivity = 71.8%, temperature ¼ 692 °C, feed flow rate ¼ 1 ml/min, and water glycerol molar ratio (WGMR) 9.5:1	Zamzuri et al. (2017)
Supercritical water reforming	CCD	Glycerol concentration, reaction temperature, KOH concentration	H_2 mole fraction, CO mole fraction, CH4 mole fraction, CO2 mole fraction	H_2 of 27.9 mol% at optimum conditions of 500°C, 7 wt% glycerol concentration and 2.39 mol L-1 KOH concentration	Yang et al. (2012)
Steam reforming	BBD	Gas hourly space velocity, reaction temperature and feed ratio			Senseni et al. (2016b)

BBD: Box-Buehnken design; CCD. Central composite design.

glycerol molar ratio, and weight hourly space velocity on the yields of the gaseous products were investigated. Various optimum conditions were obtained for each of the optimization processes.

19.4 CONCLUSION

The various modeling and optimization strategies that have been investigated for hydrogen-rich syngas from thermo-catalytic and bioconversion of biodiesel-derived glycerol have been discussed. These various strategies have been proven to be effective in modeling and optimizing the various glycerol conversion processes. The use of ANNs for modeling the prediction of hydrogen-rich syngas has demonstrated the importance of artificial intelligence as a reliable technique for predictive modeling. This could help in saving costs expended in experimentation, and the algorithms of the modeling process could be incorporated in the event of scale-up.

The technical challenges that are usually encountered due to variation in the process parameters of the glycerol conversion process can be overcome through appropriate optimization strategies. Setting the right optimization objectives could help prevent material wastage, minimize energy utilization, and maximize profitability. The individual and the interaction effects of the process parameters could help determine how each of the process parameters interact and their corresponding effects on the process output. Modeling and optimization of a chemical process are key strategies that can be employed in scaling-up production.

REFERENCES

Aarts, A. C. T. 2010. *Methodology of Mathematical Modeling*. Eindhoven: Eindhoven University of Technology.

Adeniyi, A. G. and J. O. Ighalo. 2019. Study of process factor effects and interactions in synthesis gas production via a simulated model for glycerol steam reforming. *Chemical Product and Process Modeling* 14:34.

Adhikari, S., S. D. Fernando, and A. Haryanto. 2009. Kinetics and reactor modeling of hydrogen production from glycerol via steam reforming process over Ni/CeO_2 catalysts. *Chemical Engineering & Technology* 32: 541– 547.

Alsaffar, M. A., B. V. Ayodele, and S. I. Mustapa. 2020. Scavenging carbon deposition on alumina supported cobalt catalyst during renewable hydrogen-rich syngas production by methane dry reforming using artificial intelligence modeling technique. *Journal of Cleaner Production* 247: 119168.

Arif, N. N. M., S. Z. Abidin, and O. Y. Osazuwa, et al. 2019. Hydrogen production via CO_2 dry reforming of glycerol over Re-Ni/CaO catalysts. *International Journal of Hydrogen Energy* 44: 20857–20871.

Ayodele, B. V., S. I. Mustapa, T. A. R. B. T. Abdullah, and S. F. Salleh. 2019. A mini-review on hydrogen-rich syngas production by thermo-catalytic and bioconversion of biomass and its environmental implications. *Frontiers in Energy Research* 7: 118.

Ayodele, B. V., T. A. R. B. T. Abdullah, M. A. Alsaffar, S. I. Mustapa, and S. F. Salleh. 2020. Recent advances in renewable hydrogen production by thermo-catalytic conversion of biomass-derived glycerol: Overview of prospects and challenges. *International Journal of Hydrogen Energy* 45: 18160–18185.

Bobadilla, L. F., A. Penkova, and A. Alvarez, et al. 2015. Glycerol steam reforming on bimetallic NiSn/CeO_2-MgO-Al_2O_3 catalysts: Influence of the support, reaction parameters and deactivation/regeneration processes. *Applied Catalysis A: General* 492: 38–47.

Byrd, A. J., K. K. Pant, and R. B. Gupta. 2008. Hydrogen production from glycerol by reforming in supercritical water over Ru/Al_2O_3 catalyst. *Fuel* 87: 2956–2960.

Caglar, O. Y., C. D. Demirhan, and A. K. Avci. 2015. Modeling and design of a microchannel reformer for efficient conversion of glycerol to hydrogen. *International Journal of Hydrogen Energy* 40: 7579–7585.

Cheng, C. K., S. Y. Foo, and A. A. Adesina. 2011. Steam reforming of glycerol over Ni/Al_2O_3 catalyst. *Catalysis Today* 178: 25–33.

Chookaew, T., S. O-Thong, and P. Prasertsan. 2014. Statistical optimization of medium components affecting simultaneous fermentative hydrogen and ethanol production from crude glycerol by thermotolerant *Klebsiella* sp. TR17. *International Journal of Hydrogen Energy* 39: 751–760.

Dieuzeide, M. L., M. Laborde, N. Amadeo, C. Cannilla, and G. Bonura. 2015. Hydrogen production by glycerol steam reforming : How Mg doping affects the catalytic behaviour of Ni/Al_2O_3 catalysts. *International Journal of Hydrogen Energy* 41: 157–166.

Estahbanati, M. R. K., M. Feilizadeh, and M. C. Iliuta. 2017. Photocatalytic valorization of glycerol to hydrogen: Optimization of operating parameters by artificial neural network. *Applied Catalysis B: Environmental* 209: 483–492.

Galera, S. and F. J. G. Ortiz. 2015. Techno-economic assessment of hydrogen and power production from supercritical water reforming of glycerol. *Fuel* 144:307–316.

Garlapati, V. K., U. Shankar, and A. Budhiraja. 2016. Bioconversion technologies of crude glycerol to value added industrial products. *Biotechnology Reports* 9: 9–14.

Ghasemzadeh, K, F. Ahmadnejad, A. Aghaeinejad-Meybodi, and A. Basile. 2018. Hydrogen production by a Pd-Ag membrane reactor during glycerol steam reforming: ANN modeling study. *International Journal of Hydrogen Energy* 43: 7722–7730.

Ghosh, D., I. F. Sobro, and P. C. Hallenbeck. 2012. Stoichiometric conversion of biodiesel derived crude glycerol to hydrogen: Response surface methodology study of the effects of light intensity and crude glycerol and glutamate concentration. *Bioresource Technology* 106: 154–160.

Hajjaji, N., A. Chahbani, Z. Khila, and M. N. Pons. 2014a. A comprehensive energy-exergy-based assessment and parametric study of a hydrogen production process using steam glycerol reforming. *Energy* 64: 473–483.

Hajjaji, N., I. Baccar, and M. N. Pons. 2014b. Energy and exergy analysis as tools for optimization of hydrogen production by glycerol autothermal reforming. *Renewable Energy* 71: 368–380.

Jin, H., S. Guo, L. Guo, and C. Cao. 2016. A mathematical model and numerical investigation for glycerol gasification in supercritical water with a tubular reactor. *Journal of Supercritical Fluids* 107: 526–533.

Li, M., Y. Li, S. Peng, G. Lu, and S. Li. 2009. Photocatalytic hydrogen generation using glycerol wastewater over Pt/TiO_2. *Frontiers of Chemistry in China* 4:32–38.

Liu, S. K. and Y. C. Lin. 2014. Generation of syngas through autothermal partial oxidation of glycerol over $LaMnO_3$- and $LaNiO_3$-coated monoliths. *Catalysis Today* 237: 62–70.

Manfro, R. L., T. P. M. D. Pires, N. F.P. Rireiro, and M. M. V. M. Souza. 2013. Aqueous-phase reforming of glycerol using Ni-Cu catalysts prepared from hydrotalcite-like precursors. *Catalysis Science & Technology* 3: 1278–1287.

Mangayil, R., T. Aho, M. Karp, and V. Santala. 2015. Improved bioconversion of crude glycerol to hydrogen by statistical optimization of media components. *Renewable Energy* 75: 583–589.

Mardani, A., D. Streimikiene, F. Cavallaro, N. Loganathan, and M. Khoshnoudi. 2019. Carbon dioxide (CO_2) emissions and economic growth: A systematic review of two decades of research from 1995 to 2017. *Science of the Total Environment* 649: 31–49.

Markocic, E., B. Kramberger, and J. G. van Bennekom, et al. 2013. Glycerol reforming in supercritical water; a short review. *Renewable and Sustainable Energy Reviews* 23: 40–48.

Mekhilef, S., S. Siga, and R. Saidur 2011. A review on palm oil biodiesel as a source of renewable fuel. *Renewable and Sustainable Energy Reviews* 15: 1937–1949.

Monteiro, M. R., C. L. Kugelmeier, R. S. Pinheiro, M. Otavio, and A. da Silva Cesar. 2018. Glycerol from biodiesel production : Technological paths for sustainability. *Renewable and Sustainable Energy Reviews* 88: 109–122.

Muhammad, A. F. S., A. Awad, and R. Saidur, et al. 2018. Recent advances in cleaner hydrogen productions via thermo-catalytic decomposition of methane: Admixture with hydrocarbon. *International Journal of Hydrogen Energy* 43: 18713–18734.

OECD/FAO. 2012. *OECD-FAO Agricultural Outlook 2012*, 132–246. Paris: OECD Publishing.

Ortiz, F. J. G., P. Ollero, A. Serrera, and S. Galera. 2013. Optimization of power and hydrogen production from glycerol by supercritical water reforming. *Chemical Engineering Journal* 218: 309–318.

Pakhare, D. and J. Spivey. 2014. A review of dry (CO_2) reforming of methane over noble metal catalysts. *Chemical Society Reviews* 43: 7813–7837.

Quispe, C. A. G. and C. J. R. Coranado, and J. A. Carvalho. 2013. Glycerol: Production, consumption, prices, characterization and new trends in combustion. *Renewable and Sustainable Energy Reviews* 27: 475–493.

Reddy, S. N., S. Nanda, and J. A. Kozinski. 2016. Supercritical water gasification of glycerol and methanol mixtures as model waste residues from biodiesel refinery. *Chemical Engineering Research and Design* 113: 17–27.

Rennard, D. C., J. S. Kruger, and L. D. Schmidt. 2009. Autothermal catalytic partial oxidation of glycerol to syngas and to non-equilibrium products. *Chem SusChem* 2: 89–98.

Roslan, N. A., S. Z. Abidin, A. Ideris, and D. V. N. Vo. 2020. A review on glycerol reforming processes over Ni-based catalyst for hydrogen and syngas productions. *International Journal of Hydrogen Energy* 45.18466–18489.

Seadira, T., G. Sadanandam, T. A. Ntho, et al. 2017. Hydrogen production from glycerol reforming : Conventional and green production. *Reviews in Chemical Engineering* 34: 695–726.

Senseni, A. Z., F. Meshkani, and M. Rezaei. 2016a. Steam reforming of glycerol on mesoporous nanocrystalline Ni/Al_2O_3 catalysts for H_2 production. *International Journal of Hydrogen Energy* 41: 20137–20146.

Senseni, A. Z., S. M. S. Fattahi, M. Rezaei, and F. Meshkani. 2016b. A comparative study of experimental investigation and response surface optimization of steam reforming of glycerol over nickel nano-catalysts. *International Journal of Hydrogen Energy* 41: 10178–10192.

Seretis, A. and P. Tsiakaras. 2016. Crude bio-glycerol aqueous phase reforming and hydrogenolysis over commercial SiO2-Al_2O_3 nickel catalyst. *Renewable Energy* 97: 373–379.

Shahirah, M. N. N., S. Abdullah, J. Gimbun, Y. H. Ng, and C. K. Cheng. 2016. A study on the kinetics of syngas production from glycerol over alumina-supported samarium-nickel catalyst. *International Journal of Hydrogen Energy* 41: 10568-10577.

Shahirah, M. N. N., J. Gimbun, A. Ideris, M. R. Khan, and K. C. Cheng. 2017. Catalytic pyrolysis of glycerol into syngas over ceria-promoted. *Renewable Energy* 107: 223–234.

Silva, J. M., M. A. Soria, and L. M. Madeira. 2015. Steam reforming of glycerol for hydrogen production: Modeling study. *International Journal of Hydrogen Energy* 41: 1408–1418.

Tan, H. W., A. R. A. Aziz, and M. K. Aroua. 2013. Glycerol production and its applications as a raw material: A review. *Renewable and Sustainable Energy Reviews* 27: 118–127.

Unlu, D. and N. D. Hilmioglu. 2020. Application of aspen plus to renewable hydrogen production from glycerol by steam reforming. *International Journal of Hydrogen Energy* 45: 3509–3515.

Vaiano, V., M. A. Lara, and G. Iervolino, et al. 2018. Photocatalytic H_2 production from glycerol aqueous solutions over fluorinated Pt-TiO_2 with high {001} facet exposure. *Journal of Photochemistry and Photobiology A: Chemistry* 365: 52–59.

Vaidya, P. D. and A. E. Rodrigues. 2009. Glycerol reforming for hydrogen production: A review. *Chemical Engineering & Technology* 32: 1463–1469.

Yang, F., M. A. Hanna, and R. Sun. 2012. Value-added uses for crude glycerol-a byproduct of biodiesel production. *Biotechnology for Biofuels* 5: 1–10.

Zamzuri, N. H., R. Mat, N. A. S. Amin, A. Talebian-Kiakalaieh. 2017. Hydrogen production from catalytic steam reforming of glycerol over various supported nickel catalysts. *International Journal of Hydrogen Energy* 42: 9087–9098.

20 Propanediol Production from Glycerol

A Review of the Research

Ozcan Konur

CONTENTS

20.1 INTRODUCTION

Crude oils have been primary sources of energy and fuels, such as petrodiesel fuels. However, significant public concerns about the sustainability, price fluctuations, and adverse environmental impact of crude oils have emerged since the 1970s (Ahmadun et al., 2009; Atlas, 1981; Babich and Moulijn, 2003; Kilian, 2009; Perron, 1989). Thus, biooils (Bridgwater et al., 1999; Bridgwater and Peacocke, 2000; Czernik and Bridgwater, 2004; Mohan et al., 2006; Zhang et al., 2007) and biooil-based biodiesel fuels (Chisti, 2007; Hill et al., 2006; Lotero et al., 2005; Ma and Hanna, 1999) have emerged as alternatives to crude oils and crude oil-based petrodiesel fuels in recent decades. Nowadays, although petrodiesel fuels are still used extensively, biodiesel fuels are being used increasingly in the transportation and power sectors (Konur, 2021a–ag).

There has also been great public concern about the by-product of biodiesel fuels, glycerol, in recent years as waste materials are produced in large volumes (Behr et al., 2008; da Silva et al., 2009; Dasari et al., 2005; Pagliaro et al., 2007; Thompson and He, 2006; Wang et al., 2001). In the light of these concerns, glycerol has been converted into biofuels, such as biohydrogen and biochemicals, for example propanediols (PDs), 1,3-propanediol (1,3-PD), and 1,2-propanediol (1,2-PD).

However, for the efficient progression of the research in this field, it is necessary to develop efficient incentive structures for the primary stakeholders and to inform these stakeholders about the research (Konur, 2000, 2002a–c, 2006a–b, 2007a–b; North, 1991a–b).

Although there have been a number of reviews and book chapters on propanediol production from glycerol (Kubiak et al., 2012; Maervoet et al., 2011; Martin et al., 2013; Nakagawa et al., 2014; Wang et al., 2015; Willke and Vorlop, 2008), there has been no review of the 25-most-cited articles in this field. Thus, this chapter reviews these most-cited articles by highlighting their key findings on propanediol production from glycerol as a case study of biochemical production from glycerol. Then, it discusses these key findings.

20.2 MATERIALS AND METHODOLOGY

The search for the literature was carried out in the 'Web of Science' (WOS) database in February 2020. It contains the 'Science Citation Index-Expanded' (SCI-E), the Social Sciences Citation Index' (SSCI), the 'Book Citation Index-Science' (BCI-S), the 'Conference Proceedings Citation Index-Science' (CPCI-S), the 'Emerging Sources Citation Index' (ESCI), the 'Book Citation Index-Social Sciences and Humanities' (BCI-SSH), the 'Conference Proceedings Citation Index-Social Sciences and Humanities' (CPCI-SSH), and the 'Arts and Humanities Citation Index' (A&HCI).

The keywords for the search of the literature are collated from the screening of abstract pages for the first 1,000 highly cited papers in glycerol. These keywords sets are provided in the Appendix of the related book chapter (Konur, 2021m).

The 25-most-cited articles are selected for this chapter and the key findings of the review are presented and discussed briefly.

20.3 RESULTS

20.3.1 Catalytic Propanediol Production

Nakagawa et al. (2010) study the direct hydrogenolysis of glycerol into 1,3-propanediol (1,3-PD) over a Rhenium (Re)-modified Iridium (Ir) catalyst in a paper with 247 citations. They find that Re-oxide-modified and supported Ir nanoparticles on silica catalyzed the direct hydrogenolysis of glycerol to 1,3-PD in an aqueous media. The selectivity to 1,3-PD at an initial stage reached 67%. The yield reached 38% at an 81% conversion of glycerol. 1,3-PD was produced by the attack of an active hydrogen species on iridium metal to a 1-glyceride species formed on the oxidized Re cluster.

Miyazawa et al. (2007) study the glycerol hydrogenolysis to 1,2-propanediol (1,2-PD) catalyzed by a heat-resistant ion-exchange resin combined with Ru/C in a paper with 202 citations. They observe that the combination of Ru/C with Amberlyst ion-exchange resins was effective for the hydrogenolysis of glycerol to 1,2-PD. The difference between Amberlyst 15 and 70 in performance was small at 393 K; however, it became larger at a higher reaction temperature. This can be related to the highest operating temperature of the resins and can be explained by the poisoning of Ru/C with the sulfur compounds originated from the thermal decomposition of the resins. The Ru/C+ heat-resistant Amberlyst 70, whose highest operating temperature is 463 K, showed higher conversion, selectivity to 1,2-PD, and stability at 453 K.

Amada et al. (2011) study the reaction mechanism of glycerol hydrogenolysis to 1,3-PD over an Ir-ReO_x/SiO_2 catalyst in a paper with 194 citations. They focus on the catalytic performance, structure, reaction kinetics, and reactivity trends of various substrates over the catalysts with different amounts of Re. They find that the conversion in the glycerol hydrogenolysis increased by increasing the amount of Re, up to Re/Ir = 2, and the high selectivity to 1,3-PD (ca. 60%) was almost independent of the Re amount. The average size of the Ir metal particle gradually decreased by increasing the amount of Re. The Ir metal surface was partially covered with an ReO_x cluster, regardless of the Re amount. The reaction order on H_2 pressure over Ir-ReO_x/SiO_2 (Re/Ir = 1) was one, suggesting that one active hydrogen species was produced from one hydrogen molecule. A low reaction order on glycerol concentration represented the strong interaction between glycerol and the catalyst surface. This catalyst was also applicable to the selective hydrogenolysis of the C–O bond neighboring a $-CH_2OH$ group. They confirm the direct reaction mechanism for the formation of 1.3-PD from glycerol via a 2,3-dihydroxypropoxide species.

Gandarias et al. (2010) study Pt supported on amorphous silico alumina (Pt/ASA) as a catalyst for glycerol hydrogenolysis to 1,2-PD under mild operating conditions (493 K and 45 bar H_2 pressure) in a paper with 157 citations. Glycerol hydrogenolysis also took place in experiments performed under N_2 pressure due to hydrogen available from glycerol aqueous phase reforming. As both acid and metallic sites are involved in this process, they apply a study, including activity tests and different characterization techniques, to this catalytic system (ASA support and a Pt/ASA catalyst) in order to obtain a deeper understanding of their interactions.

Kurosaka et al. (2008) study the production of 1,3-PD by hydrogenolysis of glycerol catalyzed by Pt/WO_3/ZrO_2 in a paper with 157 citations. They obtain yields of up to 24%. The catalytic activities and the selectivity toward 1,3-PD were remarkably affected by the type of support, loaded noble metal (NM), and the preparation/impregnation procedure. The active site of catalyst for the formation of 1,3-PD may be the Pt over WO_3 supported on ZrO_2.

Yuan et al. (2010) study glycerol hydrogenolysis to 1,2-PD on Cu/MgO catalysts in a paper with 137 citations. They prepared CuO/MgO catalysts by impregnation and coprecipitation at 180°C and 3.0 MPa H_2. They find that the Cu (15)/MgO catalyst prepared by coprecipitation had the best activity. The conversion of glycerol and the selectivity of 1,2-PD over Cu (15)/MgO reached 72.0% and 97.6%, respectively. And the conversion of glycerol was further increased to 82.0% when a small amount of NaOH was added in the reaction mixture. The activity of the prepared catalysts depended strongly on the particle sizes of both Cu and MgO. Catalysts that have smaller sized Cu and MgO particles are more active for glycerol hydrogenolysis.

Alhanash et al. (2008) study the hydrogenolysis of glycerol to 1,2-PD over Ru in a paper with 135 citations. They observe that Ru-doped (5 wt%) acidic heteropoly salt $Cs_{2.5}H_{0.5}[PW_{12}O_{40}]$ (CsPW) is an active bifunctional catalyst for the one-pot hydrogenolysis of glycerol to 1,2-PD in liquid phase, providing 96% selectivity to 1,2-PD at a 21% glycerol conversion at 150°C and an unprecedented low hydrogen pressure of 5 bar. A rhodium catalyst, 5%Rh/CsPW, although less active, shows considerable selectivity to 1,3-PD (7.1%), with 1,2-PD being the main product (65%).

Akiyama et al. (2009) study the vapor-phase reaction of glycerol over copper metal catalysts at an ambient hydrogen pressure in a paper with 133 citations. They convert glycerol into 1,2-PD through dehydration into hydroxyacetone, followed by hydrogenation into 1,2-PD. The yield of 1,2-PD was limited up to 80% at a constant temperature of 190°C because of a trade-off problem between the dehydration and the hydrogenation. Dehydration needs relatively high reaction temperatures, whereas hydrogenation favors low temperatures and a high hydrogen concentration. They developed an efficient process during which glycerol was converted into 1,2-PD with a yield higher than 96% in the hydrogen flow at gradient temperatures; the dehydration into hydroxyacetone was catalyzed at ca. 200°C, and the following hydrogenation into 1,2-PD was completed at ca. 120°C. The developed process controls the thermodynamic equilibrium of second-step hydrogenation.

Zhu et al. (2013) study glycerol hydrogenolysis to 1,2-PD in a paper with 131 citations. They prepared a series of Cu/SiO_2 catalysts with various B_2O_3 loadings for glycerol hydrogenolysis via the precipitation-gel method followed by impregnation with boric acid. They observe that the addition of B_2O_3 to Cu/SiO_2 can greatly restrain the growth of copper particles and promote the dispersion of copper species upon calcination, reduction, and reaction, which results in enhanced catalytic activity and stability. The optimal $3CuB/SiO_2$ reached complete conversion with 98.0% 1,2-PD selectivity. The strong correlation between the 1,2-PD yield and the Cu surface area provided direct evidence that the active Cu species were the primary active sites for glycerol hydrogenolysis.

Balaraju et al. (2008) study the hydrogenolysis of glycerol to 1,2-PD over Cu-ZnO catalysts in a paper with 126 citations. They prepared a series of Cu-ZnO catalysts with varying Cu to Zn weight ratio by the co-precipitation method and evaluate them for hydrogenolysis of glycerol. They observe that the catalyst with Cu to Zn ratio of 50:50 was highly active under relatively low H_2 pressure. The catalysts were highly selective towards 1,2-PD (>93%). The glycerol conversion depended upon the bifunctional nature of the catalyst, which required both acidic sites and a metal surface. The presence of a sufficient amount of the glycerol with a small particle size of ZnO and Cu were required for the high conversion of glycerol and selectivity to 1,2-PD.

Nakagawa et al. (2012) study the hydrogenolysis of glycerol to 1,3-PD over $Ir\text{-}ReO_x/SiO_2$ and a solid acid co-catalyst in a paper with 125 citations. Considering reusability and activity, they observe that H-ZSM-5 was the most suitable solid co-catalyst. The property of an $Ir\text{-}ReO_x/SiO_2$+H-ZSM-5 system, including kinetics and selectivity trends in various reaction conditions, was similar to the case of $Ir\text{-}ReO_x/SiO_2+H_2SO_4$. The catalyst stability, activity, and maximum yield of 1,3-PD of $Ir\text{-}ReO_x/SiO_2$+H-ZSM-5 were slightly lower than for $Ir\text{-}ReO_x/SiO_2+H_2SO_4$. Added acid may protonate the surface of the ReOx cluster to increase the number of hydroxorhenium sites, which activates glycerol by the formation of glyceride species.

20.3.2 Microbial Propanediol Production

Papanikolaou et al. (2008) study the production of 1,3-PD along with citric acid and single cell oils in a paper with 272 citations. Using *Clostridium butyricum* F2b, they produced 47.1 gL^{-1} of 1,3-PD in batch anaerobic cultures, while the substrate uptake rate, expressed in $gL^{-1}h^{-1}$, increased with the increase in glycerol concentration in the medium. In continuous cultures, they studied microbial behavior in transitory states after the addition of 1,3-PD in the chemostat vessel. They find that microbial growth was not affected by the high 1,3-PD (which was added in the chemostat vessel) concentration, while butyric and acetic acid concentrations were increased. In a two-stage continuous culture, they produced 43.5 gL^{-1} of 1,3-PD with a total volumetric productivity of 1.33 $gL^{-1}h^{-1}$.

Homann et al. (1990) study the fermentation of glycerol to 1,3-PD by *Klebsiella* and *Citrobacter* strains in a paper with 217 citations. They enriched glycerol-fermenting anaerobes with glycerol at low and high concentrations in order to obtain strains that produce 1,3-PD. They selected six isolates for more detailed characterization and identified four of them as *Citrobacter freundii*, one as *Klebsiella oxytoca*, and one as *K. pneumoniae*. The *Citrobacter* strains formed 1.3-PD and acetate and almost no by-products, while the *Klebsiella* strains produced varying amounts of ethanol in addition, and accordingly less 1,3-PD. Enterobacterial strains of the genera *Enterobacter*, *Klebsiella*, and *Citrobacter* from culture collections showed similar product patterns except for one group which formed limited amounts of ethanol, but no PD. They grew seven strains in pH-controlled batch cultures to determine the parameters necessary to evaluate their capacity for 1,3-PD production. They found that *K. pneumoniae* DSM 2026 exhibited the highest final concentration (61 g/L) and the best productivity (1.7 g/L h), whereas *C. freundii Zu* and K2 achieved only 35 g/L and 1.4 g/L h, respectively. The *Citrobacter* strains on the other hand gave somewhat better yields which were very close to the theoretical optimum of 65 mol %.

Menzel et al. (1997) study the production of 1,3-PD from the continuous fermentation of glycerol by *Klebsiella pneumonia* in a paper with 202 citations. They obtained a final PD concentration of 35.2–48.5 gL^{-1} and a volumetric productivity of 4.9–8.8 $gL^{-1}h^{-1}$ at dilution rates of between 0.1 and 0.25 h^{-1}. These results correspond to about 80–96% of the theoretical maxima under ideal conditions (with no ethanol and hydrogen formation). The highest PD concentration achieved was close to the maximum PD concentration (50–60 gL^{-1}) found in batch and fed-batch cultures. The productivity of the continuous culture was, however, about 2.0–3.5 times higher.

Mu et al. (2006) study the production of 1,3-PD by *Klebsiella pneumoniae* using crude glycerol in a paper with 199 citations. They observed that the 1,3-PD concentration of 51.3 g/L^{-1} on crude glycerol from alkali-catalyzed methanolysis of soybean oil was comparable to that of 53 g/L^{-1} on crude glycerol derived from a lipase-catalyzed process. The production of 1.7 $gL^{-1}h^{-1}$ on crude glycerol was comparable to that of 2 $gL^{-1}h^{-1}$ on pure glycerol. They conclude that the crude glycerol could be directly converted to 1,3-PD without any prior purification.

Papanikolaou et al. (2000) study the production of 1,3-PD from glycerol by a *Clostridium butyricum* strain in a paper with 174 citations. They performed batch and continuous cultures of a *Clostridium butyricum* strain on industrial glycerol. For both types of cultures, they found that the conversion yield obtained was around 0.55 g of 1,3-PD formed per 1 g of glycerol consumed, whereas the highest 1,3-PD concentration, achieved during single-stage continuous cultures, was 35–48 gL^{-1}. Moreover, the strain presented a strong tolerance to the inhibitory effect of 1,3-PD, even at high concentrations of this substance in the chemostat (e.g. 80 gL^{-1}). 1,3-PD was associated with cell growth, whereas acetate and butyrate seemed to be non-growth-associated products. At low and medium dilution rates (up to 0.1 h^{-1}), butyrate production was favored, whereas at higher rates acetate production increased. The maximum 1,3-PD volumetric productivity obtained was 5.5 $gL^{-1}h^{-1}$. They also performed a two-stage continuous fermentation. The first stage presented high 1,3-PD volumetric productivity, whereas the second stage (with a lower dilution rate) served to further increase the final product concentration. High 1,3-PD concentrations were achieved (41–46 gL^{-1}), with a maximum volumetric productivity of 3.4 $gL^{-1}h^{-1}$. A cell concentration decrease was reported between the second and the first fermenter.

Biebl et al. (1998) study the fermentation of glycerol to 1,3-PD, together with 2,3-butanediol, by *Klebsiella pneumonia* under conditions of uncontrolled pH in a paper with 155 citations. Formation of 2,3-butanediol starts with a delay of some hours and is accompanied by reuse of the acetate that was formed in the first period. The fermentation was demonstrated with the type strain *K. pneumoniae*, but growth was better with the more acid-tolerant strain GT1, which was isolated from the environment. In continuous cultures in which the pH was lowered stepwise from 7.3 to 5.4, 2,3-butanediol formation started at pH 6.6 and reached a maximum yield at pH 5.5, whereas formation of acetate and ethanol declined in this pH range. 2,3-butanediol and acetoin were also found among the products in chemostat cultures grown at pH 7 under conditions of glycerol excess, though only with low yields. At any of the pH values tested, excess glycerol in the culture enhanced the butanediol yield. Both effects are seen as a consequence of product inhibition, the undissociated acid being a stronger trigger than the less toxic diols and acid anions.

Biebl et al. (1992) study the glycerol conversion to 1,3-PD by clostridia in a paper with 149 citations. They obtain *Clostridium butyricum* from pasteurized mud and soil samples. The most active strain, SH1=DSM 5431, was able to convert up to 110 g/L of glycerol to 56 g/L of 1,3-PD in 29 h. A few *Clostridium* strains from culture collections (3 out of 16 of the *C. butyricum* group) and some isolates of *Kutzner* from cheese samples were also able to ferment glycerol, but the final concentration and productivity of 1,3-PD was lower than in strain SH1. Strain SH1 grew well in a pH range between 6.0 and 7.5, with a weak optimum at 6.5, and was stimulated by sparging with N_2. They obtain best overall productivity in a fed-batch culture with a starting concentration of 5% glycerol. In all fermentations, the yield of 1,3-PD in relation to glycerol was higher than expected from 'nicotinamide adenine dinucleotide' (NADH) production by acid formation. On the other hand, H_2 production was lower than expected, if, per mole of acetyl coenzyme A, one mole of H_2 is released. They highlight a substantial transfer of reducing potential from ferredoxin to NAD, which finally results in increased 1,3-PD production.

Barbirato et al. (1998) study 1,3-PD production by fermentation in a paper with 137 citations. They use *Klebsiella pneumoniae*, *Citrobacter freundii*, *Enterobacter agglomerans*, and *Clostridium butyricum* by batch fermentation and compare low and high glycerol contents. They observed an important metabolic flexibility with the enterobacteria in contrast to *C. butyricum*. Fermentation of glycerol by the enterobacterial species revealed the occurrence of an inhibitory phenomenon. They assign this to the accumulation in the fermentation medium of a strongly inhibitory compound, 3-hydroxypropionaldehyde, the only intermediate of the 1,3-PD metabolic pathway. This phenomenon significantly decreased biological activity and was dependent on the culture's pH conditions. This was not observed with *C. butyricum*. By comparing the efficiency of the fermentation among the four bacteria, they choose *C. butyricum* for an applied study consisting of the bioconversion of glycerol containing industrial wastewaters. With glycerin coming from the ester production and from wine stillage, they observed a high efficiency of conversion to 1,3-PD. They purified the 1,3-PD produced by liquid/liquid extraction with a recovery yield of about 100%.

Gonzalez-Pajuelo et al. (2005) study the production of 1,3-PD from glycerol in a paper with 131 citations. To obtain a better 'vitamin B-12-free' biological process, they developed a metabolic engineering strategy with *Clostridium acetobittylicum*. They then introduced the 1,3-PD pathway from *C. butyricum* on a plasmid in several mutants of *C. acetobutylicum* altered in product formation. The DG1(pSPD5) recombinant strain was the most efficient one. Chemostat cultures of this strain grown on glucose alone produced only acids (acetate, butyrate, and lactate) and a high level of hydrogen. In contrast, when glycerol was metabolized in a chemostat culture, 1,3-PD became the major product, the specific rate of acid formation decreased, and a very low level of hydrogen was observed. In a fed-batch culture, the DG1(pSPD5) strain produced 1,3-PD at a higher concentration (1,104 mM) and productivity than the natural producer *C. butyricum* VPI 3266. Furthermore, they also used this strain successfully for the very long term continuous production of 1,3-PD at high volumetric productivity (3 $gL^{-1}h^{-1}$) and titer (788 mM).

Barbirato et al. (1996) study the inhibitory metabolite of glycerol fermentation to 1,3-PD by enterobacterial species in a paper with 130 citations. Glycerol fermentation by *Enterobacter agglomerans* revealed that both growth and 1,3-PD production ceased after consumption of about 430 mM glycerol, irrespective of the initial glycerol content. They assign this phenomenon to the production of 3-hydroxypropionaldehyde. They also observed accumulation during glycerol fermentation with two other enterobacterial species, i.e. *Klebsiella pneumoniae* and *Citrobacter freundii*.

Schutz and Radler (1984) study the production of 1,3-PD from glycerol by *Lactobacillus brevis* and *Lactobacillus buchneri* in a paper with 129 citations. Three strains of *Lactobacillus brevis* and one strain of *Lactobacillus buchneri* grew very poorly on glucose. They observed good growth on glucose plus glycerol, while glucose was fermented to acetate or ethanol, lactate and CO_2, glycerol which was dehydrated to 3-hydroxypropanal and subsequently reduced to 1,3-PD. Cell extracts of *L. brevis* and *L. buchneri* grown on glucose plus glycerol contained a B12-dependent glycerol dehydratase and a 1,3-PD dehydrogenase. Glycerol was not metabolized when used as the only substrate. Fructose as sole carbon source was partially reduced to mannitol. The joint fermentation of fructose and glycerol yielded 1,3-PD from glycerol. Ribose was fermented but did not support glycerol fermentation.

Extracts from ribose grown cells did not contain glycerol dehydratase or 1,3-PD dehydrogenase. Besides glycerol the following diols were metabolized as cosubstrates with glucose: 1,2-PD, ethylene glycol, and butanediol-2.3, yielding propanol-1, ethanol, and butanol-2, respectively. Washed cells of two *L. brevis* strains, B18 and B20, formed 1,3-PD and 1,2-PD from glycerol, the third strain B 22 formed only 1,2-PD from glycerol in the absence of glucose.

Gonzalez-Pajuelo et al. (2004) evaluate the growth inhibition of *Clostridium butyricum* VPI 3266 by raw glycerol in a paper with 122 citations. They observed that *C. butyricum* presents the same tolerance to raw and to commercial glycerol when both are of a similar grade, i.e. above 87% (w/v). They observed a 39% increase of growth inhibition in the presence of 100 gl^{-1} of a lower grade raw glycerol (65% w/v). Furthermore, they observed 1,3-PD production from two raw glycerol types (65% w/v and 92% w/v), without any prior purification, in batch and continuous cultures, on a synthetic medium. There were no significant differences in *C. butyricum* fermentation patterns on raw and commercial glycerol as the sole carbon source. In every case, the 1,3-PD yield was around 0.60 mol/mol glycerol consumed.

Biebl (1991) studies the fermentation of glycerol to 1,3-PD, acetate, and butyrate by *Clostridium butyricum* with respect to growth inhibition by the accumulating products in a paper with 118 citations. He grew the clostridia in a pH-auxostat culture at low cell density and product concentration and near the maximum growth rate. He then added the products individually to the medium in increasing concentrations and used the resulting depression of the growth rate as a quantitative estimate of product inhibition. Under these conditions, he observed that growth was totally inhibited at concentrations of 60 g/L for 1,3-PD, 27 g/L for acetic acid, and 19 g/L for butyric acid at pH 6.5. He found an appreciable inhibition by glycerol only above a concentration of 80 g/L. In a pH-auxostat without added products but with high cell density as well as in batch cultures, the product proportions were different. The 1,3-PD concentration may approach the value of complete inhibition while the concentrations of acetic and butyric acids remained below these values by at least one order of magnitude. He concludes that 1,3-PD was the first range inhibitor in this fermentation.

Forsberg (1987) studies the production of 1,3-PD from glycerol by *Clostridium acetobutylicum* and other *Clostridium* species in a paper with 117 citations. He used four strains of *Clostridium acetobutylicum*, six of *C. butylicum*, two of *C. beijerinckii*, one of *C. kainantoi*, and three of *C. butyricum*. He observed that during growth of *C. butylicum* B593 in a chemostat culture at pH 6.5, 61% of the glycerol fermented was converted to 1,3-PD. When the pH was decreased to 4.9, growth and 1,3-PD production were substantially reduced.

20.4 DISCUSSION

Table 20.1 provides information on the research fronts in this field. As this table shows the primary research fronts of ‘catalytic propanediol production’ and ‘microbial propanediol production’ comprise 44 and 56% of these papers, respectively.

TABLE 20.1
Research Fronts

	Research Front	Papers (%)
1	Catalytic propanediol production	44
2	Microbial propanediol production	56

20.4.1 Catalytic Propanediol Production

Nakagawa et al. (2010) study the direct hydrogenolysis of glycerol into 1,3-PD over an re-modified Ir catalyst in a paper with 247 citations. Miyazawa et al. (2007) study glycerol hydrogenolysis to 1,2-PD, catalyzed by a heat-resistant ion-exchange resin combined with Ru/C in a paper with 202 citations. Amada et al. (2011) study the reaction mechanism of glycerol hydrogenolysis to 1,3-PD over an Ir-ReO_x/SiO_2 catalyst in a paper with 194 citations.

Gandarias et al. (2010) study Pt supported on amorphous silico alumina as a catalyst for glycerol hydrogenolysis to 1,2-PD under mild operating conditions (493 K and 45 bar H_2 pressure) in a paper with 157 citations. Kurosaka et al. (2008) study the production of 1,3-PD by hydrogenolysis of glycerol, catalyzed by Pt/WO_3/ZrO_2 in a paper with 157 citations. Yuan et al. (2010) study glycerol hydrogenolysis to 1,2-PD on Cu/MgO catalysts in a paper with 137 citations.

Alhanash et al. (2008) study the hydrogenolysis of glycerol to 1,2-PD over Ru in a paper with 135 citations. Akiyama et al. (2009) study the vapor-phase reaction of glycerol over copper metal catalysts at an ambient hydrogen pressure in a paper with 133 citations. Zhu et al. (2013) study glycerol hydrogenolysis to 1,2-PD in a paper with 131 citations.

Balaraju et al. (2008) study the hydrogenolysis of glycerol to 1,2-PD over Cu-ZnO catalysts in a paper with 126 citations. Nakagawa et al. (2012) study the hydrogenolysis of glycerol to 1,3-PD over Ir-ReO_x/SiO_2 and a solid acid co-catalyst in a paper with 125 citations.

These prolific studies highlight propanediol production from glycerol using chemical catalysts.

20.4.2 Microbial Propanediol Production

Papanikolaou et al. (2008) study the production of 1,3-PD along with citric acid and single cell oils using *Clostridium butyricum* in a paper with 272 citations. Homann et al. (1990) study the fermentation of glycerol to 1,3-PD by *Klebsiella* and *Citrobacter* strains in a paper with 217 citations. Menzel et al. (1997) study the production of 1,3-PD from the continuous fermentation of glycerol by *Klebsiella pneumonia* in a paper with 202 citations.

Mu et al. (2006) study the production of 1,3-PD by *Klebsiella pneumoniae* using crude glycerol in a paper with 199 citations. Papanikolaou et al. (2000) study the production of 1,3-PD from glycerol by a *Clostridium butyricum* strain in a paper with 174 citations. Biebl et al. (1998) study the fermentation of glycerol to 1,3-PD together with 2,3-butanediol by *Klebsiella pneumonia* under conditions of uncontrolled pH in a paper with 155 citations.

Biebl et al. (1992) study glycerol conversion to 1,3-PD by clostridia in a paper with 149 citations. Barbirato et al. (1998) study 1,3-PD production by fermentation using *Klebsiella pneumoniae*, *Citrobacter freundii*, *Enterobacter agglomerans*, and *Clostridium butyricum* in a paper with 137 citations. Gonzalez-Pajuelo et al. (2005) study the production of 1,3-PD from glycerol using *Clostridium acetobittylicum* in a paper with 131 citations.

Barbirato et al. (1996) study the inhibitory metabolite of glycerol fermentation to 1,3-PD by enterobacterial species such as *Enterobacter agglomeran* in a paper with 130 citations. Schutz and Radler (1984) study the production of 1,3-PD from glycerol by *Lactobacillus brevis* and *Lactobacillus buchneri* in a paper with 129 citations. Gonzalez-Pajuelo et al. (2004) evaluate the growth inhibition of *Clostridium butyricum* VPI 3266 by raw glycerol in a paper with 122 citations.

Biebl (1991) studies the fermentation of glycerol to 1,3-PD, acetate, and butyrate by *Clostridium butyricum* with respect to growth inhibition by the accumulating products in a paper with 118 citations. Forsberg (1987) studies the production of 1,3-PD from glycerol by *Clostridium acetobutylicum* and other *Clostridium* species in a paper with 117 citations.

These prolific studies highlight propanediol production from glycerol using microbes.

20.5 CONCLUSION

This chapter has presented the key findings of the 25-most-cited article papers in this field. Table 20.1 provides information on the research fronts in this field. As this table shows the primary research fronts of 'catalytic propanediol production' and 'microbial propanediol production' comprise 44 and 56% of these papers, respectively.

These prolific studies on two different research fronts provide valuable evidence on catalytic and microbial propanediol production. Propanediol production from glycerol by these separate methods brings substantial benefits to the society where the glycerol is bioremediated as a form of waste treatment (Busch and Stumm, 1968; Finstein et al., 1983) and a substantial amount of biochemicals are produced, helping to reduce the cost of biodiesel production (Apostolakou et al., 2009; Marchetti et al., 2008).

It is recommended that similar studies are carried out for each research front as well.

ACKNOWLEDGMENTS

The contribution of the highly cited researchers in this field is greatly acknowledged.

REFERENCES

Ahmadun, F. R., A. Pendashteh, and L. C. Abdullah, et al. 2009. Review of technologies for oil and gas produced water treatment. *Journal of Hazardous Materials* 170:530–551.

Akiyama, M., S. Sato, R. Takahashi, K. Inui, and M. Yokota. 2009. Dehydration-hydrogenation of glycerol into 1,2-propanediol at ambient hydrogen pressure. *Applied Catalysis A-General* 371:60–66.

Alhanash, A., E. F. Kozhevnikova, and I. V. Kozhevnikov. 2008. Hydrogenolysis of glycerol to propanediol over Ru: Polyoxometalate bifunctional catalyst. *Catalysis Letters* 120:307–311.

Amada, Y., Y. Shinmi, and S. Koso, et al. 2011. Reaction mechanism of thecerol hydrogenolysis to 1,3-propanediol over Ir-ReO_x/SiO_2 catalyst. *Applied Catalysis B-Environmental* 105:117–127.

Apostolakou, A. A., I. K. Kookos, C. Marazioti, and K. C. Angelopoulos. 2009. Techno-economic analysis of a biodiesel production process from vegetable oils. *Fuel Processing Technology* 90:1023–1031.

Atlas, R. M. 1981. Microbial degradation of petroleum hydrocarbons: An environmental perspective. *Microbiological Reviews* 45:180–209.

Babich, I. V. and J. A. Moulijn. 2003. Science and technology of novel processes for deep desulfurization of oil refinery streams: A review. *Fuel* 82: 607–631.

Balaraju, M., V. Rekha, and P. S. S. Prasad, R. B. N. Prasad, and N. Lingaiah. 2008. Selective hydrogenolysis of glycerol to 1,2 propanediol over Cu-ZnO catalysts. *Catalysis Letters* 126: 119–124.

Barbirato, F., J. P. Grivet, P. Soucaille, and A. Bories. 1996.3-hydroxypropionaldehyde, an inhibitory metabolite of glycerol fermentation to 1,3-propanediol by enterobacterial species. *Applied and Environmental Microbiology* 62: 1448–1451.

Barbirato, F., E. H. Himmi, T. Conte, and A. Bories. 1998.1,3-Propanediol production by fermentation: An interesting way to valorize glycerin from the ester and ethanol industries *Industrial Crops and Products* 7: 281–289.

Behr, A., J. Eilting, K. Irawadi, J. Leschinski, and F. Lindner. 2008. Improved utilisation of renewable resources: New important derivatives of glycerol. *Green Chemistry* 10:13–30.

Biebl, H. 1991. Glycerol fermentation of 1,3-propanediol by *Clostridium butyricum*: Measurement of product inhibition by use of a pH-auxostat. *Applied Microbiology and Biotechnology* 35: 701–705.

Biebl, H., S. Marten, H. Hippe, and W. D. Deckwer. 1992. Glycerol conversion to 1,3-propanediol by newly isolated clostridia. *Applied Microbiology and Biotechnology* 36: 592–597.

Biebl, H., A. P. Zeng, K. Menzel, and W. D. Deckwer. 1998. Fermentation of glycerol to 1,3-propanediol and 2,3-butanediol by *Klebsiella pneumonia*. *Applied Microbiology and Biotechnology* 50: 24–29.

Bridgwater, A. V. and G. V. C. Peacocke. 2000. Fast pyrolysis processes for biomass. *Renewable & Sustainable Energy Reviews* 4: 1–73.

Bridgwater, A. V., D. Meier, and D. Radlein. 1999. An overview of fast pyrolysis of biomass. *Organic Geochemistry* 30:1479–1493.

Busch, P. L. and W. Stumm. 1968. Chemical interactions in the aggregation of bacteria bioflocculation in waste treatment. *Environmental Science & Technology* 2: 49–53.

Chisti, Y. 2007. Biodiesel from microalgae. *Biotechnology Advances* 25: 294–306.

Czernik, S. and A. V. Bridgwater. 2004. Overview of applications of biomass fast pyrolysis oil. *Energy & Fuels* 18: 590–598.

Da Silva, G. P., M. Mack, and J. Contiero. 2009. Glycerol: A promising and abundant carbon source for industrial microbiology. *Biotechnology Advances* 27: 30–39.

Dasari, M. A., P. P. Kiatsimkul, W. R. Sutterlin, and G. J. Suppes. 2005. Low-pressure hydrogenolysis of glycerol to propylene glycol. *Applied Catalysis A: General* 281: 225–231.

Finstein, M. S., F. C. Miller, P. F. Strom, S. T. MacGregor, and K. M. Psarianos. 1983. Composting ecosystem management for waste treatment. *Bio/technology* 1: 347–353.

Forsberg, C. W. 1987. Production of 1,3-propanediol from glycerol by *Clostridium acetobutylicum* and other *Clostridium* species. *Applied and Environmental Microbiology* 53: 639–643.

Gandarias, I., P. L. Arias, J. Requies, M. B. Guemez, and J. L. G. Fierro. 2010. Hydrogenolysis of glycerol to propanediols over a Pt/ASA catalyst: The role of acid and metal sites on product selectivity and the reaction mechanism. *Applied Catalysis B-Environmental* 97: 248–256.

Gonzalez-Pajuelo, M., J. C. Andrade, and I. Vasconcelos. 2004. Production of 1,3-propanediol by *Clostridium butyricum* VPI 3266 using a synthetic medium and raw glycerol. *Journal of Industrial Microbiology & Biotechnology* 31: 442–446.

Gonzalez-Pajuelo, M., I. Meynial-Salles, and F. Mendes, et al. 2005. Metabolic engineering of *Clostridium acetobutylicum* for the industrial production of 1,3-propanediol from glycerol. *Metabolic Engineering* 7: 329–336.

Hill, J., E. Nelson, D. Tilman, S. Polasky, and D. Tiffany. 2006. Environmental, economic, and energetic costs and benefits of biodiesel and ethanol biofuels. *Proceedings of the National Academy of Sciences of the United States of America* 103: 11206–11210.

Homann, T., C. Tag, H. Biebl, W. D. Deckwer, and B. Schink. 1990. Fermentation of glycerol to 1,3-propanediol by *Klebsiella* and *Citrobacter* strains. *Applied Microbiology and Biotechnology* 33: 121–126.

Kilian, L. 2009. Not all oil price shocks are alike: Disentangling demand and supply shocks in the crude oil market. *American Economic Review* 99: 1053–1069.

Konur, O. 2000. Creating enforceable civil rights for disabled students in higher education: An institutional theory perspective. *Disability & Society* 15:1041–1063.

Konur, O. 2002a. Access to Nursing Education by disabled students: Rights and duties of nursing programs. *Nurse Education Today* 22: 364–374.

Konur, O. 2002b. Assessment of disabled students in higher education: Current public policy issues. *Assessment and Evaluation in Higher Education* 27: 131–152.

Konur, O. 2002c. Access to employment by disabled people in the UK: Is the Disability Discrimination Act working? *International Journal of Discrimination and the Law* 5: 247–279.

Konur, O. 2006a. Participation of children with dyslexia in compulsory education: Current public policy issues. *Dyslexia* 12: 51–67.

Konur, O. 2006b. Teaching disabled students in Higher Education. *Teaching in Higher Education* 11: 351–363.

Konur, O. 2007a. A judicial outcome analysis of the Disability Discrimination Act: A windfall for the employers? *Disability & Society* 22: 187–204.

Konur, O. 2007b. Computer-assisted teaching and assessment of disabled students in higher education: The interface between academic standards and disability rights. *Journal of Computer Assisted Learning* 23: 207–219.

Konur, O., ed. 2021a. *Handbook of Biodiesel and Petrodiesel Fuels: Science, Technology, Health, and Environment.* Boca Raton, FL: CRC Press.

Konur, O., ed. 2021b. *Handbook of Biodiesel and Petrodiesel Fuels: Science, Technology, Health, and Environment. Volume 1. Biodiesel Fuels: Science, Technology, Health, and Environment.* Boca Raton, FL: CRC Press.

Konur, O., ed. 2021c. *Handbook of Biodiesel and Petrodiesel Fuels: Science, Technology, Health, and Environment. Volume 2. Biodiesel Fuels based on the Edible and Nonedible Feedstocks, Wastes, and Algae: Science, Technology, Health, and Environment.* Boca Raton, FL: CRC Press.

Konur, O., ed. 2021d. *Handbook of Biodiesel and Petrodiesel Fuels: Science, Technology, Health, and Environment. Volume 3. Petrodiesel Fuels: Science, Technology, Health, and Environment.* Boca Raton, FL: CRC Press.

Konur, O. 2021e. Biodiesel and petrodiesel fuels: Science, technology, health, and environment. In *Handbook of Biodiesel and Petrodiesel Fuels: Science, Technology, Health, and Environment. Volume 1. Biodiesel Fuels: Science, Technology, Health, and Environment*, ed. O. Konur. Boca Raton, FL: CRC Press.

Konur, O. 2021f. Biodiesel and petrodiesel fuels: A scientometric review of the research. In *Handbook of Biodiesel and Petrodiesel Fuels: Science, Technology, Health, and Environment. Volume 1. Biodiesel Fuels: Science, Technology, Health, and Environment*, ed. O. Konur. Boca Raton, FL: CRC Press.

Konur, O. 2021g. Biodiesel and petrodiesel fuels: A review of the research. In *Handbook of Biodiesel and Petrodiesel Fuels: Science, Technology, Health, and Environment. Volume 1. Biodiesel Fuels: Science, Technology, Health, and Environment*, ed. O. Konur. Boca Raton, FL: CRC Press.

Konur, O. 2021h Nanotechnology applications in the diesel fuels and the related research fields: A review of the research. In *Handbook of Biodiesel and Petrodiesel Fuels: Science, Technology, Health, and Environment. Volume 1. Biodiesel Fuels: Science, Technology, Health, and Environment*, ed. O. Konur. Boca Raton, FL: CRC Press.

Konur, O. 2021i. Biooils: A scientometric review of the research. In *Handbook of Biodiesel and Petrodiesel Fuels: Science, Technology, Health, and Environment. Volume 1. Biodiesel Fuels: Science, Technology, Health, and Environment*, ed. O. Konur. Boca Raton, FL: CRC Press.

Konur, O. 2021j. Characterization and properties of biooils: A review of the research. In *Handbook of Biodiesel and Petrodiesel Fuels: Science, Technology, Health, and Environment. Volume 1. Biodiesel Fuels: Science, Technology, Health, and Environment*, ed. O. Konur. Boca Raton, FL: CRC Press.

Konur, O. 2021k. Biomass pyrolysis and pyrolysis oils: A review of the research. In *Handbook of Biodiesel and Petrodiesel Fuels: Science, Technology, Health, and Environment. Volume 1. Biodiesel Fuels: Science, Technology, Health, and Environment*, ed. O. Konur. Boca Raton, FL: CRC Press.

Konur, O. 2021l. Biodiesel fuels: A scientometric review of the research. In *Handbook of Biodiesel and Petrodiesel Fuels: Science, Technology, Health, and Environment. Volume 1. Biodiesel Fuels: Science, Technology, Health, and Environment*, ed. O. Konur. Boca Raton, FL: CRC Press.

Konur, O. 2021m. Glycerol: A scientometric review of the research. In *Handbook of Biodiesel and Petrodiesel Fuels: Science, Technology, Health, and Environment. Volume 1. Biodiesel Fuels: Science, Technology, Health, and Environment*, ed. O. Konur. Boca Raton, FL: CRC Press. 23.1.2020

Konur, O. 2021n. Propanediol production from glycerol: A review of the research. In *Handbook of Biodiesel and Petrodiesel Fuels: Science, Technology, Health, and Environment. Volume 1. Biodiesel Fuels: Science, Technology, Health, and Environment*, ed. O. Konur. Boca Raton, FL: CRC Press. 17.2.2020

Konur, O. 2021o. Edible oil-based biodiesel fuels: A scientometric review of the research. *In Handbook of Biodiesel and Petrodiesel Fuels: Science, Technology, Health, and Environment. Volume 2. Biodiesel Fuels based on the Edible and Nonedible Feedstocks, Wastes, and Algae: Science, Technology, Health, and Environment,* ed. O. Konur. Boca Raton, FL: CRC Press.

Konur, O. 2021p. Palm oil-based biodiesel fuels: A review of the research. In *Handbook of Biodiesel and Petrodiesel Fuels: Science, Technology, Health, and Environment. Volume 2. Biodiesel Fuels based on the Edible and Nonedible Feedstocks, Wastes, and Algae*, ed. O. Konur. Boca Raton, FL: CRC Press.

Konur, O. 2021q. Rapeseed oil-based biodiesel fuels: A review of the research. In *Handbook of Biodiesel and Petrodiesel Fuels: Science, Technology, Health, and Environment. Volume 2. Biodiesel Fuels based on the Edible and Nonedible Feedstocks, Wastes, and Algae*, ed. O. Konur. Boca Raton, FL: CRC Press.

Konur, O. 2021r. Nonedible oil-based biodiesel fuels: A scientometric review of the research. In *Handbook of Biodiesel and Petrodiesel Fuels: Science, Technology, Health, and Environment. Volume 2. Biodiesel Fuels based on the Edible and Nonedible Feedstocks, Wastes, and Algae: Science, Technology, Health, and Environment*, ed. O. Konur. Boca Raton, FL: CRC Press. 19.1.2020

Konur, O. 2021s. Waste oil-based biodiesel fuels: A scientometric review of the research. In *Handbook of Biodiesel and Petrodiesel Fuels: Science, Technology, Health, and Environment. Volume 2. Biodiesel Fuels based on the Edible and Nonedible Feedstocks, Wastes, and Algae: Science, Technology, Health, and Environment*, ed. O. Konur. Boca Raton, FL: CRC Press.

Konur, O. 2021t. Algal biodiesel fuels: A scientometric review of the research. In *Handbook of Biodiesel and Petrodiesel Fuels: Science, Technology, Health, and Environment. Volume 2. Biodiesel Fuels based on the Edible and Nonedible Feedstocks, Wastes, and Algae: Science, Technology, Health, and Environment*, ed. O. Konur. Boca Raton, FL: CRC Press.

Konur, O. 2021u. Algal biomass production for biodiesel production: A review of the research. In *Handbook of Biodiesel and Petrodiesel Fuels: Science, Technology, Health, and Environment. Volume 2. Biodiesel Fuels based on the Edible and Nonedible Feedstocks, Wastes, and Algae*, Ed. O. Konur. Boca Raton, FL: CRC Press. 23.2.2020

Konur, O. 2021v. Algal biomass production in wastewaters for biodiesel production: A review of the research. In *Handbook of Biodiesel and Petrodiesel Fuels: Science, Technology, Health, and Environment. Volume 2. Biodiesel Fuels based on the Edible and Nonedible Feedstocks, Wastes, and Algae*, ed. O. Konur. Boca Raton, FL: CRC Press. 23.2.2020

Konur, O. 2021x. Algal lipid production for biodiesel production: A review of the research. In *Handbook of Biodiesel and Petrodiesel Fuels: Science, Technology, Health, and Environment. Volume 2. Biodiesel Fuels based on the Edible and Nonedible Feedstocks, Wastes, and Algae*, Ed. O. Konur. Boca Raton, FL: CRC Press. 15.5.2020

Konur, O. 2021y. Crude oils: A scientometric review of the research. In *Handbook of Biodiesel and Petrodiesel Fuels: Science, Technology, Health, and Environment. Volume 3. Petrodiesel Fuels: Science, Technology, Health, and Environment*, ed. O. Konur. Boca Raton, FL: CRC Press.

Konur, O. 2021z. Petrodiesel fuels: A scientometric review of the research. In *Handbook of Biodiesel and Petrodiesel Fuels: Science, Technology, Health, and Environment. Volume 3. Petrodiesel Fuels: Science, Technology, Health, and Environment*, ed. O. Konur. Boca Raton, FL: CRC Press.

Konur, O. 2021aa. Bioremediation of petroleum hydrocarbons in the contaminated soils: A review of the research. In *Handbook of Biodiesel and Petrodiesel Fuels: Science, Technology, Health, and Environment. Volume 3. Petrodiesel Fuels: Science, Technology, Health, and Environment*, ed. O. Konur. Boca Raton, FL: CRC Press.

Konur, O. 2021ab. Desulfurization of diesel fuels: A review of the research. In *Handbook of Biodiesel and Petrodiesel Fuels: Science, Technology, Health, and Environment. Volume 3. Petrodiesel Fuels: Science, Technology, Health, and Environment*, ed. O. Konur. Boca Raton, FL: CRC Press.

Konur, O. 2021ac. Diesel fuel exhaust emissions: A scientometric review of the research. In *Handbook of Biodiesel and Petrodiesel Fuels: Science, Technology, Health, and*

Environment. Volume 3. Petrodiesel Fuels: Science, Technology, Health, and Environment, ed. O. Konur. Boca Raton, FL: CRC Press.

Konur, O. 2021ad. The adverse health and safety impact of diesel fuels: A scientometric review of the research. In *Handbook of Biodiesel and Petrodiesel Fuels: Science, Technology, Health, and Environment. Volume 3. Petrodiesel Fuels: Science, Technology, Health, and Environment*, ed. O. Konur. Boca Raton, FL: CRC Press.

Konur, O. 2021ae. Respiratory illnesses caused by the diesel fuel exhaust emissions: A review of the research. In *Handbook of Biodiesel and Petrodiesel Fuels: Science, Technology, Health, and Environment. Volume 3. Petrodiesel Fuels: Science, Technology, Health, and Environment*, ed. O. Konur. Boca Raton, FL: CRC Press.

Konur, O. 2021af. Cancer caused by the diesel fuel exhaust emissions: A review of the research. In *Handbook of Biodiesel and Petrodiesel Fuels: Science, Technology, Health, and Environment. Volume 3. Petrodiesel Fuels: Science, Technology, Health, and Environment*, ed. O. Konur. Boca Raton, FL: CRC Press.

Konur, O. 2021ag. Cardiovascular and other illnesses caused by the diesel fuel exhaust emissions: A review of the research. In *Handbook of Biodiesel and Petrodiesel Fuels: Science, Technology, Health, and Environment. Volume 3. Petrodiesel Fuels: Science, Technology, Health, and Environment*, ed. O. Konur. Boca Raton, FL: CRC Press.

Kubiak, P., K. Leja, and K. Myszka, et al. 2012. Physiological predisposition of various *Clostridium* species to synthetize 1,3-propanediol from glycerol. *Process Biochemistry* 47: 1308–1319.

Kurosaka, T., H. Maruyama, I. Naribayashi, and Y. Sasaki. 2008. Production of 1,3-propanediol by hydrogenolysis of glycerol catalyzed by Pt/WO_3/ZrO_2. *Catalysis Communications* 9: 1360–1363.

Lotero, E., Y. J. Liu, and D. E. Lopez, et al. 2005. Synthesis of biodiesel via acid catalysis. *Industrial & Engineering Chemistry Research* 44: 5353–5363.

Ma, F. R. and M. A. Hanna. 1999. Biodiesel production: A review. *Bioresource Technology* 70:1–15.

Maervoet, V. E. T., M. de Mey, J. Beauprez, S. de Maeseneire, and W. K. Soetaert. 2011. Enhancing the microbial conversion of glycerol to 1,3-propanediol using metabolic engineering. *Organic Process Research & Development* 15:189–202.

Marchetti, J. M., V. U. Miguel, and A. F. Errazu. 2008. Techno-economic study of different alternatives for biodiesel production. *Fuel Processing Technology* 89: 740–748.

Martin, A., U. Armbruster, I. Gandarias, and P. L. Arias. 2013. Glycerol hydrogenolysis into propanediols using *in situ* generated hydrogen: A critical review. *European Journal of Lipid Science and Technology* 115: 9–27.

Menzel, K., A. P. Zeng, and W. D. Deckwer. 1997. High concentration and productivity of 1,3-propanediol from continuous fermentation of glycerol by *Klebsiella pneumonia. Enzyme and Microbial Technology* 20: 82–86.

Miyazawa, T., S. Koso, K. Kunimori, and K. Tomishige. 2007. Glycerol hydrogenolysis to 1,2-propanediol catalyzed by a heat-resistant ion-exchange resin combined with Ru/C. *Applied Catalysis A-General* 329: 30–35.

Mohan, D., C. U. Pittman, and P. H. Steele. 2006. Pyrolysis of wood/biomass for bio-oil: A critical review. *Energy & Fuels* 20: 848–889.

Mu, Y., H. Teng, D. J. Zhang, W. Wang, and Z. L. Xiu. 2006. Microbial production of 1,3-propanediol by *Klebsiella pneumoniae* using crude glycerol from biodiesel preparations. *Biotechnology Letters* 28: 1755–1759.

Nakagawa, Y., Y. Shinmi, S. Koso, and K. Tomishige. 2010. Direct hydrogenolysis of glycerol into 1,3-propanediol over rhenium-modified iridium catalyst. *Journal of Catalysis* 272: 191–194.

Nakagawa, Y., X. H. Ning, Y. Amada, and K. Tomishige. 2012. Solid acid co-catalyst for the hydrogenolysis of glycerol to 1,3-propanediol over Ir-ReO_x/SiO_2. *Applied Catalysis A-General* 433: 128–134.

Nakagawa, Y., M. Tamura, and K. Tomishige. 2014. Catalytic materials for the hydrogenolysis of glycerol to 1,3-propanediol. *Journal of Materials Chemistry A* 2: 6688–6702.

North, D. C. 1991a. *Institutions, Institutional Change and Economic Performance*. Cambridge, Mass.: Cambridge University Press.

North, D.C. 1991b. Institutions. *Journal of Economic Perspectives* 5: 97–112.

Pagliaro, M., R. Ciriminna, H. Kimura, M. Rossi, and C. Della Pina. 2007. From glycerol to value-added products. *Angewandte Chemie International Edition* 46: 4434–4440.

Papanikolaou, S., P. Ruiz-Sanchez, B. Pariset, F. Blanchard, and M. Fick. 2000. High production of 1,3-propanediol from industrial glycerol by a newly isolated *Clostridium butyricum* strain. *Journal of Biotechnology* 77: 191–208.

Papanikolaou, S., S. Fakas, and M. Fick, et al. 2008. Biotechnological valorisation of raw glycerol discharged after bio-diesel (fatty acid methyl esters) manufacturing process: Production of 1,3-propanediol, citric acid and single cell oil. *Biomass & Bioenergy* 32: 60–71.

Perron, P. 1989. The great crash, the oil price shock, and the unit root hypothesis. *Econometrica: Journal of the Econometric Society* 57: 1361–1401.

Schutz, H. and Radler, F, 1984. Anaerobic reduction of glycerol to propanediol-1.3 by *Lactobacillus brevis* and *Lactobacillus buchneri. Systematic and Applied Microbiology* 5:169–178.

Thompson, J. C. and B. B. He. 2006. Characterization of crude glycerol from biodiesel production from multiple feedstocks. *Applied Engineering in Agriculture* 22: 261–265.

Wang, Z., J. Zhuge, H. Fang, and B. A. Prior. 2001. Glycerol production by microbial fermentation: A review. *Biotechnology Advances* 19: 201–223.

Wang, Y. L., J. X. Zhou, and X. W. Guo. 2015. Catalytic hydrogenolysis of glycerol to propanediols: A review. *RSC Advances* 5: 74611–74628.

Willke, T. and K. Vorlop. 2008. Biotransformation of glycerol into 1,3-propanediol. *European Journal of Lipid Science and Technology* 110: 83140.

Yuan, Z. L., J. H. Wang, and L. N. Wang, et al. 2010. Biodiesel derived glycerol hydrogenolysis to 1,2-propanediol on Cu/MgO catalysts. *Bioresource Technology* 101: 7088–7092.

Zhang, Q., J. Chang, T. J. Wang, and Y. Xu. 2007. Review of biomass pyrolysis oil properties and upgrading research. *Energy Conversion and Management* 48: 87–92.

Zhu, S. H., X. Q. Gao, and Y. L. Zhu, et al. 2013. Promoting effect of boron oxide on Cu/SiO_2 catalyst for glycerol hydrogenolysis to 1,2-propanediol. *Journal of Catalysis* 303: 70–79.

Index

C

F

N

O

P

R

S

T

U

V

W

X

Y

Z

Printed in the United States
by Baker & Taylor Publisher Services